Matrix Algorithms

Matrix Algorithms

Volume I: Basic Decompositions

G. W.
Stewart
University of Maryland
College Park, Maryland

Society for Industrial and Applied Mathematics

Philadelphia

Copyright ©1998 by the Society for Industrial and Applied Mathematics.

10 9 8 7 6 5 4 3 2 1

All rights reserved. Printed in the United States of America. No part of this book may be reproduced, stored, or transmitted in any manner without the written permission of the publisher. For information, write to the Society for Industrial and Applied Mathematics, 3600 University City Science Center, Philadelphia, PA 19104-2688.

Library of Congress Cataloging-in-Publication Data

Stewart, G. W. (Gilbert W.)
    Matrix algorithms / G.W. Stewart.
       p. cm.
    Includes bibliographical references and index.
    Contents: v. 1. Basic decompositions
    ISBN 0-89871-414-1 (v. 1 : pbk.)
    1. Matrices. I. Title.
  QA188.S714 1998
512.9'434--dc21   98-22445

0-89871-414-1 (Volume I)
0-89871-418-4 (set)

siam is a registered trademark.

CONTENTS

Algorithms   xiii

Notation   xv

Preface   xvii

1 Matrices, Algebra, and Analysis   1
- 1 Vectors   2
  - 1.1 Scalars   2
    Real and complex numbers. Sets and Minkowski sums.
  - 1.2 Vectors   3
  - 1.3 Operations with vectors and scalars   5
  - 1.4 Notes and references   7
    Representing vectors and scalars. The scalar product. Function spaces.
- 2 Matrices   7
  - 2.1 Matrices   8
  - 2.2 Some special matrices   9
    Familiar characters. Patterned matrices.
  - 2.3 Operations with matrices   13
    The scalar-matrix product and the matrix sum. The matrix product. The transpose and symmetry. The trace and the determinant.
  - 2.4 Submatrices and partitioning   17
    Submatrices. Partitions. Northwest indexing. Partitioning and matrix operations. Block forms.
  - 2.5 Some elementary constructions   21
    Inner products. Outer products. Linear combinations. Column and row scaling. Permuting rows and columns. Undoing a permutation. Crossing a matrix. Extracting and inserting submatrices.
  - 2.6 LU decompositions   23
  - 2.7 Homogeneous equations   25

2.8 Notes and references 26
    Indexing conventions. Hyphens and other considerations.
    Nomenclature for triangular matrices. Complex symmetric
    matrices. Determinants. Partitioned matrices. The
    LU decomposition.

3 Linear Algebra 28
  3.1 Subspaces, linear independence, and bases 28
      Subspaces. Linear independence. Bases. Dimension.
  3.2 Rank and nullity 33
      A full-rank factorization. Rank and nullity.
  3.3 Nonsingularity and inverses 36
      Linear systems and nonsingularity. Nonsingularity and inverses.
  3.4 Change of bases and linear transformations 39
      Change of basis. Linear transformations and matrices.
  3.5 Notes and references 42
      Linear algebra. Full-rank factorizations.

4 Analysis 42
  4.1 Norms 42
      Componentwise inequalities and absolute values. Vector norms.
      Norms and convergence. Matrix norms and consistency. Operator
      norms. Absolute norms. Perturbations of the identity. The
      Neumann series.
  4.2 Orthogonality and projections 55
      Orthogonality. The QR factorization and orthonormal bases.
      Orthogonal projections.
  4.3 The singular value decomposition 61
      Existence. Uniqueness. Unitary equivalence. Weyl's theorem and
      the min-max characterization. The perturbation of singular
      values. Low-rank approximations.
  4.4 The spectral decomposition 70
  4.5 Canonical angles and the CS decomposition 73
      Canonical angles between subspaces. The CS decomposition.
  4.6 Notes and references 75
      Vector and matrix norms. Inverses and the Neumann series. The
      QR factorization. Projections. The singular value decomposition.
      The spectral decomposition. Canonical angles and the
      CS decomposition.

5 Addenda 77
  5.1 Historical 77
      On the word matrix. History.
  5.2 General references 78
      Linear algebra and matrix theory. Classics of matrix

computations. Textbooks. Special topics. Software. Historical sources.

2 Matrices and Machines **81**
  1 Pseudocode 82
    1.1 Generalities 82
    1.2 Control statements 83
        The if statement. The for statement. The while statement. Leaving and iterating control statements. The goto statement.
    1.3 Functions 85
    1.4 Notes and references 86
        Programming languages. Pseudocode.
  2 Triangular Systems 87
    2.1 The solution of a lower triangular system 87
        Existence of solutions. The forward substitution algorithm. Overwriting the right-hand side.
    2.2 Recursive derivation 89
    2.3 A "new" algorithm 90
    2.4 The transposed system 92
    2.5 Bidiagonal matrices 92
    2.6 Inversion of triangular matrices 93
    2.7 Operation counts 94
        Bidiagonal systems. Full triangular systems. General observations on operations counts. Inversion of a triangular matrix. More observations on operation counts.
    2.8 BLAS for triangular systems 99
    2.9 Notes and references 100
        Historical. Recursion. Operation counts. Basic linear algebra subprograms (BLAS).
  3 Matrices in Memory 101
    3.1 Memory, arrays, and matrices 102
        Memory. Storage of arrays. Strides.
    3.2 Matrices in memory 104
        Array references in matrix computations. Optimization and the BLAS. Economizing memory — Packed storage.
    3.3 Hierarchical memories 109
        Virtual memory and locality of reference. Cache memory. A model algorithm. Row and column orientation. Level-two BLAS. Keeping data in registers. Blocking and the level-three BLAS.
    3.4 Notes and references 119
        The storage of arrays. Strides and interleaved memory. The BLAS. Virtual memory. Cache memory. Large memories and matrix problems. Blocking.

4 Rounding Error . . . 121
4.1 Absolute and relative error . . . 121
Absolute error. Relative error.
4.2 Floating-point numbers and arithmetic . . . 124
Floating-point numbers. The IEEE standard. Rounding error. Floating-point arithmetic.
4.3 Computing a sum: Stability and condition . . . 129
A backward error analysis. Backward stability. Weak stability. Condition numbers. Reenter rounding error.
4.4 Cancellation . . . 136
4.5 Exponent exceptions . . . 138
Overflow. Avoiding overflows. Exceptions in the IEEE standard.
4.6 Notes and references . . . 141
General references. Relative error and precision. Nomenclature for floating-point numbers. The rounding unit. Nonstandard floating-point arithmetic. Backward rounding-error analysis. Stability. Condition numbers. Cancellation. Exponent exceptions.

3 Gaussian Elimination — 147
1 Gaussian Elimination . . . 148
1.1 Four faces of Gaussian elimination . . . 148
Gauss's elimination. Gaussian elimination and elementary row operations. Gaussian elimination as a transformation to triangular form. Gaussian elimination and the LU decomposition.
1.2 Classical Gaussian elimination . . . 153
The algorithm. Analysis of classical Gaussian elimination. LU decompositions. Block elimination. Schur complements.
1.3 Pivoting . . . 165
Gaussian elimination with pivoting. Generalities on pivoting. Gaussian elimination with partial pivoting.
1.4 Variations on Gaussian elimination . . . 169
Sherman's march. Pickett's charge. Crout's method. Advantages over classical Gaussian elimination.
1.5 Linear systems, determinants, and inverses . . . 174
Solution of linear systems. Determinants. Matrix inversion.
1.6 Notes and references . . . 180
Decompositions and matrix computations. Classical Gaussian elimination. Elementary matrix. The LU decomposition. Block LU decompositions and Schur complements. Block algorithms and blocked algorithms. Pivoting. Exotic orders of elimination. Gaussian elimination and its variants. Matrix inversion. Augmented matrices. Gauss–Jordan elimination.
2 A Most Versatile Algorithm . . . 185

   2.1 Positive definite matrices 185
         Positive definite matrices. The Cholesky decomposition. The
         Cholesky algorithm.
   2.2 Symmetric indefinite matrices 190
   2.3 Hessenberg and tridiagonal matrices 194
         Structure and elimination. Hessenberg matrices. Tridiagonal
         matrices.
   2.4 Band matrices 202
   2.5 Notes and references 207
         Positive definite matrices. Symmetric indefinite systems. Band
         matrices.

3 The Sensitivity of Linear Systems 208
   3.1 Normwise bounds 209
         The basic perturbation theorem. Normwise relative error and the
         condition number. Perturbations of the right-hand side. Artificial
         ill-conditioning.
   3.2 Componentwise bounds 217
   3.3 Backward perturbation theory 219
         Normwise backward error bounds. Componentwise backward
         error bounds.
   3.4 Iterative refinement 221
   3.5 Notes and references 224
         General references. Normwise perturbation bounds. Artificial
         ill-conditioning. Componentwise bounds. Backward perturbation
         theory. Iterative refinement.

4 The Effects of Rounding Error 225
   4.1 Error analysis of triangular systems 226
         The results of the error analysis.
   4.2 The accuracy of the computed solutions 227
         The residual vector.
   4.3 Error analysis of Gaussian elimination 229
         The error analysis. The condition of the triangular factors. The
         solution of linear systems. Matrix inversion.
   4.4 Pivoting and scaling 235
         On scaling and growth factors. Partial and complete pivoting.
         Matrices that do not require pivoting. Scaling.
   4.5 Iterative refinement 242
         A general analysis. Double-precision computation of the residual.
         Single-precision computation of the residual. Assessment of
         iterative refinement.
   4.6 Notes and references 245
         General references. Historical. The error analyses. Condition of

the L- and U-factors. Inverses. Growth factors. Scaling. Iterative refinement.

4 The QR Decomposition and Least Squares — 249
1 The QR Decomposition ... 250
1.1 Basics ... 250
Existence and uniqueness. Projections and the pseudoinverse. The partitioned factorization. Relation to the singular value decomposition.
1.2 Householder triangularization ... 254
Householder transformations. Householder triangularization. Computation of projections. Numerical stability. Graded matrices. Blocked reduction.
1.3 Triangularization by plane rotations ... 270
Plane rotations. Reduction of a Hessenberg matrix. Numerical properties.
1.4 The Gram–Schmidt algorithm ... 277
The classical and modified Gram–Schmidt algorithms. Modified Gram–Schmidt and Householder triangularization. Error analysis of the modified Gram–Schmidt algorithm. Loss of orthogonality. Reorthogonalization.
1.5 Notes and references ... 288
General references. The QR decomposition. The pseudoinverse. Householder triangularization. Rounding-error analysis. Blocked reduction. Plane rotations. Storing rotations. Fast rotations. The Gram–Schmidt algorithm. Reorthogonalization.

2 Linear Least Squares ... 292
2.1 The QR approach ... 293
Least squares via the QR decomposition. Least squares via the QR factorization. Least squares via the modified Gram–Schmidt algorithm.
2.2 The normal and seminormal equations ... 298
The normal equations. Forming cross-product matrices. The augmented cross-product matrix. The instability of cross-product matrices. The seminormal equations.
2.3 Perturbation theory and its consequences ... 305
The effects of rounding error. Perturbation of the normal equations. The perturbation of pseudoinverses. The perturbation of least squares solutions. Accuracy of computed solutions. Comparisons.
2.4 Least squares with linear constraints ... 312
The null-space method. The method of elimination. The weighting method.

		2.5 Iterative refinement 320

- 2.5 Iterative refinement 320
- 2.6 Notes and references 323
  Historical. The QR approach. Gram–Schmidt and least squares. The augmented least squares matrix. The normal equations. The seminormal equations. Rounding-error analyses. Perturbation analysis. Constrained least squares. Iterative refinement.

3 Updating 326
- 3.1 Updating inverses 327
  Woodbury's formula. The sweep operator.
- 3.2 Moving columns 333
  A general approach. Interchanging columns.
- 3.3 Removing a column 337
- 3.4 Appending columns 338
  Appending a column to a QR decomposition. Appending a column to a QR factorization.
- 3.5 Appending a row 339
- 3.6 Removing a row 341
  Removing a row from a QR decomposition. Removing a row from a QR factorization. Removing a row from an R-factor (Cholesky downdating). Downdating a vector.
- 3.7 General rank-one updates 348
  Updating a factorization. Updating a decomposition.
- 3.8 Numerical properties 350
  Updating. Downdating.
- 3.9 Notes and references 353
  Historical. Updating inverses. Updating. Exponential windowing. Cholesky downdating. Downdating a vector.

5 Rank-Reducing Decompositions   357

1 Fundamental Subspaces and Rank Estimation 358
- 1.1 The perturbation of fundamental subspaces 358
  Superior and inferior singular subspaces. Approximation of fundamental subspaces.
- 1.2 Rank estimation 363
- 1.3 Notes and references 365
  Rank reduction and determination. Singular subspaces. Rank determination. Error models and scaling.

2 Pivoted Orthogonal Triangularization 367
- 2.1 The pivoted QR decomposition 368
  Pivoted orthogonal triangularization. Bases for the fundamental subspaces. Pivoted QR as a gap-revealing decomposition. Assessment of pivoted QR.
- 2.2 The pivoted Cholesky decomposition 375

2.3 The pivoted QLP decomposition 378
    The pivoted QLP decomposition. Computing the pivoted
    QLP decomposition. Tracking properties of the
    QLP decomposition. Fundamental subspaces. The matrix $\hat{Q}$ and
    the columns of X. Low-rank approximations.
2.4 Notes and references 385
    Pivoted orthogonal triangularization. The pivoted Cholesky
    decomposition. Column pivoting, rank, and singular values.
    Rank-revealing QR decompositions. The QLP decomposition.

3 Norm and Condition Estimation 387
    3.1 A 1-norm estimator 388
    3.2 LINPACK-style norm and condition estimators 391
        A simple estimator. An enhanced estimator. Condition estimation.
    3.3 A 2-norm estimator 397
    3.4 Notes and references 399
        General. LINPACK-style condition estimators. The 1-norm
        estimator. The 2-norm estimator.

4 UTV decompositions 400
    4.1 Rotations and errors 401
    4.2 Updating URV decompositions 402
        URV decompositions. Incorporation. Adjusting the gap.
        Deflation. The URV updating algorithm. Refinement. Low-rank
        splitting.
    4.3 Updating ULV decompositions 412
        ULV decompositions. Updating a ULV decomposition.
    4.4 Notes and references 416
        UTV decompositions.

References   **417**

Index   **441**

ALGORITHMS

Chapter 2. Matrices and Machines
- 1.1 Party time 82
- 2.1 Forward substitution 88
- 2.2 Lower bidiagonal system 93
- 2.3 Inverse of a lower triangular matrix 94
- 4.1 The Euclidean length of a 2-vector 140

Chapter 3. Gaussian Elimination
- 1.1 Classical Gaussian elimination 155
- 1.2 Block Gaussian elimination 162
- 1.3 Gaussian elimination with pivoting 166
- 1.4 Gaussian elimination with partial pivoting for size 168
- 1.5 Sherman's march 171
- 1.6 Pickett's charge east 173
- 1.7 Crout's method 174
- 1.8 Solution of $AX = B$ 175
- 1.9 Solution of $A^\mathrm{T} X = B$ 175
- 1.10 Inverse from an LU decomposition 179
- 2.1 Cholesky decomposition 189
- 2.2 Reduction of an upper Hessenberg matrix 197
- 2.3 Solution of an upper Hessenberg system 198
- 2.4 Reduction of a tridiagonal matrix 200
- 2.5 Solution of a tridiagonal system 201
- 2.6 Cholesky decomposition of a positive definite tridiagonal matrix . . . 201
- 2.7 Reduction of a band matrix 207

Chapter 4. The QR Decomposition and Least Squares
- 1.1 Generation of Householder transformations 257
- 1.2 Householder triangularization 259
- 1.3 Projections via the Householder decomposition 261
- 1.4 UTU representation of $\prod_i (I - u_i u_i^\mathrm{T})$ 268
- 1.5 Blocked Householder triangularization 269
- 1.6 Generation of a plane rotation 272
- 1.7 Application of a plane rotation 273

1.8	Reduction of an augmented Hessenberg matrix by plane rotations	275
1.9	Column-oriented reduction of an augmented Hessenberg matrix	276
1.10	The classical Gram–Schmidt algorithm	278
1.11	The modified Gram–Schmidt algorithm: column version	279
1.12	The modified Gram–Schmidt algorith: row version	279
1.13	Classical Gram–Schmidt orthogonalization with reorthogonalization	287
2.1	Least squares from a QR decomposition	295
2.2	Hessenberg least squares	296
2.3	Least squares via modified Gram–Schmidt	298
2.4	Normal equations by outer products	301
2.5	Least squares by corrected seminormal equations	305
2.6	The null space method for linearly constrained least squares	313
2.7	Constrained least squares by elimination	316
2.8	Constrained least squares by weights	320
2.9	Iterative refinement for least squares (residual system)	321
2.10	Solution of the general residual system	323
3.1	Updating $Ax = b$ to $(A - uv^{\mathrm{T}})y = b$	329
3.2	The sweep operator	331
3.3	QR update: exchanging columns	335
3.4	QR update: removing columns	338
3.5	Append a column to a QR decomposition	339
3.6	Append a row to a QR decomposition	341
3.7	Remove the last row from a QR decomposition	343
3.8	Remove the last row from a QR factorization	345
3.9	Cholesky downdating	347
3.10	Downdating the norm of a vector	348
3.11	Rank-one update of a QR factorization	350

Chapter 5. Rank-Reducing Decompositions

2.1	Pivoted Householder triangularization	369
2.2	Cholesky decomposition with diagonal pivoting	377
2.3	The pivoted QLP decomposition	379
3.1	A 1-norm estimator	390
3.2	A simple LINPACK estimator	392
3.3	An estimator for $\|L^{-1}\|_2$	395
3.4	A 2-norm estimator	399
4.1	URV updating	409
4.2	URV refinement	411

NOTATION

$\mathbb{R}$	The set of real numbers	2
$\mathbb{C}$	The set of complex numbers	2
Re z	The real part of z	2
Im z	The imaginary part of z	2
$|z|$	The absolute value or modulus of z	2
$\bar{z}$	The conjugate of z	2
arg z	The argument of z	2
$\mathbb{R}^n$	The set of real n-vectors	4
$\mathbb{C}^n$	The set of complex n-vectors	4
$\mathbf{e}$	The vector of ones	4
$\mathbf{e}_i$	The ith unit vector	5
$\mathbb{R}^{m \times n}$	The set of real $m \times n$ matrices	8
$\mathbb{C}^{m \times n}$	The set of complex $m \times n$ matrices	8
I_n, I	The identity matrix of order n	9
A^T	The transpose of A	14
A^H	The conjugate transpose of A	15
trace(A)	The trace of A	16
det(A)	The determinant of A	16
$\mathcal{F}$	The cross operator	23
$\mathcal{X} \oplus \mathcal{Y}$	The direct sum of $\mathcal{X}$ and $\mathcal{Y}$	29
span($\mathcal{X}$)	The span of $\mathcal{X}$	29
dim($\mathcal{X}$)	The dimension of the subspace $\mathcal{X}$	32
$\mathcal{R}(A)$	The column space of A	34
rank(A)	The rank of A	34
$\mathcal{N}(A)$	The null space of A	35
null(A)	The nullity of A	35
A^{-1}	The inverse of A	38
A^{-T}	The inverse transpose of A	38
A^{-H}	The inverse conjugate transpose of A	38
X^{I}	A left inverse of X	39
$A \leq B, A \geq B$, etc.	Componentwise matrix inequalities	43
$|A|$	The absolute value of A	43
$\|\cdot\|$	A vector norm	44

$\\|\cdot\\|_1, \\|\cdot\\|_2, \\|\cdot\\|_\infty$	The vector 1-, 2-, and ∞-norms	44
$\\|\cdot\\|$	A matrix norm	48
$\\|\cdot\\|_F$	The Frobenius norm	49
$\\|\cdot\\|_1, \\|\cdot\\|_2, \\|\cdot\\|_\infty$	The matrix 1-, 2-, and ∞-norms	51
$\theta(x,y)$	The angle between x and y	56
$x \perp y$	The vector x is orthogonal to y	56
$\mathcal{X}_\perp$	The orthogonal complement of $\mathcal{X}$	59
$P_\mathcal{X}, P_X$	The projection onto $\mathcal{X}, \mathcal{R}(X)$	60
$P_\mathcal{X}^\perp, P_\perp$	The projection onto the orthogonal complement of $\mathcal{X}$	60
$\inf(X)$	The smallest singular value of X	64
$\sigma_i(X)$	The ith singular value of X	67
$\Theta(\mathcal{X},\mathcal{Y})$	The canonical angles between $\mathcal{X}$ and $\mathcal{Y}$	73
$O(n)$	Big O notation	95
fladd, flmlt	A floating-point addition, multiplication	96
fldiv, flsqrt	A floating-point divide, square root	96
flam	A floating-point addition and multiplication	96
flrot	An application of a plane rotation	96
ϵ_M	The rounding unit	128
$\text{fl}(a)$	The rounded value of a	127
$\text{fl}(a \circ b)$	The operation $a \circ b$ computed in floating point	128
ϵ'_M	The adjusted rounding unit	131
$\leftrightarrow$	The exchange operator	165
$\kappa(A)$	The condition number of A (square)	211
$\kappa(X)$	The condition number of X (rectangular)	283

PREFACE

This book, *Basic Decompositions*, is the first volume in a projected five-volume series entitled *Matrix Algorithms*. The other four volumes will treat eigensystems, iterative methods for linear systems, sparse direct methods, and special topics, including fast algorithms for structured matrices.

My intended audience is the nonspecialist whose needs cannot be satisfied by black boxes. It seems to me that these people will be chiefly interested in the methods themselves — how they are derived and how they can be adapted to particular problems. Consequently, the focus of the series is on algorithms, with such topics as rounding-error analysis and perturbation theory introduced impromptu as needed. My aim is to bring the reader to the point where he or she can go to the research literature to augment what is in the series.

The series is self-contained. The reader is assumed to have a knowledge of elementary analysis and linear algebra and a reasonable amount of programming experience — about what you would expect from a beginning graduate engineer or an undergraduate in an honors program. Although strictly speaking the individual volumes are not textbooks, they are intended to teach, and my guiding principle has been that if something is worth explaining it is worth explaining fully. This has necessarily restricted the scope of the series, but I hope the selection of topics will give the reader a sound basis for further study.

The focus of this and part of the next volume will be the computation of matrix decompositions — that is, the factorization of matrices into products of simpler ones. This decompositional approach to matrix computations is relatively new: it achieved its definitive form in the early 1960s, thanks to the pioneering work of Alston Householder and James Wilkinson. Before then, matrix algorithms were addressed to specific problems — the solution of linear systems, for example — and were presented at the scalar level in computational tableaus. The decompositional approach has two advantages. First, by working at the matrix level it facilitates the derivation and analysis of matrix algorithms. Second, by deemphasizing specific problems, the approach turns the decomposition into a computational platform from which a variety of problems can be solved. Thus the initial cost of computing a decomposition can pay for itself many times over.

In this volume we will be chiefly concerned with the LU and the QR decompositions along with certain two-sided generalizations. The singular value decomposition also plays a large role, although its actual computation will be treated in the second volume of this series. The first two chapters set the stage not only for the present volume but for the whole series. The first is devoted to the mathematical background — matrices, vectors, and linear algebra and analysis. The second chapter discusses the realities of matrix computations on computers.

The third chapter is devoted to the LU decomposition — the result of Gaussian elimination. This extraordinarily flexible algorithm can be implemented in many different ways, and the resulting decomposition has innumerable applications. Unfortunately, this flexibility has a price: Gaussian elimination often quivers on the edge of instability. The perturbation theory and rounding-error analysis required to understand why the algorithm works so well (and our understanding is still imperfect) is presented in the last two sections of the chapter.

The fourth chapter treats the QR decomposition — the factorization of a matrix into the product of an orthogonal matrix and an upper triangular matrix. Unlike the LU decomposition, the QR decomposition can be computed two ways: by the Gram–Schmidt algorithm, which is old, and by the method of orthogonal triangularization, which is new. The principal application of the decomposition is the solution of least squares problems, which is treated in the second section of the chapter. The last section treats the updating problem — the problem of recomputing a decomposition when the original matrix has been altered. The focus here is on the QR decomposition, although other updating algorithms are briefly considered.

The last chapter is devoted to decompositions that can reveal the rank of a matrix and produce approximations of lower rank. The issues stand out most clearly when the decomposition in question is the singular value decomposition, which is treated in the first section. The second treats the pivoted QR decomposition and a new extension, the QLP decomposition. The third section treats the problem of estimating the norms of matrices and their inverses — the so-called problem of condition estimation. The estimators are used in the last section, which treats rank revealing URV and ULV decompositions. These decompositions in some sense lie between the pivoted QR decomposition and the singular value decomposition and, unlike either, can be updated.

Many methods treated in this volume are summarized by displays of pseudocode (see the list of algorithms following the table of contents). These summaries are for purposes of illustration and should not be regarded as finished implementations. In the first place, they often leave out error checks that would clutter the presentation. Moreover, it is difficult to verify the correctness of algorithms written in pseudocode. In most cases, I have checked the algorithms against MATLAB implementations. Unfortunately, that procedure is not proof against transcription errors.

A word on organization. The book is divided into numbered chapters, sections, and subsections, followed by unnumbered subsubsections. Numbering is by section, so that (3.5) refers to the fifth equations in section three of the current chapter. References to items outside the current chapter are made explicitly — e.g., Theorem 2.7, Chapter 1.

Initial versions of the volume were circulated on the Internet, and I received useful comments from a number of people: Lawrence Austin, Alekxandar S. Bozin, Andrew H. Chan, Alan Edelman, Lou Ehrlich, Lars Elden, Wayne Enright, Warren Ferguson, Daniel Giesy, Z. Han, David Heiser, Dirk Laurie, Earlin Lutz, Andrzej Mackiewicz, Andy Mai, Bart Truyen, Andy Wolf, and Gehard Zielke. I am particularly indebted to Nick Higham for a valuable review of the manuscript and to Cleve Moler for some incisive (what else) comments that caused me to rewrite parts of Chapter 3.

The staff at SIAM has done their usual fine job of production. I am grateful to Vickie Kearn, who has seen this project through from the beginning, to Mary Rose Muccie for cleaning up the index, and especially to Jean Keller-Anderson whose careful copy editing has saved you, the reader, from a host of misprints. (The ones remaining are my fault.)

Two chapters in this volume are devoted to least squares and orthogonal decompositions. It is not a subject dominated by any one person, but as I prepared these chapters I came to realize the pervasive influence of Åke Björck. His steady stream of important contributions, his quiet encouragment of others, and his definitive summary, *Numerical Methods for Least Squares Problems,* have helped bring the field to a maturity it might not otherwise have found. I am pleased to dedicate this volume to him.

G. W. Stewart
College Park, MD

1

MATRICES, ALGEBRA, AND ANALYSIS

There are two approaches to linear algebra, each having its virtues. The first is abstract. A vector space is defined axiomatically as a collection of objects, called vectors, with a sum and a scalar-vector product. As the theory develops, matrices emerge, almost incidentally, as scalar representations of linear transformations. The advantage of this approach is generality. The disadvantage is that the hero of our story, the matrix, has to wait in the wings.

The second approach is concrete. Vectors and matrices are defined as arrays of scalars — here arrays of real or complex numbers. Operations between vectors and matrices are defined in terms of the scalars that compose them. The advantage of this approach for a treatise on matrix computations is obvious: it puts the objects we are going to manipulate to the fore. Moreover, it is truer to the history of the subject. Most decompositions we use today to solve matrix problems originated as simplifications of quadratic and bilinear forms that were defined by arrays of numbers.

Although we are going to take the concrete approach, the concepts of abstract linear algebra will not go away. It is impossible to derive and analyze matrix algorithms without a knowledge of such things as subspaces, bases, dimension, and linear transformations. Consequently, after introducing vectors and matrices and describing how they combine, we will turn to the concepts of linear algebra. This inversion of the traditional order of presentation allows us to use the power of matrix methods to establish the basic results of linear algebra.

The results of linear algebra apply to vector spaces over an arbitrary field. However, we will be concerned entirely with vectors and matrices composed of real and complex numbers. What distinguishes real and complex numbers from an arbitrary field of scalars is that they posses a notion of limit. This notion of limit extends in a straightforward way to finite-dimensional vector spaces over the real or complex numbers, which inherit this topology by way of a generalization of the absolute value called the norm. Moreover, these spaces have a Euclidean geometry — e.g., we can speak of the angle between two vectors. The last section of this chapter is devoted to exploring these analytic topics.

1. VECTORS

Since we are going to define matrices as two-dimensional arrays of numbers, called scalars, we could regard a vector as a degenerate matrix with a single column, and a scalar as a matrix with one element. In fact, we will make such identifications later. However, the words "scalar" and "vector" carry their own bundles of associations, and it is therefore desirable to introduce and discuss them independently.

1.1. SCALARS

Although vectors and matrices are represented on a computer by floating-point numbers — and we must ultimately account for the inaccuracies this introduces — it is convenient to regard matrices as consisting of real or complex numbers. We call these numbers scalars.

Real and complex numbers

The set of real numbers will be denoted by $\mathbb{R}$. As usual, $|x|$ will denote the absolute value of $x \in \mathbb{R}$.

The set of complex numbers will be denoted by $\mathbb{C}$. Any complex number z can be written in the form
$$z = x + iy,$$
where x and y are real and i is the principal square root of -1. The number x is the *real part* of z and is written Re z. The number y is the *imaginary part* of z and is written Im z. The *absolute value*, or *modulus*, of z is $|z| = \sqrt{x^2 + y^2}$. The conjugate $x - iy$ of z will be written $\bar{z}$. The following relations are useful:

1. $2\operatorname{Re} z = z + \bar{z}$,
2. $2\operatorname{Im} z = z - \bar{z}$,
3. $|z|^2 = z\bar{z}$.

If $z \neq 0$ and we write the quotient $z/|z| = c + is$, then $c^2 + s^2 = 1$. Hence for a unique angle θ in $[0, 2\pi)$ we have $c = \cos\theta$ and $s = \sin\theta$. The angle θ is called the *argument* of z, written arg z. From Euler's famous relation
$$e^{i\theta} = \cos\theta + i\sin\theta,$$
we have the *polar representation* of a nonzero complex number:
$$z = |z|e^{i\arg z}.$$

The parts of a complex number are illustrated in Figure 1.1.

Scalars will be denoted by lower-case Greek or Latin letters.

SEC. 1. VECTORS

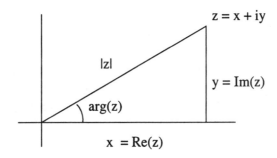

Figure 1.1: A complex number

Sets and Minkowski sums

Sets of objects will generally be denoted by script letters. For example,

$$\mathcal{C} = \{z \colon |z| = 1\}$$

is the unit circle in the complex plane. We will use the standard notation $\mathcal{X} \cup \mathcal{Y}$, $\mathcal{X} \cap \mathcal{Y}$, and $\mathcal{X} \setminus \mathcal{Y}$ for the union, intersection, and difference of sets.

If a set of objects has operations these operations can be extended to subsets of objects in the following manner. Let $\circ$ denote a binary operation between objects, and let $\mathcal{X}$ and $\mathcal{Y}$ be subsets. Then $\mathcal{X} \circ \mathcal{Y}$ is defined by

$$\mathcal{X} \circ \mathcal{Y} = \{x \circ y \colon x \in \mathcal{X}, y \in \mathcal{Y}\}.$$

The extended operation is called the *Minkowski operation*. The idea of a Minkowski operation generalizes naturally to operations with multiple operands lying in different sets.

For example, if $\mathcal{C}$ is the unit circle defined above, and $\mathcal{B} = \{-1, 1\}$, then the Minkowski sum $\mathcal{B} + \mathcal{C}$ consists of two circles of radius one, one centered at -1 and the other centered at 1.

1.2. VECTORS

In three dimensions a directed line segment can be specified by three numbers x, y, and z as shown in Figure 1.2. The following definition is a natural generalization of this observation.

Definition 1.1. *A* VECTOR x *of* DIMENSION n *or* n-VECTOR *is an array of n scalars of the form*

$$x = \begin{pmatrix} x_1 \\ x_2 \\ \vdots \\ x_n \end{pmatrix}.$$

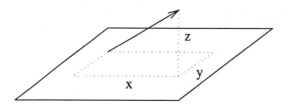

Figure 1.2: A vector in 3-Space

alpha	α	a	kappa	κ	k	sigma	ς	q
beta	β	b	lambda	λ	l, ℓ	tau	τ	t
gamma	γ	c, g	mu	μ	m	upsilon	υ	u
delta	δ	d	nu	ν	n, v	phi	ϕ	f
epsilon	ϵ	e	xi	ξ	x	chi	χ	
zeta	ζ	z	omicron	o	o	psi	ψ	
eta	η	y, h	pi	π	p	omega	ω	w
theta	θ		rho	ρ	r			
iota	ι	i	sigma	σ	s			

Figure 1.3: The Greek alphabet and Latin equivalents

We also write

$$x = (x_1, x_2, \ldots, x_n)^{\mathrm{T}}.$$

The scalars x_i are called the COMPONENTS of x. The set of n-vectors with real components will be written $\mathbb{R}^n$. The set of n-vectors with real or complex components will be written $\mathbb{C}^n$. These sets are called REAL and COMPLEX n-SPACE.

In addition to allowing vectors with more than three components, we have allowed the components to be complex. Naturally, a real vector of dimension greater than three cannot be represented graphically in the manner of Figure 1.2, and a nontrivial complex vector has no such representation. Nonetheless, most facts about vectors can be illustrated by drawings in real 2-space or 3-space.

Vectors will be denoted by lower-case Latin letters. In representing the components of a vector, we will generally use an associated lower-case Latin or Greek letter. Thus the components of the vector b will be b_i or possibly β_i. Since the Latin and Greek alphabets are not in one-one correspondence, some of the associations are artificial. Figure 1.3 lists the ones we will use here. In particular, note the association of ξ with x and η with y.

The *zero vector* is the vector whose components are all zero. It is written 0, whatever its dimension. The vector whose components are all one is written e. The vector

Sec. 1. Vectors

whose ith component is one and whose other components are zero is written e_i and is called the ith *unit vector*.

In summary,

$$0 = \begin{pmatrix} 0 \\ \vdots \\ 0 \\ \vdots \\ 0 \end{pmatrix}, \qquad e = \begin{pmatrix} 1 \\ \vdots \\ 1 \\ \vdots \\ 1 \end{pmatrix}, \qquad e_i = i \begin{pmatrix} 0 \\ \vdots \\ 1 \\ \vdots \\ 0 \end{pmatrix}.$$

1.3. Operations with Vectors and Scalars

Vectors can be added and multiplied by scalars. These operations are performed componentwise as specified in the following definition.

Definition 1.2. *Let x and y be n-vectors and α be a scalar. The* SUM *of x and y is the vector*

$$x + y = \begin{pmatrix} x_1 + y_1 \\ x_2 + y_2 \\ \vdots \\ x_n + y_n \end{pmatrix}.$$

The SCALAR-VECTOR PRODUCT αx *is the vector*

$$\alpha x = \begin{pmatrix} \alpha x_1 \\ \alpha x_2 \\ \vdots \\ \alpha x_n \end{pmatrix}.$$

The following properties are easily established from the definitions of the vector sum and scalar-vector product.

Theorem 1.3. *Let x, y, and z be n-vectors and α and β be scalars. Then*

1. $x + y = y + x$,
2. $(x + y) + z = x + (y + z)$,
3. $x + 0 = x$,
4. $x + (-1)x = 0$,
5. $(\alpha \beta) x = \alpha(\beta x)$,
6. $(\alpha + \beta) x = \alpha x + \beta x$,
7. $\alpha(x + y) = \alpha x + \alpha y$,
8. $1 \cdot x = x$.

(1.1)

The properties listed above insure that a sum of products of the form

$$\alpha_1 x_1 + \alpha_2 x_2 + \cdots + \alpha_m x_m$$

is unambiguously defined and independent of the order of summation. Such a sum of products is called a *linear combination* of the vectors $x_1, x_2, \ldots, x_m$.

The properties listed in Theorem 1.3 are sufficient to define a useful mathematical object called a *vector space* or *linear space*. Specifically, a vector space consists of a field $\mathcal{F}$ of objects called scalars and a set of objects $\mathcal{X}$ called vectors. The vectors can be combined by a sum that satisfies properties (1.1.1) and (1.1.2). There is a distinguished element $0 \in \mathcal{X}$ satisfying (1.1.3), and for every x there is a vector $-x$ such that $x + (-x) = 0$. In addition there is a scalar-vector product satisfying (1.1.8).

Vector spaces can be far more general than the spaces $\mathbb{R}^n$ and $\mathbb{C}^n$ of real and complex n-vectors. Here are three examples of increasing generality.

Example 1.4. *The following are vector spaces under the natural operations of summation and multiplication by a scalar.*

1. *The set $\mathcal{P}_n$ of polynomials of degree not greater than n*
2. *The set $\mathcal{P}_\infty$ of polynomials of any degree*
3. *The set $\mathcal{C}[0,1]$ of all real functions continuous on $[0,1]$*

- The first example is really our friend $\mathbb{C}^{n+1}$ in disguise, since the polynomial $\alpha_0 z^0 + \alpha_1 z^1 + \cdots + \alpha_n z^n$ can be identified with the $(n+1)$-vector $(\alpha_0, \alpha_1, \ldots, \alpha_n)^{\mathrm{T}}$ in such a way that sums and scalar-vector products in the two spaces correspond.

   Any member of $\mathcal{P}_n$ can be written as a linear combination of the monomials z^0, $z^1, \ldots, z^n$, and no fewer will do the job. We will call such a set of vectors a *basis* for the space in question (see §3.1).

- The second example cannot be identified with $\mathbb{C}^n$ for any n. It is an example of an infinite-dimensional vector space. However, any element of $\mathcal{P}_\infty$ can be written as the *finite* sum of monomials.

- The third example, beloved of approximation theorists, is also an infinite-dimensional space. But there is no countably infinite set of elements such that any member $\mathcal{C}[0,1]$ can be written as a finite linear combination of elements of the set. The study of such spaces belongs to the realm of functional analysis.

———

Given rich spaces like $\mathcal{C}[0,1]$, little spaces like $\mathbb{R}^n$ may seem insignificant. However, many numerical algorithms for continuous problems begin by reducing the problem to a corresponding finite-dimensional problem. For example, approximating a member of $\mathcal{C}[0,1]$ by polynomials of bounded degree immediately places us in a finite-dimensional setting. For this reason vectors and matrices are important in almost every branch of numerical analysis.

1.4. Notes and references

Representing vectors and scalars

There are many conventions for representing vectors and matrices. A common one is to represent vectors by bold lower-case letters and their components by the same letter subscripted and in ordinary type. It has the advantage that bold Greek letters can be used as vectors while their components can be represented by the corresponding nonbold letters (so that probabilists can have their π and eat it too). It has the disadvantage that it does not combine well with handwriting — on a blackboard for example. An alternative, popularized among numerical analysts by Householder [189], is to use lower-case Latin letters for vectors and lower-case Greek letters exclusively for scalars. The scheme used here is a hybrid, in which the status of lower-case Latin letters is ambiguous but always resolvable from context.

The scalar product

The scalar-vector product should not be confused with the *scalar product* of two vectors x and y (also known as the *inner product* or *dot product*). See (2.9).

Function spaces

The space $\mathcal{C}[0,1]$ is a distinguished member of a class of infinite-dimensional spaces called *function spaces*. The study of these spaces is called functional analysis. The lack of a basis in the usual sense is resolved by introducing a norm in which the space is closed. For example, the usual norm for $\mathcal{C}[0,1]$ is defined by

$$\|f\| = \max_{x \in [0,1]} |f(x)|.$$

Convergence in this norm corresponds to uniform convergence on $[0,1]$, which preserves continuity. A basis for a function space is any linearly independent set such that any element of the space can be approximated arbitrarily closely in the norm by a finite linear combination of the basis elements. For example, since any continuous function in $[0,1]$ can be uniformly approximated to any accuracy by a polynomial of sufficiently high degree — this is the Weierstrass approximation theorem [89, §6.1] — the polynomials form a basis for $\mathcal{C}[0,1]$. For introductions to functional analysis see [72, 202].

2. Matrices

When asked whether a programming language supports matrices, many people will think of two-dimensional arrays and respond, "Yes." Yet matrices are more than two-dimensional arrays — they are arrays with operations. It is the operations that cause matrices to feature so prominently in science and engineering.

2.1. MATRICES

Matrices and the matrix-vector product arise naturally in the study of systems of equations. An $m \times n$ system of linear equations

$$
\begin{aligned}
a_{11}x_1 + a_{12}x_2 + \cdots + a_{1n}x_n &= b_1 \\
a_{21}x_1 + a_{22}x_2 + \cdots + a_{2n}x_n &= b_2 \\
&\;\;\vdots \\
a_{m1}x_1 + a_{m2}x_2 + \cdots + a_{mn}x_n &= b_m
\end{aligned}
\qquad (2.1)
$$

can be written compactly in the form

$$
\sum_{j=1}^{n} a_{ij}x_j = b_i, \qquad i = 1, 2, \ldots, m. \qquad (2.2)
$$

However, matrices provide an even more compact representation. If we define arrays A, x, and b by

$$
A = \begin{pmatrix} a_{11} & a_{12} & \cdots & a_{1n} \\ a_{21} & a_{22} & \cdots & a_{2n} \\ \vdots & \vdots & & \vdots \\ a_{m1} & a_{m2} & \cdots & a_{mn} \end{pmatrix}, \quad x = \begin{pmatrix} x_1 \\ x_2 \\ \vdots \\ x_n \end{pmatrix}, \quad \text{and} \quad b = \begin{pmatrix} b_1 \\ b_2 \\ \vdots \\ b_m \end{pmatrix},
$$

and define the product Ax by the left-hand side of (2.2), then (2.1) is equivalent to

$$Ax = b.$$

Nothing could be simpler.

With the above example in mind, we make the following definition.

Definition 2.1. *An $m \times n$ MATRIX A is an array of scalars of the form*

$$
A = \begin{pmatrix} a_{11} & a_{12} & \cdots & a_{1n} \\ a_{21} & a_{22} & \cdots & a_{2n} \\ \vdots & \vdots & & \vdots \\ a_{m1} & a_{m2} & \cdots & a_{mn} \end{pmatrix}.
$$

The scalars a_{ij} are called the ELEMENTS *of A. The set of $m \times n$ matrices with real elements is written $\mathbb{R}^{m \times n}$. The set of $m \times n$ matrices with real or complex components is written $\mathbb{C}^{m \times n}$.*

The indices i and j of the elements a_{ij} of a matrix are called respectively the *row index* and the *column index*. Typically row and column indices start at one and work their way up by increments of one. In some applications, however, matrices begin with zero or even negative indices.

SEC. 2. MATRICES

Matrices will be denoted by upper-case Latin and Greek letters. We will observe the usual correspondences between the letter denoting a matrix and the letter denoting its elements (see Figure 1.3).

We will make no distinction between a 1×1 matrix a 1-vector and a scalar and likewise for $n\times 1$ matrices and n-vectors. A $1\times n$ matrix will be called an n-dimensional *row vector*.

2.2. SOME SPECIAL MATRICES

This subsection is devoted to the taxonomy of matrices. In a rough sense the division of matrices has two aspects. First, there are commonly occurring matrices that interact with matrix operations in special ways. Second, there are matrices whose nonzero elements have certain patterns. We will treat each in turn.

Familiar characters

- **Void matrices.** A *void matrix* is a matrix with no rows or no columns (or both). Void matrices are convenient place holders in degenerate matrix partitions (see §2.4).

- **Square matrices.** An $n\times n$ matrix A is called a *square matrix*. We also say that A is of *order n*.

- **The zero matrix.** A matrix whose elements are zero is called a *zero matrix*, written 0.

- **Identity matrices.** The matrix I_n of order n defined by

$$\iota_{ij} = \begin{cases} 1 & \text{if } i = j, \\ 0 & \text{if } i \neq j \end{cases}$$

is called the *identity matrix*. The ith column of the identity matrix is the ith unit vector $\mathbf{e}_i$: symbolically,

$$I_n = (\mathbf{e}_1 \ \mathbf{e}_2 \ \cdots \ \mathbf{e}_n). \tag{2.3}$$

When context makes the order clear, we will drop the subscript and simply write I for the identity matrix.

- **Permutation matrices.** Let $\mathcal{I} = \{i_1, i_2, \ldots, i_n\}$ be a permutation of the integers $1, 2, \ldots, n$. The matrix

$$P_\mathcal{I} = (\mathbf{e}_{i_1} \ \mathbf{e}_{i_2} \ \cdots \ \mathbf{e}_{i_n})$$

is called a *permutation matrix*. Thus a permutation matrix is just an identity with its columns permuted. Permutation matrices can be used to reposition rows and columns of matrices (see §2.5).

The permutation obtained by exchanging columns i and j of the identity matrix is called the (i,j)-*exchange matrix*. Exchange matrices are used to interchange rows and columns of other matrices.

Patterned matrices

An important theme in matrix computations is the reduction of matrices to ones with special properties, properties that make the problem at hand easy to solve. Often the property in question concerns the distribution of zero and nonzero elements in the matrix. Although there are many possible distributions, a few are ubiquitous, and we list them here.

- **Diagonal matrices.** A square matrix D is *diagonal* if

$$i \neq j \implies \delta_{ij} = 0.$$

In other words, a matrix is diagonal if its off-diagonal elements are zero. To specify a diagonal matrix with diagonal elements $\delta_1, \delta_2, \ldots, \delta_n$, we write

$$D = \text{diag}(\delta_1, \delta_2, \ldots, \delta_n).$$

If a matrix is called D, Λ, or Σ in this work there is a good chance it is diagonal.

The following convention, due to J. H. Wilkinson, is useful in describing patterns of zeros in a matrix. The symbol 0 stands for a zero element. The symbol X stands for an element that may or may not be zero (but probably is not). In this notation a 5×5 diagonal matrix can be represented as follows:

$$\begin{pmatrix} X & 0 & 0 & 0 & 0 \\ 0 & X & 0 & 0 & 0 \\ 0 & 0 & X & 0 & 0 \\ 0 & 0 & 0 & X & 0 \\ 0 & 0 & 0 & 0 & X \end{pmatrix}.$$

We will call such a representation a *Wilkinson diagram*.

An extension of this convention is useful when more than one matrix is in play. Here 0 stands for a zero element, while any lower-case letter stands for a potential nonzero. In this notation, a diagonal matrix might be written

$$\begin{pmatrix} d & 0 & 0 & 0 & 0 \\ 0 & d & 0 & 0 & 0 \\ 0 & 0 & d & 0 & 0 \\ 0 & 0 & 0 & d & 0 \\ 0 & 0 & 0 & 0 & d \end{pmatrix}.$$

- **Triangular matrices.** A square matrix U is *upper triangular* if

$$i > j \implies u_{ij} = 0. \tag{2.4}$$

Sec. 2. Matrices

In other words, an upper triangular matrix has the form

$$\begin{pmatrix} X & X & X & X & X \\ 0 & X & X & X & X \\ 0 & 0 & X & X & X \\ 0 & 0 & 0 & X & X \\ 0 & 0 & 0 & 0 & X \end{pmatrix}.$$

Upper triangular matrices are often called U or R.

A square matrix L is *lower triangular* if

$$i < j \implies \ell_{ij} = 0. \tag{2.5}$$

A lower triangular matrix has the form

$$\begin{pmatrix} X & 0 & 0 & 0 & 0 \\ X & X & 0 & 0 & 0 \\ X & X & X & 0 & 0 \\ X & X & X & X & 0 \\ X & X & X & X & X \end{pmatrix}.$$

Lower triangular matrices tend to be called L.

A matrix does not have to be square to satisfy (2.4) or (2.5). An $m \times n$ matrix with $m \leq n$ that satisfies (2.4) is *upper trapezoidal*. If $m \leq n$ and it satisfies (2.5) it is *lower trapezoidal*. Why these matrices are called trapezoidal can be seen from their Wilkinson diagrams.

A triangular matrix is *strictly triangular* if its diagonal elements are zero. If its diagonal elements are one, it is *unit triangular*. The same terminology applies to trapezoidal matrices.

- **Cross diagonal and triangular matrices.** A matrix is *cross diagonal*, *cross upper triangular*, or *cross lower triangular* if it is (respectively) of the form

$$\begin{pmatrix} 0 & 0 & 0 & 0 & X \\ 0 & 0 & 0 & X & 0 \\ 0 & 0 & X & 0 & 0 \\ 0 & X & 0 & 0 & 0 \\ X & 0 & 0 & 0 & 0 \end{pmatrix}, \quad \begin{pmatrix} X & X & X & X & X \\ X & X & X & X & 0 \\ X & X & X & 0 & 0 \\ X & X & 0 & 0 & 0 \\ X & 0 & 0 & 0 & 0 \end{pmatrix}, \quad \text{or} \quad \begin{pmatrix} 0 & 0 & 0 & 0 & X \\ 0 & 0 & 0 & X & X \\ 0 & 0 & X & X & X \\ 0 & X & X & X & X \\ X & X & X & X & X \end{pmatrix}.$$

These cross matrices are obtained from their more placid relatives by reversing the orders of their rows and columns. We will call any matrix form obtained in this way a cross form.

- **Hessenberg matrices.** A matrix A is *upper Hessenberg* if

$$i > j+1 \implies a_{ij} = 0.$$

An upper Hessenberg matrix is zero below its first subdiagonal:

$$\begin{pmatrix} X & X & X & X & X \\ X & X & X & X & X \\ 0 & X & X & X & X \\ 0 & 0 & X & X & X \\ 0 & 0 & 0 & X & X \end{pmatrix}.$$

A *lower Hessenberg matrix* is zero above its first superdiagonal:

$$\begin{pmatrix} X & X & 0 & 0 & 0 \\ X & X & X & 0 & 0 \\ X & X & X & X & 0 \\ X & X & X & X & X \\ X & X & X & X & X \end{pmatrix}.$$

- **Band matrices.** A matrix is *tridiagonal* if it is both lower and upper Hessenberg:

$$\begin{pmatrix} X & X & 0 & 0 & 0 \\ X & X & X & 0 & 0 \\ 0 & X & X & X & 0 \\ 0 & 0 & X & X & X \\ 0 & 0 & 0 & X & X \end{pmatrix}.$$

It acquires its name from the fact that it consists of three diagonals: a superdiagonal, a main diagonal, and a subdiagonal.

A matrix is *lower bidiagonal* if it is lower triangular and tridiagonal; that is, if it has the form

$$\begin{pmatrix} X & 0 & 0 & 0 & 0 \\ X & X & 0 & 0 & 0 \\ 0 & X & X & 0 & 0 \\ 0 & 0 & X & X & 0 \\ 0 & 0 & 0 & X & X \end{pmatrix}.$$

An *upper bidiagonal* matrix is both upper triangular and tridiagonal.

Diagonal, tridiagonal, and bidiagonal matrices are examples of band matrices. A matrix B is a *band matrix* with *lower band width* p and *upper band width* q if

$$i > j + p \implies b_{ij} = 0 \quad \text{and} \quad i < j - q \implies b_{ij} = 0. \tag{2.6}$$

The *band width* of B is $p+q+1$.

In terms of diagonals, a band matrix with lower band width p and upper band width q has p subdiagonals below the principal diagonal and q superdiagonals above the principal diagonal. The band width is the total number of diagonals.

2.3. OPERATIONS WITH MATRICES

In this subsection we will introduce the matrix operations and functions that turn matrices from lifeless arrays into vivacious participants in an algebra of their own.

The scalar-matrix product and the matrix sum

The scalar-matrix product and the matrix sum are defined in the same way as their vector analogues.

Definition 2.2. *Let λ be a scalar and A and B be $m \times n$ matrices. The* SCALAR-MATRIX PRODUCT *of λ and A is the matrix*

$$\lambda A = \begin{pmatrix} \lambda a_{11} & \lambda a_{12} & \cdots & \lambda a_{1n} \\ \lambda a_{21} & \lambda a_{22} & \cdots & \lambda a_{2n} \\ \vdots & \vdots & & \vdots \\ \lambda a_{m1} & \lambda a_{m2} & \cdots & \lambda a_{mn} \end{pmatrix}.$$

The SUM *of A and B is the matrix*

$$A + B = \begin{pmatrix} a_{11}+b_{11} & a_{12}+b_{12} & \cdots & a_{1n}+b_{1n} \\ a_{21}+b_{21} & a_{22}+b_{22} & \cdots & a_{2n}+b_{2n} \\ \vdots & \vdots & & \vdots \\ a_{m1}+b_{m1} & a_{m2}+b_{m2} & \cdots & a_{mn}+b_{mn} \end{pmatrix}.$$

The matrix sum is defined only for matrices having the same dimensions. Such matrices are said to be *conformable with respect to summation*, or when the context is clear simply *conformable*. Obviously the matrix sum is associative [i.e., $(A + B) + C = A + (B + C)$] and commutative [i.e., $A + B = B + A$]. The identity for summation is the conforming zero matrix.

These definitions make $\mathbb{R}^{m \times n}$ a real mn-dimensional vector space. Likewise the space $\mathbb{C}^{m \times n}$ is a complex mn-dimensional vector space. Thus any general results about real and complex vector spaces hold for $\mathbb{R}^{m \times n}$ and $\mathbb{C}^{m \times n}$.

The matrix product

The matrix-matrix product is a natural generalization of the matrix-vector product defined by (2.1). One motivation for its definition is the following. Suppose we have two linear systems

$$Ax = b \quad \text{and} \quad By = x.$$

Then y and b are related by a linear system $Cy = b$, where the coefficients matrix C can be obtained by substituting the scalar formulas for the components of $x = By$ into the scalar form of the equation $Ax = b$. It turns out that

$$c_{ij} = \sum_k a_{ik} b_{kj}. \tag{2.7}$$

On the other hand, if we symbolically substitute By for x in the first equation we get the equation

$$ABy = x.$$

Thus, the matrix product should satisfy $AB = C$, where the elements of C are given by (2.7). These considerations lead to the following definition.

Definition 2.3. *Let A be an $\ell \times m$ matrix and B be a $m \times n$ matrix. The product of A and B is the $\ell \times n$ matrix C whose elements are*

$$c_{ij} = \sum_{k=1}^{m} a_{ik} b_{kj}, \quad i = 1, \ldots, \ell, \, j = 1, \ldots, n.$$

For the product AB to be defined the number of columns of A must be equal to the number of rows of B. In this case we say that A and B are *conformable with respect to multiplication*. The product has the same number of rows as A and the same number of columns as B.

It is easily verified that if $A \in \mathbb{C}^{m \times n}$ then

$$I_m A = A I_n = A.$$

Thus the identity matrix is an identity for matrix multiplication.

The matrix product is associative [i.e., $(AB)C = A(BC)$] and distributes over the matrix sum [i.e., $A(B+C) = AB + AC$]. But it is not commutative. Commutativity can fail in three ways. First, if $\ell \neq n$ in the above definition, the product BA is not defined. Second, if $\ell = n$ but $m \neq n$, then AB is $n \times n$ but BA is $m \times m$, and the two products are of different orders. Thus we can have commutativity only if A and B are square and of the same order. But even here commutativity can fail, as almost any randomly chosen pair of matrices will show. For example,

$$\begin{pmatrix} 2 & 6 \\ 0 & 6 \end{pmatrix} \begin{pmatrix} 9 & 5 \\ 3 & 8 \end{pmatrix} = \begin{pmatrix} 36 & 58 \\ 18 & 48 \end{pmatrix} \neq \begin{pmatrix} 18 & 84 \\ 6 & 66 \end{pmatrix} = \begin{pmatrix} 9 & 5 \\ 3 & 8 \end{pmatrix} \begin{pmatrix} 2 & 6 \\ 0 & 6 \end{pmatrix}.$$

The failure to respect the noncommutativity of matrix products accounts for the bulk of mistakes made by people encountering matrices for the first time.

Since we have agreed to make no distinction between vectors and matrices with a single column, the above definition also defines the matrix-vector product Ax, which of course reduces to (2.1).

The transpose and symmetry

The final operation switches the rows and column of a matrix.

Definition 2.4. *Let A be an $m \times n$ matrix. The* TRANSPOSE *of A is the $n \times m$ matrix*

$$A^{\mathrm{T}} = \begin{pmatrix} a_{11} & a_{21} & \cdots & a_{m1} \\ a_{12} & a_{22} & \cdots & a_{m2} \\ \vdots & \vdots & & \vdots \\ a_{1n} & a_{2n} & \cdots & a_{mn} \end{pmatrix}.$$

The CONJUGATE TRANSPOSE *of A is the matrix*

$$A^{\mathrm{H}} = \begin{pmatrix} \bar{a}_{11} & \bar{a}_{21} & \cdots & \bar{a}_{m1} \\ \bar{a}_{12} & \bar{a}_{22} & \cdots & \bar{a}_{m2} \\ \vdots & \vdots & & \vdots \\ \bar{a}_{1n} & \bar{a}_{2n} & \cdots & \bar{a}_{mn} \end{pmatrix}.$$

By our conventions, vectors inherit the above definition of transpose and conjugate transpose. The transpose x^{T} of an n-vector x is an n-dimensional row vector.

The transpose and the conjugate transpose of a real matrix are the same. For a complex matrix they are different, and the difference is significant. For example, the number

$$x^{\mathrm{H}} x = \bar{x}_1 x_1 + \bar{x}_2 x_2 + \cdots + \bar{x}_n x_n = |x_1|^2 + |x_2|^2 + \cdots + |x_n|^2$$

is a nonnegative number that is the natural generalization of the square of the Euclidean length of a 3-vector. The number $x^{\mathrm{T}} x$ has no such interpretation for complex vectors, since it can be negative, complex, or even zero for nonzero x. For this reason, the simple transpose is used with complex vectors and matrices only in special applications.

The transpose and conjugate transpose interact nicely with matrix addition and multiplication. The proof of the following theorem is left as an exercise.

Theorem 2.5. *Let A and B be matrices. If $A + B$ is defined, then*

$$(A + B)^{\mathrm{T}} = A^{\mathrm{T}} + B^{\mathrm{T}}.$$

If AB is defined, then

$$(AB)^{\mathrm{T}} = B^{\mathrm{T}} A^{\mathrm{T}}.$$

The same holds for the conjugate transpose.

Matrices that are invariant under transposition occur very frequently in applications.

Definition 2.6. *A matrix A of order n is* SYMMETRIC *if $A = A^{\mathrm{T}}$. It is* HERMITIAN *if $A = A^{\mathrm{H}}$. The matrix A is* SKEW SYMMETRIC *if $A = -A^{\mathrm{T}}$ and* SKEW HERMITIAN *if $A = -A^{\mathrm{H}}$.*

Symmetric matrices are so called because they are symmetric about their diagonals:

$$a_{ij} = a_{ji}.$$

Hermitian matrices satisfy

$$a_{ij} = \bar{a}_{ji},$$

from which it immediately follows that the diagonal elements of a Hermitian matrix are real. The diagonals of a real skew symmetric matrix are zero, and the diagonals of a skew Hermitian matrix are pure imaginary. Any real symmetric matrix is Hermitian, but a complex symmetric matrix is not.

The trace and the determinant

In addition to the four matrix operations defined above, we mention two important functions of a square matrix. The first is little more than notational shorthand.

Definition 2.7. *Let A be of order n. The* TRACE *of A is the number*

$$\mathrm{trace}(A) = a_{11} + a_{22} + \cdots + a_{nn}.$$

The second function requires a little preparation. Let $\mathcal{I} = (i_1, i_2, \ldots, i_n)$ be a permutation of the integers $\{1, 2, \ldots, n\}$. The function

$$\varphi(\mathcal{I}) = \prod_{j=1}^{n-1} \prod_{k=j+1}^{n} (i_k - i_j)$$

is clearly nonzero since it is the product of differences of distinct integers. Thus we can define

$$\mathrm{sign}(\mathcal{I}) = \begin{cases} 1 & \text{if } \varphi(\mathcal{I}) > 0, \\ -1 & \text{if } \varphi(\mathcal{I}) < 0. \end{cases}$$

With this notation, we can make the following definition.

Definition 2.8. *The* DETERMINANT *of A is the number*

$$\det(A) = \sum_{(i_1, i_2, \ldots, i_n)} \mathrm{sign}[(i_1, i_2, \ldots, i_n)] a_{1i_1} a_{2i_2} \cdots a_{ni_n},$$

where $(i_1, i_2, \ldots, i_n)$ ranges over all permutations of the integers $1, 2, \ldots, n$.

The determinant has had a long and honorable history in the theory of matrices. It also appears as a volume element in multidimensional integrals. However, it is not much used in the derivation or analysis of matrix algorithms. For that reason, we will not develop its theory here. Instead we will list some of the properties that will be used later.

Theorem 2.9. *The determinant has the following properties (here we introduce terminology that will be defined later).*

1. $\det(A^{\mathrm{H}}) = \overline{\det(A)}$.
2. *If A is of order n, then $\det(\mu A) = \mu^n(A)$.*
3. $\det(AB) = \det(A)\det(B)$.
4. $\det(A^{-1}) = \det(A)^{-1}$.
5. *If A is block triangular with diagonal blocks $A_{11}, A_{22}, \ldots, A_{kk}$, then*

   $$\det(A) = \det(A_{11})\det(A_{22})\cdots\det(A_{kk}).$$

6. $\det(A)$ *is the product of the eigenvalues of A.*
7. $|\det(A)|$ *is the product of the singular values of A. (See §4.3.)*

2.4. SUBMATRICES AND PARTITIONING

One of the most powerful tools in matrix algebra is the ability to break a matrix into parts larger than scalars and express the basic matrix operations in terms of these parts. The parts are called submatrices, and the act of breaking up a matrix into submatrices is called partitioning.

Submatrices

A submatrix of a matrix A is a matrix formed from the intersection of sets of rows and columns of A. For example, if A is a 4×4 matrix, the matrices

$$\begin{pmatrix} a_{22} & a_{24} \\ a_{32} & a_{34} \end{pmatrix} \quad \text{and} \quad \begin{pmatrix} a_{22} & a_{23} \\ a_{32} & a_{33} \end{pmatrix}$$

are submatrices of A. The second matrix is called a contiguous submatrix because it is in the intersection of contiguous rows and columns; that is, its elements form a connected cluster in the original matrix. A matrix can be partitioned in many ways into contiguous submatrices. The power of such partitionings is that matrix operations may be used in the interior of the matrix itself.

We begin by defining the notion of a submatrix.

Definition 2.10. *Let $A \in \mathbb{C}^{m \times n}$ matrix. Let $1 \leq i_1 < i_2 < \cdots < i_p \leq m$ and $1 \leq j_1 < j_2 < \cdots < j_q \leq n$. Then the matrix*

$$B = \begin{pmatrix} a_{i_1 j_1} & a_{i_1 j_2} & \cdots & a_{i_1 j_q} \\ a_{i_2 j_1} & a_{i_2 j_2} & \cdots & a_{i_2 j_q} \\ \vdots & \vdots & & \vdots \\ a_{i_p j_1} & a_{i_p j_2} & \cdots & a_{i_p j_q} \end{pmatrix}$$

consisting of the elements in the intersection of rows $1 \leq i_1 < i_2 < \cdots < i_p \leq m$ and columns $1 \leq j_1 < j_2 < \cdots < j_q \leq n$ is a SUBMATRIX *A. The* COMPLEMENTARY SUBMATRIX *is the submatrix corresponding to the complements of the sets $\{i_1, i_2, \ldots, i_p\}$ and $\{j_1, j_2, \ldots, j_q\}$. If we have $i_{k+1} = i_k+1$ ($k = 1, \ldots, p-1$) and $j_{k+1} = j_k+1$ ($k = 1, \ldots, q-1$), then B is a* CONTIGUOUS SUBMATRIX. *If $p = q$ and $i_k = j_k$ ($k = 1, \ldots, p$), then B is a* PRINCIPAL SUBMATRIX. *If $i_p = p$ and $j_q = q$, then B is a* LEADING SUBMATRIX. *If, on the other hand, $i_1 = m-p+1$ and $j_1 = n-q+1$, then B is a* TRAILING SUBMATRIX.

Thus a principal submatrix is one formed from the same rows and columns. A leading submatrix is a submatrix in the northwest corner of A. A trailing submatrix lies in the southeast corner. For example, in the following Wilkinson diagram

$$\begin{pmatrix} \ell & \ell & \ell & x & x & x \\ \ell & \ell & \ell & x & x & x \\ \ell & \ell & \ell & t & t & t \\ x & x & x & t & t & t \end{pmatrix}$$

the 3×3 matrix whose elements are ℓ is a leading principal submatrix and the 2×3 submatrix whose element are t is a trailing submatrix.

Partitions

We begin with a definition.

Definition 2.11. *Let $A \in \mathbb{C}^{m \times n}$. A* PARTITIONING *of A is a representation of A in the form*

$$A = \begin{pmatrix} A_{11} & A_{12} & \cdots & A_{1q} \\ A_{21} & A_{12} & \cdots & A_{1q} \\ \vdots & \vdots & & \vdots \\ A_{p1} & A_{p2} & \cdots & A_{pq} \end{pmatrix},$$

where $A_{ij} \in \mathbb{C}^{m_i \times n_j}$ are contiguous submatrices, $m_1 + \cdots + m_p = m$, and $n_1 + \cdots + n_q = n$. The elements A_{ij} of the partition are called BLOCKS.

By this definition The blocks in any one column must all have the same number of columns. Similarly, the blocks in any one row must have the same number of rows.

A matrix can be partitioned in many ways. We will write

$$A = (a_1 \ a_2 \ \cdots \ a_n),$$

where a_j is the jth column of A. In this case A is said to be *partitioned by columns*. [We slipped in a partition by columns in (2.3).] A matrix can also be *partitioned by rows*:

$$A = \begin{pmatrix} a_1^{\mathrm{T}} \\ a_2^{\mathrm{T}} \\ \vdots \\ a_m^{\mathrm{T}} \end{pmatrix},$$

where a_i^{T} is the ith row of A. Again and again we will encounter the 2×2 partition

$$A = \begin{pmatrix} A_{11} & A_{12} \\ A_{21} & A_{22} \end{pmatrix},$$

particularly in the form where A_{11} is a scalar:

$$A = \begin{pmatrix} \alpha_{11} & a_{12}^{\mathrm{T}} \\ a_{21} & A_{22} \end{pmatrix}.$$

Northwest indexing

The indexing conventions we have used here are natural enough when the concern is with the partition itself. However, it can lead to conflicts of notation when it comes to describing matrix algorithms. For example, if A is of order n and in the partition

$$A = \begin{pmatrix} A_{11} & a_{12} \\ a_{21}^T & \alpha_{22} \end{pmatrix}$$

the submatrix A_{11} is of order $n-1$, then the element we have designated by α_{22} is actually the (n,n)-element of A and must be written as such in any algorithm. An alternate convention that avoids this problem is to index the blocks of a partition by the position of the element in the northwest corner of the blocks. With this convention the above matrix becomes

$$A = \begin{pmatrix} A_{11} & a_{1n} \\ a_{n1}^T & \alpha_{nn} \end{pmatrix}.$$

We will call this convention *northwest indexing* and say that the partition has been *indexed to the northwest*.

Partitioning and matrix operations

The power of matrix partitioning lies in the fact that partitions interact nicely with matrix operations. For example, if

$$A = \begin{pmatrix} A_{11} & A_{12} \\ A_{21} & A_{22} \end{pmatrix}$$

and

$$B = \begin{pmatrix} B_{11} & B_{12} \\ B_{21} & B_{22} \end{pmatrix},$$

then

$$AB = \begin{pmatrix} A_{11}B_{11} + A_{12}B_{21} & A_{11}B_{12} + A_{12}B_{22} \\ A_{21}B_{11} + A_{22}B_{21} & A_{21}B_{12} + A_{22}B_{22} \end{pmatrix},$$

provided that the dimensions of the partitions allow the indicated products and sums. In other words, the partitioned product is formed by treating the submatrices as scalars and performing an ordinary multiplication of 2×2 matrices. This idea generalizes. The proof of the following theorem is left as an exercise.

Theorem 2.12. *Let*

$$A = \begin{pmatrix} A_{11} & A_{12} & \cdots & A_{1q} \\ A_{21} & A_{22} & \cdots & A_{2q} \\ \vdots & \vdots & & \vdots \\ A_{p1} & A_{p2} & \cdots & A_{pq} \end{pmatrix}$$

and
$$B = \begin{pmatrix} B_{11} & B_{12} & \cdots & B_{1s} \\ B_{21} & B_{22} & \cdots & B_{2s} \\ \vdots & \vdots & & \vdots \\ B_{r1} & B_{r2} & \cdots & B_{rs} \end{pmatrix},$$
where $A_{ij} \in \mathbb{C}^{k_i \times \ell_j}$ and $B_{ij} \in \mathbb{C}^{m_i \times n_i}$. Then
$$A^T = \begin{pmatrix} A_{11}^T & A_{21}^T & \cdots & A_{p1}^T \\ A_{12}^T & A_{22}^T & \cdots & A_{p2}^T \\ \vdots & \vdots & & \vdots \\ A_{1q}^T & A_{2q}^T & \cdots & A_{pq}^T \end{pmatrix},$$
and the same equation holds with the transpose replaced by the conjugate transpose. If $p = r$, $q = s$, $k_i = m_i$, and $\ell_j = n_j$, then
$$A + B = \begin{pmatrix} A_{11} + B_{11} & A_{12} + B_{12} & \cdots & A_{1q} + B_{1q} \\ A_{21} + B_{21} & A_{22} + B_{22} & \cdots & A_{2q} + B_{2q} \\ \vdots & \vdots & & \vdots \\ A_{p1} + B_{p1} & A_{p2} + B_{p2} & \cdots & A_{pq} + B_{pq} \end{pmatrix}.$$
If $q = r$ and $\ell_i = m_i$, then
$$AB = \begin{pmatrix} \sum_{i=1}^{q} A_{1i} B_{i1} & \sum_{i=1}^{q} A_{1i} B_{i2} & \cdots & \sum_{i=1}^{q} A_{1i} B_{is} \\ \sum_{i=1}^{q} A_{2i} B_{i1} & \sum_{i=1}^{q} A_{2i} B_{i2} & \cdots & \sum_{i=1}^{q} A_{2i} B_{is} \\ \vdots & \vdots & & \vdots \\ \sum_{i=1}^{q} A_{pi} B_{i1} & \sum_{i=1}^{q} A_{pi} B_{i2} & \cdots & \sum_{i=1}^{q} A_{pi} B_{is} \end{pmatrix}.$$

The restrictions on the dimensions of the matrices in the above theorem insure conformity. The general principal is to treat the submatrices as scalars and perform the operations. However, in transposition the individual submatrices must also be transposed. And in multiplying partitioned matrices, keep in mind that the matrix product is not commutative.

Block forms

The various forms of matrices — diagonal matrices, triangular matrices, etc. — have block analogues. For example, a matrix A is *block upper triangular* if it can be partitioned in the form
$$A = \begin{pmatrix} A_{11} & A_{12} & A_{13} & \cdots & A_{1m} \\ 0 & A_{22} & A_{23} & \cdots & A_{21m} \\ 0 & 0 & A_{13} & \cdots & A_{1m} \\ \vdots & \vdots & \vdots & & \vdots \\ 0 & 0 & 0 & \cdots & A_{mm} \end{pmatrix}, \tag{2.8}$$
where the diagonal blocks A_{ii} are square.

2.5. SOME ELEMENTARY CONSTRUCTIONS

The following is a potpourri of elementary constructions that will appear throughout this work. They are good illustrations of the ideas introduced above.

Inner products

Given two n-vectors x and y, the *inner product* of x and y is the scalar

$$y^H x = \bar{y}_1 x_1 + \bar{y}_2 x_1 + \cdots + \bar{y}_n x_n. \tag{2.9}$$

When x and y are real 3-vectors of length one, the inner product is the cosine of the angle between x and y. This observation provides one way of extending the definition of angle to more general settings [see (4.18) and Definition 4.35].

The inner product is also known as the *scalar product* or the *dot product*.

Outer products

Given an n-vector x and an m-vector y, the *outer product* of x and y is the $m \times n$ matrix

$$xy^H = \begin{pmatrix} x_1 \bar{y}_1 & x_1 \bar{y}_2 & \cdots & x_1 \bar{y}_n \\ x_2 \bar{y}_1 & x_2 \bar{y}_2 & \cdots & x_2 \bar{y}_n \\ \vdots & \vdots & & \vdots \\ x_m \bar{y}_1 & x_m \bar{y}_2 & \cdots & x_m \bar{y}_n \end{pmatrix}.$$

The outer product is a special case of a full-rank factorization to be treated later (see Theorem 3.13).

Linear combinations

The linear combination

$$y = \alpha_1 x_1 + \alpha_2 x_2 + \cdots + \alpha_k x_k$$

has a useful matrix representation. Let $X = (x_1 \ x_2 \ \cdots \ x_k)$ and form a vector $a = (\alpha_1 \ \alpha_2 \ \cdots \ \alpha_k)^T$ from the coefficients of the linear transformation. Then it is easily verified that

$$y = Xa.$$

In other words:

> *The product of a matrix and a vector is a linear combination of the columns of the matrix. The coefficients of the linear combination are the components of the vector.*

If A is a $k\times\ell$ matrix partitioned by columns, then

$$XA = X(a_1\ a_2\ \cdots\ a_\ell) = (Xa_1\ Xa_2\ \cdots\ Xa_\ell).$$

Thus:

> The columns of XA are linear combinations of the columns of X. The coefficients of the linear combination for the jth column Xa are the elements of the jth column of A.

Column and row scaling

Let $A\in\mathbb{C}^{m\times n}$, and let

$$D = \mathrm{diag}(\delta_1, \delta_2, \ldots, \delta_n).$$

If A is partitioned by columns,

$$AD = (\delta_1 a_1\ \delta_2 a_2\ \cdots\ \delta_n a_n).$$

In other words:

> The columns of the product AD are the original columns scaled by the corresponding diagonal elements of D.

Likewise:

> The rows of DA are the original rows scaled by the diagonal elements of D.

Permuting rows and columns

Let $A\in\mathbb{C}^{m\times n}$. Then Ae_j is easily seen to be the jth column of A. It follows that if

$$P = (\mathbf{e}_{j_1}\ \mathbf{e}_{j_2}\ \cdots\ \mathbf{e}_{j_n})$$

is a permutation matrix, then

$$AP = (a_{j_1}\ a_{j_2}\ \cdots\ a_{j_n}).$$

In other words:

> Postmultiplying a matrix by a permutation matrix permutes the columns of the matrix into the order of the permutation.

Likewise:

> Premultiplying a matrix by the transpose of a permutation matrix permutes the rows of the matrix into the order of the permutation.

Undoing a permutation

It is easy to see that:

If P is a permutation matrix, then $P^\mathrm{T} P = PP^\mathrm{T} = I$.

Consequently, having interchanged columns by computing $B = AP$, we can undo the interchanges by computing $A = BP^\mathrm{T}$.

Crossing a matrix

Let $\mathcal{F} \in \mathbb{R}^{n \times n}$ be defined by

$$\mathcal{F} = (\mathbf{e}_n \ \mathbf{e}_{n-1} \ \cdots \ \mathbf{e}_1). \tag{2.10}$$

Then it is easily verified that if T is a triangular matrix then $\mathcal{F}T$ and $T\mathcal{F}$ are cross triangular. More generally, $\mathcal{F}x$ is the vector obtained by reversing the order of the components of x. We will call $\mathcal{F}$ the *cross operator*.

Extracting and inserting submatrices

Sometimes it is necessary to have an explicit formula for a submatrix of a given matrix A.

Let $A \in \mathbb{C}^{m \times n}$ and let $1 \leq i_1 < i_2 < \cdots < i_p \leq m$ and $1 \leq j_1 < j_2 < \cdots < j_q \leq n$. Let

$$E = (\mathbf{e}_{i_1} \ \mathbf{e}_{i_2} \ \cdots \ \mathbf{e}_{i_p}) \quad \text{and} \quad F = (\mathbf{e}_{j_1} \ \mathbf{e}_{j_2} \ \cdots \ \mathbf{e}_{j_q}).$$

Then $E^\mathrm{T} AF$ is the submatrix in the intersection of rows $\{i_1, \ldots, i_p\}$ and columns $\{j_1, \ldots, j_q\}$ of A. Moreover, if $B \in \mathbb{C}^{p \times q}$, then forming $A + EBF^\mathrm{T}$ replaces the submatrix $E^\mathrm{T} AF$ with $E^\mathrm{T} AF + B$.

2.6. LU DECOMPOSITIONS

A matrix decomposition is a factorization of a matrix into a product of simpler matrices. Decompositions are useful in matrix computations because they can simplify the solution of a problem. For example, if a matrix can be factored into the product of lower and upper triangular matrices, the solution of a linear system involving that matrix reduces to the solution of two triangular systems. The existence of such an LU decomposition is the point of the following theorem.

Theorem 2.13 (Lagrange, Gauss, Jacobi). Let $A \neq 0$ be an $m \times n$ matrix. Then there are permutation matrices P and Q and an integer $k \leq \min\{m, n\}$ such that

$$P^\mathrm{T} AQ = LU, \tag{2.11}$$

where

1. L is an $m \times k$ unit lower trapezoidal matrix,
2. U is a $k \times n$ upper trapezoidal matrix with nonzero diagonal elements. (2.12)

This factorization is called a PIVOTED LU DECOMPOSITION.

Proof. The proof is by induction on m, i.e., the number of rows in A.

For $m = 1$, let α_{1j} be a nonzero element of A. Let $P = 1$, and let Q be the permutation obtained by interchanging columns one and j of the identity matrix. Then the $(1,1)$-element of $P^T A Q$ is nonzero. Hence if we take $L = 1$ and $U = P^T A Q$, then L and U satisfy (2.11) and (2.12).

Now let $m > 1$. Let $\hat{P}$ and $\hat{Q}$ be permutations such that the $(1,1)$-element of $\hat{P}^T A \hat{Q}$ is nonzero. Partition

$$\hat{P}^T A \hat{Q} = \begin{pmatrix} \alpha & d^T \\ c & B \end{pmatrix},$$

and let

$$\ell = \begin{pmatrix} 1 \\ \alpha^{-1} c \end{pmatrix} \quad \text{and} \quad u^T = (\alpha \ d^T).$$

Then it is easily verified that

$$\hat{P}^T A \hat{Q} - \ell u^T = \begin{pmatrix} 0 & 0 \\ 0 & B - \alpha^{-1} c d^T \end{pmatrix}.$$

If $B - \alpha^{-1} c d^T = 0$, then we may take $L = \ell$, $U = u^T$, $P = \hat{P}$, and $Q = \hat{Q}$. Otherwise by the induction hypothesis, the matrix $B - \alpha^{-1} c d^T$ has an LU decomposition

$$\check{P}^T (B - \alpha^{-1} c d^T) \check{Q} = \check{L} \check{U}. \tag{2.13}$$

If we set

$$P = \hat{P} \begin{pmatrix} 1 & 0 \\ 0 & \check{P} \end{pmatrix} \quad \text{and} \quad Q = \hat{Q} \begin{pmatrix} 1 & 0 \\ 0 & \check{Q} \end{pmatrix}$$

and

$$L = \begin{pmatrix} 1 & 0 \\ \alpha^{-1} \check{P}^T c & \check{L} \end{pmatrix} \quad \text{and} \quad U = \begin{pmatrix} \alpha & d^T \check{Q} \\ 0 & \check{U} \end{pmatrix},$$

then P, Q, L, and U satisfy (2.11) and (2.12). ∎

Three comments on this theorem.

- Up to the permutations P and Q, the LU decomposition is unique. We will defer the proof until Chapter 3 (Theorem 1.5, Chapter 3), which is devoted to the LU decomposition and its variants.

SEC. 2. MATRICES

- The proof of Theorem 2.13 is constructive in that it presents an algorithm for computing LU decompositions. Specifically, interchange rows and columns of A so that its $(1,1)$-element is nonzero. Then with A partitioned as in (2.13), form $B - \alpha^{-1}cd^T$ and apply the procedure just sketched recursively. This process is called Gaussian elimination.

- The integer k is unique, but the proof does not establish this fact. For a proof see Theorem 3.13.

2.7. HOMOGENEOUS EQUATIONS

A central problem in matrix computations is to solve the linear system

$$Ax = b$$

or, when A has more rows than columns, at least compute an x such that Ax is a good approximation to b. In either case the solution will be unique if and only if the *homogeneous equation* or *system*

$$Av = 0 \qquad (2.14)$$

has only the solution $v = 0$. For if $Av = 0$ with $v \neq 0$, then $A(x+v) = Ax$, and $x+v$ solves the problem whenever x does. Conversely if $Ax = Ay$ for $x \neq y$, the vector $v = x - y$ is a nontrivial solution of (2.14).

If $A \in \mathbb{C}^{m \times n}$ and $m \geq n$, the homogeneous equation (2.14) has a nontrivial solution only in special circumstances [see (3.14.1)]. If $m < n$, the system (2.14) is said to be *underdetermined*. Because its right-hand side is zero it always has a nontrivial solution, as the proof of following theorem shows.

Theorem 2.14. *An underdetermined homogeneous system has a nontrivial solution.*

Proof. If $A = 0$, any nonzero vector v is a solution of (2.14). Otherwise let $P^T A Q = LU$ be an LU decomposition of A. Suppose that $Uw = 0$, where $w \neq 0$. Then with $v = Qw$, we have

$$Av = AQw = PLUw = 0.$$

Thus the problem becomes one of finding a nontrivial solution of the system $Uw = 0$.

Because U is upper trapezoidal, we can write it in the form

$$U = \begin{pmatrix} v & u^T \\ 0 & U_* \end{pmatrix}.$$

Moreover, $v \neq 0$. By an obvious induction step (we leave the base case as an exercise), the system $U_* w_* = 0$ has a nontrivial solution. Let $w^T = (-v^{-1} u^T w_* \; w_*^T)$. Then

$$Uw = \begin{pmatrix} v & u^T \\ 0 & U_* \end{pmatrix} \begin{pmatrix} -v^{-1} u^T w_* \\ w_* \end{pmatrix} = \begin{pmatrix} -u^T w_* + u^T w_* \\ U_* w_* \end{pmatrix} = 0. \quad \blacksquare$$

2.8. Notes and references

For general references on matrices and their applications see the addenda to this chapter.

Indexing conventions

The reason matrix indices begin at one in this work and in most books and articles on matrix computations is that they all treat matrix algorithms independently of their applications. Scientists and engineers, on the other hand, have no difficulty coming up with unusual indexing schemes to match their applications. For example, queueing theorists, whose queues can be empty, generally start their matrices with a $(0,0)$-element.

Hyphens and other considerations

Adjectives and adverbs tend to pile up to the left of the word "matrix"; e.g., upper triangular matrix. Strict orthographic convention would have us write "upper-triangular matrix," something nobody does since there is no such thing as an upper matrix. In principle, a block upper triangular matrix could be simply a partitioned upper triangular matrix, but by convention a block upper triangular matrix has the form (2.8). This convention breaks down when there is more than one structure to block. Anyone writing about block Toeplitz Hessenberg matrices should picture it in a display.

Nomenclature for triangular matrices

The conventional notation for upper and lower triangular matrices comes from English and German. The use of L and U to denote lower and upper triangular matrices is clear enough. But a German tradition calls these matrices left and right triangular. Hence "L" stands also for the German *links* meaning left. The use of "R" to mean an upper triangular matrix comes from the German *rechts* meaning right.

Complex symmetric matrices

Real symmetric and Hermitian matrices have nice properties that make them a numerical pleasure to work with. Complex symmetric matrices are not as easy to handle. Unfortunately, they arise in real life applications—from the numerical treatment of the Helmholtz equation, for example.

Determinants

Most linear algebra texts treat determinants in varying degrees. For a historical survey, Muir's *Theory of Determinants in the Historical Order of Development* [238] is unsurpassed. His shorter *Treatise on the Theory of Determinants* [237] contains everything you wanted to know about determinants—and then some.

Partitioned matrices

Partitioning is a powerful tool for proving theorems and deriving algorithms. A typical example is our derivation of the LU decomposition. An early example is Schur's proof that any matrix is unitarily similar to a triangular matrix [274, 1909]. However, the technique came to be widely used only in the last half of this century. It is instructive to compare the treatment of matrix algorithms in Dwyer's *Linear Computations* [112, 1951], which looks backward to the days of hand computation, with the treatment in Householder's *Principles of Numerical Analysis* [187, 1953], which looks forward to digital computers.

The northwest indexing convention is, I think, new. It has the additional advantage that if the dimensions of a partitioned matrix are known, the dimensions of its blocks can be determined by inspection.

The LU decomposition

The LU decomposition was originally derived as a decomposition of quadratic and bilinear forms. Lagrange, in the very first paper in his collected works [205, 1759], derives the algorithm we call Gaussian elimination, using it to find out if a quadratic form is positive definite. His purpose was to determine whether a stationary point of a function was actually a minimum. Lagrange's work does not seem to have influenced his successors.

The definitive treatment of decomposition is due to Gauss, who introduced it in his treatment of the motion of heavenly bodies [130, 1809] as a device for determining the precision of least squares estimates and a year after [131, 1810] as a numerical technique for solving the normal equations. He later [134, 1823] described the algorithm as follows. Here Ω is a residual sum of squares which depends on the unknown parameters x, y, z, etc.

> *Specifically, the function Ω can be reduced to the form*
>
> $$\frac{u^0 u^0}{\mathcal{A}^0} + \frac{u' u'}{\mathcal{B}'} + \frac{u'' u''}{\mathcal{C}''} + \frac{u''' u'''}{\mathcal{D}'''} + \text{etc.} + M,$$
>
> *in which the divisors $\mathcal{A}^0$, $\mathcal{B}'$, $\mathcal{C}''$, $\mathcal{D}'''$, etc. are constants and u^0, u', u'', u''', etc. are linear functions of x, y, z, etc. However, the second function u' is independent of x; the third u'' is independent of x and y; the fourth u''' is independent of x, y, and z, and so on. The last function $u^{(\pi-1)}$ depends only on the last of the unknowns x, y, z, etc. Moreover, the coefficients $\mathcal{A}^0$, $\mathcal{B}'$, $\mathcal{C}''$, etc. multiply x, y, z, etc. in u^0, u', u'', etc. respectively. Given this reduction, we may easily find x, y, z, etc. in reverse order after setting $u^0 = 0$, $u' = 0$, $u'' = 0$, $u''' = 0$, etc.*

The relation to the LU decomposition is that the coefficients of Gauss's x, y, z, etc. in the functions u^0, u', u'', etc. are proportional to the rows of U. For more details see [306].

Both Lagrange and Gauss worked with symmetric matrices. The extension to general matrices is due to Jacobi [191, 1857, posthumous], who reduced a bilinear form in the spirit of Lagrange and Gauss.

3. LINEAR ALGEBRA

The vector spaces $\mathbb{R}^n$ and $\mathbb{C}^n$ have an algebraic structure and an analytic structure. The latter is inherited from the analytic properties of real and complex numbers and will be treated in §4, where norms and limits are introduced. The algebraic structure is common to all finite-dimensional vector spaces, and its study is called *linear algebra*. The purpose of this section is to develop the fundamentals of linear algebra. For definiteness, we will confine ourselves to $\mathbb{R}^n$, but, with the exception of (3.11), the results hold for any finite-dimensional vector space.

3.1. SUBSPACES, LINEAR INDEPENDENCE, AND BASES

A subspace is a nonempty set of vectors that is closed under addition and multiplication by a scalar. In $\mathbb{R}^n$ a basic fact about subspaces is that they can be represented by a finite set of vectors called a basis. In this subsection we will show how this is done.

Subspaces

Any linear combination of vectors in a vector space remains in that vector space; i.e., vector spaces are closed under linear combinations. Subsets of a vector space may or may not have this property. For example, the usual (x, y)-plane in $\mathbb{R}^3$, defined by

$$\{x = \alpha \mathbf{e}_1 + \beta \mathbf{e}_2 \colon \alpha, \beta \in \mathbb{R}\},$$

is closed under linear combinations. On the other hand, the octant

$$\{x \colon \xi_i \geq 0, i = 1, 2, 3\}$$

is not closed under linear combinations, since the difference of two vectors with nonnegative components may have negative components. More subtly, $\mathbb{R}^n$ regarded as a subset of $\mathbb{C}^n$ is not closed under linear combinations, since the product of a real nonzero vector and a complex scalar has complex components.

Subsets closed under linear combinations have a name.

Definition 3.1. *A nonempty subset* $\mathcal{X} \subset \mathbb{R}^n$ *is a* SUBSPACE *if*

$$x_1, x_2, \ldots, x_k \in \mathcal{X} \implies \alpha_1 x_1 + \alpha_2 x_2 + \cdots + \alpha_k x_k \in \mathcal{X}$$

for any scalars $\alpha_1, \alpha_2, \ldots, \alpha_k$.

The first thing to note about subspaces is that they are themselves vector spaces. Thus the general results of linear algebra apply equally to subspaces and the spaces that contain them.

SEC. 3. LINEAR ALGEBRA

Subspaces have an algebra of their own. The proof of the following theorem is left as an exercise.

Theorem 3.2. *Let $\mathcal{X}$ and $\mathcal{Y}$ be subspaces of $\mathbb{R}^n$. Then the following are subspaces.*

1. $\{0\}$
2. $\mathcal{X} \cap \mathcal{Y} = \{z: z \in \mathcal{X} \text{ and } z \in \mathcal{Y}\}$
3. $\mathcal{X} + \mathcal{Y} = \{x + y: x \in \mathcal{X} \text{ and } y \in \mathcal{Y}\}$

Since for any subspace $\mathcal{X}$ we have $\{0\} + \mathcal{X} = \mathcal{X}$, the subspace consisting of only the zero vector acts as an additive identity. If we regard the operation of intersection as a sort of multiplication, then $\{0\}$ is an annihilator under multiplication, as it should be.

If $\mathcal{X}$ and $\mathcal{Y}$ are subspaces of $\mathbb{R}^n$ and $\mathcal{X} \cap \mathcal{Y} = \{0\}$, we say that the subspaces are *disjoint*. Note that disjoint subspaces are not disjoint sets, since they both contain the zero vector. The sum of disjoint subspaces $\mathcal{X}$ and $\mathcal{Y}$ is written $\mathcal{X} \oplus \mathcal{Y}$ and is called the *direct sum* of $\mathcal{X}$ and $\mathcal{Y}$.

The set of all linear combinations of a set of vectors $\mathcal{X}$ is easily seen to be a subspace. Hence the following definition.

Definition 3.3. *Let $\mathcal{X} \subset \mathbb{R}^n$. The set of all linear combinations of members of $\mathcal{X}$ is a subspace $\mathcal{Y}$ called the* SPAN *of $\mathcal{X}$. We write*

$$\mathcal{Y} = \text{span}(\mathcal{X}).$$

The space spanned by the vectors $x_1, x_2, \ldots, x_k$ is also written $\text{span}(x_1, x_2, \ldots, x_k)$.

In particular,

$$\mathbb{R}^n = \text{span}(\mathbf{e}_1, \mathbf{e}_2, \ldots, \mathbf{e}_n), \tag{3.1}$$

since for any $x \in \mathbb{R}^n$

$$x = \xi_1 \mathbf{e}_1 + \xi_2 \mathbf{e}_2 + \cdots + \xi_n \mathbf{e}_n. \tag{3.2}$$

Linear independence

We have just observed that the unit vectors span $\mathbb{R}^n$. Moreover, no proper subset of the unit vectors spans $\mathbb{R}^n$. For if one of the unit vectors is missing from (3.2), the corresponding component of x is zero. A minimal spanning set such as the unit vectors is called a basis. Before we begin our treatment of bases, we introduce a far reaching definition.

Definition 3.4. *The vectors $x_1, x_2, \ldots, x_k$ are* LINEARLY INDEPENDENT *if*

$$\alpha_1 x_1 + \alpha_2 x_2 + \cdots + \alpha_k x_k = 0 \implies \alpha_1 = \alpha_2 = \cdots = \alpha_k = 0.$$

Otherwise they are LINEARLY DEPENDENT.

Let us consider the implications of this definition.

- In plain words the definition says that a set of vectors is linearly independent if and only if no nontrivial linear combination of members of the set is zero. In terms of matrices, the columns of $X = (x_1 \; x_2 \; \cdots \; x_k)$ are linearly independent if and only if

$$Xa = 0 \implies a = 0.$$

This matrix formulation of linear independence will be widely used in what follows.

- Any set containing the zero vector is linearly dependent. In particular, a set consisting of a single vector x is independent if and only if $x \neq 0$.

- If $x_1, x_2, \ldots, x_k$ are linearly dependent, then one of them can be expressed as a linear combination of the others. For there are constants α_i, not all zero, such that

$$\alpha_1 x_1 + \alpha_2 x_2 + \cdots + \alpha_k x_k = 0. \tag{3.3}$$

If, say, $\alpha_j \neq 0$, then we may solve (3.3) for x_j in the form $x_j = \alpha_j^{-1} \sum_{i \neq j} \alpha_i x_i$. In particular, if $x_1, x_2, \ldots, x_{k-1}$ are independent, then j can be taken equal to k. For if $\alpha_k = 0$, in (3.3), then $x_1, x_2, \ldots, x_{k-1}$ are linearly dependent.

- If a vector can be expressed as a linear combination of a set of linearly independent vectors $x_1, x_2, \ldots, x_k$, then that expression is unique. For if

$$\alpha_1 x_1 + \alpha_2 x_2 + \cdots + \alpha_k x_k = \beta_1 x_1 + \beta_2 x_2 + \cdots + \beta_k x_k,$$

then $\sum_i (\alpha_1 - \beta_i) x_i = 0$, and by the linear independence of the x_i we have $\alpha_i - \beta_i = 0$ ($i = 1, \ldots, k$).

- A particularly important example of a set of linearly independent vectors is the columns of a lower trapezoidal matrix L whose diagonal elements are nonzero. For suppose $La = 0$, with $a \neq 0$. Let a_i be the first nonzero component of a. Then writing out the ith equation from the relation $La = 0$, we get

$$0 = \alpha_1 \ell_{i1} + \alpha_2 \ell_{i2} + \cdots + \alpha_i \ell_{ii} = \alpha_i \ell_{ii},$$

the last equality following from the fact that $\alpha_1, \alpha_2, \ldots, \alpha_{i-1}$ are all zero. Since $\ell_{ii} \neq 0$, we must have $\alpha_i = 0$, a contradiction. This result is also true of upper and lower triangular matrices.

Bases

A basis for a subspace is a set of linearly independent vectors that span the subspace.

Definition 3.5. *Let $\mathcal{X}$ be a subspace of $\mathbb{R}^n$. A set of vectors $\{b_1, b_2, \ldots, b_k\}$ is a BASIS for $\mathcal{X}$ if*

1. *$b_1, b_2, \ldots, b_k$ are linearly independent,*

SEC. 3. LINEAR ALGEBRA 31

2. $\text{span}(b_1, b_2, \ldots, b_k) = \mathcal{X}$.

From (3.1) we see that the unit vectors e_i, which are clearly independent, form a basis for $\mathbb{R}^n$. Here is a useful generalization.

Example 3.6. *The space $\mathbb{R}^{m \times n}$ regarded as a vector space has the basis*

$$\{E_{ij} = \mathbf{e}_i \mathbf{e}_j^\mathrm{T} : i = 1, \ldots, m; j = 1, \ldots, n\}.$$

For the matrices E_{ij} are clearly linearly independent and

$$A = \sum_{i=1}^{m} \sum_{j=1}^{n} \alpha_{ij} E_{ij}.$$

If $\mathcal{B} = \{b_1, b_2, \ldots, b_k\}$ is a basis for $\mathcal{X}$ and $B = (b_1 \; b_2 \; \cdots \; b_k)$, then any member $x \in \mathcal{X}$ can be written uniquely in the form $x = Ba$. This characterization is so useful that it will pay to abuse nomenclature and call the matrix B along with the set $\mathcal{B}$ a basis for $\mathcal{X}$.

We want to show that every subspace has a basis. An obvious way is to start picking vectors from the subspace, throwing away the dependent ones and keeping the ones that are independent. The problem is to assure that this process will terminate. To do so we have to proceed indirectly by first proving a theorem about bases before proving that bases exist.

Theorem 3.7. *Let $\mathcal{X}$ be a subspace of $\mathbb{R}^n$. If $\{b_1, b_2, \ldots, b_k\}$ is a basis for $\mathcal{X}$, then any collection of $k+1$ or more vectors in $\mathcal{X}$ are linearly dependent.*

Proof. Let $B = (b_1 \; b_2 \; \cdots \; b_k)$ and $C = (c_1 \; c_2 \; \cdots \; c_\ell)$, where $\ell > k$. Then each column of C is a linear combination of the columns of B, say $c_i = Bv_i$, where $v_i \in \mathbb{R}^k$. If we set $V = (v_1 \; v_2 \; \cdots \; v_k)$, then $C = BV$. But V has more columns than rows. Hence by Theorem 2.14 there is a nonzero vector w such that $Vw = 0$. It follows that

$$Cw = BVw = B \cdot 0 = 0$$

and the columns of C are linearly dependent. ∎

An important corollary of this theorem is the following.

Corollary 3.8. *If $\mathcal{B}$ and $\mathcal{B}'$ are bases for the subspace $\mathcal{X}$, then they have the same number of elements.*

For if they did not, the larger set would be linearly dependent. In particular, since the n unit vectors form a basis for $\mathbb{R}^n$, any basis of $\mathbb{R}^n$ has n elements.

We are now in a position to show that every subspace has a basis. In particular, we can choose a basis from any spanning subset — and even specify some of the vectors.

Theorem 3.9. *Let $\mathcal{X}$ be a nontrivial subspace of $\mathbb{R}^n$ that is spanned by the set $\mathcal{B}$. Suppose $b_1, b_2, \ldots, b_\ell \in \mathcal{B}$ are linearly independent. Then there is a subset*

$$\{b_1, b_2, \ldots, b_\ell, b_{\ell+1}, \ldots, b_k\} \subset \mathcal{B}$$

that forms a basis for $\mathcal{X}$.

Proof. Let $\mathcal{B}_\ell = \{b_1, b_2, \ldots, b_\ell\}$. Note that ℓ may be zero, in which case $\mathcal{B}_0$ is the empty set.

Suppose now that for $i \geq \ell$ we have constructed a set $\mathcal{B}_i \subset \mathcal{B}$ of i linearly independent vectors. If there is some vector $b_{i+1} \in \mathcal{B}$ that is not a linear combination of the members of $\mathcal{B}_i$, adjoin it to $\mathcal{B}_i$ to get a new set of linearly independent vectors $\mathcal{B}_{i+1}$. Since $\mathbb{R}^n$ cannot contain more than n linearly independent vectors, this process of adjoining vectors must stop with some $\mathcal{B}_k$, where $k \leq n$.

We must now show that any vector $x \in \mathcal{X}$ can be expressed as a linear combination of the members of $\mathcal{B}_k$. Since $\mathcal{B}$ spans $\mathcal{X}$, the vector x may be expressed as a linear combination of the members of $\mathcal{B}$: say

$$x = \alpha_1 b_{i_1} + \alpha_1 b_{i_1} + \cdots + \alpha_m b_{i_m} \equiv Ba.$$

But by construction, $B = (b_1 \ b_2 \ \cdots \ b_k)V$ for some $k \times m$ matrix V. Hence

$$x = (b_1 \ b_2 \ \cdots \ b_k)(Va)$$

expresses x as a linear combination of the b_i. ∎

The theorem shows that we can not only construct a basis for a subspace $\mathcal{X}$ but we can start from any linearly independent subset of $\mathcal{X}$ and extend it to a basis. In particular, suppose we start with a basis $\{x_1, x_2, \ldots, x_k\}$ for a subspace $\mathcal{X}$ and extend it to a basis $\{x_1, x_2, \ldots, x_k, y_1, y_2, \ldots, y_{n-k}\}$ for $\mathbb{R}^n$ itself. The space $\mathcal{Y} = \text{span}(y_1, y_2, \ldots, y_{n-k})$ is disjoint from $\mathcal{X}$, and

$$\mathcal{X} + \mathcal{Y} = \mathbb{R}^n. \tag{3.4}$$

Two spaces satisfying (3.4) are said to be *complementary*. We thus have the following result.

Corollary 3.10. *Any subspace of $\mathbb{R}^n$ has a complement.*

Dimension

Since any nontrivial subspace of $\mathbb{R}^n$ has a basis and all bases for it have the same number of elements, we may make the following definition.

Definition 3.11. *Let $\mathcal{X} \subset \mathbb{R}^n$ be a nontrivial subspace. Then the DIMENSION of $\mathcal{X}$ is the number of elements in a basis for $\mathcal{X}$. We write $\dim(\mathcal{X})$ for the dimension of $\mathcal{X}$. By convention the dimension of the trivial subspace $\{0\}$ is zero.*

Sec. 3. Linear Algebra

Thus the dimension of $\mathbb{R}^n$ is n — a fact which seems obvious but, as we have seen, takes some proving.

The dimension satisfies certain relations

Theorem 3.12. *For any subspaces $\mathcal{X}$ and $\mathcal{Y}$ of $\mathbb{R}^n$,*

$$\dim(\mathcal{X} \cap \mathcal{Y}) \leq \dim(\mathcal{X}), \dim(\mathcal{Y})$$

and

$$\dim(\mathcal{X} + \mathcal{Y}) = \dim(\mathcal{X}) + \dim(\mathcal{Y}) - \dim(\mathcal{X} \cap \mathcal{Y}). \tag{3.5}$$

Proof. We will prove the second equality, leaving the first inequality as an exercise.

Let $\dim(\mathcal{X} \cap \mathcal{Y}) = j$, $\dim(\mathcal{X}) = k$, and $\dim(\mathcal{Y}) = \ell$. Let $A \in \mathbb{R}^{n \times j}$ be a basis for $\mathcal{X} \cap \mathcal{Y}$ and extend it to a basis $(A \ B)$ for $\mathcal{X}$. Note that $B \in \mathbb{R}^{n \times (k-j)}$. Similarly let $C \in \mathbb{R}^{n \times (\ell - j)}$ be such that $(A \ C)$ is a basis for $\mathcal{Y}$. Then clearly the columns of $(A \ B \ C)$ span $\mathcal{X} + \mathcal{Y}$. But the columns of $(A \ B \ C)$ are linearly independent. To see this note that if

$$Au + Bv + Cw = 0 \tag{3.6}$$

is a nontrivial linear combination then we must have $Bv \neq 0$. For if $Bv = 0$, then $v = 0$ and $Au + Cw = 0$. By the independence of the columns of $(A \ C)$ we would then have $u = 0$ and $w = 0$, and the combination (3.6) would be trivial. Hence we may assume that $Bv \neq 0$. But since

$$Bv = -Au - Cw \in \mathcal{Y},$$

and since $Bv \in \mathcal{X}$, it follows that $Bv \in \mathcal{X} \cap \mathcal{Y}$, contradicting the definition of B.

Thus $(A \ B \ C)$ is a basis for $\mathcal{X} + \mathcal{Y}$, and hence the number of columns of the matrix $(A \ B \ C)$ is the dimension of $\mathcal{X} + \mathcal{Y}$. But $(A \ B \ C)$ has

$$j + (k - j) + (\ell - j) = k + \ell - j = \dim(\mathcal{X}) + \dim(\mathcal{Y}) - \dim(\mathcal{X} \cap \mathcal{Y})$$

columns. ∎

In what follows we will most frequently use (3.5) in the two weaker forms

$$\dim(\mathcal{X} + \mathcal{Y}) \leq \dim(X) + \dim(Y) \tag{3.7}$$

and

$$\dim(\mathcal{X} \oplus \mathcal{Y}) = \dim(X) + \dim(Y).$$

3.2. Rank and Nullity

Matrices have useful subspaces associated with them. For example, the column space of a matrix is the space spanned by the columns of the matrix. In this subsection we are going to establish the properties of two such subspaces, the columns space and the null space. But first we begin with a useful matrix factorization.

A full-rank factorization

If a matrix has linearly dependent columns, some of them are redundant, and it is natural to seek a more economical representation. For example, the $m \times n$ matrix

$$A = (\beta_1 a \ \beta_2 a \ \cdots \ \beta_n a),$$

whose columns are proportional to one another, can be written in the form

$$A = ab^{\mathrm{T}},$$

where $b^{\mathrm{T}} = (\beta_1 \ \beta_2 \ \cdots \ \beta_n)$. The representation encodes the matrix economically using $m+n$ scalars instead of the mn scalars required by the more conventional representation.

The following theorem shows that any matrix has an analogous representation.

Theorem 3.13. *Let $A \in \mathbb{R}^{m \times n}$. Then A has a* FULL-RANK FACTORIZATION *of the form*

$$A = XY^{\mathrm{T}},$$

where $X \in \mathbb{R}^{m \times k}$ and $Y \in \mathbb{R}^{n \times k}$ have linearly independent columns. If $A = \hat{X}\hat{Y}^{\mathrm{T}}$ is any other full-rank factorization, then $\hat{X}$ and $\hat{Y}$ also have k columns.

Proof. Let $P^{\mathrm{T}}AQ = LU$ be an LU decomposition of A (see Theorem 2.13). Since L and U^{T} are lower trapezoidal with nonzero diagonal elements, their columns are linearly independent. But then so are the columns of $X = PL$ and $Y = QU$. Moreover, $A = XY^{\mathrm{T}}$, so that XY^{T} is a full-rank factorization of A.

If $A = XY^{\mathrm{T}}$ is a full-rank factorization of A, then the columns of X form a basis for the space spanned by the columns of A. Thus the X-factors of all full-rank factorizations have the same number of columns. By conformity, the Y-factors also have the same number of columns. ∎

Rank and nullity

A consequence of the existence of full-rank factorizations is that the spaces spanned by the columns and rows of a matrix have the same dimension. We call that dimension the rank of the matrix.

Definition 3.14. *Let $A \in \mathbb{R}^{m \times n}$. The* COLUMN SPACE *of A is the subspace*

$$\mathcal{R}(A) = \{Ax : x \in \mathbb{R}^n\}.$$

The ROW SPACE *of A is the space $\mathcal{R}(A^{\mathrm{T}})$. The* RANK *of A is the integer*

$$\mathrm{rank}(A) \stackrel{\mathrm{def}}{=} \dim[\mathcal{R}(A)] = \dim[\mathcal{R}(A^{\mathrm{T}})].$$

The rank satisfies two inequalities.

SEC. 3. LINEAR ALGEBRA

Theorem 3.15. *The rank of the sum of two matrices satisfies*

$$\text{rank}(A + B) \leq \text{rank}(A) + \text{rank}(B). \tag{3.8}$$

The rank of the product satisfies

$$\text{rank}(AB) \leq \min\{\text{rank}(A), \text{rank}(B)\}. \tag{3.9}$$

Proof. Since $\mathcal{R}(A+B)$ is contained in $\mathcal{R}(A) + \mathcal{R}(B)$, it follows from (3.7) that the rank satisfies (3.8). Since the row space of AB is contained in the row space of A, we have $\text{rank}(AB) \leq \text{rank}(A)$. Likewise, since the column space of AB is contained in the column space of A, we have $\text{rank}(AB) \leq \text{rank}(B)$. Together these inequalities imply (3.9). ∎

We have observed that a solution of the linear system $Ax = b$ is unique if and only if the homogeneous equation $Ax = 0$ has no nontrivial solutions [see (2.14)]. It is easy to see that the set of all solutions of $Ax = 0$ forms a subspace. Hence the following definition.

Definition 3.16. *Let $A \in \mathbb{R}^{m \times n}$. The set*

$$\mathcal{N}(A) = \{x: Ax = 0\}$$

is called the NULL SPACE *of A. The dimension of $\mathcal{N}(A)$ is called the* NULLITY *of A and is written*

$$\text{null}(A) = \dim[\mathcal{N}(A)].$$

A nonzero vector in the null space of A — that is, a nonzero vector x satisfying $Ax = 0$ — is called a NULL VECTOR *of A.*

The null space determines how the solutions of linear systems can vary. Specifically:

If the system $Ax = b$ has a solution, say x_0, then any solution lies in the set (3.10)

$$x_0 + \mathcal{N}(A) = \{x_0 + x: x \in \mathcal{N}(A)\}.$$

Thus the nullity of A in some sense measures the amount of nonuniqueness in the solutions of linear systems involving A.

The basic facts about the null space are summarized in the following theorem.

Theorem 3.17. *Let $A \in \mathbb{R}^{m \times n}$. Then*

$$\mathbb{R}^n = \mathcal{N}(A) \oplus \mathcal{R}(A^T) \tag{3.11}$$

and

$$\text{rank}(A) + \text{null}(A) = n. \tag{3.12}$$

Proof. Let rank(A) = k and let $A = XY^T$ be a full-rank factorization of A. Then $\mathcal{R}(A^T) = \mathcal{R}(Y)$ and $\mathcal{N}(A) = \mathcal{N}(Y^T)$.

Let $V = (v_1 \; v_2 \; \cdots \; v_\ell)$ be a basis for $\mathcal{N}(A)$. Then the v_i are independent of the columns of Y. For suppose $v_i = Ya$ for some nonzero a. Since $v_i \neq 0$ and $v_i^T Y = 0$, it follows that

$$0 < v_i^T v_i = v_i^T Y a = 0, \tag{3.13}$$

a contradiction. Hence the columns of $(Y \; V)$ span a $(k+\ell)$-dimensional space.

We will now show that $k+\ell = n$. If not, the system

$$\begin{pmatrix} Y^T \\ V^T \end{pmatrix} v = 0$$

is underdetermined and has a nontrivial solution. Since $Y^T v = 0$, the solution v is in $\mathcal{N}(A)$. Since $V^T v = 0$, the vector v is independent of the columns of V, which contradicts the fact that the columns of V span $\mathcal{N}(A)$.

Thus $\mathcal{R}(A^T)$ and $\mathcal{N}(A)$ are disjoint subspaces whose direct sum spans an n-dimensional subspace of $\mathbb{R}^n$. But the only n-dimensional subspace of $\mathbb{R}^n$ is $\mathbb{R}^n$ itself. This establishes (3.11).

Equation (3.12) is equivalent to the equation $k+\ell = n$. ∎

Two comments.

- As the inequality in (3.13) suggests, the proof depends in an essential way on the fact that the components of vectors in $\mathbb{R}^n$ are real. (The proof goes through in $\mathbb{C}^n$ if the transpose is replaced by the conjugate transpose.) In fact, (3.11) is not true of finite-dimensional vector spaces over general fields, though (3.12) is.

- The theorem shows that $\mathcal{R}(A^T)$ and $\mathcal{N}(A)$ are complementary subspaces. In fact, the theorem offers an alternative proof that every subspace has a complement. For if the columns of X span a subspace $\mathcal{X}$, then $\mathcal{N}(X^T)$ is a complementary subspace. In fact, it is an orthogonal complement — see (4.25).

3.3. NONSINGULARITY AND INVERSES

We now turn to one of the main concerns of this volume — the solution of linear systems $Ax = b$. We have already established that a solution is unique if and only if null(A) = 0 [see (3.10)]. Here we will be concerned with the existence of solutions.

Linear systems and nonsingularity

We begin with a trivial but useful characterization of the existence of solutions of the system $Ax = b$. Specifically, for any x the vector $Ax \in \mathcal{R}(A)$. Conversely any vector in $\mathcal{R}(A)$ can be written in the form Ax. Hence:

A solution of the system $Ax = b$ exists if and only if $b \in \mathcal{R}(A)$.

Sec. 3. Linear Algebra

To proceed further it will help to make some definitions.

Definition 3.18. *A matrix* $A \in \mathbb{R}^{m \times n}$ *is of* FULL RANK *if*

$$\text{rank}(A) = \min\{m, n\}.$$

If $m \geq n$, A *is of* FULL COLUMN RANK. *If* $m \leq n$, A *is of* FULL ROW RANK. *If* $m = n$, A *is* NONSINGULAR.

If A *is not of full rank,* A *is* RANK DEGENERATE *or simply* DEGENERATE. *If* A *is square and degenerate, then* A *is* SINGULAR.

Most matrices want to be of full rank. Even when they start off degenerate, the slightest perturbation will usually eliminate the degeneracy. This is an important consideration in a discipline where matrices usually come equipped with errors in their elements.

Example 3.19. *The matrix*

$$A = \begin{pmatrix} 1 & 1 \\ 2 & 2 \end{pmatrix}$$

is singular. But a nonzero perturbation, however small, in any one element will make it nonsingular.

If we are willing to accept that full-rank matrices are more likely to occur than degenerate ones, we can make some statements — case by case.

1. If $m > n$, the matrix $(A \; b)$ will generally have full column rank, and hence b will not lie in $\mathcal{R}(A)$. Thus overdetermined systems usually have no solutions. On the other hand, $\text{null}(A)$ will generally be zero; and in this case when a solution exists it is unique.
2. If $m < n$, the matrix A will generally be of full row rank, and hence of rank m. In this case $\mathcal{R}(A) = \mathbb{R}^m$, and a solution exists. However, $\text{null}(A) > 0$, so no solution is unique. (3.14)
3. If $m = n$, the matrix A will generally be nonsingular. In this case a solution exists and is unique.

A warning. The above statements, correct though they are, should not lull one into thinking errors in a matrix can make the difficulties associated with degeneracies go away. On the contrary, the numerical and scientific problems associated with near degeneracies are subtle and not easy to deal with. These problems are treated more fully in §1, Chapter 5.

Nonsingularity and inverses

Square systems are unusual because the same condition — nonsingularity — that insures existence also insures uniqueness. It also implies the existence of an inverse matrix.

Theorem 3.20. *Let A be nonsingular of order n. Then there is a unique matrix A^{-1}, called the* INVERSE *of A, such that*

$$A^{-1}A = AA^{-1} = I_n. \tag{3.15}$$

Proof. Since rank$(A) = n$, the equations

$$Ax_i = e_i, \quad i = 1, 2, \ldots, n,$$

each have unique solutions x_i. If we set $X = (x_1 \; x_2 \; \cdots \; x_n)$, then $AX = I$. Similarly, by considering the systems

$$A^T y_i = e_i, \quad i = 1, 2, \ldots, n,$$

we may determine a unique matrix Y such that $Y^T A = I$. But

$$Y^T = Y^T I = Y^T(AX) = (Y^T A)X = IX = X.$$

Hence the matrix $A^{-1} \equiv Y^T = X$ satisfies (3.15). ∎

If A is nonsingular, then so is A^T, and $(A^T)^{-1} = (A^{-1})^T$. A convenient shorthand for the inverse of a transpose is A^{-T}. If A is complex, we write A^{-H} for the inverse conjugate transpose.

The existence of an inverse is just one characterization of nonsingularity. Here are a few more. It is an instructive exercise to establish their equivalence. (The last three characterizations presuppose more background than we have furnished to date.)

Theorem 3.21. *Let A be of order n. Then A is nonsingular if and only if it satisfies any one of the following conditions.*

1. rank$(A) = n$.
2. null$(A) = 0$.
3. For any vector b, the system $Ax = b$ has a solution.
4. If a solution of the system $Ax = b$ exists, it is unique.
5. For all x, $Ax = 0 \Rightarrow x = 0$.
6. The columns (rows) of A are linearly independent.
7. There is a matrix A^{-1} such that $A^{-1}A = AA^{-1} = I$.
8. det$(A) \neq 0$.
9. The eigenvalues of A are nonzero.
10. The singular values of A are nonzero.

From item 5 in the above it easily follows that:

The product of square matrices A and B is nonsingular if and only if A and B are nonsingular. In that case

$$(AB)^{-1} = B^{-1}A.$$

Unfortunately there are no simple general conditions for the existence of the inverse of the sum $A+B$ of two square matrices. However, in special cases we can assert the existence of such an inverse and even provide a formula; see (3.4), Chapter 4.

The inverse is in many respects the driving force behind matrix algebra. For example, it allows one to express the solution of a linear system $Ax = b$ succinctly as $x = A^{-1}b$. For this reason, disciplines that make heavy use of matrices load their books and papers with formulas containing inverses. Although these formulas are full of meaning to the specialist, they seldom represent the best way to compute. For example, to write $x = A^{-1}b$ is to suggest that one compute x by inverting A and multiplying the inverse by b — which is why specialists in matrix computations get frequent requests for programs to calculate inverses. But there are faster, more stable algorithms for solving linear systems than this invert-and-multiply algorithm. (For more on this point see §1.5, Chapter 3, and Example 4.11, Chapter 3.)

3.4. CHANGE OF BASES AND LINEAR TRANSFORMATIONS

The equation

$$x = \xi_1 \mathbf{e}_1 + \xi_2 \mathbf{e}_2 + \cdots + \xi_n \mathbf{e}_n$$

represents a vector $x \in \mathbb{R}^n$ as a sum of the unit vectors. This unit basis occupies a distinguished position because the coefficients of the representation are the components of the vector. In some instances, however, we may need to work with another basis. In this subsection, we shall show how to switch back and forth between bases.

Change of basis

First a definition.

Definition 3.22. *Let X be a basis for a subspace $\mathcal{X}$ in $\mathbb{R}^n$, and let $x = Xu$. Then the components of u are the* COMPONENTS OF x WITH RESPECT TO THE BASIS X.

By the definition of basis every $x \in \mathcal{X}$ can be represented in the form Xu and hence has components with respect to X. But what precisely are these components? The following theorem supplies the wherewithal to answer this question.

Theorem 3.23. *Let $X \in \mathbb{R}^{n \times k}$ have full column rank. Then there is a matrix X^{I} such that*

$$X^{\mathrm{I}} X = I.$$

Proof. The proof mimics the proof of Theorem 3.20. Since $\mathrm{rank}(X) = k$, the columns of X^{T} span $\mathbb{R}^k$. Hence the equations

$$X^{\mathrm{T}} y_i = \mathbf{e}_i, \qquad i = 1, 2, \ldots, k, \tag{3.16}$$

have solutions. If we set $X^{\mathrm{I}} = (y_1 \; y_2 \; \cdots \; y_k)^{\mathrm{T}}$, then equation (3.16) implies that $X^{\mathrm{I}} X = I$. ∎

The matrix X^{I} is called a *left inverse* of X. It is not unique unless X is square. For otherwise the systems (3.16) are underdetermined and do not have unique solutions.

The solution of the problem of computing the components of x with respect to X is now simple. If $x = Xu$ and X^{I} is a left inverse of X, then $u = X^{\mathrm{I}} x$ contains the components of x with respect to X. It is worth noting that although X^{I} is not unique, the vector $u = X^{\mathrm{I}} x$ is unique for any $x \in \mathcal{R}(X)$.

Now suppose we change to another basis $\hat{X}$. Then any vector $x \in \mathcal{X}$ can be expressed in the form $\hat{X} \hat{u}$. The relation of $\hat{u}$ to u is contained in the following theorem, in which we repeat some earlier results.

Theorem 3.24. *Let $\mathcal{X} \in \mathbb{R}^n$ be a subspace, and let X and $\hat{X}$ be bases for $\mathcal{X}$. Let X^{I} be a left inverse of X. Then*

$$P = X^{\mathrm{I}} \hat{X} \tag{3.17}$$

is nonsingular and

$$\hat{X} = XP.$$

If $x = Xu$ is in $\mathcal{X}$, then the components of

$$u = X^{\mathrm{I}} x \tag{3.18}$$

are the components of x with respect to X. Moreover, $P^{-1} X^{\mathrm{I}}$ is a left inverse of $\hat{X}$. Hence if $x = \hat{X} \hat{u}$,

$$\hat{u} = P^{-1} X^{\mathrm{I}} x = P^{-1} u.$$

Proof. We have already established (3.18). Since X is a basis and the columns of $\hat{X}$ lie in $\mathcal{X}$, we must have $\hat{X} = XP$ for some P. On premultiplying this relation by X^{I}, we get (3.17). The matrix P is nonsingular, since otherwise there would be a nonzero vector v such that $Pv = 0$. Then $\hat{X} v = XPv = 0$, contradicting the fact that $\hat{X}$ is a basis. The rest of the proof is a matter of direct verification. ∎

Linear transformations and matrices

Under a change of basis, the components of a vector change. We would expect the elements of a matrix likewise to change. But before we can say how they change, we must decide just what a matrix represents. Since it is easy to get bogged down in the details, we will start with a special case.

Let $A \in \mathbb{R}^{m \times n}$. With each vector $x \in \mathbb{R}^n$ we can associate a vector $f_A(x) \in \mathbb{R}^m$ defined by

$$f_A(x) = Ax.$$

SEC. 3. LINEAR ALGEBRA

The function is linear in the sense that

$$f_A(\alpha x + \beta y) = \alpha f_A(x) + \beta f_A(y).$$

Conversely, given a linear function $f: \mathbb{R}^n \to \mathbb{R}^m$, we can construct a corresponding matrix as follows. Let $a_i = f(\mathbf{e}_i)$ $(i = 1, 2, \ldots, n)$, and let $A = (a_1 \; a_2 \; \cdots \; a_n)$. If $x \in \mathbb{R}^n$, then

$$\begin{aligned} f(x) &= f(\xi_1 \mathbf{e}_1 + \cdots + \xi_n \mathbf{e}_n) \\ &= \xi_1 f(\mathbf{e}_1) + \cdots + \xi_1 f(\mathbf{e}_n) \\ &= \xi_1 a_1 + \cdots + \xi_n a_n \\ &= Ax. \end{aligned}$$

Thus the matrix-vector product Ax reproduces the action of the linear transformation f, and it is natural to call A the *matrix representation of f*.

Now suppose we change bases to Y in $\mathbb{R}^m$ and X in $\mathbb{R}^n$. Let $x = Xu$ and $f(x) = Yv$. What is the relation between u and v?

Let $X = (x_1 \; x_2 \; \cdots \; x_n)$ be partitioned by columns, and define

$$f(X) = (f(x_1) \; f(x_2) \; \cdots \; f(x_n)).$$

Then by the linearity of f we have $f(Xu) = f(X)u$. Now Y is square and nonsingular, so that $v = Y^{-1} f(X) u$. But $f(X) = AX$. Hence

$$v = (Y^{-1} A X) u.$$

We call the matrix $Y^{-1} A X$ the *representation of the linear transformation f with respect to the bases X and Y*. It transforms the vector of components of x with respect to X into the components of $f(x)$ with respect to Y.

Many matrix algorithms proceed by multiplying a matrix by nonsingular matrices until they assume a simple form. The material developed here gives us another way of looking at such transformations. Specifically, the replacement of A by $Y^{-1}AX$ amounts to changing from the natural basis of unit vectors to bases X and Y on the domain and range of the associate function f_A. The program of simplifying matrices by pre- and postmultiplications amounts to finding coordinate systems in which the underlying linear transformation has a simple structure.

Here is a particularly important example.

Example 3.25. *Let A be of order n. If we choose a basis X for $\mathbb{R}^n$, then the matrix representing the transformation corresponding to A in the new basis is $X^{-1}AX$. Such a transformation is called a* SIMILARITY TRANSFORMATION *and is important in the algebraic eigenvalue problem.*

The main result concerns linear transformations between subspaces. It is unnecessary to specify where the subspaces lie — they could be in the same space or in spaces of different dimensions. What is important is that the subspaces have bases. The proof of the following theorem follows the lines developed above and is left as an exercise.

Theorem 3.26. *Let $\mathcal{X}$ and $\mathcal{Y}$ be subspaces and let $f: \mathcal{X} \to \mathcal{Y}$ be linear. Let X and Y be bases for $\mathcal{X}$ and $\mathcal{Y}$, and let X^{I} and Y^{I} be left inverses for X and Y. For any $x \in \mathcal{X}$ let $u = X^{\mathrm{I}}x$ and $v = Y^{\mathrm{I}}f(x)$ be the components of x and $f(x)$ with respect to X and Y. Then*

$$v = [Y^{\mathrm{I}}f(X)]u.$$

We call the matrix $Y^{\mathrm{I}}f(X)$ the REPRESENTATION OF THE LINEAR TRANSFORMATION f WITH RESPECT TO THE BASES X AND Y.

If $\hat{X} = XP$ and $\hat{Y} = XQ$ are new bases for $\mathcal{X}$ and $\mathcal{Y}$, then P and Q are nonsingular and the matrix representing f with respect to the new bases is $Q^{-1}Y^{\mathrm{I}}f(X)P$.

3.5. NOTES AND REFERENCES

Linear algebra

The material in this section is the stuff of elementary linear algebra textbooks, somewhat compressed for a knowledgeable audience. For references see the addenda to this chapter.

Full-rank factorizations

Although the principal application of full-rank factorizations in this section is to characterize the rank of a matrix, they are ubiquitous in matrix computations. One of the reasons is that if the rank of a matrix is low a full-rank factorization provides an economical representation. We derived a full-rank factorization from the pivoted LU decomposition, but in fact they can be calculated from many of the decompositions to be treated later — e.g., the pivoted QR decomposition or the singular value decomposition. The tricky point is to decide what the rank is in the presence of error. See §1, Chapter 5, for more.

4. ANALYSIS

We have already pointed out that vectors and matrices regarded simply as arrays are not very interesting. The addition of algebraic operations gives them life and utility. But abstract linear algebra does not take into account the fact that our matrices are defined over real and complex numbers, numbers equipped with analytic notions such as absolute value and limit. The purpose of this section is to transfer these notions to vectors and matrices. We will consider three topics — norms, orthogonality and projections, and the singular value decomposition.

4.1. NORMS

Vector and matrix norms are natural generalizations of the absolute value of a number — they measure the magnitude of the objects they are applied to. As such they

Sec. 4. Analysis

can be used to define limits of vectors and matrices, and this notion of limit, it turns out, is independent of the particular norm used to define it. In this subsection we will introduce matrix and vector norms and describe their properties. The subsection concludes with an application to the perturbation of matrix inverses.

Componentwise inequalities and absolute values

The absolute value of a real or complex number satisfies the following three conditions.

1. Definiteness: $\xi \neq 0 \implies |\xi| > 0$
2. Homogeneity: $|\alpha \xi| = |\alpha||\xi|$
3. Triangle inequality: $|\xi + \eta| \leq |\xi| + |\eta|$

There are two ways of generalizing this notion to vectors and matrices. The first is to define functions on, say, $\mathbb{R}^n$ that satisfy the three above conditions (with ξ and η regarded as vectors and α remaining a scalar). Such functions are called norms, and they will be the chief concern of this subsection. However, we will first introduce a useful componentwise definition of absolute value.

The basic ideas are collected in the following definition.

Definition 4.1. *Let $A, B \in \mathbb{R}^{m \times n}$. Then $A \geq B$ if $\alpha_{ij} \geq \beta_{ij}$ and $A > B$ if $\alpha_{ij} > \beta_{ij}$. Similar definitions hold for the relations "<" and "$\leq$". If $A > 0$, then A is* POSITIVE. *If $A \geq 0$, then A is* NONNEGATIVE. *The* ABSOLUTE VALUE *of A is the matrix $|A|$ whose elements are $|\alpha_{ij}|$.*

There are several comments to be made about this definition.

• Be warned that the notation $A < B$ is sometimes used to mean that $B - A$ is positive definite (see §2.1, Chapter 3, for more on positive definite matrices).

• Although the above definitions have been cast in terms of matrices, they also apply to vectors.

• The relation $A > B$ means that *every* element of A is greater than the corresponding element of B. To say that $A \geq B$ with strict inequality in at least one element one has to say something like $A \geq B$ and $A \neq B$.

• If $A \neq 0$, the most we can say about $|A|$ is that $|A| \geq 0$. Thus the absolute value of a matrix is not, strictly speaking, a generalization of the absolute value of a scalar, since it is not definite. However, it is homogeneous and satisfies the triangle inequality.

• The absolute value of a matrix interacts nicely with the various matrix operations. For example,

$$|AB| \leq |A||B|,$$

a property called *consistency*. Again, if A is nonsingular, then $|A|$e is positive. In what follows we will use such properties freely, leaving the reader to supply their proofs.

Vector norms

As we mentioned in the introduction to this subsection, norms are generalizations of the absolute value function.

Definition 4.2. *A* VECTOR NORM *or simply a* NORM ON $\mathbb{C}^n$ *is a function* $\|\cdot\| : \mathbb{C}^n \to \mathbb{R}$ *satisfying the following conditions:*

1. $x \neq 0 \implies \|x\| > 0$,
2. $\|\alpha x\| = |\alpha| \|x\|$,
3. $\|x + y\| \leq \|x\| + \|y\|$.

Thus a vector norm is a definite, homogeneous function on $\mathbb{C}^n$ that satisfies the *triangle inequality* $\|x + y\| \leq \|x\| + \|y\|$. Vector norms on $\mathbb{R}^n$ are defined analogously.

The triangle inequality for vector norms has a useful variant

$$\|x - y\| \geq |\|x\| - \|y\||. \tag{4.1}$$

Another useful fact is that if $x \neq 0$ then

$$\left\| \frac{x}{\|x\|} \right\| = 1.$$

The process of dividing a nonzero vector by its norm to turn it into a vector of norm one is called *normalization*.

There are infinitely many distinct vector norms. For matrix computations, three are especially important.

Theorem 4.3. *The following three function on $\mathbb{C}^n$ are norms:*

1. $\|x\|_1 \stackrel{\text{def}}{=} \sum_i |x_i|$,
2. $\|x\|_2 \stackrel{\text{def}}{=} \sqrt{\sum_i |x_i|^2} = \sqrt{x^H x}$,
3. $\|x\|_\infty \stackrel{\text{def}}{=} \max_i |x_i|$.

The norms $\|\cdot\|_1$ *and* $\|\cdot\|_\infty$ *satisfy*

$$|x^H y| \leq \|x\|_1 \|y\|_\infty, \tag{4.2}$$

and for any x there is a y for which equality is attained—and vice versa. Moreover,

$$|x^H y| \leq \|x\|_2 \|y\|_2 \tag{4.3}$$

with equality if and only if x and y are linearly dependent.

Proof. The fact that $\|\cdot\|_1$ and $\|\cdot\|_\infty$ are norms satisfying (4.2) is left as an exercise.

The only nontrivial part of proving that $\|\cdot\|_2$ is a norm is to establish the triangle inequality. We begin by establishing (4.3). If x or y is nonzero, then the result is

SEC. 4. ANALYSIS   45

trivial. Otherwise, by normalizing x and y we may assume without loss of generality that $\|x\|_2 = \|y\|_2 = 1$. Moreover, by dividing y by a scalar of absolute value one, we may assume that $x^H y = |x^H y|$. Thus we must show that $x^H y \leq 1$. But

$$
\begin{aligned}
0 \leq \|(x-y)\|_2^2 &= (x-y)^H(x-y) \\
&= x^H x + y^H y - 2x^H y = 2 - 2x^H y.
\end{aligned}
\tag{4.4}
$$

Hence $x^H y \leq 1$, and the inequality (4.3) follows.

To prove the statement about the dependence of x and y, note that equality can be attained in (4.3) only if equality is attained in (4.4). In this case, $\|x - y\|_2 = 0$ and $x = y$, so that x and y are dependent.

Conversely if x and y are dependent, $x = \alpha y$ where $|\alpha| = 1$. Then $x^H y = \alpha x^H x = \alpha$, which implies that α is real and positive and hence is equal to one. Thus equality is attained in (4.3).

The proof of the triangular inequality for $\|\cdot\|_2$ is now elementary:

$$
\begin{aligned}
\|x+y\|_2^2 &= (x+y)^H(x+y) \\
&= x^H x + y^H y + x^H y + y^H x \\
&\leq \|x\|_2^2 + \|y\|_2^2 + 2|x^H y| \\
&\leq \|x\|_2^2 + \|y\|_2^2 + 2\|x\|_2\|y\|_2 \\
&= (\|x\|_2 + \|y\|_2)^2. \quad \blacksquare
\end{aligned}
$$

The norms defined in Theorem 4.3 are called the 1-, 2-, and ∞-norms. They are special cases of the *Hölder norms* defined for $0 \leq p \leq \infty$ by

$$\|x\|_p = \left(\sum_i |x_i|^p\right)^{\frac{1}{p}}$$

(the case $p = \infty$ is treated as a limit). The 1-norm is sometimes called the *Manhattan norm* because it is the distance you would have to traverse on a rectangular grid to get from one point to another. The 2-norm is also called the *Euclidean norm* because in real 2- or 3-space it is the Euclidean length of the vector x. All three norms are easy to compute.

Pairs of norms satisfying (4.2) with equality attainable are called *dual norms*. The inequality (4.3), which is called the *Cauchy inequality*, says that the 2-norm is self-dual. This fact is fundamental to the Euclidean geometry of $\mathbb{C}^n$, as we will see later.

Given a norm, it is easy to generate more. The proof of the following theorem is left as an exercise.

Theorem 4.4. *Let $\|\cdot\|$ be a norm on $\mathbb{C}^n$, and let $A \in \mathbb{C}^{n \times n}$ be nonsingular. Then the function $\mu_A(x)$ defined by*

$$\mu_A(x) = \|Ax\|$$

is a norm. If A is positive definite (Definition 2.1, Chapter 3), then function $\nu_A(x)$ defined by

$$\nu_A(x) = \sqrt{x^\mathrm{T} A x}$$

is a norm.

Norms and convergence

There is a natural way to extend the notion of limit from $\mathbb{C}$ to $\mathbb{C}^n$. Let

$$\{x_k = (\xi_1^{(k)}\ \xi_2^{(k)}\ \cdots\ \xi_n^{(k)})^\mathrm{T}\}_1^\infty$$

be a sequence of vectors in $\mathbb{C}^n$ and let $x \in \mathbb{C}^n$. Then we may write $x_k \to x$ provided

$$\lim_{k \to \infty} \xi_i^{(k)} = \xi_i, \qquad i = 1, 2, \ldots, n.$$

Such convergence is called *componentwise convergence*.

There is another way to define convergence in $\mathbb{C}^n$. For any sequence $\{\xi_k\}_1^\infty$ in $\mathbb{C}$ we have $\lim_k \xi_k = \xi$ if and only if $\lim_k |\xi_k - \xi| = 0$. It is therefore reasonable to define convergence in $\mathbb{C}^n$ by choosing a norm $\|\cdot\|$ and saying that $x_k \to x$ if $\lim_k \|x_k - x\| = 0$. This kind of convergence is called *normwise convergence*.

There is no compelling reason to expect the two notions of convergence to be equivalent. In fact, for infinite-dimensional vectors spaces they are not, as the following example shows.

Example 4.5. Let ℓ_∞ be the set of all infinite row vectors x satisfying

$$\|x\|_\infty \equiv \sup_i |\xi_i| < \infty.$$

It is easy to verify that $\|\cdot\|_\infty$ is a norm on ℓ_∞.

Now consider the infinite sequence whose first four terms are illustrated below:

$$\begin{aligned} x_1 &= (1\ 0\ 0\ 0\ \ldots), \\ x_2 &= (0\ 1\ 0\ 0\ \ldots), \\ x_3 &= (0\ 0\ 1\ 0\ \ldots), \\ x_4 &= (0\ 0\ 0\ 1\ \ldots). \end{aligned}$$

Clearly this sequence converges to the zero vector componentwise, since each component converges to zero. But $\|x_k - 0\|_\infty = 1$ for each k. Hence the sequence does not converge to zero in the ∞-norm.

Not only can pointwise and normwise convergence disagree, but different norms can generate different notions of convergence. Fortunately, we will only be dealing with finite-dimensional spaces, in which all notions of convergence coincide. The

SEC. 4. ANALYSIS

problem with establishing this fact is not one of specific norms. It is easy to show, for example, that the 1-, 2-, and ∞-norms all define the same notion of convergence and that it is the same as componentwise convergence. The problem is that we have infinitely many norms, and one of them might be a rogue. To eliminate this possibility we are going to prove that all norms are equivalent in the sense that they can be used to bound each other. For example, it is easy to see that

$$\|x\|_2 \leq \|x\|_1 \leq \sqrt{n}\|x\|_2. \tag{4.5}$$

The following theorem generalizes this result.

Theorem 4.6. *Let μ and ν be norms on $\mathbb{C}^n$. Then there are positive constants σ and τ such that*

$$\sigma\mu(x) \leq \nu(x) \leq \tau\mu(x), \qquad x \in \mathbb{C}^n.$$

Proof.

It is sufficient to prove the theorem for the case where ν is the 2-norm. (Why?) We will begin by establishing the upper bound on μ.

Let $\kappa = \max_i \mu(\mathbf{e}_i)$. Since $x = \sum_i \xi_i \mathbf{e}_i$,

$$\mu(x) \leq \sum_i |\xi_i|\mu(\mathbf{e}_i) \leq \kappa\|x\|_1 \leq \kappa\sqrt{n}\|x\|_2,$$

the last inequality following from (4.5). Hence, with $\sigma = 1/(\kappa\sqrt{n})$, we have $\sigma\mu(x) \leq \|x\|_2$.

An immediate consequence of the bound is that $\mu(x)$ as a function of x is continuous in the 2-norm. Specifically, from (4.1)

$$|\nu(x) - \nu(y)| \leq \nu(x-y) \leq \sigma^{-1}\|x-y\|_2.$$

Hence $\lim_{\|y-x\|_2 \to 0} |\nu(x) - \nu(y)| \leq \sigma^{-1} \lim_{\|y-x\|_2 \to 0} \|x-y\|_2 = 0$, which is the definition of continuity.

Now let $S = \{x \colon \|x\|_2 = 1\}$. Since S is closed and bounded and μ is continuous, μ assumes a minimum at some point $x_{\min}$ on S. Let $\tau = 1/\mu(x_{\min})$. Then

$$1 = \|x\|_2 \leq \tau\mu(x), \qquad x \in S. \tag{4.6}$$

Hence by the homogeneity of $\|\cdot\|$ and μ, (4.6) holds for all $x \in \mathbb{C}^n$. ∎

The equivalence of norms assures us that in $\mathbb{C}^n$ we can use any convenient norm to establish a continuity result or a perturbation theorem, and an equivalent result will hold for any other norm. As a practical matter, however, this may not be much comfort.

Example 4.7. *On $\mathbb{R}^2$ let $\|x\|_{\text{bad}} = \sqrt{\xi_1^2 + 10^{-100}\xi_2^2}$. Then convergence to, say, $\mathbf{e}$ is the same in $\|\cdot\|_{\text{bad}}$ as in the 2-norm. But if one is interested in the accuracy of the second component of a vector whose components are near one, then $\|\cdot\|_2$ does the job, whereas $\|\cdot\|_{\text{bad}}$ fails miserably. For example,*

$$\left\|\begin{pmatrix}1\\2\end{pmatrix} - \begin{pmatrix}1\\1\end{pmatrix}\right\|_2 = 1 \quad \text{but} \quad \left\|\begin{pmatrix}1\\2\end{pmatrix} - \begin{pmatrix}1\\1\end{pmatrix}\right\|_{\text{bad}} = 10^{-50}.$$

Matrix norms and consistency

The approach of defining a vector norm as a definite, homogeneous function satisfying the triangle inequality can be extended to matrices.

Definition 4.8. A MATRIX NORM ON $\mathbb{C}^{m\times n}$ *is a function* $\|\cdot\| : \mathbb{C}^{m\times n} \to \mathbb{R}$ *satisfying the following conditions:*

1. $A \neq 0 \implies \|A\| > 0$,
2. $\|\alpha A\| = |\alpha|\|A\|$, (4.7)
3. $\|A + B\| \leq \|A\| + \|B\|$.

All the properties of vector norms are equally true of matrix norms. In particular, all matrix norms are equivalent and define the same notion of limit, which is also the same as elementwise convergence.

A difficulty with this approach to matrix norms is that it does not specify how matrix norms interact with matrix multiplication. To compute upper bounds, we would like a multiplicative analogue of the triangle inequality:

$$\|AB\| \leq \|A\|\|B\|. \qquad (4.8)$$

However, the conditions (4.7) do not imply (4.8). For example, if we attempt to generalize the infinity norm in a natural way by setting

$$\|A\| = \max_{i,j} |a_{ij}|, \qquad (4.9)$$

then

$$\left\|\begin{pmatrix} 1 & 1 \\ 1 & 1 \end{pmatrix}\begin{pmatrix} 1 & 1 \\ 1 & 1 \end{pmatrix}\right\| = 2$$

but

$$\left\|\begin{pmatrix} 1 & 1 \\ 1 & 1 \end{pmatrix}\right\|\left\|\begin{pmatrix} 1 & 1 \\ 1 & 1 \end{pmatrix}\right\| = 1 \cdot 1 = 1.$$

The relation (4.8) is called consistency (with respect to multiplication). Although that equation implies that A is square (since A, B, and AB all must have the same dimensions), consistency can be defined whenever the product AB is defined.

Definition 4.9. *Let* $\|\cdot\|_{ln}$, $\|\cdot\|_{lm}$, *and* $\|\cdot\|_{mn}$ *be norms on* $\mathbb{C}^{l\times n}$, $\mathbb{C}^{l\times m}$, $\mathbb{C}^{m\times n}$. *Then these norms are* CONSISTENT *if*

$$\|AB\|_{ln} \leq \|A\|_{lm}\|B\|_{mn}$$

for all $A \in \mathbb{C}^{l\times m}$ *and* $B \in \mathbb{C}^{m\times n}$.

Sec. 4. Analysis

Since we have agreed to identify $\mathbb{C}^{n\times 1}$ with $\mathbb{C}^n$, the above definition also serves to define consistency between matrix and vector norms.

An example of a consistent matrix norm is the widely used Frobenius norm.

Definition 4.10. *The* Frobenius norm *is the function* $\|\cdot\|_F$ *defined for any matrix by*

$$\|A\|_F = \sqrt{\sum_{i,j} |a_{ij}|^2}.$$

The Frobenius norm is defined in analogy with the vector 2-norm and reduces to it when the matrix in question has only one column. Just as $\|x\|_2^2$ can be written in the form $x^H x$, so can the square of the Frobenius norm be written as

$$\|A\|_F^2 = \text{trace}(A^H A) = \text{trace}(AA^H). \tag{4.10}$$

Since the diagonal elements of $A^T A$ are the squares of the 2-norms of the columns of A, we have

$$\|A\|_F^2 = \sum_j \|a_j\|_2^2,$$

where a_j is the jth column of A. There is a similar expression in terms of the rows of A.

The Cauchy inequality can be written as a consistency relation in the Frobenius norm:

$$\|x^H y\|_F \leq \|x\|_F \|y\|_F.$$

The proof of the following theorem begins with this special case and elevates it to general consistency of the Frobenius norm.

Theorem 4.11. *Whenever the product AB is defined,*

$$\|AB\|_F \leq \|A\|_F \|B\|_F.$$

Proof. We will first establish the result for the matrix-vector product $y = Ax$. Let $A^H = (a_1 \ a_2 \ \cdots \ a_m)$ be a partitioning of A by rows, so that $y_i = a_i^H x$. Then

$$\|y\|_F^2 = \sum_i |a_i^H x|^2 \leq \sum_i \|a_i\|_2^2 \|x\|_2^2 = \|x\|_2^2 \sum_i \|a_i\|_2 = \|A\|_F^2 \|x\|^2,$$

the inequality following from the Cauchy inequality. Now let $B = (b_1 \ \cdots \ b_n)$ be partitioned by columns. Then

$$\|AB\|_F^2 = \sum_j \|Ab_j\|_2^2 \leq \sum_j \|A\|_F^2 \|b_j\|_2^2 = \|A\|_F^2 \|B\|_F^2. \quad \blacksquare$$

It sometimes happens that we have a consistent matrix norm, say defined on $\mathbb{C}^{n\times n}$, and require a consistent vector norm. The following theorem provides one.

Theorem 4.12. *Let $\|\cdot\|$ be a consistent matrix norm on $\mathbb{C}^{n\times n}$. Then there is a vector norm that is consistent with $\|\cdot\|$.*

Proof. Let v be nonzero and define $\|\cdot\|_v$ by

$$\|x\|_v = \|xv^{\mathrm{T}}\|.$$

Then it is easily verified that $\|\cdot\|_v$ is a vector norm. But

$$\|Ax\|_v = \|Axv^{\mathrm{T}}\| \leq \|A\|\|xv^{\mathrm{T}}\| = \|A\|\|x\|_v. \quad\blacksquare$$

Operator norms

The obvious generalizations of the usual vector norms to matrices are not guaranteed to yield consistent matrix norms, as the example of the ∞-norm shows [see (4.9)]. However, there is another way to turn vector norms into matrix norms, one that always results in consistent norms. The idea is to regard the matrix in question as an operator on vectors and ask how much it changes the size of a vector.

For definiteness, let ν be a norm on $\mathbb{C}^n$, and let A be of order n. For any vector x with $\nu(x) = 1$ let $\rho_x = \nu(Ax)$, so that ρ_x measures how much A expands or contracts x in the norm ν. Although ρ_x varies with x, it has a well-defined maximum. This maximum defines a norm, called the operator norm subordinate to the vector norm $\|\cdot\|$.

Before we make a formal definition, an observation is in order. Most of the norms we work with are generic — that is, they are defined generally for spaces of all dimensions. Although norms on different spaces are different mathematical objects, it is convenient to refer to them by a common notation, as we have with the 1-, 2-, and ∞-norms. We shall call such a collection a *family of norms*. In defining operator norms, it is natural to work with families, since the result is a new family of matrix norms defined for matrices of all dimensions. This is the procedure we adopt in the following definition.

Definition 4.13. *Let ν be a family of vector norms. Then the* OPERATOR NORM SUB-ORDINATE TO ν *or* GENERATED BY ν *is defined by*

$$\|A\|_\nu = \max_{\nu(x)=1} \nu(Ax).$$

Although we have defined operator norms for a family of vector norms there is nothing to prevent us from restricting the definition to one or two spaces — e.g., to $\mathbb{C}^n$.

The properties of operator norms are summarized in the following theorem.

Theorem 4.14. *Let $\|\cdot\|_\nu$ be an operator norm subordinate to a family of vector norms ν. Then $\|\cdot\|_\nu$ is a consistent family of matrix norms satisfying*

$$\|I\|_\nu = 1. \tag{4.11}$$

SEC. 4. ANALYSIS 51

The operator norm is consistent with the generating vector norm. Moreover, if $\nu(\xi) = |\xi|$, then $\|a\|_\nu = \nu(a)$.

Proof. We must first verify that $\|\cdot\|$ is a norm. Definiteness and homogeneity are easily verified. For the triangle inequality we have

$$\begin{aligned}\|A+B\|_\nu &= \max_{\nu(x)=1} \nu[(A+B)x] \\ &= \max_{\nu(x)=1} \nu(Ax + Bx) \\ &\leq \max_{\nu(x)=1} [\nu(Ax) + \nu(Bx)] \\ &\leq \max_{\nu(x)=1} \nu(Ax) + \max_{\nu(x)=1} \nu(Bx) \\ &= \|A\|_\nu + \|B\|_\nu.\end{aligned}$$

For consistency, first note that by the definition of an operator norm we have a fortiori $\nu(Ax) \leq \|A\|_\nu \nu(x)$. Hence

$$\begin{aligned}\|AB\| &= \max_{\nu(x)=1} \nu[(AB)x] \\ &= \max_{\nu(x)=1} \nu[A(Bx)] \\ &\leq \max_{\nu(x)=1} \|A\|_\nu \nu(Bx) \\ &= \|A\|_\nu \max_{\nu(x)=1} \nu(Bx) \\ &= \|A\|_\nu \|B\|_\nu.\end{aligned}$$

For (4.11) we have

$$\|I\|_\nu = \max_{\nu(x)=1} \nu(Ix) = \max_{\nu(x)=1} \nu(x) = 1.$$

Finally, suppose that $\nu(\xi) = |\xi|$. Then

$$\|a\|_\nu = \max_{\nu(\xi)=1} \nu(a\xi) = \max_{|\xi|=1} \nu(a\xi) = \nu(a),$$

so that the operator norm reproduces the generating vector norm. ∎

Turning now to the usual norms, we have the following characterization of the 1 and ∞ operator norms.

Theorem 4.15. *The 1- and ∞-norms may be characterized as follows:*

1. $\|A\|_1 = \max_j \sum_i |a_{ij}|$,
2. $\|A\|_\infty = \max_i \sum_j |a_{ij}|$.

The proof is left as an exercise.

Since the vector 1- and ∞-norms satisfy $\|\xi\| = |\xi|$, the operator norms and the vector norms coincide for vectors, as can be verified directly from the characterizations. Hence there is no need to introduce new notation for the operator norm. The matrix 1-norm is also called the *row sum norm*, since it is the maximum of the 1-norms of the rows of A. Similarly, the matrix ∞-norm is also called the *column sum norm*. These norms are easy to compute, which accounts for their widespread use.

Although the Frobenius norm is consistent with the vector 2-norm, it is not the same as the operator 2-norm — as can be seen from the fact that for $n > 1$ we have $\|I_n\|_F = \sqrt{n} \neq 1$. The matrix 2-norm, which is also called the *spectral norm*, is not easy to compute; however, it has many nice properties that make it valuable in analytical investigations. Here is a list of some of the properties. The proofs are left as exercises. (See §4.3 and §4.4 for singular values and eigenvalues.)

Theorem 4.16. *The 2-norm has the following properties.*

1. $\|A\|_2 = \max_{\|x\|_2 = \|y\|_2 = 1} |y^H A x| =$ the largest singular value of A.
2. $\|A\|_2^2 = \max_{\|x\|_2 = 1} x^H (A^H A) x =$ the largest eigenvalue of $A^H A$.
3. $\|A\|_2 = \|A^H\|_2$.
4. $\|A\|_2 \leq \|A\|_F$, *with equality if and only if* $\mathrm{rank}(A) = 1$.
5. $\|A\|_2^2 \leq \|A\|_1 \|A\|_\infty$.

Absolute norms

It stands to reason that if the elements of a vector x are less in absolute value than the elements of a vector y then we should have $\|x\| \leq \|y\|$. Unfortunately, there are easy counterexamples to this appealing conjecture. For example, the function

$$\|x\| = x_1^2 - \frac{1}{2} x_1 x_2 + x_2^2$$

is a norm, but

$$\left\|\begin{pmatrix} 1 \\ 0 \end{pmatrix}\right\| = 1 \quad \text{and} \quad \left\|\begin{pmatrix} 1 \\ 0.1 \end{pmatrix}\right\| = 0.96.$$

Since norms that are monotonic in the elements are useful in componentwise error analysis (see §3, Chapter 3), we make the following definition.

Definition 4.17. *A norm* $\|\cdot\|$ *is* ABSOLUTE *if*

$$|x| \leq |y| \implies \|x\| \leq \|y\| \tag{4.12}$$

or equivalently if

$$\|x\| = \| |x| \|. \tag{4.13}$$

Here are some comments on this definition.

- The equivalence of (4.12) and (4.13) is not trivial.
- The vector 1-, 2-, and ∞-norms are clearly absolute.

SEC. 4. ANALYSIS

- We may also speak of absolute matrix norms. The matrix 1-, ∞-, and Frobenius norms are absolute. Unfortunately, the matrix 2-norm is not absolute. However, it does satisfy the relation

$$|A| \leq B \implies \|A\|_2 \leq \||A|\|_2 \leq \|B\|_2.$$

Perturbations of the identity

The basic matrix operations—multiplication by a scalar, matrix addition, and matrix multiplication—are continuous. The matrix inverse on the other hand is not obviously continuous and needs further investigation. We will begin with perturbations of the identity matrix. The basic result is contained in the following theorem.

Theorem 4.18. *Let $\|\cdot\|$ be a consistent matrix norm on $\mathbb{C}^{n \times n}$. For any matrix P of order n, if*

$$\|P\| < 1,$$

then $I - P$ is nonsingular. Moreover,

$$\|(I - P)^{-1} X\| \leq \frac{\|X\|}{1 - \|P\|} \tag{4.14}$$

and

$$\|(I - P)^{-1} - I\| \leq \frac{\|P\|}{1 - \|P\|}. \tag{4.15}$$

Proof. By Theorem 4.12 there is a vector norm, which we will also write $\|\cdot\|$, that is consistent with the matrix norm $\|\cdot\|$. Now let $x \neq 0$. Then

$$\|(I - P)x\| = \|x - Px\| \geq \|x\| - \|Px\|$$
$$\geq \|x\| - \|P\|\|x\| = (1 - \|P\|)\|x\| > 0.$$

Hence by Theorem 3.21 A is nonsingular.

To establish (4.14), set $G = (I - P)^{-1} X$. Then $X = (I - P)G = G - PG$. Hence $\|X\| \geq \|G\| - \|P\|\|G\|$, and (4.14) follows on solving for $\|G\|$.

To establish (4.15), set $H = (I - P)^{-1} - I$. Then on multiplying by $I - P$, we find that $H - PH = P$. Hence $\|P\| \geq \|H\| - \|P\|\|H\|$, and (4.15) follows on solving for $\|H\|$. ∎

Three comments.

- The results are basically a statement of continuity for the inverse of a perturbation of the identity. If we take $X = I$ in (4.14), then it says that $\|(I - P)^{-1}\|$ is near $\|I\|$

in proportion as P is small. The inequality (4.15) says that $(I - P)^{-1}$ is itself near $I^{-1} = I$.

- The result can be extended to a perturbation $A - E$ of a nonsingular matrix A. Write $A - E = A(I - A^{-1}E)$, so that $(A - E)^{-1} = (I - A^{-1}E)A^{-1}$. Thus we have the following corollary.

Corollary 4.19. If
$$\|A^{-1}E\| \leq 1,$$
then
$$\|(A + E)^{-1}\| \leq \frac{\|A^{-1}\|}{1 - \|A^{-1}E\|}.$$

Moreover, $(A + E)^{-1} - A^{-1} = [(I - A^{-1}E) - I]A^{-1}$, so that
$$\|(A + E)^{-1} - A^{-1}\| \leq \frac{\|A^{-1}\|\|A^{-1}E\|}{1 - \|A^{-1}E\|}.$$

The above bounds remain valid when $\|A^{-1}E\|$ is replaced by $\|EA^{-1}\|$.

This corollary is closely related to results on the perturbation of solutions of linear systems presented in §3.1, Chapter 3.

- If $\|\cdot\|$ is a family of consistent norms, then (4.14) continues to hold for any matrix X for which the product PX is defined. The most frequent application is when X is a vector, in which case the bound assumes the form
$$\|(I - P)^{-1}x\| \leq \frac{\|x\|}{1 - \|P\|}.$$

The Neumann series

In some instances it is desirable to have accurate approximations to $(I - P)^{-1}$. We can obtain such approximations by a generalization of the geometric series called the Neumann series. To derive it, suppose $I - P$ is nonsingular and consider the identity
$$(I - P)(I + P + P^2 + \cdots + P^k) = I - P^{k+1}.$$

Multiplying this identity by $(I - P)^{-1}$ and subtracting $(I - P)^{-1}$ from both sides, we get
$$(I + P + P^2 + \cdots P^k) - (I - P)^{-1} = -(I - P)^{-1}P^{k+1}. \quad (4.16)$$

Thus if $I - P$ is nonsingular and the powers P^k converge to zero, the *Neumann series*
$$I + P + P^2 + P^3 + \cdots$$
converges to $(I - P)^{-1}$. In fact, even more is true.

Sec. 4. Analysis

Theorem 4.20. *Let $P \in \mathbb{C}^n$ and suppose that $\lim_{k \to \infty} P^k = 0$. Then $I - P$ is nonsingular and*

$$(I - P)^{-1} = \sum_{k=0}^{\infty} P^k.$$

A sufficient condition for $P^k \to 0$ is that $\|P\| < 1$ in some consistent norm, in which case

$$\|(I + P + P^2 + \cdots + P^k) - (I - P)^{-1}\| \leq \frac{\|P\|^{k+1}}{1 - \|P\|}. \tag{4.17}$$

Proof. Suppose that $P^k \to 0$, but $I - P$ is singular. Then there is a nonzero x such that $(I - P)x = 0$ or $Px = x$. Hence $P^k x = x$, and $I - P^k$ is singular for all k. But since $P^k \to 0$, for some k we must have $\|P\|_\infty < 1$, and by Theorem 4.18 $I - P^k$ is nonsingular—a contradiction.

Since $I - P$ is nonsingular, the convergence of the Neumann series follows on taking limits in (4.16).

If $\|P\| < 1$, where $\|\cdot\|$ is a consistent norm, then $\|P^k\| \leq \|P\|^k \to 0$, and the Neumann series converges. The error bound (4.17) follows on taking norms and applying (4.14). ∎

The following corollary will be useful in deriving componentwise bounds for linear systems.

Corollary 4.21. *If $|P|^k \to 0$, then $(I - |P|)^{-1}$ is nonnegative and*

$$|(I - P)^{-1}| \leq (I - |P|)^{-1}.$$

Proof. Since $|P^k| \leq |P|^k$, P^k approaches zero along with $|P|^k$. The nonnegativity of $(I - |P|)^{-1}$ and the inequality now follow on taking limits in the inequality

$$|I + P + P^2 + \cdots + P^k| \leq I + |P| + |P|^2 \cdots |P|^k. \quad \blacksquare$$

4.2. Orthogonality and Projections

In real 2- or 3-space any pair of vectors subtend a well defined angle, and hence we can speak of vectors being at right angles to one another. Perpendicular vectors enjoy advantages not shared by their oblique cousins. For example, if they are the sides of a right triangle, then by the theorem of Pythagoras the sum of squares of their lengths is the square of the length of the hypotenuse. In this subsection we will generalize the notion of perpendicularity and explore its consequences.

Orthogonality

In classical vector analysis, it is customary to write the Cauchy inequality in the form $|x^T y| = \cos\theta \|x\|_2 \|y\|_2$. In real 2- or 3-space it is easy to see that θ is the angle between x and y. This suggests that we use the Cauchy inequality to extend the notion of an angle between two vectors to $\mathbb{C}^n$.

Definition 4.22. Let $x, y \in \mathbb{C}^n$ be nonzero. Then the ANGLE $\theta(x,y)$ BETWEEN x AND y is defined by

$$\theta(x, y) = \cos^{-1} \frac{|x^H y|}{\|x\|_2 \|y\|_2}. \tag{4.18}$$

If $x^H y = 0$ (whether or not x or y is nonzero), we say that x and y are ORTHOGONAL and write

$$x \perp y.$$

Thus for nonzero vectors orthogonality generalizes the usual notion of perpendicularity. By our convention any vector is orthogonal to the zero vector.

The *Pythagorean equality*, mentioned above, generalizes directly to orthogonal vectors. Specifically,

$$x \perp y \implies \|x + y\|_2^2 = \|x\|_2^2 + \|y\|_2^2.$$

In fact,

$$\begin{aligned} \|x + y\|_2^2 &= (x+y)^H (x+y) \\ &= x^H x + y^H y + 2\operatorname{Re}(x^H y) \\ &= x^H x + y^H y \\ &= \|x\|_2^2 + \|y\|_2^2. \end{aligned}$$

We will encounter orthogonal vectors most frequently as members of orthonormal sets of vectors or matrices whose columns or rows are orthonormal.

Definition 4.23. Let $u_1, u_2, \ldots, u_k \in \mathbb{C}^n$. Then the vectors u_i are ORTHONORMAL if

$$u_i^H u_j = \begin{cases} 0 & \text{if } i \neq j, \\ 1 & \text{if } i = j. \end{cases}$$

Equivalently if $U = (u_1 \; u_2 \; \cdots \; u_k)$ then

$$U^H U = I_k. \tag{4.19}$$

Any matrix U satisfying (4.19) will be said to be ORTHONORMAL. A square orthonormal matrix is said to be ORTHOGONAL if it is real and UNITARY if it is complex.

Some comments on this definition.

- The term orthonormal applied to a matrix is not standard. But it is very useful. The distinction between orthogonal and unitary matrices parallels the distinction between symmetric and Hermitian matrices. A real unitary matrix is orthogonal.

- The product of an orthonormal matrix and a unitary matrix is orthonormal.

Sec. 4. Analysis

- From (4.19) it follows that the columns of an orthonormal matrix are linearly independent. In particular, an orthonormal matrix must have at least as many rows as columns.

- If U is unitary (or orthogonal), then its inverse is its conjugate transpose:

$$U^H U = U U^H = I. \qquad (4.20)$$

- For any $n \times p$ orthonormal matrix U we have $U^T U = I_p$. Hence from Theorem 4.16 and (4.10) we have

$$\|U\|_2 = 1 \quad \text{and} \quad \|U\|_F = \sqrt{p}.$$

- If U is orthonormal and UA is defined, then

$$\|UA\|_p = \|A\|_p, \qquad p = 2, F.$$

Similarly if AU^H is defined, then

$$\|AU^H\|_p = \|A\|_p, \qquad p = 2, F.$$

Because of these relations, the 2-norm and the Frobenius norm are said to be *unitarily invariant*.

—

Orthogonal (and unitary) matrices play an important role in matrix computation. There are two reasons. First, because of (4.20) orthogonal transformations are easy to undo. Second, the unitary invariance of the 2- and Frobenius norms makes it easy to reposition errors in a formula without magnifying them. For example, suppose an algorithm transforms a matrix A by an orthogonal transformation Q, and in the course of computing QA we introduce a small error, so that what we actually compute is $B = QA + E$. If we set $F = Q^T E$, then we have the relation $Q(A + F) = B$; i.e., the computed matrix B is the result of an exact application of Q to a perturbation $A + F$ of A. Now this casting of the error back onto the original data could be done with any nonsingular Q — simply define $F = Q^{-1}B$. But if Q is orthogonal, then $\|F\|_F = \|E\|_F$, so that the error is not magnified by the process (For more on this kind of backward error analysis, see §4.3, Chapter 2.)

The QR factorization and orthonormal bases

The nice properties of orthonormal matrices would not be of great value if orthonormal matrices themselves were in short supply. We are going to show that an orthonormal matrix can be obtained from any matrix having linearly independent columns. We actually prove a more general result.

Theorem 4.24 (QR factorization). Let $X \in \mathbb{C}^{n \times p}$ have rank p. Then X can be written uniquely in the form

$$X = QR, \qquad (4.21)$$

where Q is an $n \times p$ orthonormal matrix and R is upper triangular matrix with positive diagonal elements.

Proof. The proof is by induction on p. For $p = 1$, take $Q = X/\|X\|_2$ and $R = \|X\|_2$.
For general p we seek Q and R in the partitioned form

$$(X_1 \ x_p) = (Q_1 \ q_p) \begin{pmatrix} R_{11} & r_{1p} \\ 0 & \rho_{pp} \end{pmatrix}.$$

This is equivalent to the two equations:

1. $X_1 = Q_1 R_{11}$,
2. $x_p = Q_1 r_{1p} + q_p \rho_{pp}$. $\qquad (4.22)$

The first equation simply asserts the existence of the factorization for X_1, which exists and is unique by the induction hypothesis.

Let

$$r_{1p} = Q^H x_p \quad \text{and} \quad \hat{q}_p = x_p - Q_1 r_{1p}. \qquad (4.23)$$

Since $Q_1^H Q_1 = I$,

$$Q_1^H \hat{q}_p = Q_1^H x_p - Q_1^H Q_1 r_{1p} = r_{1p} - r_{1p} = 0.$$

Hence $\hat{q}_p$ is orthogonal to the columns of Q_1. Consequently, if $\hat{q}_p \neq 0$, we can determine ρ_{pp} and q_p from (4.22.2) by setting

$$\rho_{pp} = \|\hat{q}_p\|_2 \quad \text{and} \quad q_p = \hat{q}_p / \rho_{pp}. \qquad (4.24)$$

To show that $\hat{q}_p$ is nonzero, note that R_{11} has positive diagonal elements and hence is nonsingular (Theorem 2.1, Chapter 2). Thus from (4.22.1), $Q_1 = X_1 R_{11}^{-1}$. Hence

$$\hat{q}_p = x_p - X_1(R_{11}^{-1} r_{1p}).$$

The right-hand side of this relation is a nontrivial linear combination of the columns of X, which cannot be zero because X has full column rank.

The uniqueness of the factorization follows from the uniqueness of the factorization $X_1 = Q_1 R_{11}$, and the fact that formulas (4.23) and (4.24) uniquely determine r_{1p}, r_{pp}, and q_p. ∎

- The factorization whose existence is established by the theorem is called the *QR factorization* of X. This factorization is one of the most important tools in matrix computations.

SEC. 4. ANALYSIS

- From the relation $X = QR$ and the nonsingularity of R it follows that $\mathcal{R}(Q) = \mathcal{R}(X)$—in other words, the columns of Q form an orthonormal basis for the column space of X. Since any subspace has a basis that can be arranged in a matrix, it follows that:

    Every subspace has an orthonormal basis.

- Let $X_k = (x_1\ x_2\ \cdots\ x_k)$ and $Q_k = (q_1\ q_2\ \cdots\ q_k)$, and let R_k be the leading principal submatrix of R of order k. Then from the triangularity of R it follows that $X_k = Q_k R_k$. In other words, the first k columns of Q form an orthonormal basis for the space spanned by the first k columns of X.

- If $\mathcal{X}$ is a subspace and if X is a basis for $\mathcal{X}$, then we can extend that basis to a basis $(X\ Y)$ for $\mathbb{C}^n$. The QR factorization of $(X\ Y)$ gives an orthonormal basis Q for $\mathbb{C}^n$ whose first k columns are an orthonormal basis for X. The last $n-k$ columns of Q are a basis for a complementary subspace whose vectors are orthogonal to $\mathcal{X}$. This space is called the *orthogonal complement* of $\mathcal{X}$. Thus we have shown that:

    Every subspace has an orthogonal complement. (4.25)

We will write $\mathcal{X}_\perp$ for the orthogonal complement of a subspace $\mathcal{X}$. It is worth noting that the existence of orthogonal complements is also implied by (3.11).

- Looking at the above construction in a different way, suppose that X is an orthonormal basis for $\mathcal{X}$. Then the first k columns of Q are the columns of X. Consequently:

    If X is orthonormal, then there is an orthonormal matrix Y such that $(X\ Y)$ is unitary.

In particular, if $X = x$ is a vector, it follows that:

If x is nonzero, then there is a unitary matrix whose first column is $x/\|x\|_2$. (4.26)

This result is useful both in theory and practice. In fact, in §1, Chapter 4, we will show how to use Householder transformations to efficiently construct the required matrix.

- The proof of the existence of the QR factorization is constructive. The resulting algorithm is called the *classical Gram–Schmidt algorithm*. Be warned that the algorithm can be quite unstable; however, it has the theoretical advantage that it can be used in arbitrary inner-product spaces. We will return to the Gram–Schmidt algorithm in §1.4, Chapter 4.

Although there are infinitely many orthonormal bases for a nontrivial subspace of $\mathbb{C}^n$, they are all related by unitary transformations, as the following theorem shows.

Theorem 4.25. Let $\mathcal{X} \subset \mathbb{R}^n$ be a subspace and let X and $\hat{X}$ be two orthonormal bases for $\mathcal{X}$. Then $X^H \hat{X}$ is unitary and

$$\hat{X} = X(X^H \hat{X}).$$

Proof. Because X and $\hat{X}$ span the same space, there is a unique matrix U such that $\hat{X} = XU$. Now

$$I = \hat{X}^H \hat{X} = U^H X^H X U = U^H U.$$

Hence U is unitary. Moreover,

$$X^H \hat{X} = X^H X U = U. \quad \blacksquare$$

Orthogonal projections

Imagine an eagle flying over a desert, so high that it is a mere point to the eye. The point on the desert that is nearest the eagle is the point immediately under the eagle. Replace the eagle by a vector and the desert by a subspace, and the corresponding nearest point is called the projection of the vector onto the subspace.

To see how projections are computed, let $\mathcal{X} \subset \mathbb{C}^n$ be a subspace and let $z \in \mathbb{C}^n$. Let Q be an orthonormal basis for $\mathcal{X}$ and define

$$P_\mathcal{X} = QQ^H.$$

Then we can write z in the form

$$z = P_\mathcal{X} + (I - P_\mathcal{X})z \equiv z_\mathcal{X} + z_\perp. \tag{4.27}$$

Clearly $z_\mathcal{X} = Q(Q^H z) \in \mathcal{X}$, since it is a linear combination of columns of Q. Moreover,

$$Q^H z_\perp = Q^H (I - QQ^H) z = (Q^H - Q^H QQ^H) z = (Q^H - Q^H) z = 0.$$

Hence $z_\perp$ lies in the orthogonal complement $\mathcal{X}_\perp$ of $\mathcal{X}$.

The decomposition (4.27) is unique. For if $z = \hat{z}_\mathcal{X} - \hat{z}_\perp$ were another such decomposition, we would have

$$0 = (z_\mathcal{X} - \hat{z}_\mathcal{X}) + (z_\perp - \hat{z}_\perp).$$

But $(z_\mathcal{X} - \hat{z}_\mathcal{X}) \in \mathcal{X}$ while $(z_\perp - \hat{z}_\perp) \in \mathcal{X}_\perp$. Consequently, they are both zero.

The vector $z_\mathcal{X} = P_\mathcal{X} z$ is called *the orthogonal projection of z onto $\mathcal{X}$*. The vector $z_\perp = (I - P_\mathcal{X})z$ is called the *orthogonal projection of z onto the orthogonal complement of $\mathcal{X}$*. We write $P_\mathcal{X}^\perp$ for the projection matrix onto the orthogonal complement of $\mathcal{X}$. When $\mathcal{X}$ is clear from context, we write simply $P_\perp$.

The operation of projecting a vector is clearly linear. It therefore has a unique matrix representation, which in fact is $P_\mathcal{X}$. We call $P_\mathcal{X}$ the *orthogonal projection matrix*

SEC. 4. ANALYSIS

onto $\mathcal{X}$ — or when it is clear that an operator and not a vector is meant, simply the orthogonal projection onto $\mathcal{X}$. The projection matrix, being unique, does not depend on the choice of an orthogonal basis Q.

The projection matrix $P_\mathcal{X}$ satisfies

$$P_\mathcal{X}^2 = P_\mathcal{X} \quad \text{and} \quad P_\mathcal{X}^\mathrm{H} = P_\mathcal{X};$$

i.e., it is idempotent and Hermitian. It is an interesting exercise to verify that all Hermitian, idempotent matrices are orthogonal projections.

We can obtain another very useful expression for $P_\mathcal{X}$. Let X be a basis for $\mathcal{X}$, and let $X = QR$ be its QR factorization [see (4.21)]. Then

$$X^\mathrm{H} X = R^\mathrm{H} Q^\mathrm{H} Q R = R^\mathrm{H} R \quad \text{and} \quad Q = X R^{-1}.$$

It follows that

$$P_\mathcal{X} = QQ^\mathrm{H} = X R^{-1} R^{-\mathrm{H}} X = X(R^\mathrm{H} R)^{-1} X^\mathrm{H} = X(X^\mathrm{H} X)^{-1} X^\mathrm{H}.$$

As was suggested at the beginning of this subsection, the orthogonal projection of a vector onto a subspace is the point in the subspace nearest the vector. The following theorem gives a precise statement of this assertion.

Theorem 4.26. *Let $\mathcal{X} \subset \mathbb{C}^n$ be a subspace and let $z \in \mathbb{C}^n$. Then the unique solution of the problem*

$$\begin{aligned} \text{minimize} \quad & \|z - x\|_2 \\ \text{subject to} \quad & x \in \mathcal{X} \end{aligned}$$

is the orthogonal projection of z onto x.

Proof. Let $x \in \mathcal{X}$. Since $P_\mathcal{X}(z-x) \perp P_\perp(z-x)$, we have by the Pythagorean equality

$$\begin{aligned} \|z-x\|_2^2 &= \|P_\mathcal{X}(z-x)\|_2^2 + \|P_\perp(z-x)\|_2^2 \\ &= \|P_\mathcal{X} z - x\|_2^2 + \|P_\perp z\|_2^2. \end{aligned} \quad (4.28)$$

The second term in the right-hand side of (4.28) is independent of x. The first term is minimized precisely when $x = P_\mathcal{X} z$. ∎

4.3. THE SINGULAR VALUE DECOMPOSITION

Another way of looking at the QR factorization is as a reduction to triangular form by unitary transformations. Specifically, let $X = Q_X R$ be a QR factorization of X and let $Q = (Q_X \; Q_\perp)$ be unitary. Then

$$Q^\mathrm{H} X = \begin{pmatrix} R \\ 0 \end{pmatrix}.$$

In other words there is a unitary matrix Q that reduces X to upper triangular form. (This reduction is called the *QR decomposition*; see §1.1, Chapter 4.)

The *singular value decomposition* is a unitary reduction to diagonal form. The degrees of freedom needed for the additional simplification come from operating on the matrix on both the left and the right, i.e., transforming X to $U^{\mathrm{T}}XV$, where U and V are unitary. This subsection is concerned with the singular value decomposition—its existence and properties.

Existence

The singular value decomposition can be established by a recursive argument, similar in spirit to the proof of Theorem 2.13, which established the existence of the LU decomposition.

Theorem 4.27. Let $X \in \mathbb{C}^{n \times p}$, where $n \geq p$. Then there are unitary matrices U and V such that

$$U^{\mathrm{H}} X V = \begin{pmatrix} \Sigma \\ 0 \end{pmatrix},$$

where

$$\Sigma = \mathrm{diag}(\sigma_1, \sigma_2, \ldots, \sigma_p)$$

with

$$\sigma_1 \geq \sigma_2 \geq \cdots \geq \sigma_p \geq 0.$$

Proof. The proof is by recursive reduction of X to diagonal form. The base case is when X is a vector x. If $x = 0$ take $U = I$ and $V = 1$. Otherwise take U to be any unitary matrix whose first column is $x/\|x\|_2$ [see (4.26)] and let $V = (1)$. Then

$$U^{\mathrm{T}} x V = \begin{pmatrix} \|x\|_2 \\ 0 \end{pmatrix},$$

which is in the required form.

For the general case, let u and v be vectors of 2-norm one such that

$$Xv = \|X\|_2 u \equiv \sigma u.$$

Let $(u \; \hat{U})$ and $(v \; \hat{V})$ be unitary [see (4.26)]. Then

$$(u \; \hat{U})^{\mathrm{H}} X (v \; \hat{V}) = \begin{pmatrix} \sigma u^{\mathrm{H}} u & u^{\mathrm{H}} X \hat{V} \\ \sigma \hat{U}^{\mathrm{H}} u & \hat{U}^{\mathrm{H}} X \hat{V} \end{pmatrix} \equiv \begin{pmatrix} \sigma & w^{\mathrm{H}} \\ 0 & \hat{X} \end{pmatrix}.$$

We claim that $w = 0$. For if not, we have

$$(1 + \epsilon^2 \|w\|_2^2)^{-\frac{1}{2}} \left\| \begin{pmatrix} \sigma & w^{\mathrm{H}} \\ 0 & \hat{X} \end{pmatrix} \begin{pmatrix} 1 \\ \epsilon w \end{pmatrix} \right\|_2 \geq \frac{1 + \epsilon \|w\|_2}{\sqrt{1 + \epsilon^2 \|w\|_2^2}}$$

$$= \sigma + \epsilon \|w\|_2 + O(\epsilon^2).$$

SEC. 4. ANALYSIS

If follows that if ϵ is sufficiently small, then $\|X\|_2 > \sigma$, which contradicts the definition of σ.

Now by the induction hypothesis there are unitary matrices $\check{U}$ and $\check{V}$ such that

$$\check{U}^H \hat{X} \check{V} = \begin{pmatrix} \check{\Sigma} \\ 0 \end{pmatrix},$$

where $\check{\Sigma}$ is a diagonal matrix with nonincreasing, nonnegative diagonal elements. Set

$$U = (u \;\; \hat{U}\check{U}) \quad \text{and} \quad V = (v \;\; \hat{V}\check{V}).$$

Then U and V are unitary and

$$U^H X V = \begin{pmatrix} \sigma & 0 \\ 0 & \check{U}^H \hat{X} \check{V} \end{pmatrix} = \begin{pmatrix} \sigma & 0 \\ 0 & \check{\Sigma} \\ 0 & 0 \end{pmatrix},$$

a matrix which has the form required by the theorem. ∎

Here are some comments on this theorem.

- The proof of the theorem has much in common with the proof of the existence of LU factorizations (Theorem 2.13). However, unlike that proof, the proof of the existence of the singular value decomposition is not constructive, since it presupposes one has on hand a vector that generates the norm of X.

- The numbers σ_i are called the *singular values* of X. The columns of

$$U = (u_1 \; u_2 \; \cdots \; u_n) \quad \text{and} \quad V = (v_1 \; v_2 \; \cdots \; v_p)$$

are called *left* and *right singular vectors of* X. They satisfy

$$X v_i = \sigma_i u_i, \qquad i = 1, 2, \ldots, p,$$

and

$$X^H u_i = \sigma_i v_i, \qquad i = 1, 2, \ldots, p.$$

Moreover,

$$X^H u_i = 0, \qquad i = p+1, p+2, \ldots, n.$$

- It often happens that $n \gg p$, in which case maintaining the $n \times n$ matrix U can be burdensome. As an alternative we can set

$$U_p = (u_1 \; u_2 \; \cdots \; u_p) \tag{4.29}$$

in which case
$$X = U_p \Sigma V^{\mathrm{H}}.$$

This form of the decomposition is sometimes called the *singular value factorization*.

- The singular value decomposition provides an elegant full-rank factorization of X. Suppose $\sigma_k > 0 = \sigma_{k+1} = \cdots = \sigma_p$. Set

$$U_k = (u_1\ u_2\ \cdots\ u_k),$$
$$V_k = (v_1\ v_2\ \cdots\ v_k),$$
$$\Sigma_k = \mathrm{diag}(\sigma_1, \sigma_2, \ldots, \sigma_k).$$

Then it is easily verified that

$$X = U_k \Sigma_k V_k^{\mathrm{H}}$$

is a full-rank factorization of X. Since the rank of the factorization is k, we have the following important relation.

The rank of a matrix is the number of its nonzero singular values.

- By the proof of Theorem 4.27 we know that:

The 2-norm of a matrix is its largest singular value.

From the definition of the 2-norm (and also the proof of the existence of the singular value decomposition) it follows that

$$\sigma_1 = \max_{\|v\|_2 = 1} \|Xv\|_2 = \|X\|_2. \tag{4.30}$$

Later it will be convenient to have a notation for the smallest singular value of X. We will denote it by

$$\inf(X) \stackrel{\mathrm{def}}{=} \sigma_p(X). \tag{4.31}$$

- Since the Frobenius norm is unitarily invariant, it follows that

$$\|X\|_{\mathrm{F}}^2 = \|U^{\mathrm{H}} X V\|_{\mathrm{F}}^2 = \|\Sigma\|_{\mathrm{F}}^2 = \sum_{i=1}^p \sigma_i^2.$$

In other words:

The square of the Frobenius norm of a matrix is the sum of squares of its singular values.

SEC. 4. ANALYSIS

The characterizations of the spectral and the Frobenius norms in terms of singular values imply that

$$\|X\|_2 \leq \|X\|_F \leq \sqrt{p}\|X\|_2.$$

They also explain why, in practice, the Frobenius norm and the spectral norm often tend to be of a size. The reason is that in the sum of squares $\sigma_1^2 + \sigma_2^2 + \cdots + \sigma_p^2$ if σ_2 is just a little bit less than σ_1, the squaring makes the influence of σ_2 and the subsequent singular values negligible. For example, suppose $p = 101$, $\sigma_1 = 1$, and the remaining singular values are 0.1. Then

$$\|X\|_F^2 = 1 + 100 \cdot 0.01 = 2.$$

Thus $\|X\|_F = \sqrt{2} \cong 1 = \|X\|_2$.

Uniqueness

The singular value decomposition is one of the many matrix decompositions that are "essentially unique." Specifically, any unitary reduction to diagonal form must exhibit the same singular values on the diagonal. Moreover, the singular vectors corresponding to single distinct singular values are unique up to a factor of modulus one.

Repeated singular values are a source of nonuniqueness, as the following theorem shows.

Theorem 4.28. *Let $X \in \mathbb{C}^{n \times p}$ ($n \geq p$) have the singular value decomposition*

$$U^H X V = \begin{pmatrix} \Sigma \\ 0 \end{pmatrix}. \tag{4.32}$$

Let

$$\hat{U}^H X \hat{V} = \begin{pmatrix} \hat{\Sigma} \\ 0 \end{pmatrix} \tag{4.33}$$

be another singular value decomposition. Then $\hat{\Sigma} = \Sigma$. Moreover, $\hat{V} = VQ$ where $Q = V^H \hat{V}$ is unitary and

$$\sigma_i \neq \sigma_j \implies q_{ij} = 0. \tag{4.34}$$

A similar statement is true of U provided we regard the singular vectors $u_{p+1}, \ldots, u_n$ as corresponding to zero singular values.

Proof. From (4.32) and (4.33) we have

$$V^H X^H X V = \Sigma^2 \quad \text{and} \quad \hat{V}^H X^H X \hat{V} = \hat{\Sigma}^2.$$

It follows that with $Q = V^H\hat{V}$

$$Q^H \Sigma^2 Q = \hat{\Sigma}^2. \tag{4.35}$$

To establish that $\Sigma = \hat{\Sigma}$, let λ be regarded as a variable. Taking determinants in the equation $Q^H(\lambda I - \Sigma^2)Q = \lambda I - \hat{\Sigma}^2$, we find that

$$\det{}^2(Q)(\lambda - \sigma_1^2)(\lambda - \sigma_2^2)\cdots(\lambda - \sigma_n^2) = (\lambda - \hat{\sigma}_1^2)(\lambda - \hat{\sigma}_2^2)\cdots(\lambda - \hat{\sigma}_n^2). \tag{4.36}$$

Since Q is nonsingular, $\det(Q) \neq 0$. Hence the left- and right-hand sides of (4.36) are polynomials which are proportional to each other and therefore have the same zeros counting multiplicities. But the zeros of these polynomials are respectively the numbers σ_i^2 and $\hat{\sigma}_i^2$. Since these numbers are arranged in descending order, we must have $\hat{\Sigma} = \Sigma$.

To establish (4.34), write (4.35) in the form

$$\Sigma^2 Q = Q\Sigma^2.$$

Then $\sigma_i^2 q_{ij} = \sigma_j^2 q_{ij}$, or $(\sigma_i^2 - \sigma_j^2)q_{ij} = 0$. Hence if $\sigma_i \neq \sigma_j$, then $q_{ij} = 0$. ∎

In consequence of (4.34), Q is block diagonal, each block corresponding to a repeated singular value. Write $Q = \mathrm{diag}(Q_1, Q_2, \ldots, Q_k)$, and partition

$$V = (V_1 \ V_2 \ \cdots \ V_k) \quad \text{and} \quad \hat{V} = (\hat{V}_1 \ \hat{V}_2 \ \cdots \ \hat{V}_k)$$

conformally. Then

$$\hat{V}_i = V_i Q_i, \qquad i = 1, 2, \ldots, k.$$

Thus it is the subspace spanned by the right singular vectors that is unique. The singular vectors may be taken as any orthonormal basis — V_i, $\hat{V}_i$, what have you — for that subspace. However, once the right singular vectors are chosen, the left singular vectors corresponding to nonzero singular values are uniquely determined by the relation $XV = U\Sigma$. Analogous statements can be made about U.

The nonuniqueness in the singular value decomposition is therefore quite limited, and in most applications it makes no difference. Hence we usually ignore it and speak of *the* singular value decomposition of a matrix.

Unitary equivalence

Two matrices X and Y are said to be *unitarily equivalent* if there are unitary matrices P and Q such that $Y = P^H X Q$. If X has the singular value decomposition (4.32), then

$$(P^H U)^H Y (Q^H V) = \begin{pmatrix} \Sigma \\ 0 \end{pmatrix}$$

is a singular value decomposition of Y. It follows that:

SEC. 4. ANALYSIS

Unitarily equivalent matrices have the same singular values. Their singular vectors are related by the unitary transformations connecting the two matrices.

The proof of the following result, which is an immediate consequence of the proof of Theorem 4.28, is left as an exercise.

The singular values of $X^H X$ are the squares of the singular values of X. The nonzero singular values of $X X^H$ are the squares of the nonzero singular values of X.

Weyl's theorem and the min-max characterization

In (4.30) we characterized the largest singular value of X as the maximum of $\|Xv\|_2$ over all vectors v of norm one. This characterization has a far reaching generalization — the famous min-max characterization. We will derive it as a corollary of a theorem of Weyl, which is important in its own right.

In stating our results we will use the notation $\sigma_i(X)$ to refer to the ith singular value (in descending order) of the matrix X. As above we will write

$$U_k = (u_1 \; u_2 \; \cdots \; u_k),$$
$$V_k = (v_1 \; v_2 \; \cdots \; v_k),$$
$$\Sigma_k = \mathrm{diag}(\sigma_1, \sigma_2, \ldots, \sigma_k),$$

although this time without any assumption that $\sigma_{k+1} = 0$. Note that

$$XV_k = U_k \Sigma_k \quad \text{and} \quad X^H U_k = V_k \Sigma_k.$$

In addition we will write

$$X_k = U_k \Sigma_k V_k^H. \tag{4.37}$$

Note for future reference that the only candidates for nonzero singular values of $X - X_k$ are $\sigma_{k+1}, \ldots, \sigma_p$.

Theorem 4.29 (Weyl). *Let $X, Y \in \mathbb{C}^{n \times p}$ with $n \geq p$, and let $\mathrm{rank}(Y) = k$. Then*

$$\max_{\substack{w \in \mathcal{N}(Y) \\ \|w\|_2 = 1}} \|Xw\|_2 \geq \sigma_{k+1}(X),$$

and

$$\min_{\substack{w \in \mathcal{N}(Y) \\ \|w\|_2 = 1}} \|Xw\|_2 \leq \sigma_{p-k}(X).$$

Hence

$$\sigma_1(X - Y) \geq \sigma_{k+1}(X), \tag{4.38}$$

and
$$\sigma_p(X+Y) \le \sigma_{p-k}(X).$$

Moreover, if $A = B + C$, then

$$\sigma_{i+j-1}(A) \le \sigma_i(B) + \sigma_j(C). \tag{4.39}$$

Proof. We will establish only the inequalities involving maxima, the others being established similarly.

Let RS^H be a full-rank factorization of Y. Since the matrix $S^H V_{k+1}$ has more columns than rows, it has a nontrivial null space. Let a be a vector of 2-norm one such that $S^H V_{k+1} a = 0$. Then $w = V_{k+1} a \in \mathcal{N}(Y)$. Moreover,

$$\begin{aligned}\sigma_1^2(X-Y) &\ge \|(X-Y)w\|_2^2 \\ &= \|Xw\|_2 \\ &= \|X V_{k+1} a\|_2^2 \\ &= \|U_{k+1} \Sigma_{k+1} a\|_2^2 \\ &= \sum_{i=1}^{k+1} \sigma_i^2 a_i^2 \\ &\ge \sigma_{k+1},\end{aligned}$$

the last inequality following from the fact that $\sum_i a_i^2 = 1$.

To prove (4.39), we start with $i = j = 1$. Then

$$\sigma_1(A) = \|A\|_2 = \|B+C\|_2 \le \|B\|_2 + \|C\|_2 \le \sigma_1(B) + \sigma_1(C).$$

Now let B_{i-1} and C_{j-1} be formed in analogy with (4.37). Then $\sigma_1(B - B_{i-1}) = \sigma_i(B)$ and $\sigma_1(C - C_{j-1}) = \sigma_j(C)$. Moreover, rank$(B_{i-1} + C_{j-1}) \le i+j-2$. Hence from (4.38)

$$\begin{aligned}\sigma_i(B) + \sigma_j(C) &= \sigma_1(B - B_i) + \sigma_1(C - C_j) \\ &\ge \sigma_1(A - B_{i-1} - C_{j-1}) \\ &\ge \sigma_{i+j-1}(A). \quad \blacksquare\end{aligned}$$

The min-max characterization of singular values follows immediately from this theorem.

Corollary 4.30. *The singular values of X have the following characterizations:*

$$\sigma_k(X) = \min_{\dim(\mathcal{W}) = p-k+1} \max_{\substack{w \in \mathcal{W} \\ \|w\|_2 = 1}} \|Xw\|_2, \quad k = 1, 2, \ldots, p, \tag{4.40}$$

and

$$\sigma_k(X) = \max_{\dim(\mathcal{W}) = k} \min_{\substack{w \in \mathcal{W} \\ \|w\|_2 = 1}} \|Xw\|_2, \quad k = 1, 2, \ldots, p. \tag{4.41}$$

SEC. 4. ANALYSIS

Proof. We will prove only (4.40), leaving (4.41) as an exercise.

Given a subspace $\mathcal{W}$ of dimension $p-k+1$, let Y be a matrix of rank $k-1$ whose null space is $\mathcal{W}$. Then by Weyl's theorem

$$\max_{\substack{w \in \mathcal{W} \\ \|w\|_2=1}} \max_{\substack{w \in \mathcal{N}(Y) \\ \|w\|_2=1}} \|Xw\|_2 \geq \sigma_k(X).$$

The minimum is attained when $\mathcal{W} = \text{span}(v_k, v_{k+1}, \ldots, v_p)$. ∎

The perturbation of singular values

One of the reasons the singular value decomposition is useful is that the singular values are insensitive to perturbations in the matrix. This fact is also a corollary of Weyl's theorem.

Corollary 4.31. Let $X, E \in \mathbb{C}^{n \times p}$. Then

$$|\sigma_i(X+E) - \sigma_i(X)| \leq \|E\|_2, \qquad i = 1, 2, \ldots, p. \tag{4.42}$$

Proof. In (4.39) make the substitutions $A = X+E$, $B = X$, and $C = E$. Then with $j = 1$,

$$\sigma_i(X+E) \leq \sigma_i(X) + \sigma_1(E) = \sigma_i(X) + \|E\|_2.$$

With the substitutions $A = X$, $B = X+E$, and $C = -E$, we obtain the inequality

$$\sigma_i(X) \leq \sigma_i(X+E) + \|E\|_2.$$

The two inequalities imply (4.42). ∎

The inequality (4.42) states that no singular value of a matrix can be perturbed by more than the 2-norm of the perturbation of the matrix. It is all the more remarkable, because the bounds can be attained simultaneously for all i. Simply set $E = \epsilon U_p V_p^{\text{H}}$. Then the singular values of $X+E$ are $\sigma_i + \epsilon = \sigma_i + \|E\|_2$.

Low-rank approximations

We have already observed that if $\sigma_k > \sigma_{k+1} = 0$, then X has rank k. Since in practical applications matrices are generally contaminated by errors, we will seldom encounter a matrix that is exactly defective in rank. Instead we will find that some of the singular values of the matrix in question are small.

One consequence of small singular values is that the matrix must be near one that is defective in rank. To quantify this statement, suppose that the small singular values are $\sigma_{k+1}, \sigma_{k+2}, \ldots, \sigma_p$. If we define X_k by (4.37), then the nonzero singular values of $X_k - X$ are $\sigma_{k+1}, \sigma_{k+2}, \ldots, \sigma_p$. Hence we have

$$\|X - X_k\|_2 = \sigma_{k+1} \quad \text{and} \quad \|X - X_k\|_{\text{F}} = \sqrt{\sigma_{k+1}^2 + \sigma_{k+2}^2 + \cdots + \sigma_p^2}.$$

The following theorem shows that these low-rank approximations are optimal in the 2- and Frobenius norms.

Theorem 4.32 (Schmidt–Mirsky). *For any matrix $X \in \mathbb{C}^{n \times p}$, if $Y \in \mathbb{C}^{n \times p}$ is of rank k, then*

$$\|X - Y\|_2 \geq \sigma_{k+1}(X),$$

and

$$\|X - Y\|_F^2 \geq \sigma_{k+1}(X)^2 + \sigma_{k+2}(X)^2 + \cdots + \sigma_p(X)^2.$$

Equality is attained for both norms when $Y = X_k$ is defined by (4.37).

Proof. We have already established equality for $Y = X_k$. The first inequality is just (4.38) written in terms of norms.

For the second inequality, set $A = X$, $B = X - Y$, and $C = Y$ in Theorem 4.29. Then for $j = k+1$,

$$\sigma_{k+i}(X) \leq \sigma_i(X - Y) + \sigma_{k+1}(Y) = \sigma_i(X - Y),$$

the last equality following from the fact that $\text{rank}(Y) = k$. Hence

$$\|Y - X\|_F^2 \geq \sum_{i=1}^{p-k} \sigma_i^2(X - Y) \geq \sum_{i=k+1}^{p} \sigma_i^2(X). \quad \blacksquare$$

4.4. THE SPECTRAL DECOMPOSITION

It might be hoped that symmetry would force the left and right singular vectors of a Hermitian matrix to be the same. However, they need not be, even when the matrix is a scalar. For example, the singular value decomposition of the scalar (-1) is

$$(-1)(-1)(1) = (1),$$

so that the "left singular vector" (-1) and the "right singular vector" (1) have opposite signs. However, if we relax the requirement that the singular values be positive, we can obtain a symmetric decomposition.

Specifically, let v be a vector of 2-norm one such that

$$v^H A v \text{ is maximized}, \tag{4.43}$$

and let $(v \; \hat{V})$ be unitary. Consider the matrix

$$(v \; \hat{V})^H A (v \; \hat{V}) \equiv \begin{pmatrix} \lambda & w^H \\ w & \hat{A} \end{pmatrix}.$$

We claim that $w = 0$. For otherwise

$$(1 + \epsilon^2 \|w\|_2^2)^{-1} \begin{pmatrix} 1 \\ \epsilon w \end{pmatrix}^H \begin{pmatrix} \lambda & w^H \\ w & \hat{A} \end{pmatrix} \begin{pmatrix} 1 \\ \epsilon w \end{pmatrix} > \lambda + 2\epsilon \|w\|_2^2 + O(\epsilon^2),$$

SEC. 4. ANALYSIS

which contradicts the maximality of λ. Thus

$$(v \; \hat{V})^H A(v \; \hat{V}) = \begin{pmatrix} \lambda & 0 \\ 0 & \hat{A} \end{pmatrix}.$$

We can continue the reduction to diagonal form with $\hat{A}$, as we did with the singular value decomposition. The result is the following theorem.

Theorem 4.33 (The spectral decomposition). *If $A \in \mathbb{C}^{n \times n}$ is Hermitian, there is a unitary matrix U such that*

$$U^H A U = \mathrm{diag}(\lambda_1, \lambda_2, \ldots, \lambda_n), \tag{4.44}$$

where

$$\lambda_1 \geq \lambda_2 \geq \cdots \geq \lambda_n.$$

The decomposition (4.44) is called the *spectral decomposition* of A. The numbers λ_i are called *eigenvalues* of A and the columns u_i of U are called *eigenvectors*. The pair (λ_i, u_i) is called an *eigenpair*. The members of an eigenpair satisfy the equation

$$A u_i = \lambda_i u_i, \qquad i = 1, 2, \ldots, n.$$

Many properties of the singular value decomposition hold for the spectral decomposition. The eigenvalues of a Hermitian matrix and their multiplicities are unique. The eigenvectors corresponding to a multiple eigenvalue span a unique subspace, and the eigenvectors can be chosen as any orthonormal basis for that subspace.

It is easily verified that the singular values of a Hermitian matrix are the absolute values of its eigenvalues. Of particular importance is the following result, whose proof is left as an exercise.

> The eigenvalues of $X^H X$ are the squares of the singular values of X. Their eigenvectors are the corresponding right singular vectors of X. $\hspace{2em}$ (4.45)

The invariance of singular values under unitary equivalences has a counterpart for eigenvalues of a Hermitian matrix. Specifically,

> *If $U^H A U = \Lambda$ is the spectral decomposition of A and $B = V^H A V$, where V is unitary, then $(V^H U)^H B (V^H U) = \Lambda$ is the spectral decomposition of B.*

The transformation $A \to V^H A V$ is called a *unitary similarity transformation*. The above result shows that unitary similarities preserve eigenvalues and transform the eigenvectors by the transpose of the unitary matrix of the similarity transformation.

Theorem 4.29 and its consequences are also true of the eigenvalues of Hermitian matrices. We collect these results in the following theorem. Here the ith eigenvalue in descending order of a Hermitian matrix is written $\lambda_i(A)$.

Theorem 4.34. Let A, B, and C be Hermitian of order n. Then

1. (Weyl) If $\operatorname{rank}(B) = k$, then

$$\lambda_1(A - B) \geq \lambda_{k+1}(A)$$

and

$$\lambda_n(A - B) \leq \lambda_{n-k}(A).$$

2. (Weyl) The eigenvalues of the sum $A = B + C$ satisfy

$$\lambda_{i+j-1}(A) \leq \lambda_i(B) + \lambda_j(C).$$

3. (Fischer) The eigenvalues of A have the min-max characterization

$$\lambda_k(A) = \min_{\dim(\mathcal{W}) = n-k+1} \max_{\substack{w \in \mathcal{W} \\ \|w\|_2 = 1}} w^{\mathrm{H}} A w \qquad (4.46)$$

and the max-min characterization

$$\lambda_k(A) = \max_{\dim(\mathcal{W}) = k} \min_{\substack{w \in \mathcal{W} \\ \|w\|_2 = 1}} w^{\mathrm{H}} A w. \qquad (4.47)$$

4. If E is Hermitian, then

$$|\lambda_i(A + E) - \lambda_i(A)| \leq \|E\|_2, \qquad i = 1, 2, \ldots, n.$$

5. (Cauchy interlacing theorem) If V is an $n \times (n-1)$ orthonormal matrix, then

$$\lambda_1(A) \geq \lambda_1(V^{\mathrm{H}} A V) \geq \lambda_2(A)$$
$$\geq \lambda_2(V^{\mathrm{H}} A V) \geq \cdots \geq \lambda_{n-1}(V^{\mathrm{H}} A V) \geq \lambda_n(A).$$

Proof. The proofs of the first four items are mutatis mutandis the same as the corresponding results for the singular values. The last is established as follows. For any $k \leq n-1$ let $\mathcal{W} \subset \mathbb{C}^{n-1}$ be a k-dimensional subspace for which

$$\min_{\substack{w \in \mathcal{W} \\ \|w\|_2 = 1}} w^{\mathrm{H}} V^{\mathrm{H}} A V w = \lambda_k(V^{\mathrm{H}} A V).$$

Then $V\mathcal{W} \subset \mathbb{C}^n$ is a k-dimensional subspace. Hence by (4.47),

$$\lambda_k(V^{\mathrm{H}} A V) = \min_{\substack{w \in \mathcal{W} \\ \|w\|_2 = 1}} w^{\mathrm{H}} V^{\mathrm{H}} A V w = \min_{\substack{z \in V\mathcal{W} \\ \|z\|_2 = 1}} z^{\mathrm{H}} A z \leq \lambda_k(A).$$

The fact that $\lambda_k(V^{\mathrm{H}} A V) \geq \lambda_{k+1}(A)$ follows similarly from (4.46). ∎

4.5. CANONICAL ANGLES AND THE CS DECOMPOSITION

In some applications we need to say how near two subspaces are to one another. The notion of a set of canonical angles between subspaces provides a convenient metric. In this subsection we establish the properties of canonical angles and introduce the closely related CS decomposition.

Canonical angles between subspaces

In (4.18) we saw how the angle between two vectors x and y can be defined in terms of their inner product. An analogous definition can be made for subspaces; however, instead of a single angle, we obtain a collection of angles. For convenience we will assume that the subspaces in question are of the same dimension; however, the results can easily be extended to subspaces of different dimensions.

Let $\mathcal{X}, \mathcal{Y} \subset \mathbb{C}^n$ be subspaces of dimension p, and let X and Y be orthonormal bases for $\mathcal{X}$ and $\mathcal{Y}$. Let

$$U^{\mathrm{H}}(X^{\mathrm{H}}Y)V = \Gamma = \mathrm{diag}(\gamma_1, \gamma_2, \ldots, \gamma_p) \tag{4.48}$$

be the singular value decomposition of $X^{\mathrm{H}}Y$.

The numbers γ_i are characteristic of the subspaces $\mathcal{X}$ and $\mathcal{Y}$ and do not depend on the choice of orthonormal bases. For if $\hat{X}$ is another orthonormal basis, then by Theorem 4.25 $\hat{X}^{\mathrm{H}}X$ is unitary, and $\hat{X} = X(\hat{X}^{\mathrm{H}}X)$. Hence $\hat{X}^{\mathrm{H}}Y = (\hat{X}^{\mathrm{H}}X)^{\mathrm{H}}(X^{\mathrm{H}}Y)$. Thus $X^{\mathrm{H}}Y$ and $\hat{X}^{\mathrm{H}}Y$ are unitarily equivalent and have the same singular values γ_i.

Since $\|X^{\mathrm{H}}Y\|_2 \leq \|X\|_2 \|Y\|_2 = 1$, the γ_i are not greater than one in magnitude. This along with their uniqueness justifies the following definition.

Definition 4.35. *Let $\mathcal{X}, \mathcal{Y} \subset \mathbb{C}^n$ be subspaces of dimension p, and let X and Y be orthonormal bases for $\mathcal{X}$ and $\mathcal{Y}$. Let the singular values of $X^{\mathrm{H}}Y$ be $\gamma_1, \gamma_2, \ldots, \gamma_p$. Then the* CANONICAL ANGLES θ_i BETWEEN $\mathcal{X}$ AND $\mathcal{Y}$ *are defined by*

$$\theta_i = \cos^{-1} \gamma_i, \qquad i = 1, 2, \ldots, p.$$

We write

$$\Theta(\mathcal{X}, \mathcal{Y}) = \mathrm{diag}(\theta_1, \theta_2, \ldots, \theta_p).$$

A pair of orthonormal bases X_{bi} and Y_{bi} for $\mathcal{X}$ and $\mathcal{Y}$ are said to be *biorthogonal* if $X_{\mathrm{bi}}^{\mathrm{H}} Y_{\mathrm{bi}}$ is diagonal. From (4.48) it follows that the matrices $X_{\mathrm{bi}} = XU$ and $Y_{\mathrm{bi}} = YV$ are orthonormal bases for $\mathcal{X}$ and $\mathcal{Y}$ satisfying $X_{\mathrm{bi}}^{\mathrm{H}} Y_{\mathrm{bi}} = \cos \Theta(\mathcal{X}, \mathcal{Y})$ and hence are biorthogonal. From the uniqueness properties of the singular value decomposition it follows that any such basis must be essentially unique and the diagonal elements of $X_{\mathrm{bi}}^{\mathrm{H}} Y_{\mathrm{bi}}$ must be the cosines of the canonical angles between $\mathcal{X}$ and $\mathcal{Y}$. We summarize these results in the following theorem.

Theorem 4.36. *Let $\mathcal{X}, \mathcal{Y} \subset \mathbb{C}^n$ be subspaces of dimension p. Then there are (essentially unique)* CANONICAL ORTHONORMAL BASES X *and* Y *for $\mathcal{X}$ and $\mathcal{Y}$ such that*

$$X^{\mathrm{H}}Y = \cos \Theta(\mathcal{X}, \mathcal{Y}).$$

Two subspaces are identical if and only if their canonical angles are zero. Thus a principal application of canonical angles is to determine when subspaces are near one another. Fortunately, we do not have to compute the canonical angles themselves to test the nearness of subspaces. The following theorem shows how to compute matrices whose singular values are the sines of the canonical angles. As the subspaces approach one another, these matrices approach zero, and conversely.

Theorem 4.37. *Let $\mathcal{X}, \mathcal{Y} \subset \mathbb{C}^n$ be subspaces of dimension p, and let X and Y be orthonormal bases for $\mathcal{X}$ and $\mathcal{Y}$. Let $(X \ X_\perp)$ and $(Y \ Y_\perp)$ be unitary. Then the nonzero singular values of $X^H Y_\perp$ or $X_\perp^H Y$ are the sines of the nonzero canonical angles between $\mathcal{X}$ and $\mathcal{Y}$. Alternatively, if $P_\mathcal{X}$ and $P_\mathcal{Y}$ are the orthogonal projections onto $\mathcal{X}$ and $\mathcal{Y}$, then the sines of the nonzero canonical angles between $\mathcal{X}$ and $\mathcal{Y}$ are the nonzero singular values of $P_\mathcal{X}(I - P_\mathcal{Y})$ or $(I - P_\mathcal{X})P_\mathcal{Y}$.*

Proof. Without loss of generality we may assume that X and Y are canonical bases for $\mathcal{X}$ and $\mathcal{Y}$. Then the matrix

$$X^H(Y \ Y_\perp) = (\Gamma \ S) \tag{4.49}$$

has orthonormal rows. Hence

$$I = \Gamma^2 + SS^H.$$

It follows that $SS^H = I - \Gamma^2$ is diagonal. Since the diagonal elements of Γ are the canonical cosines, the diagonal elements of SS^H are the squares of the canonical sines. Thus the nonzero singular values of S are the sines of the nonzero canonical angles.

The result for $X_\perp^H Y$ is established similarly.

To establish the result for $P_\mathcal{X}(I - P_\mathcal{Y})$ note that

$$(X \ X_\perp)^H P_\mathcal{X}(I - P_\mathcal{Y})(Y \ Y_\perp) = (X \ X_\perp)^H (XX^H)(I - Y_\perp Y_\perp^H)(Y \ Y_\perp)$$

$$= \begin{pmatrix} X^H Y_\perp & 0 \\ 0 & 0 \end{pmatrix}.$$

The nonzero singular values of this matrix are the nonzero singular values of $X^H Y_\perp$, which establishes the result. The result for $(I - P_\mathcal{X})P_\mathcal{Y}$ is established similarly. ∎

Thus although we cannot compute, say, $\|\Theta(\mathcal{X}, \mathcal{Y})\|_F$ directly, we can compute $\|\sin\Theta(\mathcal{X}, \mathcal{Y})\|_F$ by computing, say, $\|X^H Y_\perp\|_F$. The latter is just as useful as the former for assessing the nearness of subspaces.

The CS decomposition

Suppose that we have a partitioned unitary matrix

$$Q = \begin{pmatrix} Q_{11} & Q_{12} \\ Q_{21} & Q_{22} \end{pmatrix},$$

SEC. 4. ANALYSIS

where Q_{11} is of order m. If we identify this matrix with the matrix $(Y\ Y_\perp)$ in (4.49) and set $X^H = (I_m\ 0)$, then $S = Q_{21}$. It follows that if we regard the singular values of Q_{11} as cosines then the singular values of Q_{12} are sines. A similar argument shows that the singular values of Q_{21} are the same sines. Moreover, passing to Q_{22} brings us back to the cosines.

These relations are a consequence of a beautiful decomposition of a unitary matrix called the CS decomposition. The proof of the following theorem is tedious but straightforward, and we omit it.

Theorem 4.38 (The CS decomposition). *Let the be an unitary matrix Q of order n be partitioned in the form*

$$Q = \begin{pmatrix} Q_{11} & Q_{12} \\ Q_{21} & Q_{22} \end{pmatrix},$$

where Q_{11} is of order $m \leq n/2$. Then there are unitary matrices $U_1, V_1 \in \mathbb{C}^{m \times m}$ and $U_2, V_2 \in \mathbb{C}^{(n-m) \times (n-m)}$ such that

$$\begin{pmatrix} U_1^H & 0 \\ 0 & U_2^H \end{pmatrix} \begin{pmatrix} Q_{11} & Q_{12} \\ Q_{21} & Q_{22} \end{pmatrix} \begin{pmatrix} V_1 & 0 \\ 0 & V_2 \end{pmatrix} = \begin{pmatrix} \Gamma & \Sigma & 0 \\ -\Sigma & \Gamma & 0 \\ 0 & 0 & I_{n-2m} \end{pmatrix},$$

where Γ and Σ are diagonal of order m and

$$\Gamma^2 + \Sigma^2 = I.$$

In effect the theorem states that the blocks in a partitioned unitary matrix share singular vectors. An important application of the decomposition is to simplify proofs of geometric theorems. It is an instructive exercise to derive the results on canonical angles and bases using the CS decomposition.

4.6. NOTES AND REFERENCES

Vector and matrix norms

There are two approaches to norms. The essentially axiomatic approach taken here was used by Banach [14, 1922] and Wiener [340, 1922] in defining normed linear spaces. Earlier Minkowski [228, 1911] defined norms geometrically in terms of compact, convex sets containing the origin in their interior (the unit ball $\{x : \|x\| = 1\}$ is such a set). For more on this approach, see [189].

The Cauchy inequality (in scalar form) is actually due to Cauchy [57, 1821, Note II, Théorèm XVI]. It is also associated with the names Schwarz and Bunyakovski.

The spectral norm was introduced by Peano [258, 1888]. The introduction of the Frobenius norm qua norm is due to Frobenius [124, 125, 1911].

For a proof of the equivalence of the characterizations (4.12) and (4.12) see [310, Theorem I.1.2].

For a systematic treatment of vector and matrix norms and further references, see [310].

Inverses and the Neumann series

For references on the perturbation of inverses and linear systems, see §4.6, Chapter 3. The Neumann series originated as a series of powers of operators [242, 1877].

The QR factorization

In some sense the QR factorization originated with Gram [158, 1883], who orthogonalized sequences of functions, giving determinantal expressions for the resulting orthogonal sequences. Later, Schmidt [272, 1907] gave the algorithm implicitly used in the proof of Theorem 4.24, still in terms of sequences of functions. The name of the decomposition is due to Francis [123], who used it in his celebrated QR algorithm for the nonsymmetric eigenvalue problem. Rumor has it that the "Q"in QR was originally an "O" standing for orthogonal. It is a curiosity that the formulas for the Gram–Schmidt algorithm can be found in a supplement to Laplace's *Théorie Analytique des Probabilités* [211, 1820]. However, there is no notion of orthogonalization associated with the formulas.

Projections

Not all projections have to be orthogonal. Any idempotent matrix P is a projection onto $\mathcal{R}(P)$ along $\mathcal{R}(P^{\text{H}})$.

The singular value decomposition

The singular value decomposition was introduced by Beltrami [25, 1873] and independently by Jordan [193, 1874]. They both worked with quadratic forms; however, their proofs transfer naturally to matrices. Beltrami derives the decomposition from the spectral decomposition of $X^{\text{H}}X$ [see (4.45)]. Jordan showed that vectors u and v of norm one that maximize the bilinear form $u^{\text{T}}Xv$ will deflate the problem as in the proof of Theorem 4.27. The construction used here due to Golub and Van Loan [153]. The chief disadvantage of this approach is that uniqueness has to be proven explicitly (Theorem 4.28). For another approach to uniqueness see [319].

In another line of development, Schmidt [272, 1907] established a singular value decomposition for integral operators and showed that it gave optimal low-rank approximations (Theorem 4.32) in the Frobenius norm. The theorem was rediscovered in terms of matrices by Eckart and Young [113, 1936], whose names are sometimes associated with it. Mirsky [229, 1960] established the optimality in all unitarily invariant norms. The proofs given here are essentially due to Weyl [338, 1912], who established the results for the spectral decomposition and then noted that they could be adapted to the singular value decomposition.

The min-max characterization is due to Fischer [119, 1905], who proved it for pencils of quadratic forms. The name Courant, who generalized it to differential operators [76, 1920], is sometimes associated with the theorem.

For more on the history of the singular value decomposition see [303].

The spectral decomposition

The spectral decomposition, written as a linear system of the form $Au = \lambda u$, is due to Cauchy [58, 1829], who established the orthogonality of the eigenvectors and his remarkable interlacing theorem (see Theorem 4.34) for principal submatrices.

Canonical angles and the CS decomposition

The idea of canonical angles between subspaces is due to Jordan [194, 1875]. The CS decomposition, which is a lineal descendent of Jordan's work, is implicit in an important paper by Davis and Kahan [88, 1970] and in a paper by Björck and Golub [43, 1973]. The explicit form and the name is due to Stewart [290, 1977]. It was generalized to nonrectangular partitions by Paige and Saunders [249]. These decompositions are closely related to the statistician Hotelling's work on canonical correlations [185, 1936].

Paige and Wei [250] give a historical survey of canonical angles and the CS decomposition. For computational algorithms see [295] and [327].

5. Addenda

5.1. Historical

On the word matrix

According to current thinking [221], about six thousand years ago the region between the Dnieper and Ural rivers was occupied by people speaking a language known as proto-Indo-European. Fifteen hundred years later, the language had fragmented, and its speakers began to spread out across Europe and Asia in one of the most extensive linguistic invasions ever recorded. From Alaska to India, from Patagonia to Siberia, half the world's population now speak Indo-European languages.

One piece of evidence for the common origin of the Indo-European languages is the similarity of their everyday words. For example, the word for two is *dva* in Sanskrit, *duo* in Greek, *duva* in Old Church Slavonic, and *dau* in Old Irish. More to our purpose, mother is *mather* in Sanskrit, *mater* in Greek, *mati* in Old Church Slavonic, *mathir* in Old Irish — and *mater* in Latin.

Matrix is a derivative of the Latin *mater*. It originally meant a pregnant animal and later the womb. By extension it came to mean something that surrounds, supports, or sustains — for example, the material in which a fossil is embedded. In 1850 Sylvester used it to refer to a rectangular array of numbers. It acquired its present mathematical meaning in 1855 when Cayley endowed Sylvester's array with the usual matrix operations.

History

A definitive history of vectors, matrices, and linear algebra has yet to be written. Two broad traditions can be discerned. The first begins with quaternions and passes through vector analysis to tensor analysis and differential geometry. This essentially analytic theory, whose early history has been surveyed in [80], touches only lightly on the subject of this work.

The second tradition concerns the theory of determinants and canonical forms. Muir [238] gives an exhaustive history of the former in four volumes. Kline, who surveys the latter in his *Mathematical Thought from Ancient to Modern Times* [199], points out that most of the fundamental results on matrices — their canonical forms and decompositions — had been obtained before matrices themselves came into widespread use. Mathematicians had been working with linear systems and quadratic and bilinear forms before Cayley introduced matrices and matrix algebra in the 1850s [59, 60], and they continued to do so.

The relation of matrices and bilinear forms is close. With every matrix A one can associate a function

$$f(x, y) = y^{\mathrm{H}} A x = \sum_{i,j} a_{ij} y_i x_j,$$

called a *bilinear form*, that is linear in each of its variables. Conversely, each bilinear form $\sum_{i,j} a_{ij} y_i x_j$ corresponds to the matrix of its coefficients a_{ij}. Under a change of variables, say $x = P\hat{x}$ and $y = Q\hat{y}$, the matrix of the form changes to $Q^{\mathrm{H}} A P$. Thus the reduction of a matrix by transformations is equivalent to simplifying a quadratic form by a change of variables.

The first simplification of this kind is due to Lagrange [205, 1759], who showed how to reduce a quadratic form to a sum of squares, in which the kth term contains only the last $n-k+1$ variables. His purpose was to determine if the form was positive definite. Gauss [130, 131, 1809, 1810] introduced essentially the same reduction — now called Gaussian elimination — to solve systems and compute variances arising from least squares problems. Throughout the rest of the century, various reductions and canonical forms appeared in the literature: e.g., the LU decomposition by Jacobi [191, 1857], Jordan's canonical form [192, 1870], reductions of matrix pencils by Weierstrass [337, 1868] and Kronecker [204, 1890], and the singular value decomposition discovered independently by Beltrami [25, 1873] and Jordan [193, 1874]. For more on these decompositions see the notes and references to the appropriate sections and chapters.

The notion of an abstract vector space seems to be more a creature of functional analysis than of matrix theory (for a history of the former see [95]). Definitions of normed linear spaces — usually called Banach spaces — were proposed independently by Banach [14, 1922] and Wiener [340, 1922]. Less the norm, these spaces became our abstract vector spaces.

5.2. GENERAL REFERENCES

This work is a survey of matrix algorithms, not of its literature. Consequently, the notes and references subsections cite only the immediately relevant literature. How-

Linear algebra and matrix theory

There are any number of texts on abstract linear algebra. My favorite is Halmos' text *Finite-Dimensional Vector Spaces* [168]. Greub's *Linear Algebra* [159] is more technical: Marcus and Minc's *Introduction to Linear Algebra* [224] is more leisurely. In addition, there are many elementary books stressing applications, e.g., [11, 56, 215, 217, 244, 312].

Gantmacher's two-volume *Theory of Matrices* [129] is the definitive survey of matrix theory up to 1959. Other introductions to matrix theory may be found in [208, 222, 223]. Berman and Plemmons [28] give a comprehensive treatment of the theory of nonnegative matrices.

The earlier literature on matrices was algebraic in flavor with its emphasis on canonical forms and decompositions. Over the past half century the subject has expanded to include the analytic properties of matrices. Bellman's *Introduction to Matrix Analysis* [24] is the classic. Horn and Johnson's *Matrix Analysis* [182] and *Topics in Matrix Analysis* [183] deserve special mention. The second book contains a wealth of historical information.

For more specialized references on inequalities and perturbation theory, see the books by Marcus and Minc [223], Bhatia [29], Kato [196], and Stewart and Sun [310].

Classics of matrix computations

The founders of modern numerical linear algebra liked to write books, and many of them can be read with profit today. James H. Wilkinson's *Rounding Errors in Algebraic Processes* [345] is the first general exposition of the modern theory of rounding error. His *Algebraic Eigenvalue Problem* [346] contains most of what was known about dense matrix computations in 1965. Alston Householder's *Theory of Matrices in Numerical Analysis* [189] is notable for its concise unification of diverse material. Faddeev and Faddeeva give the Russian view in their *Computational Methods of Linear Algebra* [115]. Richard Varga's *Matrix Iterative Analysis* [330] is an elegant introduction to the classical iterative methods.

Textbooks

The first textbook devoted exclusively to modern numerical linear algebra was Fox's *Introduction to Numerical Linear Algebra* [122]. My own text, *Introduction to Matrix Computations* [288], published in 1973, is showing its age. Golub and Van Loan's *Matrix Computations* [153] is compendious and up to date — the standard reference. Watkins' *Fundamentals of Matrix Computations* [333], Datta's *Numerical Linear Algebra and Applications* [86], and Trefethen and Bau's *Numerical Linear Algebra* [319] are clearly written, well thought out introductions to the field. Coleman and Van Loan's *Handbook for Matrix Computations* [71] provides a useful introduction to the practicalities of the subject.

Special topics

There are a number of books on special topics in matrix computations: eigenvalue problems [82, 207, 253], generalized inverses [26, 240, 267], iterative methods [13, 17, 165, 166, 332, 353], least squares [41, 213], rounding-error analysis [177], and sparse matrices [108, 143, 247].

Software

The progenitor of matrix software collections was the series of Handbook articles that appeared in *Numerische Mathematik* and were later collected in a single volume by Wilkinson and Reinsch [349]. The lineal descendants of this effort are EISPACK [284], LINPACK [99], and LAPACK [9]. It is not an exaggeration to say that applied linear algebra and matrix computations have been transformed by the availability of Cleve Moler's MATLAB system [232] and its clones. See [255] for a useful handbook.

EISPACK, LINPACK, and LAPACK are available over the web from the NETLIB repository at

```
http://www.netlib.org/index.html
```

This repository contains many other useful, high-quality numerical routines. For a general index of numerical routines, consult the *Guide to Available Mathematical Software* (GAMS) at

```
http://math.nist.gov/gams/
```

Historical sources

Kline's *Mathematical Thought from Ancient to Modern Times* [199] contains many references to original articles. Older texts on matrix theory are often good sources of references to original papers. Particular mention should be made of the books by Mac Duffee [220], Turnbull and Aitken [322], and Wedderburn, [334]. For a view of precomputer numerical linear algebra see Dwyer's *Linear Computations* [112].

2

MATRICES AND MACHINES

Matrix algorithms — at least the ones in this series — are not museum pieces to be viewed and admired for their beauty. They are meant to be programmed and run on today's computers. However, the road from a mathematical description of an algorithm to a working implementation is often long. In this chapter we will traverse the road in stages.

The first step is to decide on the vehicle that will carry us — the language we will use to describe our algorithms. In this work we will use pseudocode, which is treated in the first section of this chapter.

The second stage is the passage from a mathematical description to pseudocode. It often happens that an algorithm can be derived and written in different ways. In the second section of this chapter, we will use the problem of solving a triangular system to illustrate the ins and outs of getting from mathematics to code. We will also show how to estimate the number of arithmetic operations an algorithm performs. Although such operation counts have limitations, they are often the best way of comparing the efficiency of algorithms — short of measuring actual performance.

The third stage is to move from code to the computer. For matrix algorithms two aspects of the computer are paramount: memory and arithmetic. In the third section, we will show how hierarchical memories affect the performance of matrix algorithms and conversely how matrix algorithms may be coded to interact well with the memory system of a computer. In the fourth section, we introduce floating-point arithmetic and rounding-error analysis — in particular, backward rounding-error analysis and its companion, perturbation theory.

The further we proceed along the road from mathematical description to implementation the more important variants of an algorithm become. What appears to be a single algorithm at the highest level splits into several algorithms, each having its advantages and disadvantages. For example, the interaction of a matrix algorithm with memory depends on the way in which a matrix is stored — something not usually specified in a mathematical description. By the time we reach rounding error, truly minute changes in an algorithm can lead to enormous differences in behavior. What is an algorithm? The answer, it seems, depends on where you're at.

82　　　　　　　　　　　　　CHAPTER 2. MATRICES AND MACHINES

Here are instructions on how to get to my house. The party starts at 7:30.

    1.   Go to the last traffic light on Kingston Pike
    2.   Turn right
    3.   Drive 5.3 miles
    4.   Turn left at the convenience store
    5.   We are the ninth house on the right

Algorithm 1.1: Party time

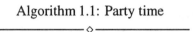

1. PSEUDOCODE

Most computer languages have all that is needed to implement algorithms for dense matrices — two-dimensional arrays, conditional statements, and looping constructs. Many also allow one to define new data structures — something that is useful in coding sparse matrix algorithms. Yet matrix algorithms are not usually presented in a standard programming language but in some form of pseudocode. The chief reason is that pseudocode allows one to abandon lexical rigor for ease of exposition. English sentences and mathematical expressions can be interleaved with programming constructs. Statements can be neatly labeled for later reference. And pseudocode provides a veneer of neutrality by appearing not to favor one language over another.

For all these reasons, we have chosen to present algorithms in pseudocode. This section is devoted to setting down the basics. The reader is assumed to be familiar with a high-level, structured programming language.

1.1. GENERALITIES

A program or code fragment is a sequence of statements, perhaps numbered sequentially. The statements can be ordinary English sentences; e.g.,

    1.   Go to the last traffic light on Kingston Pike
    2.   Turn right

When it is necessary to be formal, we will call a sequence of pseudocode an algorithm and give it a prologue explaining what it does. For example, Algorithm 1.1 describes how to get to a party.

We will use standard mathematical notation freely in our pseudocode. However, in a statement like

    1. $x = A^{-1}b$

the "=" is a replacement operator, not a mathematical assertion of equality. We might have written

SEC. 1. PSEUDOCODE

$A[\mathcal{I}, \mathcal{J}]$ The matrix in the intersection of the rows indexed by $\mathcal{I}$ and the column indexed by $\mathcal{J}$.
$A[i_1{:}i_2, j_1{:}j_2]$ The submatrix in the intersection of rows $i_1, \ldots, i_2$ and columns $j_1, \ldots, j_2$.
$A[i_1{:}i_2, j]$ The vector $(a_{i_1 j}, \ldots, a_{i_2 j})^{\mathrm{T}}$.
$A[i, j_1{:}j_2]$ The row vector vector $(a_{ij_1}, \ldots, a_{ij_2})$.
$A[:, j]$ The jth column of A.
$A[i, :]$ The ith row of A.
$A[i, j]$ The (i, j)-element of A.

Figure 1.1: Notation for submatrices

———————⋄———————

1. $b = A^{-1}b$

which is (almost) the same as

1.   Overwrite b with the solution of $Ax = b$

Many of our algorithms will involve partitioned matrices. In ordinary text it is easy enough to write statements like: Partition A in the form

$$A = \begin{pmatrix} \alpha_{11} & a_{12}^{\mathrm{T}} \\ a_{21} & A_{22} \end{pmatrix}.$$

But the same statement would be awkward in a program. Consequently we will use the conventions in Figure 1.1 to extract submatrices from a matrix. Inconsistent dimensions like $[n+1{:}n]$ represent a void vector or matrix, a convention that is useful at the beginning and end of loops [see (2.5) for an example].

1.2. CONTROL STATEMENTS

Our pseudocode has the usual elementary control statements.

The if statement

The **if** statement has the following form:

1.   **if** (first conditional expression)
2.   first block of statements
3.   **else if** (second conditional statement)
4.   second block of statements
5.   **else if** (third conditional statement)
6.   third block of statements
            . . .
7.   **else**
8.   last block of statements
9.   **end if**

Both the **else** and the **else if**s are optional. There may be no more than one **else**. The conditional statements are evaluated in order until one evaluates to true, in which case the corresponding block of statements is executed. If none of the conditional statements are true, the block of statements following the **else**, if there is one, is executed. In nested **if** statements, an **else** refers to the most recent **if** that has not been paired with an **endif** or an **else**.

The token **fi** is an abbreviation for **end if**. It is useful for one-liners:

1. **if** $(s = 0)$ **return** 0 **fi**

The for statement

The **for** statement has the following form.

1. **for** $i = j$ **to** k **by** d
2.   Block of statements
3. **end for** i

Here i is a variable, which may not be modified in the loop. The parameters j, k, and d are expressions, which will be evaluated once before the loop is executed. If the **by** part of the loop is omitted, d is assumed to be one.

For $i \leq k$ and $d > 0$, the block of statements is executed for $i = j$, $i = j+d$, $i = j+2d$, ..., $j+nd$, where n is the largest integer such that $j+nd \leq k$. Similarly, for $i \geq k$ and $d < 0$ the index i steps downward by increments of d until it falls below k. The identifier i is not required after the **end for**, but it is useful for keeping track of long loops. In fact, any appropriate token will do — e.g., the statement number of the **for** statement.

The **for** loop obeys the following useful convention.

Inconsistent loops are not executed.

For example, the following code subtracts the last $n-1$ components of an n-vector from the first component, even when $n = 1$.

1. **for** $i = 2$ **to** n
2. $x_1 = x_1 - x_i$
3. **end for**

The while statement

The **while** statement has the following form.

1. **while** (conditional expression)
2.   block of statements
3. **end while**

The while statement continues to execute the block of statements until the conditional expression evaluates to false.

Leaving and iterating control statements

The statement

    1.   **leave** <name>

causes the algorithm to leave the control statement indicated by <name>. The statement may be a **for**, **while**, or **if** statement. The name may be anything that unambiguously identifies the statement: the index of a loop, the line number of an **if** statement, or the corresponding **end if**.

The statement

    1.   **iterate** <name>

forces an new iteration of the **for** or **while** loop indicated by <name>.

The goto statement

The **goto** statement is useful in situations where the **leave** statement is inadequate. It has the form

    1.   **goto** <name>

Here name may be a statement number or a statement label. A statement label proceeds the statement and is followed by a colon:

    1.   Error: Take care of the error

1.3. FUNCTIONS

Functions and subprograms with arguments will be indicated, as customary, by preceding the code by the name of the function with its argument list. The statement **return** exits from the subprogram. It can also return a value. For example, the following function returns $\sqrt{a^2 + b^2}$, calculated in such a way as to avoid overflows and render underflows innocuous (see §4.5).

    1.   *Euclid*(a, b)
    2. $s = |a| + |b|$
    3.   **if** ($s = 0$)
    4.   **return** 0   ! Zero is a special case.
    5.   **else**
    6.   **return** $s\sqrt{(a/s)^2 + (b/s)^2}$
    7.   **end if**
    8.   **end** *Euclid*

Note the comment, which is preceded by an exclamation mark.

Parameters are passed to functions by reference, as in FORTRAN, not by value, as in C. For scalar arguments this means that any modification of an argument in a function modifies the corresponding argument in the program that invokes it.

1.4. NOTES AND REFERENCES

Programming languages

Pratt and Zelkowitz [265] give an excellent survey of programming languages. The front-running language for numerical computations has been FORTRAN77 [10] with C [197] a serious competitor. Each language has its advantages. FORTRAN handles arrays more sensibly, making it the language of choice for matrix computations. C, with its rich facilities for creating data structures, provides a more congenial environment for the manipulation of sparse matrices and for large-scale scientific programming.

Both C and FORTRAN have been extended. C++ [114, 218] is an object-oriented language that is compatible with C and will likely become the new standard. The language FORTRAN90 [227] includes the ability to define data structures that the old version lacked. If it catches on, it will reconfirm FORTRAN as the language of scientific computing. It should be noted that both extensions include powerful features that make it easy to write inefficient code.

Pseudocode

Another reason for the use of pseudocode in this book is, paradoxically, to make the algorithms a little difficult to implement in a standard language. In many of the algorithms to follow I have omitted the consistency and error tests that make their overall structure difficult to see. If these algorithms could be lifted from the text and compiled, they would no doubt find their way unpolished into the real world. In fact, implementing the algorithms, which requires line-by-line attention, is a good way to become really familiar with them.

The pseudocode used in this work shows a decided tilt toward FORTRAN in its looping construct and its passing of subprogram arguments by reference. This latter feature of FORTRAN has been used extensively to pass subarrays by the BLAS (Basic Linear Algebra Subprograms, see §3). In C one has to go to the additional trouble of creating a pointer to the subarray. But it should be added that our conventions for specifying submatrices (Figure 1.1) render the decisions to pass by reference largely moot.

The use of the colon to specify an index range (Figure 1.1) is found in array declarations FORTRAN77. It was extended to extract subarrays in MATLAB and later in FORTRAN90. The use of brackets to specify array references is in the spirit of C. It avoids loading yet another burden on the overworked parenthesis.

Twenty years ago one could be pilloried for including a **goto** statement in a language. The reason was a 1968 letter in the *Communications of the ACM* by Dijkstra titled "Go to statement considered harmful" [96]. Although others had deprecated the use of goto's earlier, Dijkstra's communication was the match that lit the fire. The argument ran that the goto's in a program could be replaced by other structured constructs — to the great improvement of the program. This largely correct view was well on the way to freezing into dogma, when Knuth in a wonderfully balanced article [200, 1974] (which contains a history of the topic and many references) showed that goto's

SEC. 2. TRIANGULAR SYSTEMS

have rare but legitimate uses.

2. TRIANGULAR SYSTEMS

The implementation of matrix algorithms is partly art, partly science. There are general principles but no universal prescriptions for their application. Consequently, any discussion of code for matrix algorithms must be accompanied by examples to bridge the gap between the general and the particular.

In this section we will use the problem of solving a lower triangular system as a running example. There are three reasons for this choice. First, it is a real problem of wide applicability. Second, it is simple enough so that the basic algorithm can be readily comprehended. Third, it is complex enough to illustrate many of the principles of sound implementation. We have chosen to work with lower triangular systems instead of upper triangular systems because the order of computations runs forward through the matrix in the former as opposed to backward in the latter. But everything we say about lower triangular systems applies mutatis mutandis to upper triangular systems.

2.1. THE SOLUTION OF A LOWER TRIANGULAR SYSTEM

Existence of solutions

It is convenient to begin with a theorem.

Theorem 2.1. *Let L be a lower triangular matrix of order n. Then L is nonsingular if and only if its diagonal elements are nonzero.*

Proof. We will use the fact that a matrix L is nonsingular if and only if the system

$$Lx = b \qquad (2.1)$$

has a solution for every b (see Theorem 3.21, Chapter 1). Let us write the system (2.1) in scalar form:

$$\begin{aligned}
b_1 &= \ell_{11}x_1 \\
b_2 &= \ell_{21}x_1 + \ell_{22}x_2 \\
b_3 &= \ell_{31}x_1 + \ell_{32}x_2 + \ell_{33}x_3 \\
&\;\cdot\;\cdot\;\cdot \\
b_n &= \ell_{n1}x_1 + \ell_{n2}x_2 + \ell_{n3}x_3 + \cdots + \ell_{nn}x_n
\end{aligned} \qquad (2.2)$$

First, suppose that the diagonal elements of L are nonzero. Then the first equation in (2.2) has the solution $x_1 = b_1/\ell_{11}$. Now suppose that we have computed $x_1, x_2, \ldots, x_{k-1}$. Then from the kth equation,

$$b_k = \ell_{k1}x_1 + \ell_{k2}x_2 + \ell_{k3}x_3 + \cdots + \ell_{kk}x_k,$$

> Let L be a nonsingular lower triangular matrix of order n. The following algorithm solves the system $Lx = b$.
>
> 1. **for** $k = 1$ **to** n
> 2. $\quad x_k = b_k$
> 3. $\quad$ **for** $j = 1$ **to** $k-1$
> 4. $\quad\quad x_k = x_k - \ell_{kj} x_j$
> 5. $\quad$ **end for** j
> 6. $\quad x_k = x_k / \ell_{kk}$
> 7. **end for** k

Algorithm 2.1: Forward substitution

---

we have

$$x_k = \frac{b_k - \ell_{k1}x_1 - \ell_{k2}x_2 - \ell_{k3}x_3 - \cdots - \ell_{k,k-1}x_{k-1}}{\ell_{kk}}. \tag{2.3}$$

Consequently, the fact that the ℓ_{kk} are all nonzero implies that equation (2.1) has a solution for any right-hand side b.

On the other hand, suppose that some diagonals of L are zero, and suppose that ℓ_{kk} is the first such diagonal. If $k = 1$, then the equation fails to have a solution whenever $b_1 \neq 0$. If $k > 1$, the quantities $x_1, x_2, \ldots, x_{k-1}$ are determined uniquely as in (2.3). If b_k is then chosen so that

$$b_k - \ell_{k1}x_1 + \ell_{k2}x_2 + \ell_{k3}x_3 + \cdots + \ell_{k-1,k-1}x_{k-1} \neq 0,$$

there is no x_k satisfying the kth equation. ∎

The forward substitution algorithm

The proof of Theorem 2.1 is not the shortest possible—it is easier to observe that $\det(L)$ is the product of the diagonal elements of L—but it is constructive in that it provides an algorithm for computing x, an algorithm that is sometimes called *forward substitution*. Algorithm 2.1 is an implementation of this process.

This algorithm is a straightforward implementation of (2.3). The loop on j computes the denominator, and the division in statement 6 completes the evaluation. By the convention on inconsistent loops, the inner loop is not executed when $k = 1$, which is just as it should be.

Overwriting the right-hand side

The solution of a triangular system is frequently just a step in the transformation of a vector. In such cases it makes little sense to create an intermediate vector x that will

SEC. 2. TRIANGULAR SYSTEMS

itself be transformed. Instead one should overwrite the vector b with the solution of the system $Lx = b$. It is easy to modify Algorithm 2.1 to do this.

1.   **for** $k = 1$ **to** n
2.   **for** $j = 1$ **to** $k-1$
3. $b_k = b_k - \ell_{kj} b_j$
4.   **end for** j
5. $b_k = b_k / \ell_{kk}$
6.   **end for** k

2.2. RECURSIVE DERIVATION

In matrix computations it is unusual to have a solution to a problem in terms of scalar formulas such as (2.3). The following derivation of Algorithm 2.1 is more representative of the way one finds new matrix algorithms.

Let the system $Lx = b$ be partitioned (with northwest indexing) in the form

$$\begin{pmatrix} L_{11} & 0 \\ \ell_{n1}^T & \lambda_{nn} \end{pmatrix} \begin{pmatrix} x_1 \\ \xi_n \end{pmatrix} = \begin{pmatrix} b_1 \\ \beta_n \end{pmatrix}. \qquad (2.4)$$

This partition is equivalent to the two equations

1. $L_{11} x_1 = b_1$,
2. $\ell_{n1}^T x_1 + \lambda_{nn} \xi_n = \beta_n$.

Since L_{11} is a lower triangular matrix of order $n-1$ we can solve the first system recursively for x_1 and then solve the second equation for ξ_n. This leads to the following recursive code.

1.   *trisolve*(L, x, b, n)
2.   **if** ($n = 0$) **return fi**
3.   *trisolve*($L[1{:}n{-}1, 1{:}n{-}1]$, $x[1{:}n{-}1]$, $b[1{:}n{-}1]$, $n{-}1$)   (2.5)
4. $x[n] = (b[n] - L[n, 1{:}n{-}1] * x[1{:}n{-}1])/L[n, n]$
5.   **end** *trisolve*

There are two things to say about this algorithm.

- We have made heavy use of the conventions in Figure 1.1 to extract submatrices and subvectors. The result is that no loop is required to compute the inner product in the formula for $x[n]$. This suggests that we can code shorter, more readable algorithms by consigning operations such as inner products to subprograms. We will return to this point when we discuss the BLAS in §3.

- Implicit in the program is the assumption that $L[n, 1{:}n{-}1] * x[1{:}n{-}1]$ evaluates to zero when $n = 1$. This is the equivalent of our convention about inconsistent **for** loops. In fact, the natural loop to compute the inner product in (2.5), namely,

1. sum = 0
2. **for** $j = 1$ **to** $n-1$
3.   sum = sum + $L[n,j]*x[j]$
4. **end for**

returns zero when $n = 1$. In what follows we will assume that degenerate statements are handled in such a way as to make our algorithms work.

—

Many matrix algorithms are derived, as was (2.5), from a matrix partition in such a way as to suggest a recursive algorithm. Another example is the recursive algorithm for computing the LU decomposition implicit in the proof of Theorem 2.13, Chapter 1. How then are we to recover a more conventional nonrecursive algorithm? A recursive matrix algorithm will typically contain a statement or sequence of statements performing a computation over a fixed range, usually from 1 or 2 to $n-1$ or n, where n is the recursion parameter—e.g., statement (2.5.4). The nonrecursive code is obtained by replacing the index n by another variable k and surrounding the statements by a loop in k that ranges between 1 and n. Whether k goes forward or backward must be determined by inspection. For example, the nonrecursive equivalent of (2.5) is

1. **for** $k = 1$ **to** n
2. $x[k] = (b[k] - L[k, 1{:}k{-}1]*x[1{:}k{-}1])/L[k,k]$ (2.6)
3. **end for**

Matrix algorithms are seldom written in recursive form. There are two plausible reasons.

1. A recursive call is computationally more expensive than iterating a **for** loop.
2. When an error occurs, it is easy to jump out of a nest of loops to an appropriate error handler. Getting out of a recursion is more difficult.

On modern computers a matrix must be rather small for the recursion overhead to count for much. Yet small matrices are often manipulated in the inner loops of application programs, and the implementer of matrix algorithms is well advised to be parsimonious whenever possible.

2.3. A "NEW" ALGORITHM

Matrices can be partitioned in many different ways, and different partitions lead to different algorithms. For example, we can write the system $Lx = b$ in the form

$$\begin{pmatrix} \lambda_{11} & 0 \\ \ell_{21} & L_{22} \end{pmatrix} \begin{pmatrix} \xi_1 \\ x_2 \end{pmatrix} = \begin{pmatrix} \beta_1 \\ b_2 \end{pmatrix}.$$

As with (2.4), this system expands into two equations:

1. $\lambda_{11}\xi_1 = \beta_1$,

Sec. 2. Triangular Systems

2. $\xi_1 \ell_{21} + L_{22} x_2 = b_2$.

Thus we can solve for ξ_1 in the form

$$\xi_1 = \frac{\beta_1}{\lambda_{11}}$$

and then go on to solve the system

$$L_{22} x_2 = b_2 - \xi_1 \ell_{21}.$$

The following algorithm implements this scheme.

1. $x = b$
2.   **for** $k = 1$ **to** n
3. $x[k] = x[k]/L[k,k]$ (2.7)
4. $x[k+1{:}n] = x[k+1{:}n] - x[k]*L[k+1{:}n, k]$
5.   **end for** k

We will call this algorithm the *axpy* algorithm after the BLAS used to implement it [see (3.9)].

This algorithm certainly appears different from (2.6). However, in one sense the two algorithms are fundamentally the same. To see this, let us focus on what happens to the component $x[k]$ of the solution. If we trace through both programs, we find that the following sequence of operations is being performed.

1. $x[k] = b[k]$
2.   **for** $j = 1$ **to** $k-1$
3. $x[k] = x[k] - L[k,j]*x[j]$ (2.8)
4.   **end for**
5. $x[k] = x[k]/L[k,k]$

Thus if we focus on only one component of the solution, we find that both algorithms perform the same operations in the same order. The difference between the two algorithms is the way operations for different components of the solution are interleaved.

In consequence, both algorithms share some properties. An obvious property is that they both have the same complexity in the sense that they both perform the same number of arithmetic operations. A rather striking property is that even in the presence of rounding error the algorithms will compute the same answer down to the very last bit, since they perform the same operations in the same order on the individual components (provided, of course, that the compiler does not do funny things with registers that work in higher precision).

An important difference in these algorithms is the order in which they access the elements of the matrix. The back-substitution algorithm is *row oriented* in the sense that the inner loop moves along the rows of the matrix. The algorithm (2.7) is *column oriented*; the inner loop moves down columns. However, these two algorithms can perform quite differently on machines with hierarchical memories (see §3.3).

2.4. THE TRANSPOSED SYSTEM

It frequently happens that one has to solve the transposed system

$$L^T x = b$$

(or equivalently the system $x^T L = b^T$). To derive an algorithm, note that the last equation of this system has the form

$$\ell_{nn} x_n = b_n,$$

which can be solved for x_n. The kth equation of the system has the form

$$b_k = \ell_{kk} x_k + \ell_{k+1,k} x_{k+1} + \cdots + \ell_{nk} x_n,$$

which can be solved for x_k provided we know $x_{k+1}, \ldots, x_n$.

These considerations lead to the following algorithm.

1. **for** $k = n$ **to** 1 **by** -1
2. $x[k] = (b[k] - L[k+1{:}n, k]^T * x[k+1{:}n])/L[k, k]$
3. **end for** i

This algorithm is the analogue of the forward substitution algorithm (*back substitution* it is called), but in changing from the original system to the transposed system it has become column oriented. The analogue for transposed systems of the of the column-oriented algorithm (2.7) is row oriented.

2.5. BIDIAGONAL MATRICES

In §2.2, Chapter 1, we listed some classes of matrices that had special structures of zero elements. One of these is the class of lower triangular matrices that has been our concern in this section. It often happens that one class is contained in another class with a richer supply of nonzero elements. For example, a bidiagonal matrix of the form

$$\begin{pmatrix} X & 0 & 0 & 0 & 0 \\ X & X & 0 & 0 & 0 \\ 0 & X & X & 0 & 0 \\ 0 & 0 & X & X & 0 \\ 0 & 0 & 0 & X & X \end{pmatrix}$$

is clearly lower triangular. Hence the algorithms we have just derived will solve bidiagonal systems. But they will spend most of their time manipulating zero elements. We can get a more efficient algorithm by restricting the computations to nonzero elements.

For example, in the relation

$$x_k = \frac{b_k - \ell_{k1} x_1 - \ell_{k2} x_2 - \ell_{k3} x_3 - \cdots - \ell_{k-1,k-1} x_{k-1}}{\ell_{kk}}$$

SEC. 2. TRIANGULAR SYSTEMS

Let L be a nonsingular lower bidiagonal matrix of order n. The following algorithm solves the system $Lx = b$.

1. $x[1] = b[1]/L[1,1]$
2. **for** $k = 2$ **to** n
3. $\quad x[k] = (b[k] - L[k, k-1]*x[k-1])/L[k, k]$
4. **end for**

Algorithm 2.2: Lower bidiagonal system

⬦

defining x_k, only $\ell_{k,k-1}$ and $\ell_{k,k}$ are nonzero. Hence we may rewrite it in the form

$$x_k = \frac{b_k - \ell_{k,k-1} x_{k-1}}{\ell_{kk}}.$$

Thus we get Algorithm 2.2. This algorithm is clearly cheaper than Algorithm 2.1. But how much cheaper? We will return to this question after we derive another algorithm.

2.6. INVERSION OF TRIANGULAR MATRICES

There is seldom any need to compute the inverse a matrix, since the product $x = A^{-1}b$ can be computed more cheaply by solving the system $Ax = b$. (We will return to this point in §1.5, Chapter 3.) Occasionally, however, the elements of an inverse have meaning for the problem at hand and it is desirable to print them out. For this reason, algorithms for matrix inversion are not entirely useless.

A general strategy for computing the inverse $X = (x_1 \; x_2 \; \cdots \; x_n)$ of a matrix A is to solve the system

$$Ax_j = \mathbf{e}_j, \qquad j = 1, 2, \ldots, n$$

(cf. the proof of Theorem 3.20, Chapter 1). However, when $A = L$ is lower triangular there are some special savings. As is often the case, the algorithm is a spin-off of a useful result.

Theorem 2.2. *The inverse of a lower (upper) triangular matrix is lower (upper) triangular.*

Proof. We will prove the result for lower triangular matrices. Partition the system $Lx_j = \mathbf{e}_j$ in the form

$$\begin{pmatrix} L_{11}^{(j)} & 0 \\ L_{21}^{(j)} & L_{22}^{(j)} \end{pmatrix} \begin{pmatrix} x_1^{(j)} \\ x_2^{(j)} \end{pmatrix} = \begin{pmatrix} 0 \\ \mathbf{e}_1 \end{pmatrix},$$

Let L be a nonsingular lower triangular matrix of order n. The following algorithm computes the inverse X of L.

1. **for** $k = 1$ **to** n
2. $\quad X[k,k] = 1/L[k,k]$
3. $\quad$ **for** $i = k{+}1$ **to** n
4. $\quad\quad X[i,k] = -L[i,k{:}i{-}1]*X[k{:}i{-}1,k]/L[i,i]$
5. $\quad$ **end for** i
6. **end for** k

Algorithm 2.3: Inverse of a lower triangular matrix

———————◇———————

where $L_{11}^{(j)}$ is $(j{-}1)\times(j{-}1)$. Then $L_{11}^{(j)} x_1^{(j)} = 0$, and hence $x_1^{(j)} = 0$. This shows that the first $j{-}1$ components of the jth column of L^{-1} are zero, and it follows that L^{-1} is lower triangular. ∎

The proof of Theorem 2.2 implies that to compute the inverse of L we need only solve the $(n{-}j{+}1)\times(n{-}j{+}1)$ systems $L_{22}^{(j)} x_2^{(j)} = \mathbf{e}_1$ for $j = 1, 2, \ldots, n$. If we use Algorithm 2.1 to solve these systems, we obtain Algorithm 2.3.

The algorithm can be modified to overwrite L with its inverse by replacing all references to X with references to L. The reader should verify that the following algorithm does the job.

1. **for** $k = 1$ **to** n
2. $\quad L[k,k] = 1/L[k,k]$
3. $\quad$ **for** $i = k{+}1$ **to** n
4. $\quad\quad L[i,k] = -L[i,k{:}i{-}1]*L[k{:}i{-}1,k]/L[k,k]$
5. $\quad$ **end for** i
6. **end for** k

(2.9)

The savings in storage can be considerable, since a lower triangular matrix of order n has at most $n(n{+}1)/2$ nonzero elements.

2.7. OPERATION COUNTS

Most matrix problems can be solved in more than one way. Of the many considerations that enter into the choice of an algorithm, three are paramount: speed, storage, and stability. Which algorithm runs faster? Which uses the least storage? Are the answers satisfactory? We will treat storage and stability in the next two sections. And in fact, we will have to reconsider the issue of speed, since speed and storage are connected. Nonetheless, we can learn useful things about the speed of an algorithm simply by counting arithmetic operations.

Sec. 2. Triangular Systems

Bidiagonal systems

Let us look first at the number of operations required to solve a bidiagonal system. For $k = 1$, the loop in Algorithm 2.2 performs a single division. For $k > 1$, it performs one multiplication, one addition (actually a subtraction), and one division. Since the loop runs from $k = 1$ to n, the entire algorithm requires

1. $n-1$ additions,
2. $n-1$ multiplications,
3. n divisions.

Full triangular systems

Let us now consider Algorithm 2.1 for the solution of a full triangular system. The body of the inner loop (statement 4) performs one addition and one multiplication. It is executed $k-1$ times as j varies from 1 to $k-1$, and k itself varies from one to n. Consequently, the algorithm requires

$$\sum_{k=1}^{n}(k-1) = \frac{n(n-1)}{2}$$

additions and multiplications. Taking into account the number of divisions, we get the following operation count:

1. $\frac{1}{2}n^2 - \frac{1}{2}n$ additions,
2. $\frac{1}{2}n^2 - \frac{1}{2}n$ multiplications,
3. n divisions.

General observations on operations counts

These examples illustrate some important points about operation counts.

- **The dominant term.** For large n, the term $\frac{1}{2}n^2$ in the expression $\frac{1}{2}n^2 - \frac{1}{2}n$ dominates the term $\frac{1}{2}n$. For example, if $n = 100$ then the ratio of the terms is one hundred to one. Consequently, it is customary to report only the dominant term — in this case $\frac{1}{2}n^2$ — in operation counts.

- **Order and order constants.** The factor of the dominant term that actually grows — in this case n^2 — is called the *order* of the algorithm. Thus the algorithm for solving a full lower triangular system is of order n^2, while the algorithm for solving a bidiagonal system is of order n. We write that they are $O(n^2)$ and $O(n)$ respectively, a convention that is called *the big O notation*.

The factor $\frac{1}{2}$ in the count $\frac{1}{2}n^2$ is called the *order constant*. It turns out that it is often easy to guess the order just by looking at an algorithm, whereas getting the order constant can be tedious. For this reason, the order constant is often omitted in reporting operation counts. As a general rule, this is not sound practice, since the order

Abbreviation	Description
fladd	a floating-point addition
flmlt	a floating-point multiplication
fldiv	a floating-point division
flsqrt	a floating-point square root
flam	an addition and a multiplication
flrot	application of a plane rotation

- As is customary with physical units, the abbreviations do not take periods or a plural in "s" (e.g., $3n$ flmlt). An exception to the latter is in unquantified usage (e.g., the count in flams).

- The abbreviations take the usual prefixes denoting powers of ten (e.g., Gflam).

- Rates in operations per second are expressed by appending "/s" (e.g., Gflam/s).

- The flam is a compound operation consisting of one addition and one multiplication.

- The flrot represents the application of a plane rotation to a 2-vector. Its value is 2 fladd + 4 flmlt.

Figure 2.1: Abbreviations and conventions for reporting operation counts

―――――――――――◊―――――――――――

constant is the only thing that distinguishes algorithms of the same order and it can have important consequences for algorithms of different order.

• **Nomenclature.** The terminology for presenting operation counts is in a state of disarray. The widely used term "flop," which was originally an acronym for floating point operation, has undergone so many changes that the substance has been practically wrung out of it (for more, see the notes and references for this section). Instead we will use the abbreviations in Figure 2.1.

Note that the flam has replaced the flop in its sense (now defunct) of a floating-point addition combined with a floating-point multiplication. Since in many matrix algorithms, additions and multiplications come roughly in pairs, we will report many of our counts in flams.

• **Complex arithmetic.** We will also use this nomenclature for complex arithmetic. However, it is important to keep in mind that complex arithmetic is more expensive

SEC. 2. TRIANGULAR SYSTEMS

than its real counterpart. For example, the calculation of

$$(a+bi) + (c+di) = (a+c) + (b+d)i$$

requires two real additions and hence is twice as expensive as a real addition. Again, the calculation

$$(a+bi)*(c+di) = (a*c - b*d) + (a*d + b*c)i$$

requires four real multiplications and two real additions and is at least four times as expensive as a real multiplication. These two examples also show that the ratio of multiplication times to addition times can be different for real and complex arithmetic.

Inversion of a triangular matrix

As a final example let us consider Algorithm 2.3 for the inversion of a lower triangular system. The inner loop of the algorithm contains the statement

$$X[i,k] = -L[i, k{:}i{-}1]*X[k{:}i{-}1, k]/L[k,k]$$

which represents an inner product of length $i-k-1$ requiring about $i-k$ flams. Since i ranges from k to n and k ranges from 1 to n, the total number of flams for the algorithm is

$$\sum_{k=1}^{n} \sum_{i=k}^{n} i-k.$$

We could use standard summation formulas to evaluate this sum, but the process is error prone. However, if we are only interested in the highest-order term in the sum, we may replace the sum by an integral:

$$\int_0^n \int_k^n i-k \, di \, dj = \frac{1}{6} n^3. \tag{2.10}$$

Note that the range of the outer integral has been adjusted to make it easy to evaluate. We can do this because a shift of one or two in the limits of a range does not change the high-order term.

More observations on operation counts

The following example gives a feel of the way execution times increase for the three orders we have encountered.

Example 2.3. *Consider three algorithms of order n, n^2, and n^3, all having an order constant of one. Here is how long they take to run on a machine that can perform 10M operations per second.*

n	$O(n)$	$O(n^2)$	$O(n^3)$
10	10^{-5} sec	10^{-4} sec	10^{-3} sec
100	10^{-4} sec	10^{-2} sec	10^{+0} sec
1000	10^{-3} sec	10^{+0} sec	10^{+3} sec = 17 min
10000	10^{-2} sec	10^{+1} sec	10^{+6} sec = 12 day

The time required by an $O(n^3)$ process increases dramatically with n. Unfortunately, most conventional matrix algorithms are $O(n^3)$, where n is the order of the matrix.

This example makes clear one reason for deprecating the invert-and-multiply algorithm for solving linear systems — at least triangular systems. The direct algorithm for solving triangular systems is $O(n^2)$, while the inversion of a triangular matrix is an $O(n^3)$ process.

Since operation counts are widely used to compare algorithms, it is important to have an idea of their merits and limitations.

- **Lower bounds.** Operation counts provide a rigorous lower bound on the time an algorithm will take — just divide the various counts by their rates for the computer in question and add. Such lower bounds can be useful. If, for example, a bound predicts that a calculation will take at least a thousand years, then it is time to consider alternatives.

- **Arithmetic is not everything.** Algorithms have overheads other than arithmetic operations, overheads we will treat in the next section. Hence running time cannot be predicted from operation counts alone. In a large number of cases, however, the running time as a function of the size of the problem is proportional to the time predicted by operation counts. Moreover, the constant of proportionality is often approximately the same over many algorithms — provided they are implemented with due respect for the machine in question.

- **Comparing algorithms of equal order.** In using operation counts to compare algorithms of the same order, it is the order constant that decides. Other things being equal, one should prefer the algorithm with the smaller order constant. But keep in mind that other things are never exactly equal, and factors of, say, two in the order constants may be insignificant. The larger the factor, the more likely there is to be a corresponding difference in performance.

- **Comparing algorithms of different order.** In principle, order constants are not needed to decide between algorithms of different order: the algorithm of lower order ultimately wins. But ultimately may never come. For example, if an $O(n^3)$ algorithm has an order constant equal to one while an $O(n^2)$ has an order constant of one thousand, then the first will be better for matrices of size less than one thousand. The size of problem for which a lower-order algorithm becomes superior to a higher-order algorithm is called the *break-even point*. Many promising algorithms have been undone by high break-even points.

—

Finally, keep in mind that there are other things than speed to consider in selecting an algorithm — numerical stability, for example. An algorithm that persistently returns bad answers is useless, even if it runs at blinding speed.

	Lower Triangular		Upper Triangular
$xelib(X,L,B)$	$X = L^{-1}B$	$xeuib(X,U,B)$	$X = U^{-1}B$
$xelitb(X,L,B)$	$X = L^{-\mathrm{T}}B$	$xeuitb(X,U,B)$	$X = U^{-\mathrm{T}}B$
$xebli(X,B,L)$	$X = BL^{-1}$	$xebui(X,B,U)$	$X = BU^{-1}$
$xeblit(X,B,L)$	$X = BL^{-\mathrm{T}}$	$xebuit(X,B,U)$	$X = BU^{-\mathrm{T}}$

- The names of these BLAS describe their functions. For example, *xelib* means "X Equals L Inverse B."

- The arguments appear in the calling sequence in the order in which they appear in the names.

- l and u may be replaced by $l1$ and $u1$, indicating that the matrices in question are unit triangular with their diagonals not stored—e.g., $xell1ib(X,L,B)$.

- B may replace X in the calling sequence, in which case the result overwrites B—e.g., $xelib(B,L,B)$.

Figure 2.2: BLAS for triangular matrices

2.8. BLAS FOR TRIANGULAR SYSTEMS

Many of the algorithms in this work must at some point solve triangular systems of one form or another. Given the expense of invert-and-multiply algorithms, it would be misleading to write something like $x = L^{-1}b$ in the pseudocode. On the other hand, to write out everything in full would focus attention on inessential details. The natural compromise is to relegate the solution to a series of subprograms. Such subprograms are called *basic linear algebra subprograms*, abbreviated BLAS.

Figure 2.2 describes the BLAS we will use for triangular systems. On the right is the calling sequence, on the left the function the subprogram performs. Note that the function has been coded into the names of these BLAS. For example, *xeuitb* can be read, "X equals U inverse transpose times B." The variable B represents a matrix whose dimensions are consistent with those of the triangular matrix. The variable X, which gets the result, must be of the same dimension as B. In fact, we can write B for X in the calling sequence, in which case B is overwritten by the result.

The BLAS are more than just a notational device. Since they are defined by what they do, their implementation can vary. In particular, special BLAS can be constructed to take advantages of special features of the machine in question. We will see more of this later.

2.9. NOTES AND REFERENCES

Historical

The first explicit algorithm for solving a lower triangular system was given by Gauss [131, 1810] as part of his elimination method for solving the normal equations of least squares. In fact, in his earlier work on calculating orbits of heavenly bodies — the famous *Theoria Motus* — Gauss [130, 1809] alludes to the possibility of inverting an entire triangular system. For more see §2.8, Chapter 1.

Recursion

Although there are sound reasons why recursion is not much used in matrix computations, at least part of the story is that at one time recursion could be quite expensive. Improved compiler techniques (e.g., see [5, 317]) have made recursive calls comparatively inexpensive, so that the overhead is negligible except for very small matrices.

Operation counts

Operation counts belong to the field of algorithms and their complexity. Two classical references are the book of Aho, Hopcroft, and Ullman [4], which treats the algorithmic aspect, and the book by Hopcroft and Ullman [181], which treats the theoretical aspects. For an encyclopedic treatment with many reference see [73].

Pat Eberlein has told me that the word "flop" was in use by 1957 at the Princeton Institute for Advanced Studies. Here is a table of the various meanings that have attached themselves to the word.

1. Flop — a floating point operation.
2. Flop — a floating point addition and multiplication.
3. Flops — plural of 1 or 2.
4. Flops — flops (1 or 2) per second.

In its transmogrifications, the meaning of "flop" has flipped from 1 to 2 and back to 1 again. Golub and Van Loan [152, p. 19] hint, ever so gently, that the chief beneficiaries of the second flip were the purveyors of flops — supercomputer manufacturers whose machines got a free boost in speed, at least in advertising copy.

The system adopted here consists of natural abbreviations. Since precision requires that heterogeneous counts be spelled out, there is no canonical term for a floating-point operation. However, the flam and the rot (short for "rotation" and pronounced "wrote") cover the two most frequently occurring cases of compound operations. The usage rules were lifted from *The New York Public Library Writer's Guide to Style and Usage* [316] and *The American Heritage College Dictionary* [74].

The technique of approximating sums by integrals, as in (2.10), is a standard trick of the trade. It provides the correct asymptotic forms, including the order constant, provided the integrand does not grow too fast.

Computational theorists and matrix algorithmists measure complexity differently. The former measure the size of their problems in terms of number of inputs, the latter

Cholesky decomposition	2,780
QR decomposition (Householder triangularization)	11,875
Eigenvalues (symmetric matrix)	13,502
The product $A*A$	16,000
QR decomposition (explicit Q)	28,975
Singular values	33,321
Eigenvalues and eigenvectors (symmetric matrix)	71,327
Eigenvalues (nonsymmetric matrix)	143,797
Singular values and vectors	146,205
Eigenvalues and eigenvectors (nonsymmetric matrix)	264,351

Figure 2.3: Operation counts for some matrix algorithms: $n = 20$

———————⋄———————

in terms of the order of the matrix. Since a matrix of order n has $m = n^2$ elements, an $O(n^3)$ matrix algorithm is an $O(m^{\frac{3}{2}})$ algorithm to a computational theorist. This places matrix algorithms somewhere between the Fourier transform, which is $O(m^2)$, and the fast Fourier transform, which is $O(m \log m)$. And a good thing too! If our algorithms were $O(m^3)$, we wouldn't live long enough to run them.

Whether you call the order n^3 or $m^{\frac{3}{2}}$, the order constants of matrix algorithms can vary dramatically. The table in Figure 2.3, containing the number operations required for some common $O(n^3)$ matrix algorithms applied to a 20×20 matrix, was compiled using MATLAB. (Thanks to Jack Dongarra for the idea.) Thus the order constant for finding the eigenvalues and eigenvectors of a nonsymmetric matrix is nearly one hundred times larger than that for finding the Cholesky decomposition. Beresford Parlett, complaining about the abuse of the big O notation, says that it plays the part of a fig leaf on a statue: it covers up things people don't want seen. The above table supports this simile.

Basic linear algebra subprograms (BLAS)

For more on the BLAS see the notes and references to §3.

3. MATRICES IN MEMORY

There are many ways to execute the algorithms of the preceding section. The calculations could be done by hand, perhaps with the help of a slide rule or a table of logarithms. They could be done with an abacus or a mechanical calculator. Each mode of computation requires special adaptations of the algorithm in question. The order in which operations are performed, the numbers that are written down, the safeguards against errors — all these differ from mode to mode.

This work is concerned with matrix computations on a digital computer. Just like any other mode of computation, digital computers place their own demands on matrix algorithms. For example, recording a number on a piece of paper is an error-prone process, whereas the probability of generating an undetected error in writing to the memory of a computer is vanishingly small. On the other hand, it is easy to mismanage the memory of a computer in such a way that the speed of execution is affected.

The theme of this section is matrices in memory. We will begin by describing how dense matrices are represented on computers, with emphasis on the overhead required to retrieve the elements of a matrix. We will then move on to a discussion of hierarchical memories.

3.1. MEMORY, ARRAYS, AND MATRICES

Memory

It is useful to be a little imprecise about what a computer memory is. We shall regard it as a linear array of objects called *words*. The notion of a word will vary with context. It may be an integer, a single-precision or double-precision floating-point number, or even a data structure.

The location of an object in memory is called its *address*. If x is the address of an object in memory, we will write $x[1]$ for the object itself and $x[2]$ for the object immediately following $x[1]$ in memory. In general $x[k+1]$ will be the kth object following $x[1]$ in memory. This convention meshes nicely with our convention for representing the components of vectors. The symbol x can stand for both a vector and the address of a vector. In each case, the components of x are represented by $x[i]$.

Storage of arrays

In high-level programming languages, matrices are generally placed in two-dimensional arrays. A $p \times q$ array A is a set of pq memory locations in the computer. An element of an array is specified by two integers i and j which lie within certain ranges. In this work we will assume $1 \leq i \leq p$ and $1 \leq j \leq q$. The syntax by which an element of an array is represented will depend on the programming language. Here we will use the convention we have already been using for matrices — the (i,j)-element of the array A is written $A[i,j]$.

A difficulty with arrays is that they are two-dimensional objects that must be stored in a one-dimensional memory. There are many ways in which this can be done, each having its own advantages for specialized applications. For general matrix computations, however, there are just two conventions.

• **Storage by rows.** Beginning at a base address a, the array is stored a row at a time, the components of each row appearing sequentially in the memory. For example, a

SEC. 3. MATRICES IN MEMORY

5×3 array A would be stored in the form

$$\begin{array}{lll} A[1,1]\leftrightarrow a[1] & A[1,2]\leftrightarrow a[2] & A[1,3]\leftrightarrow a[3] \\ A[2,1]\leftrightarrow a[4] & A[2,2]\leftrightarrow a[5] & A[2,3]\leftrightarrow a[6] \\ A[3,1]\leftrightarrow a[7] & A[3,2]\leftrightarrow a[8] & A[3,3]\leftrightarrow a[9] \\ A[4,1]\leftrightarrow a[10] & A[4,2]\leftrightarrow a[11] & A[4,3]\leftrightarrow a[12] \\ A[5,1]\leftrightarrow a[13] & A[5,2]\leftrightarrow a[14] & A[5,3]\leftrightarrow a[15] \end{array}$$

This order of storage is also called *lexicographical order* because the elements $A[i,j]$ are ordered with their first subscript varying least rapidly, just like letters in words alphabetized in a dictionary. This form of storage is also called *row major order*.

The general formula for the location of the (i,j)-element of a $p\times q$ array can be deduced as follows. The first $i-1$ rows have $(i-1)q$ elements. Consequently, the first element of the ith row is $a[(i-1)q+1]$. Since the elements of a row are stored in sequence, the jth element of the ith row must be $a[(i-1)q+j]$. Thus

$$A[i,j] \leftrightarrow a[(i-1)q+j]. \tag{3.1}$$

- **Storage by columns.** Here the array is stored a column at a time. The memory locations containing a 5×3 array are shown below.

$$\begin{array}{lll} A[1,1]\leftrightarrow a[1] & A[1,2]\leftrightarrow a[6] & A[1,3]\leftrightarrow a[11] \\ A[2,1]\leftrightarrow a[2] & A[2,2]\leftrightarrow a[7] & A[2,3]\leftrightarrow a[12] \\ A[3,1]\leftrightarrow a[3] & A[3,2]\leftrightarrow a[8] & A[3,3]\leftrightarrow a[13] \\ A[4,1]\leftrightarrow a[4] & A[4,2]\leftrightarrow a[9] & A[4,3]\leftrightarrow a[14] \\ A[5,1]\leftrightarrow a[5] & A[5,2]\leftrightarrow a[10] & A[5,3]\leftrightarrow a[15] \end{array}$$

The general correspondence is

$$A[i,j] \leftrightarrow a[(j-1)p+i]. \tag{3.2}$$

This form of storage is also called *column major order*.

Strides

When a $p\times q$ array is stored rowwise, the distance in memory between an element and the next element in the same column is q. This number is called the *stride* of the array because it is the stride you must take through memory to walk down a column. When the array is stored by columns, the stride is p and refers to the stride required to traverse a row.

The stride is confusing to people new to matrix computations. One reason is that it depends on whether the array is stored rowwise or columnwise. Thus the stride is different in C, which stores arrays rowwise, and FORTRAN, which stores them columnwise.

Another source of confusion is that matrices are frequently stored in arrays whose dimensions are larger than those of the matrix. Many programs manipulate matrices whose dimensions are unknown at the time the program is invoked. One way of handling this problem is to create a $p \times q$ array whose dimensions are larger than any matrix the program will encounter. Then any $m \times n$ matrix with $m \leq p$ and $n \leq q$ can be stored in the array — usually in the northwest corner.

Now it is clear from (3.1) and (3.2) that you have to know the stride to locate elements in an array. Consequently, if a matrix in an oversized array is passed to a subprogram, the argument list must contain not only the dimensions of the matrix but also the stride of the array. The source of confusion is that the stride is generally different from the dimensions of the matrix.

Example 3.1. *In FORTRAN arrays are stored columnwise. Consequently, a subroutine to solve a lower triangular system might begin*

```
SUBROUTINE LTSLV(N, L, LDL, B, X)
REAL L(LDL,*), B(*), X(*)
```

The parameter N *is the order of the matrix* L. *The parameter* LDL *is the stride of the array* L *containing the matrix* L. *The name is an abbreviation for "leading dimension of* L*" because the first dimension in the declaration of an array in FORTRAN is the stride of the array.*

3.2. MATRICES IN MEMORY

Although many kinds of objects can be stored in arrays, in this work we are concerned chiefly with matrices, whose algorithms make special demands on the memory. In this and the following subsections we will treat the interaction of matrix algorithms and memory.

Array references in matrix computations

Let us now see how Algorithm 2.1 for solving a triangular system looks from the point of view of a computer. For ready reference we repeat the algorithm here.

1. **for** $k = 1$ **to** n
2. $x_k = b_k$
3.   **for** $j = 1$ **to** $k-1$
4. $x_k = x_k - \ell_{kj} x_j$ (3.3)
5.   **end for** j
6. $x_k = x_k / \ell_{kk}$
7. **end for** k

We want to convert this program to one using pure memory references, so that we can see the overhead involved in the array indexing. By our conventions, vectors present no problems. The kth component of the vector x is $x[k]$. Matrices require more work. If we assume that L is stored rowwise with stride p and ℓ is the address of

Sec. 3. Matrices in Memory

ℓ_{11}, then we can use (3.1) to retrieve elements of L. The result of these alterations to (3.3) is the following program.

1.   **for** $k = 1$ **to** n
2. $x[k] = b[k]$
3.   **for** $j = 1$ **to** $k-1$
4. $x[k] = x[k] - \ell[(k-1)p+j]*x[j]$ (3.4)
5.   **end for** j
6. $x[k] = x[k]/\ell[(k-1)p+k]$
7.   **end for** k

This program shows why arithmetic operation counts alone cannot predict the running time of an algorithm. Each array reference involves a certain amount of additional computational work. For example, the reference to ℓ_{kj} in the original program translates into $\ell[(k-1)p+j]$, which requires two additions and a multiplication to compute. But the program also shows why the running time is proportional to the arithmetic operation count. Each floating-point addition and multiplication in the inner loop is accompanied by the same overhead for memory references.

Optimization and the BLAS

The efficiency of the program (3.4) can be improved. As j is incremented, the expression $(k-1)p+j$ in the inner loop simply increases by one. Thus if we precompute the value $(k-1)p$ we can simply increment in the loop. Here is pseudocode embodying this idea.

1. $i = 1$
2.   **for** $k = 1$ **to** n
3. $x[k] = b[k]$
4.   **for** $j = 1$ **to** $k-1$
5. $x[k] = x[k] - \ell[i]*x[j]$
6. $i = i+1$ (3.5)
7.   **end for** j
8. $x[k] = x[k]/\ell[i]$
9. $i = i+p-k$
10. **end for** k

This implementation is more efficient than (3.4), since the multiplication and two additions in the inner loop have been replaced by one addition to increment i. In fact, we have succeeded in eliminating all multiplications in the index calculation. The key is to observe that at the end of the loop on j the value of i is exactly $p-k$ less that the value needed for the next iteration of the loop, so that its next value can be computed by the simple addition in statement (3.5.9). However, for large matrices this trick has only negligible effect, since the savings occur outside the inner loop.

The difference between the algorithms (3.4) and (3.5) is essentially the difference between the codes produced by nonoptimizing and optimizing compilers. The nonoptimizing compiler will generate the address computation in (3.4.4), not recognizing that it represents an increment of one with each iteration of the loop. The optimizing compiler will simply generate code to increment the index.

Since time is important in computations with large matrices, it is natural to ask if there is a way to circumvent the code generated by a nonoptimizing compiler. The answer is yes. The idea is to isolate frequently occurring computations into subprograms where they can be optimized — by hand if necessary. Let's see how this works for the forward substitution algorithm.

The place to start (as always in speeding up code) is the inner loop — in this case the loop on j in (3.3). The effect of this loop is to compute the inner product

$$L[k, 1{:}k{-}1]*x[1{:}k{-}1]$$

and subtract it from x_k. Since the inner product represents the bulk of the work, that is the computation we want to isolate in a subprogram. The following function does the job.

1. $dot(n, x, y)$
2. $s = 0$
3.   **for** $k = 1$ **to** n
4. $s = s + x[k]*y[k]$
5.   **end for** k
6.   **return** s
7.   **end** dot

Now we can substitute the program *dot* for the inner loop on j:

1.   **for** $k = 1$ **to** n
2. $x[k] = (b[k] - dot(k{-}1, L[k, 1], x[1]))/A[k, k]$ (3.6)
3.   **end for** k

There are five comments to be made about *dot* and its usage.

- The subprogram uses the convention for inconsistent loops and returns zero when n is zero.

- Comparing the calling sequence $dot(k{-}1, L[k, 1], x[1])$ with *dot* itself, we see that it is the address of $L[k, 1]$ and $x[1]$ that is passed to the subprogram. This is sometimes called *call by reference*, as contrasted with *call by value* in which the value of the argument is passed (see §1.3).

- Since we have replaced doubly subscripted array references to L with singly subscripted array references to x in the subprogram *dot*, even a nonoptimizing compiler will generate efficient code. But if not, we could compile *dot* on an optimizing compiler (or code it in assembly language) and put it in a library. Since the inner product is one of the more frequently occurring matrix operations, the effort will pay for itself many times over.

SEC. 3. MATRICES IN MEMORY

- The subprogram *dot* can be written to take advantage of special features of the machine on which it will be run. For example, it can use special hardware — if it exists — to compute the inner product.

- The subprogram *dot* is not as general as it should be. To see why, imagine that L is stored by columns rather than by rows, say with stride p. Then to move across a row, the index of x in *dot* must increase by p instead of one. A revised subprogram will take care of this problem.

```
1.   dot(n, x, xstr, y, ystr)
2.      ix = 1;   iy = 1
3.      s = 0
4.      for k = 1 to n
5.         s = s + x[ix]*y[iy]
6.         ix = ix+xstr
7.         iy = iy+ystr
8.      end for k
9.      return s
10.  end dot
```

In this subprogram the index of both x and y are incremented by a stride provided by the user. In particular, to convert (3.6) to handle a matrix stored by columns with stride p, the statement (3.6.2) would be replaced by

$$x[k] = (b[k] - dot(k-1, L[k,1], p, x[1], 1))/A[k,k]$$

———

The function *dot* is representative of a class of subprograms that perform frequently occurring tasks in linear algebra. They are called the BLAS for "basic linear algebra subprograms." In particular, *dot* is called a level-one BLAS because it operates on vectors. As we shall see, there are higher-level BLAS that perform matrix-vector and matrix-matrix operations. We have already met another class of BLAS for the solution of triangular systems (see Figure 2.2).

The utility of the BLAS is universally recognized, and any attempt to produce quality matrix software must come to grips with them. Nonetheless, BLAS will not be much used to present algorithms in this work. There is no contradiction in this. The fact that we use partitioning and matrix operations to present our algorithms means that the appropriate BLAS are suggested by the pseudocode itself. For example, compare the statement

$$x[k] = (b[k] - dot(k-1, L[k,1], p, x[1], 1))/A[k,k]$$

with

$$x[k] = (b[k] - L[k, 1{:}k-1]*x[1{:}k-1])/A[k,k]$$

Economizing memory — Packed storage

In the early days of digital computing, memory was in short supply — so much so that economy of storage was often more important than speed when it came to selecting an algorithm. Now memory is plentiful. Yet it is not infinite, and matrices demand a lot of it. Therefore, techniques for economizing memory are still useful.

Triangular matrices furnish an opportunity for economizing storage, since about half of the elements are zero. Because the zero elements are not referenced in the course of solving a triangular system, there is no point in storing them. However, most programming languages do not have triangular arrays as a primitive data type. Consequently, we must store the elements in a linear array.

For definiteness let us suppose that the elements of the $n \times n$ lower triangular matrix L are stored columnwise in an array ℓ, so that we have the correspondence

$$L[1,1] \leftrightarrow \ell[1]$$
$$L[2,1] \leftrightarrow \ell[2] \quad L[2,2] \leftrightarrow \ell[4]$$
$$L[3,1] \leftrightarrow \ell[3] \quad L[3,2] \leftrightarrow \ell[5] \quad L[3,3] \leftrightarrow \ell[6]$$

This cramming of the nonzero elements of a matrix into a linear array is called a *packed representation*.

We will implement the column-oriented algorithm (2.7), which for convenience is reproduced here.

1. $x = b$
2. **for** $k = 1$ **to** n
3. $x[k] = x[k]/L[k,k]$
4. $x[k+1:n] = x[k+1:n] - x[k]*L[k+1:n, k]$
5. **end for** k

(3.7)

The implementation is in the spirit of the algorithm (3.5) in that we set up an index i that moves through the array ℓ.

1. $i = 1$
2. $x = b$
3. **for** $k = 1$ **to** n
4. $x[k] = x[k]/\ell[i]$
5. $i = i+1$
6.   **for** $j = k+1$ **to** n
7. $x[j] = x[j] - x[k]*\ell[i]$
8. $i = i+1$
9.   **end for** j
10. **end for** k

(3.8)

It is worth noting that there is no need to do an extra computation to get the index of the diagonal of L before the division. Because of the packed representation, all we need do is increment i by one.

The level-one BLAS interact nicely with packed representations. For example, the basic operation the statement (3.7.4) performs is to overwrite a vector by its sum with

SEC. 3. MATRICES IN MEMORY

a constant times another vector. Consequently, if we create a new BLAS called *axpy* (for ax plus y) that overwrites a vector y with $ax + y$, we can use it to implement the inner loop of (3.8).

Specifically, define *axpy* as follows.

1.   *axpy*(n, a, x, *xstr*, y, *ystr*)
2. $ix = 1$; $iy = 1$
3.   **for** $k = 1$ **to** n
4. $y[iy] = y[iy] + a*x[ix]$
5. $ix = ix + xstr$
6. $iy = iy + ystr$
7.   **end for** k
8.   **end** *axpy*

(3.9)

Then (3.8) can be written in the form

1. $i = 1$
2. $x = b$
3.   **for** $k = 1$ **to** n
4. $x[k] = x[k]/\ell[i]$
5.   *axpy*($n-k$, $-x[k]$, $\ell[i+1]$, 1, $x[k+1]$, 1)
6. $i = i+n-k+1$
7.   **end for** k

Most matrices with regular patterns of zeros lend themselves to a packed representation. The packing need not be in a single array. People often pack the three diagonals of a tridiagonal matrix in three linear arrays (see §2.3, Chapter 3). Nor is packing confined to matrices with zero elements. Almost half the elements of a symmetric matrix are redundant (since $a_{ij} = a_{ji}$) and do not need to be stored.

Packing is not the only way to economize storage. Overwriting is another. For example, the algorithm (2.9) overwrites a lower triangular matrix with its inverse, thus saving $O(n^2)$ words of memory. Later in this work we shall see how a matrix can be overwritten by a decomposition of itself. However, we can overwrite a matrix only if we know that we will not need it again.

3.3. HIERARCHICAL MEMORIES

People want their memories large so they can solve large problems. And they want their memories fast so they can solve large problems in a reasonable time. Unfortunately, speed and capacity work against each other. Large memories are slow and fast memories are small.

To get around this difficulty computer architects have evolved a compromise called *hierarchical memory*. The idea is that a large slow memory backs up a fast small one. The computer works through the fast memory, which, of course, cannot contain all the words in the large memory. When the computer references a word that is not in the fast memory a block containing the word is swapped from the large memory.

Most computers have more than one level of memory. Figure 3.1 exhibits a typical (though idealized) hierarchical memory. At the bottom are the registers of the central processing unit—the place where words from higher memories are manipulated. At the top is a disk containing the entire memory allocated to the machine—the virtual memory that cannot fit into the main memory below. Between the main memory and the registers is a small, fast cache memory.

Volumes have been written on hierarchical memories, and it is impossible to descend to architectural details in a work of this nature. But even a superficial description will be enough to suggest how matrix algorithms might be coded to avoid problems with memory hierarchies. We will begin this subsection by discussing virtual and cache memory. We will then turn to strategies for writing matrix algorithms that use hierarchical memories efficiently.

Virtual memory and locality of reference

Our model of a computer memory continues to be a linear array of words addressed by integers. The size of the memory is called the *address space*. The address space of many modern computers is enormous. Address spaces of 2^{32} bytes are common, and address spaces of 2^{64} are now appearing.

An address space of 2^{32} bytes represents about five hundred million double-precision floating-point words. Although such a memory could be built, it would be impractically expensive for most applications. The cure is to make do with a smaller main memory and store the rest on a backing store, usually a disk. When this is done in such a way that the process is invisible to the programmer, it called *virtual memory*.

Specifically, the address space is divided into blocks of contiguous words called *pages*. Typically a page will contain several kilobytes. Some of the pages are contained in main memory; the rest are kept on a backing store. When a program references a memory location, the hardware determines where the page containing the location lies. There are two possibilities.

1. The page is in the main memory. In this case the reference — whether a read or a write — is performed with no delay.
2. The page is not in main memory, a condition known as a *page fault*. In this case the system swaps the page in backing store with one of the pages in main memory and then honors the reference.

The problem with this arrangement is that reads and writes to backing store are more costly than references to main memory, e.g., a hundred thousand times more costly. It is therefore important to code in such a way as to avoid page faults. In computations with large matrices, some page faults are inevitable because the matrices consume so much memory. But it is easy to miscode matrix algorithms so that they cause unnecessary page faults.

The key to avoiding page faults is *locality of reference*. Locality of reference has two aspects, *locality in space* and *locality in time*.

SEC. 3. MATRICES IN MEMORY

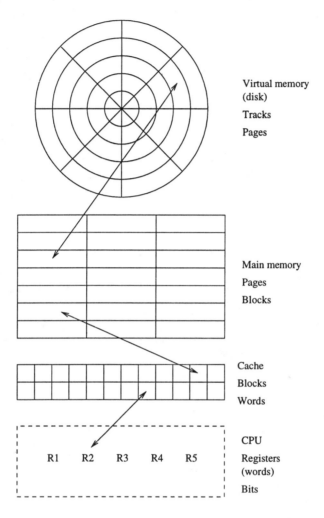

This is a representation of a hierarchical memory. At the highest level is a disk-based virtual memory. It is divided into pages of words that are swapped in and out of the main memory below. The pages of main memory are divided into blocks that are swapped with the fast cache memory. Finally words in the cache memory move between the registers of the central processing unit.

Figure 3.1: A hierarchical memory

Locality in space refers to referencing nearby locations. The rationale is that contiguous memory locations are likely to lie in the same page, so that a cluster of references to nearby locations is unlikely to generate page faults. On the other hand, locations far removed from one another will lie on different pages and referencing them one after another may cause a sequence of page faults. Thus it is desirable to arrange computations so that if a location is referenced subsequent references are to nearby locations.

To understand the notion of locality in time consider two references to a single memory location. If these references occur near each other in time — the extreme case is when they occur one right after the other — the page containing the item is likely to be still around. As the references become further separated in time, the probability of a page fault increases. Thus it is desirable to arrange computations so that repeated references to the same locations are made close together in time.

Cache memory

In recent years the speed of processors has increased faster than the speed of memory — at least memory that can be built in quantity at a reasonable cost. To circumvent this roadblock, computer architects have incorporated small, fast memories — called *cache memories* or simply *caches* — into computers.

Cache memory bears the same relation to main memory as main memory does to virtual memory, though the details and terminology differ. The cache is divided into *blocks* which contain segments from main memory. When a memory reference is made, the hardware determines if it is in the cache. If it is, the request is honored right away. If it is not — a *cache miss* this situation is called — an appropriate (and generally time-consuming) action is taken before the reference is honored.

An important difference between cache and virtual memories is that writes a cache are usually more expensive than reads. The reason is the necessity of preserving *cache coherency* — the identity of the contents of the cache and the contents of the corresponding block of memory. A coherent cache block, say one that has only been read from, can be swapped out at any time simply by overwriting it. An incoherent cache block, on the other hand, cannot be overwritten until its coherency is restored.

There are two common ways of maintaining cache coherency. The first, called *write through*, is to replicate any write to cache with a write to the corresponding location in main memory. This will cause writes — or at least a sequence of writes near each other in time — to be slower than reads. The other technique, called *write back*, is to wait for a miss and if necessary write the whole block back to memory, also a time-consuming procedure. Actually, write-through and write-back represent two extremes. Most caches have buffering that mitigates the worst behavior of both. Nonetheless, hammering on a cache with writes is a good way to slow down algorithms.

A model algorithm

We are now going to consider techniques by which we can improve the interaction of matrix algorithms with hierarchical memories. It must be stressed that this is more an art than a science. Machines and their compilers have become so diverse and so complicated that it is difficult to predict the effects of these techniques. All we can say is that they have the potential for significant speedups.

In order to present the techniques in a uniform manner, we will consider a model calculation. People acquainted with Gaussian elimination—to be treated in Chapter 3—will recognize the following fragment as a stripped down version of that algorithm. The matrix A in the fragment is of order n.

1. **for** $k = 1$ **to** $n-1$
2. $\quad A[k+1{:}n, k+1{:}n] = A[k+1{:}n, k+1{:}n]$
   $\quad\quad - A[k+1{:}n, k] * A[k, k+1{:}n]$ (3.10)
3. **end for** k

In matrix terms, if at the kth stage the array A is partitioned in the form (northwest indexing)

$$\begin{pmatrix} A_{11} & a_{1k} & A_{1,k+1} \\ a_{k1}^{\text{T}} & \alpha_{kk} & a_{k,k+1}^{\text{T}} \\ A_{k+1,1} & a_{k+1,k} & A_{k+1,k+1} \end{pmatrix},$$

then the next matrix is obtained by the substitution

$$A_{k+1,k+1} \leftarrow A_{k+1,k+1} - a_{k+1,k} a_{k,k+1}^{\text{T}}. \tag{3.11}$$

Row and column orientation

Figure 3.2 exhibits two obvious ways of implementing our model algorithm. They differ only in the order of the loops on i and j. But the effects of the difference can be great.

To see this, suppose the matrix A is stored columnwise in an $n \times n$ array with address a. Then for $k = 1$, the first algorithm, which moves along the rows of the array, makes the following sequence of memory references.

$$\begin{array}{ccccccccc}
& a[1] & \to & a[n+1] & \to & a[2n+1] & \to & \cdots & \to & a[n(n-1)+1] \\
\to & a[2] & \to & a[n+2] & \to & a[2n+2] & \to & \cdots & \to & a[n(n-1)+2] \\
\to & a[3] & \to & a[n+3] & \to & a[2n+3] & \to & \cdots & \to & a[n(n-1)+3] \\
& \vdots & & \vdots & & \vdots & & & & \vdots \\
\to & a[n] & \to & a[2n] & \to & a[3n] & & \cdots & \to & a[n^2]
\end{array}$$

If n is at all large, the references jump around memory—i.e., they do not preserve locality in space.

114   CHAPTER 2. MATRICES AND MACHINES

1. **for** $k = 1$ **to** n
2.   **for** $i = k+1$ **to** n
3.   **for** $j = k+1$ **to** n
4. $A[i,j] = A[i,j] - A[i,k]*A[k,j]$
5.   **end for** j
6.   **end for** i
7. **end for** k

Row-Oriented Algorithm

1. **for** $k = 1$ **to** n
2.   **for** $j = k+1$ **to** n
3.   **for** $i = k+1$ **to** n
4. $A[i,j] = A[i,j] - A[i,k]*A[k,j]$
5.   **end for** i
6.   **end for** j
7. **end for** k

Column-Oriented Algorithm

Figure 3.2: Row- and column-oriented algorithms

On the other hand, the second algorithm, which traverses the columns of A, makes the following sequence of memory references.

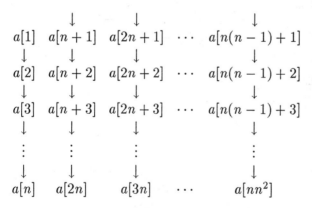

Thus the algorithm proceeds from one memory location to the next — i.e., it preserves locality in space about as well as any algorithm can.

As k increases the behavior of the algorithms becomes more complicated. But the first algorithm never jumps in memory by less than n words, while the second algorithm never jumps by more than n words and usually by only one word. If A is stored by rows instead of columns, the two algorithms reverse themselves with the second

SEC. 3. MATRICES IN MEMORY

making the large jumps.

We say that the first algorithm is *row oriented* and that the second is *column oriented*. The above reasoning shows that column orientation preserves locality of reference when the matrix in question is stored by columns; row orientation preserves locality of reference when the matrix is stored by rows. Thus our first technique for coding algorithms is:

Make the orientation of the algorithm agree with the storage scheme. (3.12)

Here are some comments on this technique.

- Although many of the simpler matrix algorithms have both row- and column-oriented versions, in more complicated algorithms we may have to compromise. Thus the assertion (3.12) should be taken, not as a strict requirement, but as an ideal to strive for.

- In days gone by, when pages were small and caches were nonexistent, orientation was used to avoid page faults. Today, they chiefly prevent cache misses. The reason is that main memories have grown so large that they can contain extremely large matrices. For example, a 64 Mbyte memory can contain an IEEE double-precision matrix of order almost 3,000. Once a matrix has been brought in main memory, it can generally be manipulated without page faults.

- Because FORTRAN, which stores arrays by column, has dominated the field of matrix computations, most packages such as LINPACK and LAPACK are column oriented. In a general way we will follow this convention, although we will not hesitate to present mixed or row-oriented algorithms if circumstances dictate.

Level-two BLAS

It is not always necessary to decide on row or column orientation at the time an algorithm is written. For example, we have seen that the basic operation of our model algorithm is the subtraction of a rank-one matrix:

$$A_{k+1,k+1} \leftarrow A_{k+1,k+1} - a_{k+1,k} a_{k,k+1}^{\mathrm{T}}$$

[see (3.11)]. Suppose we write a function *amrnk1*(A, x, y^{T}) that overwrites A with $A - xy^{\mathrm{T}}$ (the name means "A minus a rank-one matrix). Then we can write our model algorithm in the form

1. **for** $k = 1$ **to** $n-1$
2.   *amrnk1*$(A[k+1{:}n, k+1{:}n], A[k+1{:}n, k], A[k, k+1{:}n])$ (3.13)
3. **end for** k

The program *amrnk1* can then be loaded from a library of code written for the target system and language.

The program *amrnk1* is called a *level-two* BLAS. The name comes from the fact that it performs $O(n^2)$ matrix-vector operations. Other examples of level-two BLAS are the ones for triangular systems listed in Figure 2.2. Yet another example is the formation of a matrix-vector product. It turns out that the catalogue of useful matrix-vector operations is small enough that it is practical to code libraries of them for a given machine or language. Row and column orientation, as required, can then be incorporated in these libraries.

We will not make explicit use of level-two BLAS in presenting algorithms. The reason is the same as for level-one BLAS — our notation is sufficient to express the action of most level-two BLAS. Provided we present our algorithms at a high enough level, the level-two BLAS required will be obvious on inspection.

Unfortunately, the level-two BLAS are not a panacea. Algorithms for the more complicated matrix decompositions usually have row or column orientation built into them in subtle ways — for example, in the decision whether to store a transformation or its transpose. Until we agree on a common language or until all languages become ambidextrous, orientation of matrix algorithms will continue to trouble us.

Keeping data in registers

With a pure write-through cache it is as expensive to write to cache as to main memory. Now both of the programs in Figure 3.2 perform a great many writes — about $\frac{1}{3}n^3$. The number can be reduced by recasting the algorithm.

Let us put ourselves in the position of a typical element of A and ask what computations the model algorithm performs on it. There are two cases. If the element is a_{ik} with $i > k$, the algorithm performs the following computations:

$$a_{ik} \leftarrow a_{ik} - a_{i1}a_{1k} - a_{i2}a_{2j} - \cdots - a_{i,k-1}a_{k-1,k}.$$

On the other hand, if the element is a_{kj} with $k \geq j$, the algorithm performs the following computations:

$$a_{kj} \leftarrow a_{kj} - a_{k1}a_{1j} - a_{k2}a_{2j} - \cdots - a_{k,k-1}a_{k-1,j}.$$

These formulas presuppose that the a's forming the products on the left have already been processed. This will be true if, as k goes from one to $n-1$, we compute a_{ik} ($i = k+1, \ldots, n$) and a_{kj} ($j = k, \ldots, n$).

These considerations give us the following algorithm.

Sec. 3. Matrices in Memory

$$\begin{aligned}
&\text{1.} \quad \textbf{for } k = 1 \textbf{ to } n{-}1 \\
&\text{2.} \quad\quad \textbf{for } i = k{+}1 \textbf{ to } n \\
&\text{3.} \quad\quad\quad \textbf{for } j = 1 \textbf{ to } k{-}1 \\
&\text{4.} \quad\quad\quad\quad A[i,k] = A[i,k] - A[i,j]*A[j,k] \\
&\text{5.} \quad\quad\quad \textbf{end for } j \\
&\text{6.} \quad\quad \textbf{end for } i \\
&\text{7.} \quad\quad \textbf{for } j = k \textbf{ to } n \\
&\text{8.} \quad\quad\quad \textbf{for } i = 1 \textbf{ to } k{-}1 \\
&\text{9.} \quad\quad\quad\quad A[k,j] = A[k,j] - A[k,i]*A[i,k] \\
&\text{10.} \quad\quad\quad \textbf{end for } i \\
&\text{11.} \quad\quad \textbf{end for } j \\
&\text{12.} \quad \textbf{end for } k
\end{aligned}$$
(3.14)

Incidentally, this program is a stripped down version of the Crout form of Gaussian elimination (see Algorithm 1.7, Chapter 3).

The advantage of this form of the algorithm is that the reference to $A[i,k]$ in statement 4 does not change in the inner loop on i. Consequently, we can put it in a register and work with it there without having to write to cache. Similarly for the computation in statement 9. Thus the reorganized algorithm is potentially faster than either of the algorithms in Figure 3.2 when writes to cache are expensive.

Unfortunately, there are trade-offs. The program (3.14) is neither column- nor row-oriented and cannot be made so. What we gain by keeping the data in registers we may lose to the poor orientation. Moreover, there is no way to use BLAS to hide the difference between (3.14) and the algorithms in Figure 3.2. With regard to memory they are fundamentally different algorithms, and the choice between them must be at the highest level.

Blocking and the level-three BLAS

In the program (3.13) we used the level-two BLAS *amrnk1* to subtract a rank-one matrix (in exterior-product form) from a submatrix of A. Although we can implement this routine in different ways — e.g., vary its orientation — a rank-one update is too simple to give us much scope for variation. It turns out that by a process called *blocking* we can elevate the rank-one update into an update of higher rank.

To see how to do this, choose a *block size* m and partition A in the form

$$A = \begin{pmatrix} A_{11} & A_{1,m+1} \\ A_{m+1,1} & A_{m+1,m+1} \end{pmatrix}, \qquad (3.15)$$

where the indexing is to the northwest. If we process the elements in A_{11}, $A_{1,m+1}$, and $A_{m+1,1}$ in the usual way, then the effect of the first m steps of the algorithm (3.13) on $A_{m+1,m+1}$ is to overwrite $A_{m+1,m+1}$ as follows

$$A_{m+1,m+1} \leftarrow A_{m+1,m+1} - A_{m+1,1} A_{1,m+1}.$$

This overwriting is a rank-m update. After the update, we can repeat the process on the matrix $A_{m+1,m+1}$.

1. **for** $k = 1$ **to** n **by** m
2. $ku = \min\{k+m-1, n\}$
3.   **for** $l = k$ **to** $ku-1$
4.   **for** $j = l+1$ **to** n
5.   **for** $i = l+1$ **to** $\min\{j, ku\}$
6. $A[i,j] = A[i,j] - A[i,l]*A[l,j]$
7.   **end for** i
8.   **end for** j
9.   **for** $j = l+1$ **to** ku
10.   **for** $i = j+1$ **to** n
11. $A[i,j] = A[i,j] - A[i,l]*A[l,j]$
12.   **end for** i
13.   **end for** j
14.   **end for** l
15. $A[ku+1:n, ku+1:n] = A[ku+1:n, ku+1:n]$
              $- A[ku+1:n, k:ku]*A[k:ku, ku+1:n]$
16. **end for** k

Figure 3.3: A blocked algorithm

The code in Figure 3.3 implements this scheme. The code may be best understood by referring to the following figure.

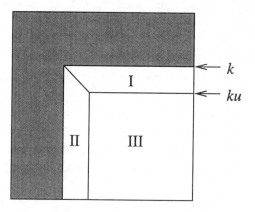

The grey region contains the elements that have been completely processed. Regions I (which contains the diagonal) and II are the blocks corresponding to A_{11}, $A_{1,m+1}$, and $A_{m+1,1}$ in the partition (3.15). They are processed in the loop on l. The loop in statement 4 processes region I. The loop in statement 9 processes region II. Region III is processed in statement 15, which if the block size m is not large accounts for most of the work in the algorithm.

If we now define a *level-three* BLAS *amrnkm*(A, X, Y) that overwrites A with $A -$

SEC. 3. MATRICES IN MEMORY

XY, we may replace statement 15 with

$$\mathit{amrnkm}(A[ku{+}1{:}n, ku{+}1{:}n],\ A[ku{+}1{:}n, k{:}ku],\ A[k{:}ku, ku{+}1{:}n])$$

The BLAS *amrnkm* can then be coded to take advantage of the features of a particular machine. The fact that *amrnkm* works with more data than *amrnkl* gives us more opportunity to economize. For example, if m is not too large we may be able to use inner products in the style of (3.14) without triggering a volley of cache misses.

The choice of the block size m is not easy. Two things limit its size. First, the overhead for processing regions I and II increases until it swamps out any benefits. Second, as we have suggested above, if m is too large we increase the probability of cache misses and page faults. The routines provided by LAPACK choose a value of 1, 32, or 64, depending on the name of the BLAS and whether the arithmetic is real or complex.

Blocking can be remarkably effective in speeding up matrix algorithms — especially the simpler ones. However, we shall not present blocked algorithms in this work. There are three reasons. First, the blocking obscures the simplicity of the basic algorithm. Second, once a matrix algorithm is well understood, it is usually an easy matter to code a blocked form. Finally, the LAPACK codes are thoroughly blocked and well commented, so that the reader can easily learn the art of blocking by studying them. For these reasons, we will present our algorithms at the level of matrix-vector operations, i.e., algorithms that can be implemented with the level-two BLAS.

3.4. NOTES AND REFERENCES

The storage of arrays

We have stressed the storage of arrays by rows and columns because that is the way it is done in high-level programming languages. But in some cases other representations may be preferable. For example, one can partition a matrix A into submatrices and store the submatrices as individual arrays, a scheme that can improve performance on computers with hierarchical memories.

Strides and interleaved memory

In many computers, memory is divided into banks which can be accessed independently (e.g., see [169, p. 305ff.]). Specifically, if the memory has m banks, then the word with address x is assigned to bank $x \bmod m$. Because the banks are independent, if $x[i]$ is referenced a subsequent reference to $x[i{+}1]$ can proceed immediately. However, a reference to $x[i{+}m]$ must wait for the access to $x[i]$ to complete. This means that one should avoid arrays with strides that are equal to the number of banks. Since banks tend to come in powers of two, one should never create an array with a stride that is a power of two. A product p of odd numbers is a good bet, since the smallest integer k for which kp is a multiple of 2^i is 2^i itself.

The BLAS

The BLAS arose in stages, as suggested by their level numbers. The original BLAS [214], formally proposed in 1979, specified only vector operations. When this level of abstraction was found to be unsatisfactory for certain vector supercomputers, notably the various CRAYs, the level-two BLAS [102, 103] for matrix-vector operations were proposed in 1988. Finally, the level-three BLAS [101, 100] were proposed in 1990 to deal with hierarchical memories.

The fact that the BLAS could enhance the performance of code generated by nonoptimizing compilers was first noted by the authors of LINPACK and was an important factor in their decision to adopt the BLAS.

A problem with generalizing the original BLAS is that each level of ascent adds disproportionately to the functions that could be called BLAS. For example, the solution of triangular systems is counted among the level-two BLAS. But then, why not include the solution of Hessenberg systems, which is also an $O(n^2)$ process. By the time one reaches the level-three BLAS, everything in a good matrix package is a candidate. The cure for this problem is, of course, a little common sense and a lot of selectivity.

Virtual memory

Virtual memory was proposed by Kilburn, Edwards, Lanigan, and Sumner in 1962 [198]. Virtual memories are treated in most books on computer architecture (e.g., [169, 317, 257, 78]). Moler [231] was the first to point out the implications of virtual memory for matrix computations.

A common misconception is that virtual memory in effect gives the user a memory the size of the address space — about 4 Gbytes for an address space of 2^{32} bytes. But on a multiuser system each user would then have to be allocated 4 Gbytes of disk, which would strain even a large system. In practice, each user is given a considerably smaller amount of virtual memory.

Cache memory

Cache memory was the creation of Maurice Wilkes [341], the leader of the project that resulted in the first effective stored program computer. A comprehensive survey may be found in [283]. Also see [78, 169, 173, 257].

Large memories and matrix problems

We have already observed in §2.9 that if $m = n^2$ is the amount of memory required to store a general matrix then the complexity of $O(n^3)$ algorithms is $O(m^{\frac{3}{2}})$, which is a superlinear function of m. This implies that to keep up with an increase in memory, processor speed (including access to memory) must increase disproportionally. Evidently, such a disproportionate increase has been the rule until quite recently. For decades the rule of thumb for matrix computations was: If you can fit it in main memory, you can afford to solve it. Only recently has the balance tipped to the other side, and now you begin to see papers in which the authors beg off running the largest pos-

Sec. 4. Rounding Error

sible problem with the excuse that it would take too long.

Blocking

It is important to distinguish between a blocked algorithm like the one in Figure 3.3 and a block algorithm in which the blocks of a partitioned matrix are regarded as (noncommuting) scalars. We will return to this point when we consider block Gaussian elimination (Algorithm 1.2, Chapter 3).

4. Rounding Error

> *As I was going up the stair*
> *I met a man who wasn't there!*
> *He wasn't there again today!*
> *I wish, I wish he'd stay away!*
>
> Hughs Mearns

Rounding error is like that man. For most people it isn't there. It isn't there as they manipulate spreadsheets, balance checking accounts, or play computer games. Yet rounding error hovers at the edge of awareness, and people wish it would go away.

But rounding error is inevitable. It is a consequence of the finite capacity of our computers. For example, if we divide 1 by 3 in the decimal system, we obtain the nonterminating fraction $0.33333\ldots$. Since we can store only a finite number of these 3's, we must round or truncate the fraction to some fixed number of digits, say 0.3333. The remaining 3's are lost, and forever after we have no way of knowing whether we are working with the fraction $1/3$ or some other number like $0.33331415\ldots$.

Any survey of matrix algorithms — or any book on numerical computation, for that matter — must come to grips with rounding error. Unfortunately, most rounding-error analyses are tedious affairs, consisting of several pages of algebraic manipulations followed by conclusions that are obvious only to the author. Since the purpose of this work is to describe algorithms, not train rounding-error analysts, we will confine ourselves to sketching how rounding error affects our algorithms. To understand the sketches, however, the reader must be familiar with the basic ideas — absolute and relative error, floating-point arithmetic, forward and backward error analysis, and perturbation theory. This section is devoted to laying out the basics.

4.1. Absolute and Relative Error

There are two common ways of measuring the degree to which a quantity b approximates another quantity a — absolute and relative error. The difference between the two is that absolute error is defined without reference to the size of the quantities involved, whereas relative error incorporates the size as a scaling factor.

Absolute error

We begin with a definition.

Definition 4.1. *Let a and b be scalars. Then the* ABSOLUTE ERROR *in b as an approximation to a is the number*

$$\epsilon = |b - a|.$$

There are three comments to be made about this definition.

- The absolute error measures the distance between an approximation to a quantity and its true value. However, it does not allow one to retrieve the true value, since the direction between it and its approximation are unknown. However, if we introduce the number $e = b - a$, then by definition $b = a + e$. We may summarize this as follows.

    If b is an approximation to a with absolute error ϵ, then there is a number $e = b - a$ such that
    1. $|e| = \epsilon$,
    2. $b = a + e$.

- The number $e = b - a$ is usually called the *error* in b, and some people would confine the use of the term "error" to this difference. Such a restriction, however, would require us to qualify any other measure of deviation, such as absolute error, even when it is clear what is meant. In this work the meaning of the word "error" will vary with the context.

- In many applications only an approximate quantity is given, while the true value is unknown. This means that we cannot know the error exactly. The problem is resolved by computing upper bounds on the absolute error. We will see many examples in what follows.

- The absolute error is difficult to interpret without additional information about the true value.

Example 4.2. *Suppose b approximates a with an absolute error of 0.01. If $a = 22.43$, then a and b agree to roughly four decimal digits. On the other hand, if $a = 0.002243$, then the error overwhelms a. In fact, we could have $b = 0.012243$, which is almost five times the size of a.*

Relative error

Example 4.2 suggests that the problem with absolute error is that it does not convey a sense of scale, i.e., of the relation of the error to the quantity being approximated. One way of expressing this relation is to take the ratio of the error to the true value. In the above example, if $a = 22.43$, this ratio is about 0.0004, which is satisfactorily small. If, on the other hand, $a = 0.002243$, the ratio is about four. These considerations lead to the following definition.

SEC. 4. ROUNDING ERROR

Definition 4.3. *Let $a \neq 0$ and b be scalars. Then the* RELATIVE ERROR *in b as an approximation to a is the number*

$$\rho = \frac{|b - a|}{|a|}.$$

Relative error is somewhat more complicated than absolute error.

- The requirement that the number a be nonzero is in keeping with our motivation of the definition of relative error. Any error, however small, is infinitely large compared with zero.

- Just as a small absolute error tells us that we must add a quantity near zero to the true value to get the approximate value, so a small relative error tells us that we must multiply by a number near one. Specifically:

If b is an approximation to a with relative error ρ, then there is a number $r = (b - a)/a$ such that

$$\begin{aligned}1.\ & |r| = \rho, \\ 2.\ & b = a(1 + r).\end{aligned} \tag{4.1}$$

Conversely, if b satisfies (4.1.2), then b is an approximation to a with relative error $|r|$.

- If b is an approximation to a with absolute error ϵ, then a can be regarded as an approximation to b with absolute error ϵ. In general, no such reciprocal relation exists for the relative error. As an extreme example, zero approximates everything except zero with relative error one. But, as we have observed above, no approximation to zero has a relative error. Nonetheless, if the relative error is small, then an approximate reciprocity exists.

Theorem 4.4. *Let b approximate a with relative error $\rho < 1$. Then b is nonzero, and*

$$\frac{|a - b|}{|b|} \leq \frac{\rho}{1 - \rho}.$$

Proof. From the definition of relative error, we have $\rho|a| = |b - a| \geq |a| - |b|$, from which it follows that $|b| \geq (1 - \rho)|a| > 0$. Hence from the definition of relative error and the last inequality, it follows that

$$\frac{\rho}{1 - \rho} = \frac{|b - a|}{(1 - \rho)|a|} \geq \frac{|a - b|}{|b|}. \qquad \blacksquare$$

Thus, in passing from the relative error in b as an approximation to a to the relative error in a as an approximation to b, we must multiply by a factor of $(1 - \rho)^{-1}$. As ρ decreases, this factor quickly becomes insignificant. If, for example, $\rho = 0.1$, the factor is about 1.1. Therefore, if the relative error is at all small, it makes no real difference which quantity is used as a normalizer. In this case one may speak of the relative error in a and b without bothering to specify which is the quantity being approximated.

- The relative error is related to the number of significant digits to which two numbers agree. Consider, for example, the following approximations to $e = 2.71828\ldots$ and their relative errors.

Approx.	R.E.
2	$3 \cdot 10^{-1}$
2.7	$7 \cdot 10^{-3}$
2.71	$3 \cdot 10^{-3}$
2.718	$1 \cdot 10^{-4}$
2.7182	$3 \cdot 10^{-5}$
2.71828	$6 \cdot 10^{-7}$

These numbers suggest the following rule of thumb.

> If the relative error of a with respect to b is ρ, then a and b agree to roughly $-\log \rho$ significant decimal digits.

The rule is only approximate and can vary by an order of magnitude either way. For example, 9.99999 and 9.99899 agree to three digits and have a relative error of about 10^{-4}. On the other hand, the numbers 1.00000 and 1.00999 also agree to three digits but have a relative error of about 10^{-2}.

The rule applies to number systems other than decimal. For binary systems the rule reads:

> If a and b have relative error of approximately 2^{-t}, then a and b agree to about t bits.

4.2. FLOATING-POINT NUMBERS AND ARITHMETIC

Anyone who wishes to do serious rounding-error analysis must grapple with the details of floating-point arithmetic. Fortunately, only the rudiments are required to understand how such analyses are done and how to interpret them. The remaining subsections are devoted to those rudiments.

Floating-point numbers

Floating-point numbers and their arithmetic are familiar to anyone who has used a hand calculator in scientific mode. For example, when I calculate $1/\pi$ on my calculator, I might see displayed

$\quad$ 3.183098 -01

This display has two components. The first is the number 3.183098, which is called the mantissa. The second is the number -01, called the exponent, which represents

SEC. 4. ROUNDING ERROR

a power of ten by which the mantissa is to be multiplied. Thus the display represents the number

$$3.183098 \cdot 10^{-1} = 0.3183098.$$

It is easy to miss an important aspect of the display. The numbers have only a finite number of digits—seven for the mantissa and two for the exponent. This is characteristic of virtually all floating-point systems. The mantissa and exponent are represented by numbers with a fixed number of digits. As we shall see, the fixed number of digits in the mantissa makes rounding error inevitable.

Although the above numbers are all represented in the decimal system, other bases are possible. In fact most computers use binary floating-point numbers.

Let us summarize these observations in a working definition of floating-point number.

Definition 4.5. *A t-digit, base-β* FLOATING-POINT NUMBER *having* EXPONENT RANGE $[e_{\min}, e_{\max}]$ *is a pair (m, e), where*

1. *m is a t-digit number in the base β with its β-point in a fixed location, and*
2. *e is an integer in the interval $[e_{\min}, e_{\max}]$.*

The number m is called the MANTISSA *of (m, e), and the number e is its* EXPONENT. *The* VALUE *of the number (m, e) is*

$$m \cdot \beta^e.$$

The number (m, e) is NORMALIZED *if the leading digit in m is nonzero.*

It is important not to take this definition too seriously; the details of floating-point systems are too varied to capture in a few lines. Instead, the above definition should be taken as a model that exhibits the important features of most floating-point systems.

On hand calculators the floating-point base is ten. On most digital computers it is two, although base sixteen occurs on some IBM computers. The location of the β-point varies. On hand calculators it is immediately to the left of the most significant digit, e.g., $(3.142, 0)$. On digital computers the binary point is located either to the left of the most significant digit, as in $(1.10010, 1)$, or to the right, as in $(.110010, 2)$.

The way in which the exponent is represented also varies. In the examples in the last paragraph, we used decimal numbers to represent the exponent, even though the second example concerned a binary floating-point number. In the IBM hexadecimal format, the exponent is represented in binary.

A floating-point number on a digital computer typically occupies one or two 32-bit words of memory. A number occupying one word is called a *single-precision* number, one occupying two words is called a *double-precision* number. Some systems provide *quadruple-precision* numbers occupying four words. The necessity of representing floating-point numbers within the confines of a fixed number of words accounts for the limits on the size of the mantissa and on the range of the exponent.

The representation of π in 4-digit, decimal floating point as $(0.314, 1)$ wastes a digit representing the leading zero. The representation $(3.142, 0)$ is more accurate. For this reason most floating-point systems automatically adjust the exponent so the mantissa is normalized, i.e., so that its leading bit is nonzero.

The IEEE standard

Since the details of floating-point systems vary, it is useful to consider one in detail. We have chosen the IEEE standard, which has been widely adopted—at least nominally.

Example 4.6 (IEEE floating-point standard). *Single-precision IEEE standard floating-point numbers have the form*

```
 0 1    8 9                          31
|σ| exp |         frac                 |
```

The small numbers above the box denote bit positions within the 32-bit word containing the number. The box labeled σ contains the sign of the mantissa. The other two boxes contain the exponent and the trailing part of the mantissa. The value of the number is

$$(-1)^\sigma 1.\text{frac} \cdot 2^{\text{exp}-127}, \qquad 1 \leq \text{exp} \leq 254.$$

The double-precision format is

```
 0 1      11 12                        63
|σ|  exp  |          frac                |
```

The value of a double precision number is

$$(-1)^\sigma 1.\text{frac} \cdot 2^{\text{exp}-1023}, \qquad 1 \leq \text{exp} \leq 2046.$$

The quantities frac and exp are not the same as the quantities m and e in Definition 4.5. Here is a summary of the differences.

- Since the leading bit in the mantissa of a normalized, binary, floating-point number is always one, it is wasteful to devote a bit to its representation. To conserve precision, the IEEE fraction stores only the part below the leading bit and recovers the mantissa via the formula $m = (-1)^\sigma \cdot 1.\text{frac}$.

- The number exp is called a *biased exponent*, since the true value e of the exponent is computed by subtracting a bias. The unbiased exponent range for single precision is $[-126, 127]$, which represents a range of numbers from roughly 10^{-38} to 10^{38}. Double precision ranges from roughly 10^{-307} to 10^{307}. In both precisions the extreme exponents (i.e., -127 and 128 in single precision) are reserved for special purposes.

- Zero is represented by $\text{exp} = 0$ (one of the reserved exponents) and $f = 0$. The sign bit can be either 0 or 1, so that the system has both a $+0$ and a -0.

SEC. 4. ROUNDING ERROR

Rounding error

Relative error is the natural mode in which to express the errors made in rounding to a certain number of digits. For example, if we round $\pi = 3.14159\ldots$ to four digits, we obtain the approximation 3.142, which has a relative error of about 10^{-4}. The exponent -4, or something nearby, is to be expected from the relation between relative error and the number of significant digits to which two numbers agree.

More generally, consider the problem of rounding a normalized binary fraction a to t digits. We can represent this fraction as

$$a = 0.1 \overbrace{xx \ldots xx}^{t-1} yzz \ldots , \qquad (4.2)$$

where the y and the z's represent the digits to be rounded. If $y = 0$ we truncate the fraction, which gives the number

$$\tilde{a} = 0.1 \overbrace{xx \ldots xx}^{t-1} . \qquad (4.3)$$

The worst possible error (which is approached when the z's are all one) is 2^{-t-1}. On the other hand, if y is one, we round up to get the number

$$\tilde{a} = 0.1 \overbrace{xx \ldots xx}^{t-1} + 2^{-t}.$$

Again the worst error (which is attained when the z's are all zero) is 2^{-t-1}. Since the smallest possible value of a is $1/2$, we make a relative error

$$\frac{|\tilde{a} - a|}{|a|} \leq \frac{2^{-t-1}}{2} = 2^{-t}. \qquad (4.4)$$

In a floating-point system rounding depends only on the mantissa and not on the exponent. To see this result let a be multiplied by 2^e, where e is an integer. Then the value of $\tilde{a}$ will also be multiplied by 2^e, and the factor 2^e will cancel out as we compute the relative error (4.4).

Let us write $\mathrm{fl}(a)$ for the rounded value of a number a. Then from the characterization (4.1), it follows that if a number a is rounded to a t-digit binary floating-point number we have

$$\mathrm{fl}(a) = a(1 + \rho), \qquad |\rho| \leq 2^{-t}. \qquad (4.5)$$

There are other ways to shoehorn a number into a finite precision word. The rounding we have just described is sometimes called "round up" because the number

$$a = 0.1 \overbrace{xx \ldots xx}^{t-1} 1000 \ldots$$

is always rounded upward. It could equally well be rounded downward — or to the nearest even number, a strategy which has a lot to recommend it. In *chopping* or *truncation* the trailing digits are simply lopped off, so that the chopped representation of (4.2) is (4.3), whatever the value of the digits $yzz\ldots$. All these methods generate bounds of the form (4.5), though possibly with a different bound on $|\rho|$.

The key players in (4.5) are an equality stating that rounding introduces a relative error and an inequality bounding how large the error can be. Abstracting from this equation, we will assume:

> For any floating-point system, there is a smallest number ϵ_M such that
>
> $$\mathrm{fl}(a) = a(1+\rho), \qquad |\rho| \leq \epsilon_M.$$
>
> The number ϵ_M is called the ROUNDING UNIT for the system in question.

Typically, the rounding unit for a t-digit, base-β floating-point system will be approximately β^{-t}. The size can vary a little depending on the details. For example, the rounding unit for chopping is generally twice the rounding unit for ordinary rounding. Although this increase is minor, we will see later that chopping has an important drawback that does not reveal itself in the bounds.

Example 4.7 (IEEE standard). *The single-precision rounding unit for IEEE floating point is about 10^{-7}. The double precision rounding unit is about 10^{-16}.*

Floating-point arithmetic

Floating-point numbers have an arithmetic that mimics the arithmetic of real numbers. The operations are usually addition, subtraction, multiplication, and division. This arithmetic is necessarily inexact. For example, the product of two four-digit numbers is typically an eight-digit number, and in a four-digit floating-point system it must be rounded back. The standard procedure is for each operation to return the correctly rounded answer. This implies the following error bounds for floating-point arithmetic.

> Let $\circ$ denote one of the arithmetic operations $+, -, \times, \div$, and let $\mathrm{fl}(a \circ b)$ denote the result of performing the operation in a floating-point system with rounding unit ϵ_M. Then
>
> $$\mathrm{fl}(a \circ b) = (a \circ b)(1+\rho), \qquad |\rho| \leq \epsilon_M. \qquad (4.6)$$

The bound (4.6) will be called the *standard bound* for floating-point arithmetic, and a floating-point system that obeys the standard bound will be called a *standard system*. The standard bound is the basis for most rounding-error analyses of matrix algorithms. Only rarely do we need to know the details of the arithmetic itself. This fact accounts for the remarkable robustness of many matrix algorithms.

Sec. 4. Rounding Error

Example 4.8 (IEEE standard). *In the IEEE standard the default rounding mode is round to nearest even. However, the standard specifies other modes, such as round toward zero, that are useful in specialized applications.*

The practice of returning the correctly rounded answer has an important implication.

Example 4.9. *If $|a + b| < \min\{|a|, |b|\}$, we say that* CANCELLATION *has occurred in the sum $a + b$. The reason for this terminology is that cancellation is usually accompanied by a loss of significant figures. For example, consider the difference*

$$
\begin{array}{r}
0.4675 \\
-\ 0.4623 \\
\hline
0.0052
\end{array}
$$

Since cancellation implies that no more than the full complement of significant figures is required to represent the result, it follows that:

> *When cancellation occurs in a standard floating-point system, the computed result is exact.*

There is a paradox here. People frequently blame cancellation for the failure of an algorithm; yet we have just seen that cancellation itself introduces no error. We will return to this paradox in the next subsection.

One final point. Many algorithms involve elementary functions of floating-point numbers, which are usually computed in software. For the purposes of rounding-error analysis, however, it is customary to regard them as primitive operations that return the correctly rounded result. For example, most rounding error-analyses assume that

$$\mathrm{fl}(\sqrt{a}) = \sqrt{a}\,(1 + \rho), \qquad |\rho| \leq \epsilon_{\mathrm{M}}.$$

4.3. Computing a sum: Stability and condition

In this subsection we will use the standard bound to analyze the computation of the sum

$$s_n = x_1 + x_2 + \cdots + x_n.$$

Simple as this computation is, it already illustrates many of the features of a full-blown rounding-error analysis.

The details of the analysis depend on the order in which the numbers are summed. For definiteness we will analyze the following algorithm.

1. $s_1 = x_1$
2.   **for** $i = 2$ **to** n
3. $s_i = s_{i-1} + x_i$
4.   **end** i

A backward error analysis

A natural approach is to write, say, $\tilde{s}_i$ for the computed values of the s_i and attempt to compute an upper bound on $|\tilde{s}_i - s_i|$ from a bound on $|\tilde{s}_{i-1} - s_{i-1}|$. This technique is called *forward rounding-error analysis*. Although such a procedure is useful in some applications, a different, indirect method, called *backward error analysis*, works better for many matrix algorithms. The idea is to let the s_i stand for the *computed* quantities and relate them to the original data. We will proceed in stages.

- **Application of the standard bound.** This is the tedious part of the analysis. From the standard bound we have

$$s_2 = \text{fl}(x_1 + x_2) = (x_1 + x_2)(1 + \epsilon_1) = x_1(1 + \epsilon_1) + x_2(1 + \epsilon_1),$$

where $|\epsilon_1| \leq \epsilon_M$. Similarly,

$$\begin{aligned}
s_3 = \text{fl}(s_2 + x_3) &= (s_2 + x_3)(1 + \epsilon_2) \\
&= x_1(1 + \epsilon_1)(1 + \epsilon_2) + \\
&\quad x_2(1 + \epsilon_1)(1 + \epsilon_2) + \\
&\quad x_3(1 + \epsilon_2).
\end{aligned}$$

Continuing in this way, we find that

$$\begin{aligned}
s_n = \text{fl}(s_{n-1} + x_n) &= (s_{n-1} + x_n)(1 + \epsilon_{n-1}) \\
&= x_1(1 + \epsilon_1)(1 + \epsilon_2)\cdots(1 + \epsilon_{n-1}) + \\
&\quad x_2(1 + \epsilon_1)(1 + \epsilon_2)\cdots(1 + \epsilon_{n-1}) + \\
&\quad x_3(1 + \epsilon_2)\cdots(1 + \epsilon_{n-1}) + \\
&\quad \vdots \\
&\quad x_{n-1}(1 + \epsilon_{n-2})(1 + \epsilon_{n-1}) + \\
&\quad x_n(1 + \epsilon_{n-1}),
\end{aligned} \qquad (4.7)$$

where $|\epsilon_i| \leq \epsilon_M$ $(i = 1, 2, \ldots, n-1)$.

The expression (4.7) is not very informative, and it will help to introduce some simplifying notation. Let the quantities η_i be defined by

$$\begin{aligned}
1 + \eta_1 &= (1 + \epsilon_1)(1 + \epsilon_2)\cdots(1 + \epsilon_{n-1}), \\
1 + \eta_2 &= (1 + \epsilon_1)(1 + \epsilon_2)\cdots(1 + \epsilon_{n-1}), \\
1 + \eta_3 &= (1 + \epsilon_2)\cdots(1 + \epsilon_{n-1}), \\
&\vdots \\
1 + \eta_{n-1} &= (1 + \epsilon_{n-2})(1 + \epsilon_{n-1}), \\
1 + \eta_n &= (1 + \epsilon_{n-1}).
\end{aligned} \qquad (4.8)$$

Then

$$\begin{aligned}
s_n = {}& x_1(1 + \eta_1) + x_2(1 + \eta_2) + x_3(1 + \eta_3) + \cdots \\
& + x_{n-1}(1 + \eta_{n-1}) + x_n(1 + \eta_n).
\end{aligned} \qquad (4.9)$$

Sec. 4. Rounding Error

- **First-order bounds.** The numbers η_i in (4.9) are called the (relative) *backward error*, and it is important to have a bound on their sizes. To see what kind of bound we can expect, consider the product

$$1 + \eta_{n-1} = (1 + \epsilon_{n-2})(1 + \epsilon_{n-1}) = 1 + (\epsilon_{n-2} + \epsilon_{n-1}) + \epsilon_{n-2}\epsilon_{n-1}. \quad (4.10)$$

Now $|\epsilon_{n-2} + \epsilon_{n-1}| \leq 2\epsilon_M$ and $|\epsilon_{n-2}\epsilon_{n-1}| \leq \epsilon_M^2$. If, say, $\epsilon_M = 10^{-16}$, then $2\epsilon_M = 2 \cdot 10^{-16}$ while $\epsilon_M^2 = 10^{-32}$. Thus the third term on the right-hand side of (4.10) is insignificant compared to the second term and can be ignored. If we do ignore it, we get

$$\eta_{n-1} \cong \epsilon_{n-2} + \epsilon_{n-1}$$

or

$$|\eta_{n-1}| \lesssim |\epsilon_{n-2}| + |\epsilon_{n-1}| \leq 2\epsilon_M.$$

In general, we should expect

$$\begin{aligned}|\eta_1| &\lesssim (n-1)\epsilon_M, \\ |\eta_i| &\lesssim (n-i+1)\epsilon_M, \qquad i = 2, 3, \ldots, n.\end{aligned} \quad (4.11)$$

- **Rigorous bounds.** To get a completely rigorous bound, we use the following result, whose proof is left as an exercise.

Theorem 4.10. *Let*

$$w = \frac{(1 + \epsilon_1) \cdots (1 + \epsilon_k)}{(1 + \epsilon_{k+1}) \cdots (1 + \epsilon_n)},$$

where $|\epsilon_i| \leq \epsilon_M$. *If* $n\epsilon_M < 1$, *then* $w = 1 + \eta$, *where*

$$|\eta| \leq \frac{n\epsilon_M}{1 - n\epsilon_M}.$$

If further

$$n\epsilon_M \leq 0.1,$$

then

$$\eta \leq n\epsilon'_M,$$

where

$$\epsilon'_M \stackrel{\text{def}}{=} \frac{\epsilon_M}{0.9} < 1.12\epsilon_M \quad (4.12)$$

is the ADJUSTED ROUNDING UNIT.

If we assume that $n\epsilon_M \leq 0.1$, then the bounds (4.11) can be written rigorously in the form

$$|\eta_1| \leq (n-1)\epsilon'_M,$$
$$|\eta_i| \leq (n-i+1)\epsilon'_M, \quad i = 2, 3, \ldots, n. \qquad (4.13)$$

The simplicity of the bounds (4.13) more than justifies the slight overestimate of the error that results from using the adjusted rounding unit. For this reason we make the following assumption.

> *In all rounding-error analyses it is tacitly assumed that the size of the problem is such that approximate bounds of the form $|\eta| \lesssim n\epsilon_M$ can be rigorously replaced by $|\eta| \leq n\epsilon'_M$, where ϵ'_M is the adjusted rounding unit defined by (4.12).*

Backward stability

The expression

$$s_n = x_1(1+\eta_1) + x_2(1+\eta_2) + x_3(1+\eta_3) + \cdots$$
$$+ x_{n-1}(1+\eta_{n-1}) + x_n(1+\eta_n)$$

has the following interpretation.

> *When the sum of n numbers is computed in floating-point arithmetic, the result is as if we had computed the exact sum of the same n numbers perturbed by small relative errors. The errors are bounded by $n-1$ times the adjusted rounding unit.*

The key observation here is that the errors have been thrown back on the original data. Algorithms for which this is true are said to be *backward stable*, or if the context is matrix computations, simply *stable*.

Backward stability provides a different way of looking at the quality of computed solutions. In practice, the data on which a computation is based is not exact but is contaminated by errors. The errors may be measurement errors, in which case they will be generally be large compared to the rounding unit. Or the data may be computed, in which case it will be contaminated with rounding error.

Thus we must think of the input to an algorithm not as a set of numbers but as a set of regions representing where the erroneous input can lie. If the backward error is less than the size of these regions, then the algorithm that performed the computations is absolved of responsibility for any inaccuracies in the answer, since the errors in the data could have produced the same answer.

To put things another way, let us regard a computational problem as a function f of its input x. We will say that a function is *well conditioned* if $f(x)$ is insensitive to small perturbations in x, that is, if $f(y)$ is near $f(x)$ when x is near y. (The precise definition of terms like "insensitive," "small," and "near" will depend on the nature of

SEC. 4. ROUNDING ERROR

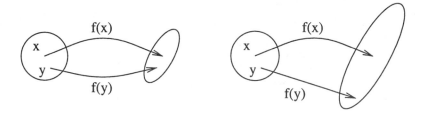

Figure 4.1: Well- and ill-conditioned problems

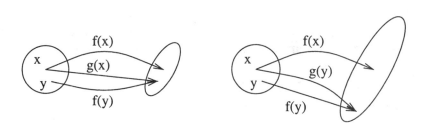

Figure 4.2: Behavior of a stable algorithm

the problem.) On the other hand, if f varies greatly with small perturbations in x we will say that it is *ill conditioned*. Examples of well- and ill-conditioned functions are illustrated in Figure 4.1. The circles on the left represent a range of values of x; the ellipses on the right represent the range of values of $f(x)$. The larger ellipse on the right represents the ill-conditioning of f.

An algorithm to solve the computational problem represented by the function f can be regarded as another function g that approximates f. Suppose that the algorithm g is backward stable and is applied to an input x. Then $g(x) = f(y)$ for some y near x. If the problem is well conditioned, then $f(x)$ must be near $f(y)$, and the computed solution $g(x)$ is accurate. On the other hand, if the problem is ill conditioned, then the computed result will generally be inaccurate. But provided the errors in the data are larger than the backward error, the answer will lie within the region of uncertainty caused by the errors in the data. These two situations are illustrated in Figure 4.2.

Weak stability

We have seen that a backward stable algorithm solves well-conditioned problems accurately. But not all algorithms that solve well-conditioned problems accurately are backward stable. For example, an algorithm might produce a solution that is as accurate as one produced by a stable algorithm, but the solution does not come from a slight perturbation of the input. This situation, which is called *weak stability*, is illustrated in Figure 4.3. The ellipse on the right represents the values of f corresponding

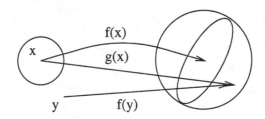

Figure 4.3: Weak stability

to the circle on the left. The large circle on the right represents the values returned by a weakly stable algorithm. Since the radius of the circle is the size of the major axis of the ellipse, the algorithm returns a value that is no less accurate than one returned by a stable algorithm. But if the value does not lie in the ellipse, it does not correspond to a data point in the circle on the left, i.e., a point near the input.

Condition numbers

As informative as a backward rounding-error analysis of an algorithm is, it does not tell us what accuracy we can expect in the computed solution. In the notation introduced above, the backward error analysis only insures that there is a y near x such that $g(x) = f(y)$. But it does not tell us how far $f(x)$ is from $f(y)$.

In fact, the problem of accuracy can be cast in general terms that has nothing to do with rounding error.

> Given a function f and two arguments x and y, bound the distance between $f(x)$ and $f(y)$ in terms of the distance between x and y.

Resolving such problems is the subject of the mathematical discipline of *perturbation theory*. The perturbation that gives rise to y from x can have any source, which need not be rounding error. For this reason we will first treat the general problem of the perturbation of a sum of numbers and then reintroduce rounding error.

We are concerned with perturbations of the sum

$$s = x_1 + x_2 + \cdots + x_n. \tag{4.14}$$

The perturbed problem is

$$\tilde{s} = x_1(1 + \eta_1) + x_2(1 + \eta_2) + \cdots + x_n(1 + \eta_n), \tag{4.15}$$

where the η_i are assumed to have a common bound

$$|\eta_i| \leq \epsilon. \tag{4.16}$$

To derive a perturbation bound, subtract (4.14) from (4.15) to get

$$\tilde{s} - s = x_1\eta_1 + x_2\eta_2 + \cdots + x_1\eta_1.$$

SEC. 4. ROUNDING ERROR

Taking absolute values and applying (4.16), we get

$$|\tilde{s} - s| \leq (|x_1| + |x_2| + \cdots + |x_n|)\epsilon.$$

Thus the relative error in the perturbed sum is

$$\frac{|\tilde{s} - s|}{|s|} \leq \kappa\epsilon, \qquad (4.17)$$

where

$$\kappa = \frac{|x_1| + |x_2| + \cdots + |x_n|}{|x_1 + x_2 + \cdots + x_n|}. \qquad (4.18)$$

The right-hand side of (4.17) is the relative error in $\tilde{s}$. The number ϵ bounds the relative error in the x's. Thus κ is a factor that mediates the passage from a bound on the perturbation of the arguments of a function to a bound on the perturbation induced in the function itself. Such a number is called a *condition number*.

Just as a backward rounding-error analysis distinguishes between satisfactory and unsatisfactory algorithms, condition numbers distinguish between easy and hard problems. For our problem the condition number is never less than one. It is equal to one when the absolute value of the sum is equal to the sum of the absolute values, something that happens whenever all the x's have the same sign. On the other hand, it is large when the sum is small compared to the x's. Thus the condition number not only bounds the error, but it provides insight into what makes a sum hard to compute.

Reenter rounding error

Let us see what the above perturbation theory says about our algorithm for summing numbers. The backward errors η_i satisfy

$$|\eta_1| \leq (n-1)\epsilon'_M,$$
$$|\eta_i| \leq (n-i+1)\epsilon'_M, \qquad i = 2, 3, \ldots, n.$$

More simply,

$$|\eta_i| \leq n\epsilon'_M.$$

Consequently, the error in the computed sum is bounded by

$$n\kappa\epsilon'_M.$$

Let us look more closely at this bound and the way in which it was derived.

In popular accounts, the accumulation of rounding error is often blamed for the failure of an algorithm. Here the accumulation of rounding error is represented by the factor $n\epsilon_M$, which grows slowly. For example, if κ is one, we cannot loose eight digits of accuracy unless n is greater than 100 million. Thus, even for large n, the condition number can be more influential than the accumulation of rounding error. In fact, a single rounding error may render an ill-conditioned problem inaccurate, as we shall see later.

The bound itself is almost invariably an overestimate. In the first place, it is derived by replacing bounds like $|\eta_i| \leq (n-i)\epsilon'_M$ with $|\eta_n| \leq n\epsilon'_M$, which exaggerates the effects of the terms added last. In fact, if we were to arrange the terms so that $x_1 \leq x_2 \leq \cdots \leq x_n$, then the larger x's will combine with the smaller η's to make the final bound an even greater overestimate.

There is another factor tending to make the bound too large. Recall that

$$1 + \eta_1 = (1 + \epsilon_1)(1 + \epsilon_2) \cdots (1 + \epsilon_{n-1})$$

[see (4.8)]. Multiplying this relation out and keeping only first-order terms in ϵ_M, we find that

$$\eta_1 \cong \epsilon_1 + \epsilon_2 + \cdots + \epsilon_{n-1}.$$

Now the worst-case bound on $|\eta_1|$ is about $(n-1)\epsilon_M$. But if we are rounding, we can expect the ϵ's to vary in sign. With negative ϵ's cancelling positive ϵ's, the sum can be much smaller than the worst-case bound. In fact, on statistical grounds, we can expect it to behave more like $\sqrt{n}\epsilon_M$. Note that if all the x_i are positive, and the floating-point system in question chops instead of rounding, all the ϵ_i will be negative, and we will see a growth proportional to n.

If the bound is an overestimate, what good is it? The above discussion provides three answers. First, even as it stands, the bound shows that rounding errors accumulate slowly. Second, by looking more closely at the derivation of the bound, we discovered that arranging the x's in ascending order of magnitude tends to make the bound a greater overestimate. Since the bound is fixed, it can only become worse if actual error becomes smaller. This suggests the following rule of thumb.

Summing a set of numbers in increasing order of magnitude tends to diminish the effects of rounding error. (4.19)

Finally, an even deeper examination of the derivation shows a fundamental difficulty with chopped arithmetic — if the numbers are all of the same sign, the chopping errors are all in the same direction and accumulate faster.

4.4. CANCELLATION

The failure of a computation is often signaled by the cancellation of significant figures as two nearly equal numbers are subtracted. However, it is seldom the *cause* of the failure. To see why, consider the sum

$$\begin{array}{r} 472635.0000 \\ +\quad 27.5013 \\ -\ 472630.0000 \\ \hline 32.5013 \end{array} \quad (4.20)$$

SEC. 4. ROUNDING ERROR

The condition number for this problem [see (4.18)] is

$$\kappa = \frac{945292.5013}{32.5013} \cong 3 \cdot 10^4.$$

Consequently, we should expect to see a loss of about four digits in attempting to compute the sum.

Our expectations are realized. If we compute the sum in six-digit, decimal arithmetic, we get first

$$\begin{array}{r} \text{fl } 472635.0000 \\ +27.5013 \\ \hline 472663.0000 \end{array} \tag{4.21}$$

and then

$$\begin{array}{r} \text{fl } 472663. \\ -472630. \\ \hline 33. \end{array} \tag{4.22}$$

The answer is accurate to only two digits.

The first thing to note about this example is that rounding error does not need to accumulate for a calculation to go bad. Here only one rounding error is committed—in (4.21). But because the problem is ill conditioned, that single rounding error is sufficient to induce great inaccuracy in the answer.

A more important point concerns the role of cancellation. It is a widespread belief that cancellation causes algorithms to fail. True believers will point to the cancellation in (4.22)—they usually cannot resist calling it *catastrophic* cancellation—as the source of the inaccuracy in the computed solution. A little reflection shows that something is wrong with this notion. Suppose we reorder the computations as follows:

$$\begin{array}{r} \text{fl } 472635. \\ -472630. \\ \hline 5. \end{array} \tag{4.23}$$

and

$$\begin{array}{r} \text{fl } 5.0000 \\ +27.5013 \\ \hline 32.5013 \end{array} \tag{4.24}$$

The cancellation in the first subtraction is every bit as catastrophic as the cancellation in (4.22). Yet the answer is exact.

There is no single paradigm for explaining the effects of cancellation. But it is frequently useful to regard cancellation as revealing a loss of information that occurred earlier in the computation—or if not in the computation, then when the input for the computation was generated. In the example above, the number 27.5013 in (4.21) could be replaced by any number in the interval [27.5000, 28.4999] and the computed sum would be unaltered. The addition has destroyed all information about the last four digits of the number 27.5013. The cancellation, without introducing any errors of its own, informs us of this fact.

Cancellation often occurs when a stable algorithm is applied to an ill-conditioned problem. In this case, there is little to be done, since the difficulty is intrinsic to the problem. But cancellation can also occur when an unstable algorithm is applied to a well-conditioned problem. In this case, it is useful to examine the computation to see where information has been lost. The exercise may result in a modification of the algorithm that makes it stable.

4.5. EXPONENT EXCEPTIONS

A floating-point *exception* occurs when an operation is undefined or produces a result out of the range of the system. Examples of exceptions are:

1. Division by zero.
2. *Overflow*, which occurs when an operation would produce a number whose exponent is too large.
3. *Underflow*, which occurs when an operation would produce a number whose exponent is too small.

Here we will treat overflow and underflow. The theme is that in some cases it is possible to eliminate overflow and render underflows harmless.

Overflow

There are two reasonable actions to take when overflow occurs.

1. Stop the calculation.
2. Return a number representing a machine infinity and continue the calculation.

For underflow there are three options.

1. Stop the calculation.
2. Return zero.
3. Return an unnormalized number or zero.

SEC. 4. ROUNDING ERROR

The following table illustrates the contrast between the two last options for four-digit decimal arithmetic with exponent range $[-99, 99]$.

Result of operation	Result returned option 2	Result returned option 3
$1.234 \cdot 10^{-99}$	$(1.234, -99)$	$(1.234, -99)$
$1.234 \cdot 10^{-100}$	0	$(0.123, -99)$
$1.234 \cdot 10^{-101}$	0	$(0.012, -99)$
$1.234 \cdot 10^{-101}$	0	$(0.001, -99)$
$1.234 \cdot 10^{-102}$	0	0

The table makes it obvious why the second option is called *flush to zero* and the third is called *gradual underflow*.

Avoiding overflows

The two ways of treating overflows have undesirable aspects. The first generates no answer, and the second generates one that is not useful except in specialized applications. Thus overflows should be avoided if at all possible. The first option for underflow is likewise undesirable; however, the second and third options may give useful results if the calculation is continued. The reason is that in many cases scaling to avoid overflows insures that underflows are harmless.

To see how this comes about, consider the problem of computing

$$c = \sqrt{a^2 + b^2} \tag{4.25}$$

in 10-digit decimal floating point with an exponent range of $[-99, 99]$. For definiteness we will assume that overflows stop the computation and that underflows flush to zero.

The natural computation suggested by (4.25) is subject to both overflows and underflows. For example, if $a = b = 10^{60}$, then a^2 and b^2 both overflow, and the computation stops. This happens in spite of the fact that the answer is $\sqrt{2}\, 10^{60}$, which is representable on our system.

Similarly, if $a = b = 10^{-60}$, their squares underflow and are set to zero. Thus the computed result is zero, which is a poor relative approximation to the true answer $\sqrt{2}\, 10^{-60}$.

The cure for the overflow is to scale. If we set

$$s = |a| + |b|,$$

then the numbers a/s and b/s are less than one in magnitude, and their squares cannot overflow. Thus we can compute c according to the formula

$$c = s\sqrt{(a/s)^2 + (b/s)^2} \tag{4.26}$$

without fear of overflow.

> The following algorithm computes $\sqrt{a^2+b^2}$ with scaling that insures that overflows cannot occur and underflows are harmless.
>
> 1. $Euclid(a, b)$
> 2. $s = |a| + |b|$
> 3.   **if** $(s = 0)$
> 4.   **return** 0   ! Zero is a special case.
> 5.   **else**
> 6.   **return** $s\sqrt{(a/s)^2 + (b/s)^2}$
> 7.   **end if**
> 8. **end** *Euclid*

Algorithm 4.1: The Euclidean length of a 2-vector

———————⋄———————

Moreover, this scaling renders underflows harmless. To see this, suppose that the number a has the largest magnitude. Then the magnitude of a/s must be greater than 0.5. Now if (b/s) underflows and is set to zero, the formula (4.26) gives the answer $|a|$. *But in this case $|a|$ is the correctly rounded answer.* For if b/s underflows, we must have $|b/a| \cong 10^{-100}$, and hence $s = |a|$. Consequently, by the Taylor series expansion $\sqrt{1+x} = 1 + 0.5x + \cdots$, we have

$$c = |a|\sqrt{1+(b/a)^2} \cong |a|[1 + .5*(b/a)^2].$$

The factor on the right represents a relative error of about 10^{-200}. Since $\epsilon_M \cong 10^{-10}$, $|a|$ is the correctly rounded value of c.

We summarize this technique in Algorithm 4.1 (which we have already seen in §1.2).

Many matrix algorithms can be scaled in such a way that overflows do not occur and underflows are not a problem. As in Algorithm 4.1, there is a price to be paid. But for most matrix algorithms the price is small compared to the total computation when n is large enough.

Exceptions in the IEEE standard

Let us conclude with a quick look at how exceptions are treated in the IEEE standard.

The IEEE standard handles arithmetic exceptions by encoding the results in the special formats that use the reserved values of the exponent. In what follows (see Example 4.6), $\underline{e}$ is the lower reserved value (corresponding to exp = 0) and $\overline{e}$ is the upper reserved value (corresponding to exp = 255 in single precision and exp = 2047 in double precision).

1. Overflow returns an infinity, which is represented by $(\pm 0, \overline{e})$. The sign is the sign one would get if overflow had not occurred.

Sec. 4. Rounding Error

2. The attempt to compute $c/0$, where $c \neq 0$ results in an infinity whose sign is the same as c.
3. Operations, such as $0/0$, whose meaning is indeterminate result in a NaN (for *not a number*). A NaN is represented by $(f, \overline{e})$, where $f \neq 0$.
4. Underflow is gradual. A denormalized number is represented by the pair $(f, \underline{e})$, where $|f| < 1$; i.e., f is denormalized. Since $(0, \underline{e})$ represents zero, gradual underflow eventually flushes to zero.

In addition, the standard specifies system interrupts on exceptions and status flags to indicate the nature of the interrupt. These can be used to take special action, e.g., stop the calculation on overflow.

4.6. Notes and references

General references

J. H. Wilkinson is rightly judged the father of modern error analysis. The first work to contain extensive analyses in floating-point arithmetic is Wilkinson's ground-breaking paper on the solution of linear systems [342]. He later published an expository book *Rounding Errors in Algebraic Processes* [345], which can still be read with profit today. His monumental *Algebraic Eigenvalue Problem* [346] contains the summation of his work on rounding error.

Books by Sterbenz [285] and Knuth [201] give much useful information. The latter includes historical material. Goldberg's survey [147] is comprehensive and contains a detailed description of IEEE arithmetic. His survey of the hardware implementation of computer arithmetic [146] is also excellent.

The *Accuracy and Stability of Numerical Algorithms* by N. J. Higham [177] is the current definitive work on rounding error.

Relative error and precision

The notion of two numbers agreeing to a certain number of significant figures is slippery. For example, in the naive view the two numbers

2.00000
1.99999

agree to no figures, yet they have a relative error of about 10^{-6}. We need to add something like: If two digits differ by one and the larger is followed by a sequence of nines while the smaller is followed by a sequence of zeros, then the digits and their trailers are in agreement.

Nomenclature for floating-point numbers

The mantissa of a floating-point number is often called the *fraction* because when the number is normalized to be less than one the mantissa is a proper fraction. But the nomenclature is misleading for other normalizations. The exponent is sometimes called the *characteristic*.

The words "mantissa" and "characteristic" come from the terminology for logarithms. For example, consider log 2000 ≅ 3.3010. The number 3 to the left of the decimal is the characteristic and the number 0.3010 is the mantissa. The characteristic is just the power of ten that multiplies the antilogarithm of the mantissa and corresponds exactly to the exponent of a floating-point number. The mantissa of a logarithm, however, is the log of the mantissa of a floating-point number. Thus both "fraction" and "mantissa" are misleading; but perhaps "mantissa" is less so, since log tables are out of fashion.

The rounding unit

The use of ϵ_M to denote the rounding unit is a tribute to the people who created the *Handbook* package of linear algebra programs [349]. It stands for "macheps" (machine epsilon), which is what the rounding unit was called in their programs. Other authors use the letter u, standing for "unit in the last place." We avoid this nomenclature because u is too useful a letter to reserve for a special purpose.

The following code is sometimes used to approximate the rounding unit.

1. $ru = 1$
2.   **while** $(1+ru \neq 1)$
3. $ru = ru/2$
4.   **end while**

The rationale is that any number just a little smaller than the rounding unit will have no effect when it is added to one. However, this code may be defeated by compilers that assign the computation to registers with extended precision (or worse yet "optimize" the test $1+ru \neq 1$ to $ru = 0$).

Nonstandard floating-point arithmetic

We have called a floating-point arithmetic that satisfies (4.6) a standard system. Unfortunately, there are nonstandard systems around which can produce sums with low relative accuracy. To see how this comes about, consider the computation of the difference $1 - 0.999999$ in six-digit decimal arithmetic. The first step is to align the operands thus:

$$\begin{array}{r} 1.000000 \\ -\ 0.999999 \\ \hline \end{array}$$

If the computation is done in *seven* digits, the computer can go on to calculate

$$\begin{array}{r} 1.000000 \\ -\ 0.999999 \\ \hline 0.000001 \end{array}$$

SEC. 4. ROUNDING ERROR

and normalize the result to the correct answer: $.100000 \cdot 10^{-6}$. However, if the computer carries *six* digits, the trailing 9 will be lost during the alignment. The resulting computation will proceed as follows

$$\begin{array}{r} 1.00000 \\ -\ 0.99999 \\ \hline 0.00001 \end{array}$$

giving a normalized answer of $.100000 \cdot 10^{-5}$. In this case, the computed answer has a relative error of ten!

The high relative error is due to the absence of a *guard digit* to preserve essential information in the course of the computation. Although the absence of a guard digit does not affect the grosser aspects of matrix calculations, it makes certain fine maneuvers difficult or impossible. At one time algorithmists had no choice but to work around computers without guard digits: there were too many of them to ignore. But as the number of such computers has declined, people have become less tolerant of those that remain, and the present consensus is that anything you can do with a standard floating-point arithmetic is legitimate.

Backward rounding-error analysis

Although von Neumann and Goldstine [331, 1947] and Turing [321, 1948] came close to performing backward rounding-error analyses, Givens [145, 1954] was the first to explicitly throw errors back on the original data. This technique was exploited by Wilkinson in the works cited above. For more see his survey "Modern error analysis" [348].

The "fl" notation is due to Wilkinson, as is the trick of simplifying error bounds by slightly increasing the rounding unit [see (4.12)].

Stability

Stability is an overworked word in numerical analysis. As used by lay people it usually means something imprecise like, "This algorithm doesn't bite." The professionals, on the other hand, have given it a number of precise but inconsistent meanings. The stability of a method for solving ordinary differential equations is very different from the stability of an iterative method for solving linear equations.

It is not clear just when the term stability in dense matrix computations acquired its present meaning of backward stability. The word does not appear in the index of *Rounding Errors in Algebraic Processes* [345, 1963] or of *The Algebraic Eigenvalue Problem* [346, 1965]. Yet by 1971 Wilkinson [348] was using it in the current sense.

The meaning of backward stability can vary according to the measure of nearness used to define it. The stability of a computed sum might be called relative, componentwise, backward stability because small relative errors are thrown back on the individual components of the input. Many classical results on stability are cast in terms of norms, which tend to smear out the error across components. For more see §3, Chapter 3.

The term "weak stability" was first published by Bunch [52, 1987], although the phrase had been floating around for some time. The first significant example of a weak stability result was Björck's analysis of the modified Gram–Schmidt algorithm for solving least squares problems [37]. It should be noted that a weakly stable algorithm may be satisfactory for a single application but fail when it is iterated [307].

Condition numbers

The notion of a condition number for matrices was introduced by Turing [321, 1948]. In his own words:

> When we come to make estimates of errors in matrix processes we shall find that the chief factor limiting the accuracy that can be obtained is "ill-conditioning" of the matrices involved. The expression "ill-conditioned" is sometimes used merely as a term of abuse applicable to matrices or equations, but seems most often to carry a meaning somewhat similar to that defined below.

He goes on to support his definition with an explicit bound.

It should not be supposed that all perturbation problems can be summarized in a single number. For example, the behavior of the solutions of linear systems under perturbations of the right-hand side is best described by two interacting numbers [see (3.9), Chapter 3].

Cancellation

Since cancellation often accompanies numerical disasters, it is tempting to conclude that a cancellation-free calculation is essentially error free. See [177, Ch. 1] for counterexamples.

To most people catastrophic cancellation means the cancellation of a large number of digits. Goldberg [147] defines it to be the cancellation of numbers that have errors in them, implying that cancellation of a single bit is catastrophic unless the operands are known exactly.

Exponent exceptions

If the scale factor s in the algorithm *Euclid* is replaced by $\max\{|a|, |b|\}$, the results may be less accurate on a hexadecimal machine. The reason is that the number

$$\sqrt{(a/s)^2 + (b/s)^2}$$

is a little bit greater than one so that the leading three bits in its representation are zero. I discovered this fact after two days of trying to figure out why an algorithm I had coded consistently returned answers about a decimal digit less accurate than the algorithm it was meant to replace. Such are the minutiae of computer arithmetic.

Sec. 4. Rounding Error

To increase the range of numbers that can be represented on a computer, Clenshaw and Olver [67] have proposed a computer arithmetic that in effect adaptively trades precision for range. The proposers argue that their arithmetic will eliminate bothersome, ad hoc scaling of the kind found in Algorithm 4.1. Demmel [91] argues that to avoid loss of precision new precautions will be needed that are equally ad hoc and just as bothersome. However, the main objection to the system is that the proposers have not shown how to implement it efficiently.

3

Gaussian Elimination

During a convivial dinner at the home of Iain Duff, the conversation turned to the following question. Suppose you know that an imminent catastrophe will destroy all the books in the world — except for one which you get to choose. What is your choice and why? There are, of course, as many answers to that question as there are people, and so we had a lively evening.

A similar question can be asked about matrix computations. Suppose that all matrix algorithms save one were to disappear. Which would you choose to survive? Now algorithms are not books, and I imagine a group of experts would quickly agree that they could not do without the ability to solve linear equations. Their algorithm of choice would naturally be Gaussian elimination — the most versatile of all matrix algorithms. Gaussian elimination is an algorithm that computes a *matrix decomposition* — in this case the factorization of a matrix A into the product LU of a lower triangular matrix L and an upper triangular matrix U. The value of having a matrix decomposition is that it can often be put to more than one use. For example, the LU decomposition can be used as follows to solve the linear system $Ax = b$. If we write the system in the form $LUx = b$, then $Ux = L^{-1}b$, and we can generate x by the following algorithm.

1. Solve the system $Ly = b$
2. Solve the system $Ux = y$

But the decomposition can also be used to solve the system $A^{T}x = b$ as follows.

1. Solve the system $U^{T}y = b$
2. Solve the system $L^{T}x = y$

It is this adaptability that makes the decompositional approach the keystone of dense matrix computations.

This chapter consists of four sections. The first is devoted to the ins and outs of Gaussian elimination when it is applied to a general matrix. However, a major virtue of Gaussian elimination is its ability to adapt itself to special matrix structures. For this reason the second section treats Gaussian elimination applied to a variety of matrices.

The third section treats the perturbation theory of linear systems. Although this section is logically independent of Gaussian elimination, it leads naturally into the fourth section, which discusses the effects of rounding error on the algorithm.

For the most part, Gaussian elimination is used with real, square matrices. Since the extension of the theory and algorithms to complex or rectangular matrices is trivial, we will make the following expository simplification.

Throughout this chapter A will be a real matrix of order n.

1. GAUSSIAN ELIMINATION

This section is concerned with Gaussian elimination for dense matrices that have no special structure. Although the basic algorithm is simple, it can be derived and implemented in many ways, each representing a different aspect of the algorithm. In the first subsection we consider the basic algorithm in four forms, each of which has its own computational consequences. The next subsection is devoted to a detailed algebraic analysis of the algorithm, an analysis which leads naturally to the topics of block elimination and Schur complements.

The basic algorithm can fail with a division by zero, and in §1.3 we show how to use row and column interchanges to remove the difficulty — a device called *pivoting*. In §1.4 we present a number of common variants of Gaussian elimination. Finally in §1.5 we show how to apply the results of Gaussian elimination to the solution of linear systems and the inversion of matrices.

1.1. FOUR FACES OF GAUSSIAN ELIMINATION

There are four closely related ways to approach Gaussian elimination:

1. Gaussian elimination as the elimination of variables in a linear system,
2. Gaussian elimination as row operations on a linear system,
3. Gaussian elimination as a transformation of a matrix to triangular form by elementary lower triangular matrices,
4. Gaussian elimination as the factorization of a matrix into the product of lower and upper triangular factors.

Although each approaches yields the same algorithm, each has its own advantages in specific applications. Moreover, the approaches can be generalized to other matrix decompositions. In fact, a large part of dense matrix computations consists of variations on the themes introduced in this subsection.

Gauss's elimination

Gauss originally derived his algorithm as a sequential elimination of variables in a quadratic form. Here we eliminate variables in a system of linear equations, but the process has the flavor of Gauss's original derivation.

SEC. 1. GAUSSIAN ELIMINATION

Let us write out the system $Ax = b$ for $n = 4$:

$$\begin{aligned}\alpha_{11}x_1 + \alpha_{12}x_2 + \alpha_{13}x_3 + \alpha_{14}x_4 &= b_1 \\ \alpha_{21}x_1 + \alpha_{22}x_2 + \alpha_{23}x_3 + \alpha_{24}x_4 &= b_2 \\ \alpha_{31}x_1 + \alpha_{32}x_2 + \alpha_{33}x_3 + \alpha_{34}x_4 &= b_3 \\ \alpha_{41}x_1 + \alpha_{42}x_2 + \alpha_{43}x_3 + \alpha_{44}x_4 &= b_4\end{aligned} \quad (1.1)$$

If we solve the first equation in this system for x_1, i.e.,

$$x_1 = \alpha_{11}^{-1}(b_1 - \alpha_{12}x_2 - \alpha_{13}x_3 - \alpha_{14}b_4),$$

and substitute this value into the last three equations, we obtain the system

$$\begin{aligned}\alpha_{11}x_1 + \alpha_{12}x_2 + \alpha_{13}x_3 + \alpha_{14}x_4 &= b_1 \\ \alpha'_{22}x_2 + \alpha'_{23}x_3 + \alpha'_{24}x_4 &= b'_2 \\ \alpha'_{32}x_2 + \alpha'_{33}x_3 + \alpha'_{34}x_4 &= b'_3 \\ \alpha'_{42}x_2 + \alpha'_{43}x_3 + \alpha'_{44}x_4 &= b'_4\end{aligned} \quad (1.2)$$

where

$$\alpha'_{ij} = \alpha_{ij} - \alpha_{i1}\alpha_{1j}/\alpha_{11} \quad \text{and} \quad b'_i = b_i - \alpha_{i1}b_1/\alpha_{11}.$$

We may repeat the process, solving the second equation in (1.2) for x_2 and substituting the results in the last two equations. The result is the system

$$\begin{aligned}\alpha_{11}x_1 + \alpha_{12}x_2 + \alpha_{13}x_3 + \alpha_{14}x_4 &= b_1 \\ \alpha'_{22}x_2 + \alpha'_{23}x_3 + \alpha'_{24}x_4 &= b'_2 \\ \alpha''_{33}x_3 + \alpha''_{34}x_4 &= b''_3 \\ \alpha''_{43}x_3 + \alpha''_{44}x_4 &= b''_4\end{aligned} \quad (1.3)$$

where

$$\alpha''_{ij} = \alpha'_{ij} - \alpha_{i2}\alpha'_{2j}/\alpha'_{22} \quad \text{and} \quad b''_i = b'_i - \alpha'_{i2}b'_2/\alpha'_{22}.$$

Finally, if we solve the third equation of the system (1.3) for x_3 and substitute it into the fourth, we obtain the system

$$\begin{aligned}\alpha_{11}x_1 + \alpha_{12}x_2 + \alpha_{13}x_3 + \alpha_{14}x_4 &= b_1 \\ \alpha'_{22}x_2 + \alpha'_{23}x_3 + \alpha'_{24}x_4 &= b'_2 \\ \alpha''_{33}x_3 + \alpha''_{34}x_4 &= b''_3 \\ \alpha'''_{44}x_4 &= b'''_4\end{aligned}$$

where

$$\alpha'''_{ij} = \alpha''_{ij} - \alpha_{i3}\alpha''_{3j}/\alpha''_{33} \quad \text{and} \quad b'''_i = b''_i - \alpha_{i3}b''_3/\alpha''_{33}.$$

This last system is upper triangular and can be solved for the x_i by any of the techniques described in §2, Chapter 2.

This process generalizes to systems of order n. The reduction to triangular form is called *Gaussian elimination*. The solution of the resulting triangular system is called *back substitution* because most texts (along with Gauss) recommend the usual back substitution algorithm.

It is instructive to cast the reduction in terms of matrices. Write the equation $Ax = b$ in the form

$$\begin{pmatrix} \alpha_{11} & a_{12}^T \\ a_{21} & A_{22} \end{pmatrix} \begin{pmatrix} \xi_1 \\ x_2 \end{pmatrix} = \begin{pmatrix} \beta_1 \\ b_2 \end{pmatrix}.$$

Then it is easily verified that after one step of Gaussian elimination the system has the form

$$\begin{pmatrix} \alpha_{11} & a_{12}^T \\ 0 & A_{22} - \alpha_{11}^{-1} a_{21} a_{12}^T \end{pmatrix} \begin{pmatrix} \xi_1 \\ x_2 \end{pmatrix} = \begin{pmatrix} \beta_1 \\ b_2 - \alpha_{11}^{-1} \beta_1 a_{21} \end{pmatrix}. \tag{1.4}$$

The matrix $A_{22} - \alpha_{11}^{-1} a_{21} a_{12}^T$ is called the *Schur complement of α_{11} in A*. Whenever a variable is eliminated from a system of equations, there is a Schur complement in the background. Later in this subsection we will show how the elimination of several variables results in a more general definition of the Schur complement (Definition 1.1).

Gaussian elimination and elementary row operations

Gaussian elimination can also be regarded as a reduction to triangular form by row operations. Specifically, in the system (1.1) let

$$\ell_{i1} = \frac{\alpha_{i1}}{\alpha_{11}}, \qquad i = 2, \ldots, 4. \tag{1.5}$$

(The numbers ℓ_{ij} will be called *multipliers*.) If for $i = 2, 3, 4$ we subtract ℓ_{i1} times the first row of the system from the ith row of the system, the result is the system (1.2). Another way of looking at this process is that the multipliers ℓ_{1i} are calculated in such a way that when ℓ_{1i} times the first row is subtracted from the ith row the coefficient of x_1 vanishes.

To continue the process, we subtract multiples of the second row of the reduced system (1.2) from the remaining rows to make the coefficient of x_2 zero. And so on.

This is an extremely productive way of viewing Gaussian elimination. If, for example, a coefficient of x_1 is already zero, there is no need to perform the corresponding row operation. This fact allows us to derive efficient algorithms for matrices with many zero elements. We will return to this view of Gaussian elimination in §2, where we consider tridiagonal, Hessenberg, and band matrices.

SEC. 1. GAUSSIAN ELIMINATION

Gaussian elimination as a transformation to triangular form

At this point it will be convenient to drop the right-hand side of our system and focus on what happens to the matrix A. Let

$$L_1 = \begin{pmatrix} 1 & 0 & 0 & 0 \\ -\ell_{21} & 1 & 0 & 0 \\ -\ell_{31} & 0 & 1 & 0 \\ -\ell_{41} & 0 & 0 & 1 \end{pmatrix}, \qquad (1.6)$$

where the numbers ℓ_{i1} are the multipliers defined by (1.5). Then it is easily verified that

$$L_1 A = \begin{pmatrix} \alpha_{11} & \alpha_{12} & \alpha_{13} & \alpha_{14} \\ 0 & \alpha'_{22} & \alpha'_{23} & \alpha'_{24} \\ 0 & \alpha'_{32} & \alpha'_{33} & \alpha'_{34} \\ 0 & \alpha'_{42} & \alpha'_{43} & \alpha'_{44} \end{pmatrix}.$$

Thus the matrix $L_1 A$ is just the matrix of the system (1.2) obtained after one step of Gaussian elimination.

The process can be continued by setting

$$L_2 = \begin{pmatrix} 1 & 0 & 0 & 0 \\ 0 & 1 & 0 & 0 \\ 0 & -\ell_{32} & 1 & 0 \\ 0 & -\ell_{42} & 0 & 1 \end{pmatrix},$$

where

$$\ell_{i2} = \frac{\alpha'_{i2}}{\alpha'_{22}}, \qquad i = 3, 4.$$

It follows that

$$L_2 L_1 A = \begin{pmatrix} \alpha_{11} & \alpha_{12} & \alpha_{13} & \alpha_{14} \\ 0 & \alpha'_{22} & \alpha'_{23} & \alpha'_{24} \\ 0 & 0 & \alpha''_{33} & \alpha''_{34} \\ 0 & 0 & \alpha''_{43} & \alpha''_{44} \end{pmatrix},$$

which corresponds to one more step of Gaussian elimination.

To finish the elimination (for $n = 4$) set

$$L_3 = \begin{pmatrix} 1 & 0 & 0 & 0 \\ 0 & 1 & 0 & 0 \\ 0 & 0 & 1 & 0 \\ 0 & 0 & -\ell_{43} & 1 \end{pmatrix},$$

where
$$\ell_{43} = \frac{\alpha''_{43}}{\alpha''_{33}}.$$
Then
$$L_3 L_2 L_1 A = \begin{pmatrix} \alpha_{11} & \alpha_{12} & \alpha_{13} & \alpha_{14} \\ 0 & \alpha'_{22} & \alpha'_{23} & \alpha'_{24} \\ 0 & 0 & \alpha''_{33} & \alpha''_{34} \\ 0 & 0 & 0 & \alpha'''_{44} \end{pmatrix} \qquad (1.7)$$
is the final triangular matrix produced by Gaussian elimination.

The matrices L_k are called *elementary lower triangular matrices*. They introduce zeros into successive columns of A in such a way that the final matrix is upper triangular. Thus Gaussian elimination is a reduction to triangular form by elementary lower triangular matrices.

This view of Gaussian elimination generalizes in two ways. In the first place, the elementary lower triangular matrices can be replaced by any class of transformations capable of introducing zeros in the appropriate places. (Of course the transformations should be simple enough to make their storage and manipulation economical.) For example, an alternative to elementary lower triangular matrices is the class of Householder transformations to be introduced in Chapter 4.

In the second place, the goal does not have to be a triangular matrix. Elementary lower triangular matrices, and especially Householder transformations, are used to reduce matrices to a variety of forms.

Gaussian elimination and the LU decomposition

If we denote the right-hand side of (1.7) by U, then
$$A = L_1^{-1} L_2^{-1} L_3^{-1} U.$$
Now the product $L = L_1^{-1} L_2^{-1} L_3^{-1}$ is easily seen to be
$$L = \begin{pmatrix} 1 & 0 & 0 & 0 \\ \ell_{21} & 1 & 0 & 0 \\ \ell_{31} & \ell_{32} & 1 & 0 \\ \ell_{41} & \ell_{42} & \ell_{43} & 1 \end{pmatrix}.$$

Hence:

> Gaussian elimination computes the LU DECOMPOSITION
>
> $$A = LU$$
>
> of A into the product of a unit lower triangular matrix and an upper triangular matrix. The elements of the upper triangular matrix are the coefficients of the final upper triangular system. The elements of L are the multipliers ℓ_{ij}. The factorization is also called the LU FACTORIZATION.

(1.8)

SEC. 1. GAUSSIAN ELIMINATION

We have already met the LU decomposition in Theorem 2.13, Chapter 1. The proof of that theorem amounts to a rederivation of Gaussian elimination from a partition of the decomposition. In brief, if A has an LU decomposition, we may partition it in the form

$$\begin{pmatrix} \alpha_{11} & a_{12}^{\mathrm{T}} \\ a_{21} & A_{22} \end{pmatrix} = \begin{pmatrix} 1 & 0 \\ \ell_{21} & L_{22} \end{pmatrix} \begin{pmatrix} \upsilon_{11} & u_{12}^{\mathrm{T}} \\ 0 & U_{22}^{\mathrm{T}} \end{pmatrix}.$$

Equivalently,

1. $\alpha_{11} = \upsilon_{11}$,
2. $a_{12}^{\mathrm{T}} = u_{12}^{\mathrm{T}}$,
3. $a_{21} = \upsilon_{11} \ell_{21}$,
4. $A_{22} = LU + \ell_{21} u_{12}^{\mathrm{T}}$.

The first two equations say that the first row of U is the same as the first row of A. The third equation, written in the form $\ell_{21} = \alpha_{11}^{-1} a_{21}$, says that the first column of L consists of the first set of multipliers. Finally, the third equation, written in the form

$$L_{22} U_{22} = A_{22} - \alpha_{11}^{-1} a_{21} a_{12}^{\mathrm{T}},$$

says that the product $L_{22} U_{22}$ is the Schur complement of α_{11}. In other words, to compute the LU decomposition of A:

1. set the first row of U equal to the first row of A
2. compute the multipliers ℓ_{i1} and store them in the first column of L (1.9)
3. apply this process recursively the Schur complement of α_{11}

The technique of partitioning a decomposition to get an algorithm is widely applicable — for example, it can be used to derive the Gram–Schmidt algorithm for orthogonalizing the columns of a matrix (§1.4, Chapter 4). By varying the partitioning one can obtain variants of the algorithm in question, and we will exploit this fact extensively in §1.4.

1.2. CLASSICAL GAUSSIAN ELIMINATION

For a general dense matrix all the above approaches to Gaussian elimination yield essentially the same algorithm — an algorithm that we will call *classical Gaussian elimination*. The purpose of this subsection is to probe more deeply into the properties of this algorithm. In particular, we will be concerned with when the algorithm can be carried to completion and what it computes along the way.

The algorithm

In presenting classical Gaussian elimination, we will regard it as a method for computing the LU decomposition of A [see (1.8)]. In most implementations of the algorithm,

$$
\begin{pmatrix} a & a & a & a & a \\ a & a & a & a & a \\ a & a & a & a & a \\ a & a & a & a & a \\ a & a & a & a & a \end{pmatrix} \xRightarrow{1} \begin{pmatrix} u & u & u & u & u \\ \ell & a & a & a & a \\ \ell & a & a & a & a \\ \ell & a & a & a & a \\ \ell & a & a & a & a \end{pmatrix} \xRightarrow{2} \begin{pmatrix} u & u & u & u & u \\ \ell & u & u & u & u \\ \ell & \ell & a & a & a \\ \ell & \ell & a & a & a \\ \ell & \ell & a & a & a \end{pmatrix} \xRightarrow{3}
$$

$$
\xRightarrow{3} \begin{pmatrix} u & u & u & u & u \\ \ell & u & u & u & u \\ \ell & \ell & u & u & u \\ \ell & \ell & \ell & a & a \\ \ell & \ell & \ell & a & a \end{pmatrix} \xRightarrow{4} \begin{pmatrix} u & u & u & u & u \\ \ell & u & u & u & u \\ \ell & \ell & u & u & u \\ \ell & \ell & \ell & u & u \\ \ell & \ell & \ell & \ell & u \end{pmatrix}
$$

Figure 1.1: The course of Gaussian elimination

the L- and U-factors overwrite the matrix A in its array. Specifically, at the first step the first row of U, which is identical with the first row of A, is already in place. The elements of $\ell_{21} = \alpha_{11}^{-1} a_{21}$ can overwrite a_{21}. (We do note need to store the element ℓ_{11}, since it is known to be one.) Symbolically, we can represent the first step as follows:

$$
\begin{pmatrix} a & a & a & a & a \\ a & a & a & a & a \\ a & a & a & a & a \\ a & a & a & a & a \\ a & a & a & a & a \end{pmatrix} \xRightarrow{1} \begin{pmatrix} u & u & u & u & u \\ \ell & a & a & a & a \\ \ell & a & a & a & a \\ \ell & a & a & a & a \\ \ell & a & a & a & a \end{pmatrix}.
$$

The procedure is then repeated on the $(n-1)\times(n-1)$ trailing submatrix — and so on. Figure 1.1 charts the course of the algorithm. The ℓ's and u's fill up the array as the a's are processed. Note how the last transformation accounts for the penultimate and ultimate row of u, all in one fell swoop. The implementation in Algorithm 1.1 is simplicity itself.

An operation count for the algorithm is easily derived. At the kth stage, the work is concentrated in statement 4, which performs $(n-k)^2$ flam. Thus the total number of flams is approximately

$$\int_0^n (n-k)^2 \, dk = \tfrac{1}{3} n^3.$$

Hence:

The operation count for Algorithm 1.1 is

$$\tfrac{1}{3} n^3 \text{ flam.}$$

(1.10)

SEC. 1. GAUSSIAN ELIMINATION

The following algorithm overwrites an $n \times n$ matrix A with its LU decomposition. The elements ℓ_{ij} ($i > j$) and u_{ij} ($i \leq j$) overwrite the corresponding elements of A. The algorithm gives an error return if a zero diagonal element is encountered.

1. **for** $k = 1$ **to** $n-1$
2.   **if** $(A[k,k] = 0)$ error **fi**
3. $A[k+1{:}n, k] = A[k+1{:}n, k]/A[k,k]$
4. $A[k+1{:}n, k+1{:}n] = A[k+1{:}n, k+1{:}n]$
      $- A[k+1{:}n, k] * A[k, k+1{:}n]$
5. **end for** k

Algorithm 1.1: Classical Gaussian elimination

───────◊───────

Analysis of classical Gaussian elimination

The diagonal elements of the array A in Algorithm 1.1 are called *pivot elements*. For the algorithm to complete the kth stage, the kth pivot element must be nonzero at the time it is used as a divisor. Obviously we cannot just look at A and determine if the algorithm will go to completion, since the diagonals of the array A change in the course of the elimination. Here we will establish conditions under which the pivot elements are nonzero. We will also find out what Gaussian elimination computes when it is terminated prematurely.

Unfortunately, the classical variant of the algorithm is not easy to analyze. The reason is that it is doing three things at once — computing L by columns, computing U by rows, and updating a Schur complement. The key to a smooth analysis is to begin with Schur complements.

The first thing to do is to generalize the notion of Schur complement, first introduced in connection with (1.4).

Definition 1.1. *Let A be partitioned in the form*

$$A = \begin{pmatrix} A_{11} & A_{12} \\ A_{21} & A_{22} \end{pmatrix},$$

and suppose that A_{11} is nonsingular. Then the SCHUR COMPLEMENT OF A_{11} IN A *is the matrix*

$$S = A_{22} - A_{21} A_{11}^{-1} A_{12}.$$

The Schur complement plays an important role in matrix computations and is well worth studying in its own right. But to keep things focused, we will only establish a single result that we need here. We will return to Schur complements later in this subsection.

Theorem 1.2. *Let*

$$A = \begin{pmatrix} A_{11} & A_{12} \\ A_{21} & A_{22} \end{pmatrix},$$

where A_{11} is nonsingular, and let S be the Schur complement of A_{11} in A. Then A is nonsingular if and only if S is nonsingular.

Proof. It is easily verified that A has a *block LU decomposition* of the form

$$\begin{pmatrix} A_{11} & A_{12} \\ A_{21} & A_{22} \end{pmatrix} = \begin{pmatrix} I & 0 \\ A_{21}A_{11}^{-1} & I \end{pmatrix} \begin{pmatrix} A_{11} & A_{12} \\ 0 & S \end{pmatrix}.$$

The first factor on the right-hand side is nonsingular, because it is lower triangular with ones on its diagonal. Hence A is nonsingular if and only if the second factor is nonsingular. This factor is block upper triangular and is therefore nonsingular if and only if its diagonal blocks are nonsingular. But by hypothesis the diagonal block A_{11} is nonsingular. Hence, A is nonsingular if and only if the second diagonal block S is nonsingular. ∎

We are now in a position to establish conditions under which Gaussian elimination goes to completion.

Theorem 1.3. *A necessary and sufficient condition for Algorithm 1.1 to go to completion is that the leading principal submatrices of A of order $1, \ldots, n-1$ be nonsingular.*

Proof. The proof of sufficiency is by induction. For $n = 1$ the algorithm does nothing, which amounts to setting $L = 1$ and $U = \alpha_{11}$.

Now assume the assertion is true of all matrices of order $n-1$. Let A be partitioned in the form

$$A = \begin{pmatrix} \alpha_{11} & a_{12}^T \\ a_{21} & A_{22} \end{pmatrix}.$$

Since $\alpha_{11} \neq 0$, the first step of Algorithm 1.1 can be performed. After this step the matrix assumes the form

$$\begin{pmatrix} \alpha_{11} & a_{12}^T \\ 0 & S \end{pmatrix},$$

where S is the Schur complement of α_{11}. Since the remaining steps of the algorithm are performed on S, by the induction hypothesis it is sufficient to prove that the leading principal submatrices of S of order $1, \ldots, n-2$ are nonsingular.

Let $A^{[k]}$ and $S^{[k]}$ denote the leading principal submatrices of A and S order k. Then $S^{[k]}$ is the Schur complement of α_{11} in $A^{[k+1]}$. (To see this, consider Algorithm 1.1 restricted to $A^{[k+1]}$.) By hypothesis $A^{[k+1]}$ is nonsingular for $k = 1, \ldots, n-2$. Hence by Theorem 1.2, $S^{[k]}$ is nonsingular for $k = 1, \ldots, n-2$. This completes the proof of sufficiency.

SEC. 1. GAUSSIAN ELIMINATION

For the necessity of the conditions, assume that the algorithm goes to completion. Then the result is an LU decomposition $A = LU$ of A. Moreover, the pivots for the elimination are the diagonal elements of v_{kk} of U. Since these elements must be nonzero for $k = 1, \ldots, n-1$, the leading principal submatrices $U^{[k]}$ are nonsingular for $k = 1, \ldots, n-1$. Since L has ones on its diagonals, the matrices $L^{[k]}$ are also nonsingular. Now by the triangularity of L and U,

$$A^{[k]} = L^{[k]} U^{[k]}.$$

Hence $A^{[k]}$ is nonsingular for $k = 1, \ldots, n-1$. ∎

There are three comments to be made about this theorem.

- Since Gaussian elimination computes the LU decomposition of A, the theorem provides sufficient conditions for the existence of the LU decomposition.

- It might seem that we have traded one unverifiable condition (the existence of nonzero pivots) for another (the nonsingularity of leading principal minors). However, in many commonly occurring cases the structure of the matrix guarantees that its leading principal submatrices are nonsingular. Such is true of positive definite matrices, diagonally dominant matrices, M-matrices, and totally positive matrices (see §4.3).

- We have noted briefly (p. 148) that Gaussian elimination can be applied to a rectangular matrix A — say $A \in \mathbb{R}^{m \times n}$. If $m \leq n$, then theorem remains true with n replaced by m. However, if $m > n$, we must require that all the leading principal submatrices of A be nonsingular. The reason is that after the $(n-1)$th step, the last $m-n$ components of L remain to be computed, a computation that requires division by v_{nn}.

- From the proof of the theorem it is clear that if the first $k < n$ principal submatrices of A are nonsingular the algorithm will at least complete the kth step.

—

The last comment raises an important question. Suppose we stop classical Gaussian elimination after the completion of the kth step. What have we computed?

The easiest way to answer this question is to focus on an individual element and ask what computations are performed on it. [For another example of this approach see (2.8), Chapter 2]. Figure 1.1 shows that the elements α_{ij} are divided into two classes: an upper class of α_{ij}'s ($i \leq j$) destined to become u's and a lower class of α_{ij}'s ($i > j$) destined to become ℓ's. The following programs show what happens to the members of each class lying in the kth row or column.

Upper Class ($j \geq k$)

U1.   **for** $i = 1$ **to** $k-1$
U2. $\alpha_{kj} = \alpha_{kj} - \ell_{ki} v_{ij}$
U3.   **end for** i
U4. $v_{kj} = \alpha_{kj}$

Lower Class ($i > k$)

L1.   **for** $j = 1$ **to** $k-1$
L2. $\alpha_{ik} = \alpha_{ik} - \ell_{ij} v_{jk}$
L3.   **end for** i
L4. $\ell_{ik} = \ell_{ik} / v_{kk}$

The computations in these little programs can be summarized in terms of matrices. Partition the first k columns of L and the first k rows of U in the form

$$\begin{pmatrix} L_{11} \\ L_{21} \end{pmatrix} \quad \text{and} \quad (U_{11} \ U_{12}). \tag{1.11}$$

Then

$$\begin{pmatrix} A_{11} & A_{12} \\ A_{21} & A_{22} \end{pmatrix} - \begin{pmatrix} L_{11} \\ L_{21} \end{pmatrix} (U_{11} \ U_{12}) = \begin{pmatrix} 0 & 0 \\ 0 & S \end{pmatrix}, \tag{1.12}$$

where the components of S are the partially processed elements of A (the a's in Figure 1.1). The correctness of this equation may be most easily seen by verifying it for the case $k = 3$.

Equation (1.12) shows that we have computed a partial LU decomposition that reproduces A in its first k rows and columns. It remains to determine what the matrix S is. To do this, multiply out (1.12) to get

1. $A_{11} = L_{11}U_{11}$,
2. $A_{12} = L_{11}U_{12}$,
3. $A_{21} = L_{21}U_{11}$,
4. $A_{22} = L_{22}U_{22} + S$.

These equations imply that $S = A_{22} - A_{21}A_{11}^{-1}A_{12}$. Thus S is the Schur complement of A_{11} in A.

These results are summarized in the following theorem.

Theorem 1.4. *Let Algorithm 1.1 be stopped at the end of the kth step, and let the contents of the first k rows and columns of A be partitioned as in (1.11). If A is partitioned conformally, then*

$$\begin{pmatrix} A_{11} & A_{12} \\ A_{21} & A_{22} \end{pmatrix} - \begin{pmatrix} L_{11} \\ L_{21} \end{pmatrix} (U_{11} \ U_{12}) = \begin{pmatrix} 0 & 0 \\ 0 & S \end{pmatrix},$$

where

$$S = A_{22} - A_{21}A_{11}^{-1}A_{12} \tag{1.13}$$

is the Schur complement of A_{22} in A.

This theorem shows that there are two ways of computing the Schur complement S. The first is to compute a nested sequence of Schur complements, as in classical Gaussian elimination. The second is to compute it in one step via (1.13). This fact is also a consequence of a general theorem on nested Schur complements (Theorem 1.7), which we will prove later in this subsection. But first some observations on LU decompositions are in order.

SEC. 1. GAUSSIAN ELIMINATION

LU decompositions

In Theorem 1.3 we gave conditions under which such an LU decomposition exists. We now turn to the uniqueness of the decomposition. To simplify the exposition, we will assume that the matrix A is nonsingular.

In discussing uniqueness it is important to keep in mind that what we have been calling an LU decomposition is just one of a continuum of factorizations of A into the product of lower and upper triangular matrices. For if $A = LU$ and D is a nonsingular diagonal matrix, then

$$A = (LD)(D^{-1}U) \tag{1.14}$$

is another such factorization. In particular, if the diagonal elements of U are nonzero, we can take $D = \text{diag}(v_{11}, \ldots, v_{nn})$, in which case the upper triangular part of the factorization (1.14) is unit triangular. To avoid confusion among the many possible factorizations we will adopt the following convention.

> *Unless otherwise stated, LU decompositions will be normalized so that L is unit triangular.* (1.15)

The basic uniqueness result is contained in the following theorem.

Theorem 1.5. *If A is nonsingular and has an LU decomposition, then the decomposition is unique.*

Proof. Let $A = LU$ and $A = \hat{L}\hat{U}$ be LU decompositions of A. Then $LU = \hat{L}\hat{U}$. Since A is nonsingular, so are its factors. Hence

$$\hat{L}^{-1}L = \hat{U}U^{-1}.$$

Now the matrix on the right-hand side of this equation is lower triangular and the matrix on the left is upper triangular. Hence they are both diagonal. By the convention (1.15) L and $\hat{L}^{-1}$ are unit lower triangular. Hence

$$\hat{L}^{-1}L = \hat{U}U^{-1} = I,$$

and $\hat{L} = L$ and $\hat{U} = U$. ∎

It should be stressed that even when A is singular, it may still have a unique LU decomposition. For example, if the leading principal submatrices of A of orders up to $n-1$ are nonsingular, then by Theorem 1.3 A has an LU decomposition, and it is easy to show that it is unique.

In general, however, a nonsingular matrix may fail to have an LU decomposition. An example is the matrix

$$A = \begin{pmatrix} 0 & 1 \\ 1 & 0 \end{pmatrix}.$$

Moreover, if a singular matrix does have an LU decomposition, it need not be unique. For example, the factorization

$$A = \begin{pmatrix} 0 & 1 \\ 0 & 1 \end{pmatrix} = \begin{pmatrix} 1 & 0 \\ \lambda & 1 \end{pmatrix} \begin{pmatrix} 0 & 1 \\ 0 & 1-\lambda \end{pmatrix}$$

is an LU decomposition of A for any value of λ.

We can say more about the way in which LU decompositions are nonunique. However, in numerical applications one tends to avoid such situations, and we will not pursue the matter here.

Block elimination

To introduce block elimination, we consider the following system:

$$\begin{pmatrix} 0 & 1 & 1 \\ 1 & 0 & 1 \\ 1 & 1 & 1 \end{pmatrix} \begin{pmatrix} \xi_1 \\ \xi_2 \\ \xi_3 \end{pmatrix} = \begin{pmatrix} 2 \\ 2 \\ 3 \end{pmatrix},$$

whose solution is the vector of all ones. Because the first pivot is zero, we cannot solve the system by eliminating the variables in their natural order. Equivalently, classical Gaussian elimination fails on this system.

The usual cure for this problem is to eliminate variables in a different order. This amounts to interchanging rows and columns of the matrix to bring a nonzero element into the pivot position. We will treat this important technique — called *pivoting* — more thoroughly in §1.3.

An alternative is to eliminate more than one variable at a time. For example, if the first two equations are solved for ξ_1 and ξ_2 in terms of ξ_3 and the results are plugged into the third equation, we get

$$\begin{pmatrix} 0 & 1 & 1 \\ 1 & 0 & 1 \\ 0 & 0 & -1 \end{pmatrix} \begin{pmatrix} \xi_1 \\ \xi_2 \\ \xi_3 \end{pmatrix} = \begin{pmatrix} 2 \\ 2 \\ -1 \end{pmatrix}.$$

This system can be solve for $\xi_3 = 1$ and then simultaneously for ξ_1 and ξ_2. This process is an example of *block elimination*.

Because block elimination is important in many of applications, in this section we will give a brief sketch of its implementation and analysis. Fortunately, if the noncommutativity of matrix multiplication is taken into account, the results for classical Gaussian elimination carry over *mutatis mutandis*.

To fix our notation, let A be partitioned in the form

$$A = \begin{pmatrix} A_{11} & A_{12} & \cdots & A_{1m} \\ A_{21} & A_{22} & \cdots & A_{2m} \\ \vdots & \vdots & & \vdots \\ A_{m1} & A_{m2} & \cdots & A_{mm} \end{pmatrix}, \qquad (1.16)$$

SEC. 1. GAUSSIAN ELIMINATION

where the diagonal blocks A_{ii} are square of order n_i, so that

$$n_1 + n_2 + \cdots + n_m = n.$$

The easiest way to approach the algorithm is to derive it as a recursive computation of a block LU decomposition in the spirit of (1.9). Specifically, repartition A in the form

$$A = \begin{pmatrix} A_{11} & A_{1*} \\ A_{*1} & A_{**} \end{pmatrix}.$$

Then A has the block LU decomposition

$$A = \begin{pmatrix} I & 0 \\ A_{*1}A_{11}^{-1} & I \end{pmatrix} \begin{pmatrix} A_{11} & A_{1*} \\ 0 & A_{**} - A_{*1}A_{11}^{-1}A_{1*} \end{pmatrix}.$$

In other words:

1. The first (block) row of U in the block LU decomposition of A is the first (block) row of A.
2. The first (block) column of L, excluding a leading identity matrix of order n_1, is $A_{*1}A_{11}^{-1}$.
3. The rest of the block decomposition can be obtained by computing the block decomposition of the Schur complement $A_{**} - A_{*1}A_{11}^{-1}A_{1*}$.

Algorithm 1.2 is an implementation of this scheme. The code parallels the code for classical Gaussian elimination (Algorithm 1.1), the major difference being the use of the indices lx and ux to keep track of the lower and upper limits of the current blocks. Three comments.

- In the scalar case, the LU decomposition is normalized so that L is unit lower triangular. Here the diagonal blocks of L are identity matrices. This means that the scalar and block LU decompositions are not the same. However, the scalar decomposition, if it exists, can be recovered from the block decomposition in the form $(LD^{-1})(DU)$, where D is a diagonal matrix formed from the U-factors of the diagonal blocks of L.

- Although we have written the algorithm in terms of inverses, in general we would use some decomposition (e.g., a pivoted LU decomposition) to implement the computations. (See §1.5 for how this is done.)

- Surprisingly, the operation count for the algorithm does not depend on the size of the blocks, provided LU decompositions are used to implement the effects of inverses in the algorithm. In fact, the count is roughly $\frac{1}{3}n^3$, the same as for scalar Gaussian elimination.

It is important to distinguish between a block algorithm, in which the elements of a partition are treated as (noncommuting) scalars, and blocked algorithms that use

Let A be partitioned as in (1.16), where the diagonal blocks A_{ii} are of order $n[i]$. The following algorithm overwrites A with its block LU decomposition

$$A = \begin{pmatrix} I & 0 & \cdots & 0 \\ L_{21} & I & \cdots & 0 \\ \vdots & \vdots & & \vdots \\ L_{m1} & L_{m2} & \cdots & I \end{pmatrix} \begin{pmatrix} U_{11} & U_{12} & \cdots & U_{1m} \\ 0 & U_{22} & \cdots & U_{2m} \\ \vdots & \vdots & & \vdots \\ 0 & 0 & \cdots & U_{mm} \end{pmatrix}.$$

The algorithm gives an error return if a singular diagonal block is encountered.

1. $nn = n[1] + \cdots + n[m]$
2. $ux = 0$
3. **for** $k = 1$ **to** $m-1$
4. $lx = ux+1$
5. $ux = lx+n[k]-1$
6.   **if** ($A[lx{:}ux, lx{:}ux]$ is singular) error **fi**
7. $A[ux+1{:}nn, lx{:}ux] = A[ux+1{:}nn, lx{:}ux] * A[lx{:}ux, lx{:}ux]^{-1}$
8. $A[ux+1{:}nn, ux+1{:}nn] = A[ux+1{:}nn, ux+1{:}nn]$
           $- A[ux+1{:}nn, lx{:}ux] * A[lx{:}ux, ux+1{:}nn]$
9. **end for** k

Algorithm 1.2: Block Gaussian elimination

the blocking strategy illustrated in Figure 3.3, Chapter 2. For more see the notes and references.

Not only is the code for block elimination quite similar to the code for classical Gaussian elimination, but the natural generalizations of the analysis are valid. If for $k = 1, \ldots, m-1$ the leading principal submatrices of A of order $n_1 + \cdots + n_k$ are nonsingular, then the algorithm goes to completion. If the algorithm is stopped after

SEC. 1. GAUSSIAN ELIMINATION

step k, then

$$\begin{pmatrix} A_{11} & \cdots & A_{1k} & \cdots & A_{1m} \\ \vdots & & \vdots & & \vdots \\ A_{k1} & \cdots & A_{kk} & \cdots & A_{km} \\ \vdots & & \vdots & & \vdots \\ A_{m1} & \cdots & A_{mk} & \cdots & A_{mm} \end{pmatrix}$$
$$- \begin{pmatrix} I & \cdots & 0 \\ \vdots & & \vdots \\ L_{k1} & \cdots & I \\ \vdots & & \vdots \\ L_{m1} & \cdots & L_{mk} \end{pmatrix} \begin{pmatrix} U_{11} & \cdots & U_{1k} & \cdots & U_{1m} \\ \vdots & & \vdots & & \vdots \\ 0 & \cdots & U_{kk} & \cdots & U_{km} \end{pmatrix}$$
$$= \begin{pmatrix} 0 & 0 \\ 0 & S \end{pmatrix},$$

where

S is the Schur complement of $\begin{pmatrix} A_{11} & \cdots & A_{1k} \\ \vdots & & \vdots \\ A_{k1} & \cdots & A_{kk} \end{pmatrix}$ in A.

Schur complements

Throughout this section we have worked piecemeal with Schur complements. We will now give a systematic statement of their more important properties. We begin with what we can derive from a 2×2 block LU decomposition. Many of the following results have already been established, and we leave the rest as an exercise. Note that they do not depend on the choice of diagonal blocks in the block LU decomposition.

Theorem 1.6. *In the partitioning*

$$A = \begin{pmatrix} A_{11} & A_{12} \\ A_{21} & A_{22} \end{pmatrix}$$

suppose that A_{11} is nonsingular. Then A has a block LU decomposition

$$\begin{pmatrix} A_{11} & A_{12} \\ A_{21} & A_{22} \end{pmatrix} = \begin{pmatrix} L_{11} & 0 \\ L_{21} & L_{22} \end{pmatrix} \begin{pmatrix} U_{11} & U_{12} \\ 0 & U_{22} \end{pmatrix}, \quad (1.17)$$

where L_{11} and U_{11} are nonsingular. Moreover, for any such decomposition

1. $L_{11}U_{11} = A_{11}$,
2. $L_{21}U_{11} = A_{21}$,
3. $L_{11}U_{12} = A_{12}$,
4. $L_{22}U_{22} = A_{22} - A_{21}A_{11}^{-1}A_{12} = A_{22} - L_{21}U_{12}$,
5. $U_{11}^{-1}U_{12} = A_{11}^{-1}A_{12}$,
6. $L_{21}L_{11}^{-1} = A_{21}A_{11}^{-1}$.

(1.18)

If in addition A is nonsingular, then so are L_{22}, U_{22}, and hence the Schur complement $A_{22} - A_{21}A_{11}^{-1}A_{12} = L_{22}U_{22}$. If we partition

$$A^{-1} = \begin{pmatrix} A_{11}^{(-1)} & A_{12}^{(-1)} \\ A_{21}^{(-1)} & A_{22}^{(-1)} \end{pmatrix},$$

then

$$A_{22}^{(-1)} = U_{22}^{-1}L_{22}^{-1} = (A_{22} - A_{21}A_{11}^{-1}A_{12})^{-1}.$$

In commenting on Theorem 1.4, we noted that the "Schur complement" computed by a sequence of k steps of classical Gaussian elimination is the same as the Schur complement of the leading principal minor of order k — a fact that is not obvious. The following theorem can be used to show that any two sequences of scalar and block elimination that terminate at the same leading principal submatrix compute the same Schur complement.

Theorem 1.7. *Let A be partitioned in the form*

$$\begin{pmatrix} A_{11} & A_{12} & A_{13} \\ A_{21} & A_{22} & A_{23} \\ A_{31} & A_{32} & A_{33} \end{pmatrix},$$

and assume that

$$A_{11} \text{ and } \begin{pmatrix} A_{11} & A_{12} \\ A_{21} & A_{22} \end{pmatrix} \text{ are nonsingular.} \tag{1.19}$$

Let

$$S = \begin{pmatrix} S_{11} & S_{12} \\ S_{21} & S_{22} \end{pmatrix}$$

be the Schur complement of A_{11} in A. Then S_{11} is nonsingular, and the Schur complement of

$$\begin{pmatrix} A_{11} & A_{12} \\ A_{21} & A_{22} \end{pmatrix}$$

in A is equal to the Schur complement of S_{11} in S.

Sec. 1. Gaussian Elimination

Proof. Consider the block LU decomposition

$$\begin{pmatrix} A_{11} & A_{12} & A_{13} \\ A_{21} & A_{22} & A_{23} \\ A_{31} & A_{32} & A_{33} \end{pmatrix} = \begin{pmatrix} L_{11} & 0 & 0 \\ L_{21} & L_{22} & 0 \\ L_{31} & L_{32} & L_{33} \end{pmatrix} \begin{pmatrix} U_{11} & U_{12} & U_{13} \\ 0 & U_{22} & U_{23} \\ 0 & 0 & U_{33} \end{pmatrix},$$

which exists by virtue of (1.19). Now from the equation

$$\begin{pmatrix} A_{11} & A_{12} \\ A_{21} & A_{22} \end{pmatrix} = \begin{pmatrix} L_{11} & 0 \\ L_{21} & L_{22} \end{pmatrix} \begin{pmatrix} U_{11} & U_{12} \\ 0 & U_{22} \end{pmatrix}$$

and (1.19), $L_{22}U_{22}$ are nonsingular. But by (1.18.4),

$$S = \begin{pmatrix} S_{11} & S_{12} \\ S_{21} & S_{22} \end{pmatrix} = \begin{pmatrix} L_{22} & 0 \\ L_{32} & L_{33} \end{pmatrix} \begin{pmatrix} U_{22} & U_{23} \\ 0 & U_{33} \end{pmatrix}.$$

Hence $S_{11} = L_{22}U_{22}$ is nonsingular, and the Schur complement of S_{11} in S is by (1.18.4) $L_{33}U_{33}$. But once again by (1.18.4), the Schur complement of

$$\begin{pmatrix} A_{11} & A_{12} \\ A_{21} & A_{22} \end{pmatrix}$$

in A is also equal to $L_{33}U_{33}$. ∎

1.3. Pivoting

Algorithm 1.1 fails if the element $A[k, k]$ in statement 3 is zero. A cure for this problem is to perform row and column interchanges to move a nonzero element from the submatrix $A[k{:}n, k{:}n]$ into $A[k, k]$, a process known as *pivoting*. Although the idea of pivoting is simple enough, it is not a trivial matter to decide which element to use as a pivot. In this subsection we discuss some generalities about pivoting. More details will be found at appropriate points in this work.

Gaussian elimination with pivoting

Pivoting is easy to incorporate into Gaussian elimination. Algorithm 1.3 is a simple modification of Algorithm 1.1. The code makes use of the exchange operator "↔", which swaps the objects on its left and right.

There is some terminology associated with this algorithm. At the kth stage of the algorithm, the element $A[k, k]$ is called the *pivot element* or simply the *pivot*. The process of performing interchanges to alter the pivot elements is called *pivoting*. Although we have introduced this terminology in connection with Gaussian elimination, many other matrix algorithms employ pivoting.

In Algorithm 1.3 the interchanges are performed on the entire array A, so that the rows of L and the columns of U are interchanged along with those of $A[k{:}n, k{:}n]$. The reason is contained in the following analogue of Theorem 1.4. (See §2.2, Chapter 1, for exchange matrices.)

```
1.  for k = 1 to n−1
2.      if ( A[k:n, k:n] = 0) return fi
3.      Find indices p_k, q_k ≥ k such that A[p_k, q_k] ≠ 0
4.      A[k, 1:n] ↔ A[p_k, 1:n]
5.      A[1:n, k] ↔ A[1:n, q_k]
6.      A[k+1:n, k] = A[k+1:n, k]/A[k, k]
7.      A[k+1:n, k+1:n] = A[k+1:n, k+1:n] − A[k+1:n, k]*A[k, k+1:n]
8.  end k
```

Algorithm 1.3: Gaussian elimination with pivoting

⎯⎯⎯⎯⎯⎯⎯⎯⎯⎯⎯⎯⎯ ◇ ⎯⎯⎯⎯⎯⎯⎯⎯⎯⎯⎯⎯⎯

Theorem 1.8. *In Algorithm 1.3 let P_k be the (k, p_k)-exchange matrix and let Q_k be the (k, q_k)-exchange matrix. Then in the notation of Theorem 1.4,*

$$P_k \cdots P_1 A Q_1 \cdots Q_k - \begin{pmatrix} L_{11} \\ L_{21} \end{pmatrix} (U_{11}\ U_{12}) = \begin{pmatrix} 0 & 0 \\ 0 & S \end{pmatrix}, \qquad (1.20)$$

where the matrix S is the Schur complement of the leading principal submatrix of order k of $P_k \cdots P_1 A Q_1 \cdots Q_k$. In particular, if we set

$$P = P_{n-1} \cdots P_2 P_1 \quad \text{and} \quad Q = Q_1 Q_2 \cdots Q_{n-1},$$

it follows from (1.20) with $k = n$ that Algorithm 1.3 computes an LU decomposition of PAQ.

The proof of this theorem is essentially a matter of tracing the effects of the interchanges through the algorithm.

Generalities on pivoting

A consequence of Theorem 1.8 is that pivoted Gaussian elimination is equivalent to making interchanges in the original matrix and performing unpivoted Gaussian elimination. For purposes of analysis, this is a useful result, since we may assume that any interchanges have been made at the outset. In practice, however, we must know something about the reduced matrix $A[k:n, k:n]$ in order to choose a pivot. Sometimes theory will guide us; but where it does not, our only recourse is to determine pivots on the fly.

The process of selecting pivots has two aspects: where pivots come from and how pivots are chosen. We will treat each in turn. Since the details of pivoting depend on the algorithm in question and its application, the following discussion is necessarily general — an overview of the territory.

• **Where pivots come from.** The most important restriction on choosing pivots is that each candidate has to be completely reduced so that it is a member of the current Schur complement. Since classical Gaussian elimination updates the entire Schur

SEC. 1. GAUSSIAN ELIMINATION

complement at each stage, every element of $A[k{:}n, k{:}n]$ is a candidate. However, other variants of Gaussian elimination postpone the reduction of some of the elements of $A[k{:}n, k{:}n]$ and thus restrict the range of choice of pivots. For examples see Algorithms 1.5, 1.6, and 1.7.

The process of choosing pivots from the entire array $A[k{:}n, k{:}n]$ is known as *complete pivoting*. Its advantage is that it gives us the widest possible choice of pivots. However, since the entire array must be searched to find a pivot, it adds a small $O(n^3)$ overhead to unpivoted Gaussian elimination.

An alternative that does not add significant overhead is to choose the pivot element from the column $A[k{:}n, k]$, a process known as *partial pivoting*. Although partial pivoting restricts our selection, it can be done with more variants of Gaussian elimination than complete pivoting. The alternative of selecting pivots from the row $A[k, n{:}k]$ is seldom done.

Schur complements in a symmetric matrix are symmetric. Consequently, Gaussian elimination preserves symmetry and, with proper organization, can factor a symmetric matrix at half the usual cost. Unfortunately, pivoting destroys symmetry. The exception is when pivots are chosen from the diagonal of $A[k{:}n, k{:}n]$. This process is known as *diagonal pivoting*. Diagonal pivoting is also required to preserve the structure of other classes of matrices — most notably, M-matrices and diagonally dominant matrices.

- **How pivots are chosen.** Although any nonzero pivot will be sufficient to advance Gaussian elimination to the next stage, in practice some pivots will be better than others. The definition of better, however, depends on what we expect from the algorithm.

The most common way of selecting a pivot from a set of candidates is to choose the one that is largest in magnitude. The process is called *pivoting for size*. There are two reasons to pivot for size.

The first reason is numerical stability. We shall see in §4 that Gaussian elimination is backward stable provided the elements of the array A do not grow too much in the course of the algorithm. Pivoting for size tends to inhibit such growth. Complete pivoting for size is unconditionally stable. Partial pivoting for size can be unstable, but real-life examples are infrequent and unusual in structure.

The second reason is to determine rank. In Theorem 2.13, Chapter 1, we used Gaussian elimination to establish the existence of a full-rank factorization of a matrix. The algorithm corresponding to this proof is a version of Algorithm 1.3 that returns at statement 2. For in that case, the current Schur complement is zero, and the first $k-1$ columns in the array A contain full-rank trapezoidal factors L_{k-1} and U_{k-1} such that $PAQ = L_{k-1}U_{k-1}$. This suggests that Gaussian elimination can be used to determine rank. For more see §2.4, Chapter 5.

Another way of choosing pivots is to preserve sparsity. A *sparse matrix* is one whose elements are mostly zero. We can often take advantage of sparsity to save time and memory in a matrix algorithm. However, most matrix algorithms tend to reduce sparsity as they proceed. For example, if A_k denotes the current matrix in Gaussian

> Let P be as in Theorem 1.8. The following algorithm computes an LU decomposition of PA.
>
> 1. **for** $k = 1$ **to** $n-1$
> 2.   **if** $(A[k{:}n, n] = 0)$ **iterate** k **fi**
> 3.   Determine p_k so that $|A[p_k, k]| \geq |A[i, k]|$ $(i = k, \ldots, n)$
> 4. $A[k, 1{:}n] \leftrightarrow A[p_k, 1{:}n]$
> 5. $A[k{+}1{:}n, k] = A[k{+}1{:}n, k]/A[k, k]$
> 6. $A[k{+}1{:}n, k{+}1{:}n] = A[k{+}1{:}n, k{+}1{:}n] - A[k{+}1{:}n, k] * A[k, k{+}1{:}n]$
> 7. **end** k

Algorithm 1.4: Gaussian elimination with partial pivoting for size

———————◇———————

elimination and $\alpha_{ik}^{(k)}$ and $\alpha_{kj}^{(k)}$ are both nonzero, then $\alpha_{ij}^{(k+1)} = \alpha_{ij}^{(k)} - \alpha_{ik}^{(k)} \alpha_{kj}^{(k)} / \alpha_{kk}^{(k)}$ will in general be nonzero—always when $\alpha_{ij}^{(k)} = 0$. This introduction of nonzero elements in a sparse matrix is called *fill-in*.

Clearly the choice of pivot influences fill-in. For example, if all the elements in the pivot row and column are nonzero, then Gaussian elimination will fill the current submatrix with nonzero elements. Consequently, most algorithms for sparse matrices use a pivoting strategy that reduces fill-in, a process called *pivoting for sparsity*. Unfortunately, pivoting for size and pivoting for sparsity can be at odds with one another, so that one must compromise between stability and sparsity.

———

A word on nomenclature. The terms complete and partial pivoting are frequently used to mean complete and partial pivoting for size. This usage is natural for dense matrices, where pivoting for size is the norm. But other applications demand other strategies. It therefore makes sense to reserve the words "complete" and "partial" to describe where the pivots are found and to add qualifiers to indicate how pivots are chosen.

Gaussian elimination with partial pivoting

Of all the pivoting strategies, by far the most common is partial pivoting for size. Consequently, we conclude this subsection with Algorithm 1.4 — arguably the most frequently used algorithm in all matrix computations.

Unlike our other versions of Gaussian elimination, this one does not return when a nonzero pivot fails to exist. Instead it goes on to the next step. It is easy to verify that the algorithm returns an LU decomposition in which the exceptional columns of L are equal to the corresponding unit vector and the corresponding diagonal of U is zero.

Sec. 1. Gaussian Elimination

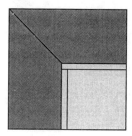

Figure 1.2: Classical Gaussian elimination

―――――◇―――――

1.4. Variations on Gaussian elimination

Classical Gaussian elimination can be thought of as an expanding LU decomposition crowding a Schur complement out of the southeast corner of an array. The situation is illustrated graphically in Figure 1.2. The dark gray area represents the part of the LU decomposition that has already been computed with L and U separated by a diagonal line. The light gray area contains the Schur complement. The thin horizontal box represents the row of U about to be computed; the vertical box represents the corresponding column of L. The way these two boxes overlap reflects the fact that the diagonal elements of L are one and are not stored.

Although the operations that classical Gaussian elimination performs on a given element are fixed, there is considerable freedom in how these operations are interleaved one with another, and each style of interleaving gives a variation on the basic algorithm. Fortunately, the important variants can be derived from diagrams like the one in Figure 1.2. The general idea is to border the part of the LU decomposition already computed with a row and a column and to extend the decomposition into the border. The algorithms differ in the shape of the region in which the decomposition has been computed. Besides classical Gaussian elimination, which is represented by Figure 1.2, there are four other algorithms, two of which are obvious variants of one another. We will treat each in turn.

Sherman's march

Figure 1.3 illustrates an algorithm we will call Sherman's march. Here (and in the other variants) no Schur complement is computed, and the white area represents untouched elements of the original matrix. A step of the algorithm proceeds from the LU decomposition of a leading principal submatrix of A and computes the LU decomposition of the leading principal submatrix of order one greater. Thus the algorithm proceeds to the southeast through the matrix, just like Sherman's procession from Chattanooga, Tennessee, to Savannah, Georgia.

The algorithm is easy to derive. Consider the LU decomposition of the leading principal submatrix of A of order k in the following partitioned form (in this subsection

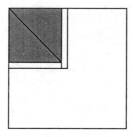

Figure 1.3: Sherman's march

all partitions are indexed to the northwest):

$$\begin{pmatrix} A_{11} & a_{1k} \\ a_{k1}^T & \alpha_{kk} \end{pmatrix} = \begin{pmatrix} L_{11} & 0 \\ \ell_{k1}^T & 1 \end{pmatrix} \begin{pmatrix} U_{11} & u_{1k} \\ 0 & v_{kk} \end{pmatrix}.$$

Then $A_{11} = L_{11}U_{11}$ is the LU decomposition of A_{11}, which we have already computed. Computing the $(1,2)$-block of this factorization, we find that

$$L_{11}u_{1k} = a_{1k}, \tag{1.21}$$

which is a triangular system that can be solved for u_{1k}. Computing the $(1,2)$-block, we get

$$\ell_{k1}^T U_{11} = a_{k1}^T, \tag{1.22}$$

another triangular system. Finally, from the $(2,2)$-block we have $\ell_{k1}^T u_{1k} + v_{kk} = \alpha_{kk}$, or

$$v_{kk} = \alpha_{kk} - \ell_{k1}^T u_{1k}.$$

Algorithm 1.5 implements the bordering method described above. The triangular systems (1.21) and (1.22) are solved by the BLAS *xellib* and *xebui* (see Figure 2.2, Chapter 2). We begin the loop at $k = 2$, since A[1,1] already contains its own LU decomposition. But with our conventions on inconsistent statements, we could equally well have begun at $k = 1$.

At this point we should reemphasize that this algorithm, and the ones to follow, are arithmetically identical with classical Gaussian elimination. If *xellib* and *xebui* are coded in a natural way, Sherman's march and classical Gaussian elimination perform exactly the same arithmetic operations on each element and for each element perform the operations in the same order. In spite of this arithmetic equivalence, Algorithm 1.5 has two important drawbacks.

First, it does not allow pivoting for size. At the kth step, the Schur complement of A_{11}, where one must look for pivots, has not been computed. For this reason the algorithm is suitable only for matrices for which pivoting is not required.

Sec. 1. Gaussian Elimination

This algorithm overwrites A with its LU decomposition.

1. **for** $k = 2$ **to** n
2. $\quad$ xellib($A[1{:}k{-}1, k]$, $A[1{:}k{-}1, 1{:}k{-}1]$, $A[1{:}k{-}1, k]$)
3. $\quad$ xebui($A[k, 1{:}k{-}1]$, $A[k, 1{:}k{-}1]$, $A[1{:}k{-}1, 1{:}k{-}1]$)
4. $\quad$ $A[k, k] = A[k, k] - A[k, 1{:}k{-}1]*A[1{:}k{-}1, k]$
5. $\quad$ **if** ($A[k, k] = 0$) error **fi**
6. **end for** k

Algorithm 1.5: Sherman's march

———◇———

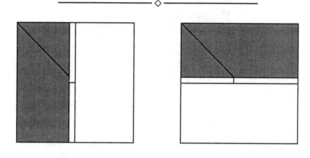

Figure 1.4: Pickett's charge

———◇———

Second, the work in the algorithm is concentrated in the solution of triangular systems. Although this does not change the operation counts, the severely sequential nature of algorithms for solving triangular systems makes it difficult to get full efficiency out of certain architectures.

Pickett's charge

Figure 1.4 illustrates the two versions of another variant of Gaussian elimination. It is called Pickett's charge because the algorithm sweeps across the entire matrix like Pickett's soldiers at Gettysburg. The charge can be to the east or to the south. We will consider the eastern version.

To derive the algorithm, partition the first k columns of the LU decomposition of A in the form

$$\begin{pmatrix} A_{11} & a_{1k} \\ A_{k1} & a_{kk} \end{pmatrix} = \begin{pmatrix} L_{11} & 0 \\ L_{k1} & \ell_{kk} \end{pmatrix} \begin{pmatrix} U_{11} & u_{1k} \\ 0 & v_{kk} \end{pmatrix}$$

and assume that L_{11}, L_{k1}, and U_{11} have already been computed. Then on computing the $(1,2)$-block, we get

$$L_{11} u_{1k} = a_{1k}, \tag{1.23}$$

which we can solve for u_{1k}. Computing the $(2,2)$-block, we get $L_{k1}u_{1k} + v_{kk}\ell_{kk} = a_{kk}$, from which we have

$$v_{kk}\ell_{kk} = a_{kk} - L_{k1}u_{1k}. \tag{1.24}$$

After the right-hand side of this relation is computed, v_{kk} is determined so that the first component of ℓ_{kk} is equal to one.

It is possible to introduce partial pivoting into this algorithm. To see how, consider the following picture.

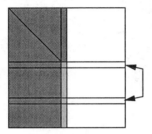

It shows the state of the array A just after the computation of $v_{kk}\ell_{kk}$, which is indicated by the lightly shaded column. Now by (1.24) this vector is equal to $a_{kk} - L_{k1}u_{1k}$, which is precisely the part of the Schur complement from which we would pick a pivot. If we choose a pivot from this column and interchange the *entire* row of the array A with the pivot row (as indicated by the arrows in the above picture), we are actually performing three distinct operations.

1. In $A[:, 1{:}k]$ we interchange the two rows of the part of the L-factor that has already been computed.
2. In $A[:, k]$ we interchange the two elements of the current Schur complement.
3. In $A[:, k{+}1{:}n]$ we interchange the two rows of A.

But these three operations are precisely the interchanges we make in Gaussian elimination with partial pivoting.

Combining these observations, we get Algorithm 1.6, in which the BLAS *xellib* is used to solve (1.23). In pivoting we perform the interchanges on the entire array A, so that the final result is an LU decomposition of the matrix A with its rows interchanged as specified by the integers p_k. The charge-to-the-south algorithm is analogous. However, if we want to pivot, we must perform column rather than row interchanges.

Crout's method

The Crout algorithm has the same pattern as classical Gaussian elimination, except that the computation of the Schur complement is put off to the last possible moment (Figure 1.5). To derive the algorithm, partition the first k columns of the LU decomposition in the form

$$\begin{pmatrix} A_{11} & a_{1k} \\ A_{k1} & a_{kk} \end{pmatrix} = \begin{pmatrix} L_{11} & 0 \\ L_{k1} & \ell_{kk} \end{pmatrix} \begin{pmatrix} U_{11} & u_{1k} \\ 0 & v_{kk} \end{pmatrix},$$

SEC. 1. GAUSSIAN ELIMINATION

This algorithm computes a LU decomposition with partial pivoting for size.

1. **for** $k = 1$ **to** n
2. $xellib(A[1{:}k{-}1, k], A[1{:}k{-}1, 1{:}k{-}1], A[1{:}k{-}1, k])$
3. $A[k{:}n, k] = A[k{:}n, k] - A[k{:}n, 1{:}k{-}1] * A[1{:}k{-}1, k]$
4.   Determine p_k so that $|A[p_k, k]| \geq |A[i, k]|$ $(i = k, \ldots, n)$
5. $A[k, 1{:}n] \leftrightarrow A[p_k, 1{:}n]$
6.   **if** $(A[k, k] \neq 0)$ $A[k{+}1{:}n, k] = A[k{+}1{:}n, k]/A[k, k]$ **fi**
7. **end for** k

Algorithm 1.6: Pickett's charge east

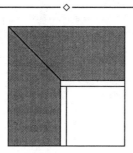

Figure 1.5: The Crout algorithm

---

where L_{11}, L_{21}, U_{11}, and u_{1k} are assumed known. Then

$$v_{kk}\ell_{kk} = a_{kk} - L_{k1}u_{1k},$$

where v_{kk} is determined so that the first component of ℓ_{kk} is one. As in Pickett's charge, we can pivot for size at this point.

Now partition the first k rows of the factorization in the form

$$\begin{pmatrix} A_{11} & a_{1k} & A_{1,k+1} \\ a_{k1}^{\mathrm{T}} & \alpha_{kk} & a_{k,k+1}^{\mathrm{T}} \end{pmatrix} = \begin{pmatrix} L_{11} & 0 \\ \ell_{k1}^{\mathrm{T}} & 1 \end{pmatrix} \begin{pmatrix} U_{11} & u_{1k} & U_{1,k+1} \\ 0 & v_{kk} & u_{k,k+1}^{\mathrm{T}} \end{pmatrix}.$$

It follows that

$$u_{k,k+1}^{\mathrm{T}} = a_{k,k+1}^{\mathrm{T}} - \ell_{k1}^{\mathrm{T}} U_{1,k+1}.$$

Thus we arrive at Algorithm 1.7. Like classical Gaussian elimination, the Crout algorithm is entirely free of triangular solves.

Advantages over classical Gaussian elimination

All these algorithms have an advantage over classical Gaussian elimination that is not immediately apparent: they make fewer alterations in the elements of the array A. For

---

1. **for** $k = 1$ **to** n
2. $\quad A[k{:}n, k] = A[k{:}n, k] - A[k{:}n, 1{:}k{-}1] * A[1{:}k{-}1, k]$
3. $\quad$ Determine p_k so that $|A[p_k, k]| \geq |A[i, k]|$ $(i = k, \ldots, n)$
4. $\quad A[k, 1{:}n] \leftrightarrow A[p_k, 1{:}n]$
5. $\quad$ **if** $(A[k, k] \neq 0)$ $A[k{+}1{:}n, k] = A[k{+}1{:}n, k] / A[k, k]$ **fi**
6. $\quad A[k, k{+}1{:}n] = A[k, k{+}1{:}n] - A[k, 1{:}k{-}1] * A[1{:}k{-}1, k{+}1{:}n]$
7. **end for** k

Algorithm 1.7: Crout's method

⸺⋄⸺

example, classical Gaussian elimination alters the (n, n)-element $n-1$ times as it updates the Schur complement. The algorithms of this subsection can be coded so that they alter it only once when it becomes v_{nn}. As we have seen in §3.3, Chapter 2, this can be a considerable advantage on machines where writing to cache is more expensive than reading from it.

1.5. LINEAR SYSTEMS, DETERMINANTS, AND INVERSES

The principle use of Gaussian elimination is to solve systems of equations. However, it is also used to calculate determinants and inverses. We will treat these three computational problems in order.

Solution of linear systems

In the introduction to this chapter we sketched algorithms for solving $Ax = b$ and $A^T x = b$. Specifically, we can use an LU decomposition of A to solve the system $Ax = b$ as follows.

$$
\begin{aligned}
&1. \quad \text{Solve the system } Ly = b \\
&2. \quad \text{Solve the system } Ux = y
\end{aligned}
\tag{1.25}
$$

We will now show how pivoting affects this algorithm.

For definiteness, suppose that we have computed an LU decomposition of A with partial pivoting. Let $P_1, P_2, \ldots, P_{n-1}$ denote the exchanges corresponding to the pivots. Then by Theorem 1.8,

$$P_{n-1} \cdots P_1 A = LU, \tag{1.26}$$

where LU is the factorization contained in the array A.

Now suppose we want to solve the system $Ax = b$. Multiplying by this system by $P_{n-1} \cdots P_1$, we have from (1.26)

$$LUx = P_{n-1} \cdots P_1 b.$$

Hence if we first perform the interchanges on b, we can proceed to solve an unpivoted system as in (1.25).

Sec. 1. Gaussian Elimination

Let the array A contain the LU decomposition of A computed with partial pivoting, and let $p_1, \ldots, p_{n-1}$ be the pivot indices. Then the following algorithm overwrites B with the solution of the system $AX = B$.

1. **for** $k = 1$ **to** $n-1$
2. $B[k,:] \leftrightarrow B[p_k,:]$
3. **end for** k
4. $xellib(B, A, B)$
5. $xeuib(B, A, B)$

Algorithm 1.8: Solution of $AX = B$

Let A be as in Algorithm 1.8. The following algorithm overwrites B by the solution of the system $A^T X = B$.

1. $xeuitb(B, A, B)$
2. $xellitb(B, A, B)$
3. **for** $k = n-1$ **to** 1 **by** -1
4. $B[k,:] \leftrightarrow B[p_k,:]$
5. **end for** k

Algorithm 1.9: Solution of $A^T X = B$

In the following implementation we extend the algorithm to solve several systems at a time. Algorithm 1.8 uses the triangular BLAS *xellib* and *xeuib* (see Figure 2.2, Chapter 2).

The treatment of pivots in the solution of the transposed system $x^T A = b^T$ is a bit trickier. In the above notation, $A^T = U^T L^T P_{n-1} \cdots P_1$ (remember $P_k^2 = I$). Hence the system can be written in the form

$$U^T L^T (P_{n-1} \cdots P_1 x) = b.$$

If we define $y = P_{n-1} \cdots P_1 x$, we can solve the system $U^T L^T y = b$ and then interchange components to get $x = P_1 \cdots P_{n-1} y$. The result is Algorithm 1.9, in which we use the BLAS *xeuitb* and *xellitb* to solve transposed triangular systems.

For all their small size, these algorithms have a lot to be said about them.

- The bulk of the work in these algorithms is concentrated in solving triangular systems. If B has only a single column, then two systems must be solved. Since it takes about $\frac{1}{2}n^2$ flam to solve a triangular system, the algorithms take about n^2 flam. More generally:

If B has ℓ columns, the operation count for Algorithm 1.8 or Algorithm 1.9 is

ℓn^2 *flam.*

- The algorithms effectively compute $A^{-1}B$ and $A^{-T}B$. The cost of these computations is the same as multiplying B by A^{-1} or by A^{-T}. Thus the algorithms represent a reasonable alternative to the invert-and-multiply algorithm. We will return to this point later.

- The algorithms overwrite the right-hand side B with the solution X. This conserves storage — which can be substantial if B is large — at the cost of forcing the user to save B whenever it is required later. An alternative would be to code, say, Algorithm 1.8 with the calling sequence *linsolve*(A, B, X), in which B is first copied to X and the solution is returned in X. The invocation *linsolve*(A, B, B) would then be equivalent to Algorithm 1.8.

- Algorithm 1.8 could be combined with one of our algorithms for computing an LU decomposition in a single program. This approach is certainly easy on the naive user, who then does not have to know that the solution of a linear system proceeds in two distinct steps: factorization and solution. But this lack of knowledge is dangerous. For example if the user is unaware that on return the array A contains a valuable factorization that can be reused, he or she is likely to recompute the factorization when another task presents itself — e.g., a subsequent solution of $A^T x = b$.

- Ideally our two algorithms for using the LU decomposition should be supplemented by two more: one to solve $XA = B$ and another to solve $XA^T = B$. Fortunately, our triangular BLAS make the coding of such algorithms an elementary exercise.

Determinants

People who work in matrix computations are often asked for programs to compute determinants. It frequently turns out that the requester wants the determinant in order to solve linear systems — usually by Cramer's rule. There is a delicious irony in this situation, for the best way to compute a determinant of a general matrix is to compute it from its LU decomposition, which, as we have seen, can be used to solve the linear system.

However, if a determinant is really needed, here is how to compute it. Since $A = P_1 \cdots P_{n-1} LU$,

$$\det(A) = \det(P_1) \cdots \det(P_n) \det(L) \det(U).$$

Now $\det(L) = 1$ because L is unit lower triangular, and $\det(U) = v_{11} \cdots v_{nn}$. Moreover, $\det(P_k)$ is 1 if P_k is the identity matrix and -1 otherwise. Thus the product of the determinant of the exchanges is 1 if the number of proper interchanges is even and -1 if the number is odd. It follows that

$$\det(A) = (-1)^{\text{number of interchanges}} v_{11} \cdots v_{nn}. \tag{1.27}$$

SEC. 1. GAUSSIAN ELIMINATION

It should be noted that the formula (1.27) can easily underflow or overflow, even when the elements of A are near one in magnitude. For example, if the v_{ii} are all ten, then the formula will overflow in IEEE single-precision arithmetic for $n > 38$. Thus a program computing the determinant should return it in coded form. For example, LINPACK returns numbers d and e such that $\det(A) = d \cdot 10^e$.

Matrix inversion

Turning now to the use of the LU decomposition to compute matrix inverses, the algorithm that comes first to mind mimics the proof of Theorem 3.20, Chapter 1, by using Algorithm 1.8 to solve the systems

$$Ax_j = e_j, \qquad j = 1, 2, \ldots, n,$$

for the columns x_j of the inverse of A. If only a few of the columns of A^{-1} are required, this is a reasonable way to compute. However, if the entire inverse is needed, we can economize on both storage and operations by generating the inverse in place directly from the factors L and U. There are many ways to do this, of which the following is one.

As above, let us suppose that the LU decomposition of A has been computed with partial pivoting. Then $P_{n-1} \cdots P_1 A = LU$

$$A^{-1} = U^{-1} L^{-1} P_{n-1} \cdots P_1. \tag{1.28}$$

In outline, our algorithm goes as follows.

1. Compute U^{-1} in place
2. Solve the system $XL = U^{-1}$ in place
3. $A = A * P_{n-1} * \cdots * P_1$

The inversion of U can be accomplished as follows. Let $S = U^{-1}$. Partition the equation $SU = I$

$$\begin{pmatrix} S_{11} & s_{1k} & S_{1,k+1} \\ 0 & \sigma_{kk} & s^T_{k,k+1} \\ 0 & 0 & S_{k+1,k+1} \end{pmatrix} \begin{pmatrix} U_{11} & u_{1k} & U_{1,k+1} \\ 0 & v_{kk} & u^T_{k,k+1} \\ 0 & 0 & U_{k+1,k+1} \end{pmatrix}$$
$$= \begin{pmatrix} I_{k-1} & 0 & 0 \\ 0 & 1 & 0 \\ 0 & 0 & I \end{pmatrix}.$$

The indexing in this partition is to the northwest. It follows that $\sigma_{kk} v_{kk} = 1$ and $S_{11} u_{1k} + v_{kk} s_{1k} = 0$. Equivalently,

1. $\sigma_{kk} = 1/v_{kk}$,
2. $s_{1k} = -\sigma_{kk} S_{11} u_{1k}$.

$$\tag{1.29}$$

These formulas for the kth column of S do not involve the first $k-1$ columns of U. Thus the columns of S can be generated in their natural order, overwriting the corresponding columns of U.

We will now show how to compute $X = U^{-1}L^{-1}$. As above, let $S = U^{-1}$, so that $XL = S$. Partition this relation in the form

$$(X_{11} \; x_{1k} \; X_{1,k+1}) \begin{pmatrix} L_{11} & 0 & 0 \\ \ell_{k1}^T & 1 & 0 \\ L_{k+1,1} & \ell_{k+1,k} & L_{k+1,k+1} \end{pmatrix} = (S_{11} \; s_{1k} \; S_{1,k+1}).$$

Then $x_{1k} + X_{1,k+1}\ell_{k+1,k} = s_{1k}$, or

$$x_{1k} = s_{1k} - X_{1,k+1}\ell_{k+1,k}.$$

Thus we can generate the columns of X in reverse order, each column overwriting the corresponding column of L and U^{-1} in the array A.

After $U^{-1}L^{-1}$ has been computed, we must perform the interchanges P_k in reverse order as in (1.28).

Algorithm 1.10 is an implementation of the method derived above. It is by far the most involved algorithm we have seen so far.

- The product $S_{11}u_{1k}$ in (1.29) is computed explicitly. In a quality implementation the task would be done by a level-two BLAS.

- The complexity of this algorithm can be determined in the usual way. The first loop on k requires the multiplication of a $(k-1)$-vector times a triangular matrix, which requires $\frac{1}{2}k^2$ flam. Consequently the total count for this loop is

$$\frac{1}{2}\int_0^n k^2 \, dk = \tfrac{1}{6}n^3.$$

The body of the second loop on k requires nk^2 flam. Hence its operation count is $\frac{1}{3}n^3$. Adding the two operation counts we get

$\frac{1}{2}n^3$ flam.

If we add this count to the $\frac{1}{3}n^3$ flam required to compute the LU decomposition in the first place, we find:

It requires

$$\tfrac{5}{6}n^3 \text{ flam} \tag{1.30}$$

to invert a matrix via the LU decomposition.

SEC. 1. GAUSSIAN ELIMINATION

Let the array A contain the LU decomposition of A computed with partial pivoting and let $p_1, \ldots, p_{n-1}$ be the pivoting indices. This algorithm overwrites the array A with A^{-1}.

      ! Invert U.
1.   for $k = 1$ to n
2. $A[k,k] = 1/A[k,k]$
3.   for $i = 1$ to $k-1$
4. $A[i,k] = -A[k,k]*(A[i,i{:}k-1]*A[i{:}k-1,k])$
5.   end for i;
6.   end for k

      ! Calculate $U^{-1}L^{-1}$.
7.   for $k = n-1$ to 1 by -1
8. $temp = A[k+1{:}n, k]$
9. $A[k+1{:}n, k] = 0$
10. $A[1{:}n, k] = A[1{:}n, k] - A[1{:}n, k+1{:}n]*temp$
11.   end for k

      ! Perform the interchanges.
12.   for $k = n-1$ to 1 by -1
13. $A[1{:}n, k] \leftrightarrow A[1{:}n, p_k]$
14.   end for k

Algorithm 1.10: Inverse from an LU decomposition

The count (1.30) has important implications for the invert-and-multiply technique for solving linear equations. Let $B \in \mathbb{R}^{n \times l}$. To solve the system $AX = B$ by invert-and-multiply, we calculate A^{-1} and then compute $X = A^{-1}B$. The latter calculation requires ln^2 flam, for a total of $\frac{5}{6}n^3 + ln^2$ flam. On the other hand, Algorithm 1.8 requires $\frac{1}{3}n^3$ flam to compute the LU decomposition of A, followed by ln^2 flam for the algorithm itself. The total is $\frac{1}{3}n^3 + ln^2$ flam. The ratio of these counts is

$$\rho = \frac{\text{flams for invert-and-multiply}}{\text{flams for inversion via LU}} = \frac{5 + 6l/n}{2 + 6l/n}.$$

The following table contains the values of ρ for various representative values of l/n.

l/n	ρ
0.00	2.5
0.25	1.9
0.50	1.6
0.75	1.5
1.00	1.3
2.00	1.2
∞	1.0

When l is small compared with n, solving directly is almost two and a half times faster than inverting and multiplying. Even when $l = n$ it is 30% faster. And of course ρ is never less than one, so that the advantage, however small, is always to the direct solution.

These ratios are a compelling reason for avoiding matrix inversion to solve linear systems. In §4 we will show that the direct solution is not only faster but it is more stable, another reason for avoiding matrix inversion.

"But," I hear someone protesting, "I really need the inverse," and I respond, "Are you sure?" Most formulas involving inverses can be reduced to the solution of linear systems. An important example is the computation of the bilinear form $\tau = y^T A^{-1} x$. The following algorithm does the job.

$$\begin{array}{ll} 1. & \text{Solve } Au = x \\ 2. & \tau = y^T u \end{array} \qquad (1.31)$$

It is worth noting that if you need the (i, j)-element of A^{-1}, all you have to do is plug $x = \mathbf{e}_j$ and $y = \mathbf{e}_i$ into (1.31).

Of course, there are applications in which the inverse is really needed. Perhaps the most important example is when a researcher wants to scan the elements of the inverse to get a feel for its structure. In such cases Algorithm 1.10 stands ready to serve.

1.6. NOTES AND REFERENCES

Decompositions and matrix computations

We have already noted in §5.1, Chapter 1, that many matrix decompositions were derived in the nineteenth century in terms of bilinear and quadratic forms. However, these decompositions had no computational significance for their originators — with the exception of Gauss's decomposition of a quadratic form. The idea of a decomposition as a platform from which other computations can be launched seems to have emerged slowly with the digital computer. It is only vaguely present in Dwyer's *Linear Computations* [112, 1951], which is devoted largely to computational tableaus. By the middle of the 1960s, it appears fully developed in the books of Householder [189, 1964] and Wilkinson [346, 1965].

Classical Gaussian elimination

Gauss's original algorithm was for positive definite systems from least squares problems. His method was equivalent to factoring $A = RDR^T$ where R is upper triangular. The original exposition of the method in Gauss's *Theoria Motus* [130, 1809] was not computationally oriented (Gauss was after the statistical precision of his least squares estimates [306]). His first layout of the algorithm [131, 1810] was, up to diagonal scaling, what we have called classical Gaussian elimination.

Elementary matrix

The terminology "elementary lower triangular matrix" for matrices like (1.6) ultimately derives from the elementary row operations found in most introductory linear algebra texts. The elementary operations are

1. Multiply a row by a (nonzero) constant,
2. Interchange two rows,
3. Subtract one row from another.

The same effect can be obtained by premultiplying by an *elementary matrix* obtained by performing the same operations on the identity. Householder [189] observed that these matrices could be written in the form $I - uv^T$ and went on to call any such matrix elementary. In particular, elementary lower triangular matrices have the form $I - \ell_k e_k^T$, where the first k components of ℓ_k are zero.

The LU decomposition

Gauss, who worked with positive definite systems, gave a symmetric decomposition that is more properly associated with the Cholesky decomposition. Jacobi [191, 1857, posthumous], factored a bilinear form $f(x, y)$ in the form

$$f(x,y) = h_1(y)g_1(x) + h_2(y)g_2(x) + \cdots + h_n(y)g_n(x),$$

in which the linear functions g_i and h_i depend only on the last $n-i+1$ components of x and y. If A is the matrix corresponding to f, the coefficients of the h_i and g_i form the columns and rows of an LU-decomposition of A. The connection of Gaussian elimination with a matrix factorization was first noted by Dwyer [111] in 1944 — one hundred and thirty-five years after Gauss published his algorithm.

Block LU decompositions and Schur complements

The first block LU decomposition is due to Schur [275, 1917, p. 217], who wrote it essentially in the form

$$\begin{pmatrix} A_{11}^{-1} & 0 \\ -A_{21}A_{11}^{-1} & I \end{pmatrix} \begin{pmatrix} A_{11} & A_{12} \\ A_{21} & A_{22} \end{pmatrix} = \begin{pmatrix} I & A_{11}^{-1}A_{12} \\ 0 & A_{22} - A_{21}A_{11}^{-1}A_{12} \end{pmatrix}.$$

Schur used the relation only to prove a theorem on determinants and did not otherwise exploit it. The name Schur complement for the matrix $A_{22} - A_{21}A_{11}^{-1}A_{12}$ is due to Haynsworth [172].

Cottle [75] and Ouellette [248] give surveys of the Schur complement with historical material.

Block algorithms and blocked algorithms

The distinction between a block algorithm and a blocked algorithm is important because the two are numerically very different creatures. The computation of a block 2×2 LU decomposition is block Gaussian elimination, and even with block pivoting it can fail in cases where ordinary Gaussian elimination with pivoting goes to completion — e.g., with the matrix

$$A = \begin{pmatrix} 0 & 1 & 0 & 0 \\ 0 & 0 & 1 & 0 \\ \hline 0 & 0 & 0 & 1 \\ 1 & 0 & 0 & 0 \end{pmatrix},$$

in which all the blocks are singular.

The systematic use of the word "blocked" in the sense used here may be found in [104]. It is unfortunate that the distinction between two very different kinds of algorithms should rest on the presence or absence of the two letters "ed," but the queue of good alternatives is empty. (Higham [177, §12.1] suggests the name "partitioned algorithm" for a blocked algorithm. But in my opinion, the word "partition" has enough to do keeping matrix elements in their place. Asking it to also keep track of arithmetic operations is too much.)

For notes on the level-three BLAS, which are used to implement blocked algorithms, see §3, Chapter 2.

Pivoting

The technique of pivoting did not arise from Gaussian elimination, which was historically a method for solving positive definite systems and did not require it to avoid division by zero. Instead the idea came from Chió's method of pivotal condensation for computing determinants [66, 1853]. In modern terminology, the idea is to choose a nonzero element a_{ij} of A and compute its Schur complement S. Then $\det(A) = (-1)^{i+j} a_{ij} \det(S)$. Thus the determinant of A can be calculated by repeating the procedure recursively on S. The element a_{ij} was called the pivot element and was selected to be nonzero. The practitioners of this method (e.g., see [6], [339]) seem not to have realized that it is related to Gaussian elimination.

The terms "partial pivoting" and "complete pivoting" are due to Wilkinson [344].

Exotic orders of elimination

There are other triangular decompositions corresponding to other orders of elimination. For example, if one starts at the southeast corner of A and introduce zeros up the columns above the diagonals, one obtains a UL decomposition. If one starts in the southwest corner and eliminates up the columns, one obtains a UL decomposition in which U and L are upper and lower cross triangular.

In one sense all these algorithms are equivalent to ordinary Gaussian elimination with pivoting, since the pivot element can be moved by row and column interchanges to the appropriate place on the diagonal. However, they access the elements of A in different orders and may behave differently on machines with hierarchical memories. For example, the LINPACK algorithm for solving symmetric indefinite systems [99, Ch. 5] computes a block UL decomposition to preserve column orientation. For more on symmetric indefinite systems see §2.2.

Another variant is to eliminate elements along the rows of the matrix. It is a surprising and useful fact that this variant gives the same LU decomposition (up to normalization of the diagonals) as classical Gaussian elimination. This is because the result of the elimination in both cases is to compute the Schur complement of α_{11}. However, partial pivoting will produce a different pivot sequence — the two algorithms look in different places for their pivots — and hence different decompositions.

Gaussian elimination and its variants

The inner-product formulas for the versions of Gaussian elimination that we have called Pickett's charge and Crout reduction (they are the same for symmetric matrices) are also due to Gauss [135, 1828]. The name Doolittle [107, 1878] is also associated with these formulas. For modern computers these variants have the advantage that they reduce the number of writes to memory. However, the formulas offered little advantage to the people who calculated by hand in the nineteenth century, since each term in the inner products had to be written down in order to add it to the others. In fact, Doolittle's contribution seems to have been to design a tableau in which the terms were expeditiously recorded (see [110]). The modern versions of the inner-product algorithms — Crout's method [79] and Dwyer's abbreviated Doolittle algorithm [110] — were devised only after mechanical calculators that could accumulate an inner product in a register became available.

All these variants of Gaussian elimination have suffered from a surplus of names. In one system [105, 246] they are categorized by the order of the indices i, j, and k in the three loops that compose the scalar form of the algorithm. Unfortunately, one person's i may be another's j, which throws the system into confusion. Another system [104] calls classical Gaussian elimination "right looking," referring to the fact that it looks ahead to compute the Schur decomposition. Pickett's charge east is called "left looking," referring to the fact that it get its data from behind the border. Crout is Crout.

The names used in this work have the following rationale. It seems ungrateful not to give Gauss his due by naming the first published form of the algorithm after him. Crout's contributions are substantial enough to attach his name to the variant he published (though a case can be made for Dwyer). Sherman's march and Pickett's charge are pure whimsy—they echo my leisure reading at the time (and they would have been different had I been reading about the campaigns of Alexander, Caesar, or Napoleon). Just for the record, Pickett charged to the east.

Matrix inversion

Algorithm 1.10 is what Higham calls Method B in his survey of four algorithms for inverting matrices [177, §13.3]. The algorithms have essentially the same numerical properties, so the choice between them must rest on other considerations—e.g., their interaction with memory.

Most modern texts on numerical linear algebra stress the fact that matrix inverses are seldom needed (e.g., [153, §3.4.11]). It is significant that although Gauss knew how to invert systems of equations he devoted most of his energies to avoiding numerical inversion [140, pp. 225–231]. However, it is important to keep things in perspective. If the matrix in question is well conditioned and easy to invert (an orthogonal matrix is the prime example), then the invert-and-multiply may be faster and as stable as computing a solution from the LU decomposition.

Augmented matrices

Let $A = LU$ be the LU decomposition of A. Given any $n \times p$ matrix B, set $C = L^{-1}B$. Then

$$(A\ B) = L(U\ C)$$

is an LU decomposition of the *augmented matrix* $(A\ B)$. Since the equation $AX = B$ is equivalent to the equation $UX = C$, we can solve linear systems by computing the LU decomposition of the expanded matrix and back-solving. When $p = 1$, this is formally equivalent to eliminating variables in the equation $Ax = b$, with row operations performed on both sides of the equality.

Gauss–Jordan elimination

Gauss–Jordan elimination is a variant of Gaussian elimination in which all the elements in a column are eliminated at each stage. A typical reduction of a 4×4 matrix would proceed as follows.

$$\begin{pmatrix} X & X & X & X \\ \hat{X} & X & X & X \\ \hat{X} & X & X & X \\ \hat{X} & X & X & X \end{pmatrix} \Longrightarrow \begin{pmatrix} X & \hat{X} & X & X \\ 0 & X & X & X \\ 0 & \hat{X} & X & X \\ 0 & \hat{X} & X & X \end{pmatrix} \Longrightarrow \begin{pmatrix} X & 0 & \hat{X} & X \\ 0 & X & \hat{X} & X \\ 0 & 0 & X & X \\ 0 & 0 & \hat{X} & X \end{pmatrix} \Longrightarrow$$

$$\Longrightarrow \begin{pmatrix} X & 0 & 0 & \hat{X} \\ 0 & X & 0 & \hat{X} \\ 0 & 0 & X & \hat{X} \\ 0 & 0 & 0 & X \end{pmatrix} \Longrightarrow \begin{pmatrix} X & 0 & 0 & 0 \\ 0 & X & 0 & 0 \\ 0 & 0 & X & 0 \\ 0 & 0 & 0 & X \end{pmatrix} \quad (1.32)$$

(Here the elements to be eliminated are given hats.) Thus the elimination reduces the matrix to diagonal form. If the same operations are applied to the right-hand side of a system, the resulting diagonal system is trivial to solve. Pivoting can be incorporated into the algorithm, but the selection must be from the Schur complement to avoid filling in zeros already introduced. The method is not backward stable, but it is weakly stable. For rounding error analyses see [262] and especially [177, §13.4], where further references will be found.

With some care, the method can be arranged so that the inverse emerges in the same array, and this has lead to elegant code for inverting positive definite matrices [22]. Combined with the expanded matrix approach, it has been used by statisticians to move variables in and out of regression problems [156]. For more see §3.1, Chapter 4.

2. A Most Versatile Algorithm

The algorithms of the last section all had the same object: to compute the LU decomposition of a general dense matrix, possibly with pivoting. However, Gaussian elimination can be adapted to matrices of special structure, and the purpose of this subsection is to show how it is done. Specifically, we will consider the use of Gaussian elimination to decompose positive definite matrices, symmetric indefinite matrices, Hessenberg and tridiagonal matrices, and general band matrices. Although these matrices have been well worked over in the literature, and excellent programs for their reduction are available, each type has something new to teach us. Positive definite matrices are one of the most important classes of matrices to appear in matrix computations. The treatment of symmetric indefinite matrices furnishes an example of a block algorithm and contains an elegant pivoting strategy. Hessenberg and tridiagonal matrices teach us how to manipulate matrices with a fixed structure. Band matrices teach us how to handle structures that depend on parameters and how to do operation counts that involve more than one parameter.

A word of warning is in order. Up to now we have been explicitly testing for zero divisors in our algorithms, as any working code must do. However, the tests clutter up the algorithms and make it more difficult to see their structure. Consequently, we will dispense with such tests, leaving it to the implementer to supply them.

2.1. Positive definite matrices

Positive definite matrices are among the most frequently encountered matrices of special structure. Because of their structure they can be factored in half the time required for an ordinary matrix and do not require pivoting for numerical stability. In this subsection we will give the algorithmic details, reserving the question of stability for later.

Positive definite matrices

We will begin with a far-reaching definition. For the moment we will drop our tacit assumption that A is real.

Definition 2.1. Let $A \in \mathbb{C}^{n \times n}$. Then A is POSITIVE DEFINITE if

1. A is Hermitian,
2. $x \neq 0 \implies x^H A x > 0$.

If $x \neq 0 \Rightarrow x^H A x \geq 0$, then A is POSITIVE SEMIDEFINITE. If equality holds for at least one nonzero x, then A is PROPERLY SEMIDEFINITE. If $x \neq 0 \Rightarrow x^H A x < 0$, we say that A is NEGATIVE DEFINITE.

The requirement that A be Hermitian reduces to symmetry for real matrices. Some people drop the symmetry requirement and call a real matrix positive definite if $x \neq 0 \Rightarrow x^T A x > 0$. We will avoid that usage in this work.

The simplest nontrivial example of a positive definite matrix is a diagonal matrix with positive diagonal elements. In particular, the identity matrix is positive definite. However, it is easy to generate more.

Theorem 2.2. Let $A \in \mathbb{C}^{n \times n}$ be positive definite, and let $X \in \mathbb{C}^{n \times p}$. Then $X^H A X$ is positive semidefinite. It is positive definite if and only if X is of full column rank.

Proof. Let $x \neq 0$ and let $y = Xx$. Then $x^H(X^H A X)x = y^H A y \geq 0$, by the positive definiteness of A. If X is of full column rank, then $y = Xx \neq 0$, and by the positive definiteness of A, $y^H A y > 0$. On the other hand if X is not of full column rank, there is a nonzero vector x such that $Xx = 0$. For this particular x, $x^H(X^H A X)x = 0$. ∎

Any principal submatrix of A can be written in the form $X^T A X$, where the columns of X are taken from the identity matrix (see §2.5, Chapter 1). Since any matrix X so formed has full column rank, it follows that:

Any principal submatrix of a positive definite matrix is positive definite.

In particular a diagonal element of a matrix is a 1×1 principal submatrix. Hence:

The diagonal elements of a positive definite matrix are positive.

If P is a permutation matrix, then P has full rank. Hence $P^T A P$ is positive definite. A transformation of the form $P^T A P$ is called a *diagonal permutation* because it rearranges the diagonal elements of A. Hence:

A diagonal permutation of a positive definite matrix is positive definite.

We now turn to the properties of positive definite matrices. One of the most important is that they can be characterized succinctly in terms of their eigenvalues.

Theorem 2.3. *A Hermitian matrix A is positive (semi)definite if and only if its eigenvalues are positive (nonnegative).*

SEC. 2. A MOST VERSATILE ALGORITHM

Proof. We will treat the definite case, leaving the semidefinite case as an exercise. Let $A = U\Lambda U^{\mathrm{H}}$ be the spectral decomposition of A (see Theorem 4.33, Chapter 1). If the eigenvalues of A are positive, then Λ is positive definite, and by Theorem 2.2, so is A. Conversely, if A is positive definite and $Au = \lambda u$, with $\|u\|_2 = 1$ then $\lambda = u^{\mathrm{H}} A u > 0$. ∎

From the facts that the determinant of a matrix is the product of the eigenvalues of the matrix and the eigenvalues of the inverse matrix are the inverses of the eigenvalues, we have the following corollary. Since the eigenvalues of A are positive, A is nonsingular. Moreover $A^{-1} = U^{\mathrm{H}}\Lambda^{-1}U$. This establishes the following corollary.

Corollary 2.4. *If A is positive semidefinite, its determinant is nonnegative. If A is positive definite, its determinant is positive. Moreover, if A is positive definite, then A is nonsingular and A^{-1} is positive definite.*

The fact that a positive definite matrix has positive eigenvalues implies that it also has a positive definite square root.

Theorem 2.5. *Let A be positive (semi)definite. Then there is a positive (semi)definite matrix $A^{\frac{1}{2}}$ such that*

$$(A^{\frac{1}{2}})^2 = A. \tag{2.1}$$

The matrix $A^{\frac{1}{2}}$ is unique.

Proof. Let $A = U\Lambda U^{\mathrm{T}}$ be the spectral decomposition of A. Then the diagonal elements λ_i of Λ are nonnegative. If we define $\Lambda^{\frac{1}{2}} = \mathrm{diag}(\lambda_1, \ldots, \lambda_n)$, then $A^{\frac{1}{2}} = U\Lambda^{\frac{1}{2}}U^{\mathrm{T}}$ satisfies (2.1). It is clearly positive semidefinite. If A is positive definite, then the numbers λ_i are positive and $A^{\frac{1}{2}}$ is positive definite.

Uniqueness is established by a rather involved argument based on the fact that subspaces spanned by eigenvectors corresponding to equal eigenvalues are unique. We omit it here. ∎

For computational purposes one of the most important facts about positive definite matrices is that they have positive definite Schur complements.

Theorem 2.6. *Let the positive (semi)definite matrix A be partitioned in the form*

$$A = \begin{pmatrix} A_{11} & A_{12} \\ A_{12}^{\mathrm{H}} & A_{22} \end{pmatrix}.$$

If A_{11} is positive definite, then its Schur complement in A is positive (semi)definite.

Proof. We will treat the positive definite case, leaving the semidefinite case as an exercise. Since A_{11} is positive definite, it is nonsingular, and hence its Schur complement is well defined. Let $x \neq 0$. Then by the positive definiteness of A and direct computation,

$$0 < (x^H A_{12}^H A_{11}^{-1} \quad -x^H) \begin{pmatrix} A_{11} & A_{12} \\ A_{12}^H & A_{22} \end{pmatrix} \begin{pmatrix} A_{11}^{-1} A_{12} x \\ -x \end{pmatrix}$$
$$= x^H (A_{22} - A_{21} A_{11}^{-1} A_{21}^H) x. \quad \blacksquare$$

The Cholesky decomposition

Since Schur complements in symmetric matrices are symmetric, we might expect that we could devise a symmetric form of Gaussian elimination in which L turned out to be U^H, say $A = R^H R$, where R is upper triangular. However, we cannot do this for just any symmetric matrix because any matrix that can be written in the form $R^H R$ must by Theorem 2.2 be positive semidefinite. But if we restrict ourselves to positive definite matrices, we have the following important result.

Theorem 2.7 (Cholesky decomposition). *If A is positive definite, then A can be factored uniquely in the form $A = R^H R$, where R is upper triangular with positive diagonal elements.*

Proof. The proof is by induction on the order n of A. For $n = 1$ take $R = \sqrt{\alpha_{11}}$.

Now assume the theorem is true for all positive definite matrices of order $n-1$. We seek a factorization of A in the partitioned form

$$\begin{pmatrix} A_{11} & a_{1n} \\ a_{1n}^H & \alpha_{nn} \end{pmatrix} = \begin{pmatrix} R_{11}^H & 0 \\ r_{1n}^H & \rho_{nn} \end{pmatrix} \begin{pmatrix} R_{11} & r_{1n} \\ 0 & \rho_{nn} \end{pmatrix}.$$

Multiplying out the equation we get the following three equations:

1. $A_{11} = R_{11}^H R_{11}$,
2. $a_{1n} = R_{11}^H r_{1n}$,
3. $\alpha_{nn} = r_{1n}^H r_{1n} + \rho_{nn}^2$.

By the induction hypothesis, R_{11} is uniquely determined by the first equation.

Since A_{11} is nonsingular, so is R_{11}, and r_{1n} is determined uniquely by the second equation in the form

$$r_{1n} = R_{11}^{-H} a_{1n}.$$

Finally from the third equation we have

$$\begin{aligned} \rho_{nn}^2 &= \alpha_{nn} - r_{1n}^H r_{1n} \\ &= \alpha_{nn} - a_{1n}^H R_{11}^{-1} R_{11}^{-H} a_{1n} \\ &= \alpha_{nn} - a_{1n}^H (R_{11}^H R_{11})^{-1} a_{1n} \\ &= \alpha_{nn} - a_{1n}^H A_{11}^{-1} a_{1n}. \end{aligned} \quad (2.2)$$

SEC. 2. A MOST VERSATILE ALGORITHM

Given a positive definite matrix stored in the upper half of the array A, this algorithm overwrites it with its Cholesky factor R.

1. **for** $k = 1$ **to** n
2.   *xeuitb*($A[1{:}k{-}1, k]$, $A[1{:}k{-}1, 1{:}k{-}1]$, $A[1{:}k{-}1, k]$)
3. $A[k, k] = \sqrt{A[k,k] - A[1{:}k{-}1, k]^\mathrm{T} * A[1{:}k{-}1, k]}$
4. **end for** k

Algorithm 2.1: Cholesky decomposition

———————◇———————

But this last number is just the Schur complement of A_{11} and hence by Theorem 2.6 is positive. Thus ρ_{nn} is uniquely determined in the form

$$\rho_{nn} = \sqrt{a_{nn} - r_{1n}^\mathrm{H} r_{1n}}. \quad \blacksquare \qquad (2.3)$$

The decomposition $A = R^\mathrm{H} R$ is called the *Cholesky decomposition* of A. There are many variants which can be obtained by scaling the diagonals of R, that is, by rewriting the decomposition in the form $\hat{R}^\mathrm{H} D \hat{R}$, where D is diagonal [see also (1.14)]. On common choice is to take $D = \mathrm{diag}(\rho_{11}, \ldots, \rho_{nn})$, so that the diagonals of $\hat{R}$ are one. In this case it is often written in the form LDL^H, where L is lower triangular. Gauss took $D = \mathrm{diag}(\rho_{11}^{-2}, \ldots, \rho_{nn}^{-2})$.

The Cholesky algorithm

The proof of existence of the Cholesky decomposition is a recursive form of Sherman's march for computing an LU decomposition. We have already observed that one cannot pivot with this kind of algorithm. However, the proof shows that positive definiteness makes pivoting unnecessary — the algorithm always goes through to completion. We will see later that pivoting is also unnecessary for numerical stability.

Algorithm 2.1 is an implementation of the algorithm suggested by the proof of Theorem 2.7. It uses the triangular BLAS *xeuitb* (see Figure 2.2, Chapter 2) to solve the system (2.3). Here are some comments.

- Only the upper half of the matrix A is stored and manipulated, and the lower half of the array A can be used for other purposes.

- Although we have assumed the upper half of the matrix A has been stored in the array A, the algorithm could be implemented for packed storage, in which the matrix has been arranged in a linear array of length $n(n+1)/2$ (see §3.2, Chapter 2).

- The complexity of the algorithm is easily computed by integration of its loops:

$$\int_{k=0}^{n} \int_{j=k}^{n} \int_{i=j}^{n} di\, dj\, dk = n^3/6.$$

Hence:

The Cholesky algorithm requires

$$\tfrac{1}{6}n^3 \text{ flam}.$$

This is the cheapest $O(n^3)$ algorithm in the repertoire.

- Although pivoting is not necessary to insure the numerical stability of Gaussian elimination applied to positive definite matrices, pivoting can be used to detect near degeneracies in rank. We will treat this case in Chapter 5, where we will show how pivoting can be incorporated into the variant of Algorithm 2.1 corresponding to classical Gaussian elimination.

2.2. SYMMETRIC INDEFINITE MATRICES

Owing to the symmetry of a positive definite matrix, the Cholesky algorithm requires only half the operations required by Gaussian elimination. The object of this subsection is to derive an equally economical algorithm for general symmetric matrices.

The chief problem in extending the Cholesky algorithm to general matrices is that there may be no positive diagonal elements on which to pivot. For example, consider the matrix

$$A = \begin{pmatrix} -2 & -6 & 4 \\ -6 & -21 & 15 \\ 4 & 15 & -13 \end{pmatrix},$$

which has no positive diagonal elements. The only way to obtain a positive pivot is to move the element 4 or the element 15 into the $(1,1)$-position, a process that obviously destroys symmetry.

In itself, this problem is not insurmountable. Since Schur complements in a symmetric matrix are symmetric, we could perform Gaussian elimination working with only the upper part of the matrix. This procedure would have the desired operation count. And we can even retain a sort of symmetry in the factorization. For if we multiply the first row of A by -1 and then perform one step of Gaussian elimination, we obtain a reduction of the form

$$A - \begin{pmatrix} 1 \\ 3 \\ -2 \end{pmatrix} (-2)(1\ 3\ -2) = \begin{pmatrix} 0 & 0 & 0 \\ 0 & -4 & 4 \\ 0 & 4 & -5 \end{pmatrix}.$$

After one more step of this procedure we obtain the factorization

$$A = \begin{pmatrix} 1 & 0 & 0 \\ 3 & 1 & 0 \\ -2 & -1 & 1 \end{pmatrix} \begin{pmatrix} -2 & 0 & 0 \\ 0 & -3 & 0 \\ 0 & 0 & -2 \end{pmatrix} \begin{pmatrix} 1 & 3 & -2 \\ 0 & 1 & -1 \\ 0 & 0 & 1 \end{pmatrix},$$

which is an $R^\mathrm{T}DR$ factorization of A in which the diagonals of R are 1. We have thus isolated the effects of the negative pivots in the matrix D.

SEC. 2. A MOST VERSATILE ALGORITHM

Unfortunately, this approach fails completely on a matrix such as

$$A = \begin{pmatrix} 0 & 1 & 2 \\ 1 & 0 & 3 \\ 2 & 3 & 0 \end{pmatrix} \qquad (2.4)$$

in which all potential pivots are zero. And even nonzero pivots, if they are sufficiently small, will cause numerical instability.

Example 2.8. *If we attempt to compute the Schur complement of the $(1,1)$-element in the matrix*

$$A = \begin{pmatrix} 10^{-6} & 1 & 2 \\ 1 & 0 & 3 \\ 2 & 3 & 0 \end{pmatrix}$$

to four decimal digits, we obtain

$$\begin{pmatrix} -1 \cdot 10^6 & -2 \cdot 10^6 \\ -2 \cdot 10^6 & -4 \cdot 10^6 \end{pmatrix}. \qquad (2.5)$$

This matrix is exactly of rank one, even though the original matrix is nonsingular. Another way of viewing this disaster is to observe that (2.5) is the matrix that would have resulted from exact computations on the singular matrix

$$\begin{pmatrix} 10^{-6} & 1 & 2 \\ 1 & 0 & 0 \\ 2 & 0 & 0 \end{pmatrix}.$$

Thus all information about the original elements in the trailing principal submatrix of order two has been lost in the passage to (2.5).

The above example is our first hint that the generation of large elements in the course of Gaussian elimination can cause numerical difficulties. We will take up this point in detail in §4. Here we will use it to derive an algorithm for symmetric indefinite systems.

The basic idea of the algorithm is to compute a block LDU decomposition in which blocks of the diagonal matrix D are of order one or two. For example, although the natural pivots in (2.4) are zero, the leading 2×2 principal submatrix is just a permutation of the identity. If we use it as a "pivot" to compute a Schur complement, we obtain a symmetric block decomposition of the form

$$A = \begin{pmatrix} 1 & 0 & 0 \\ 0 & 1 & 0 \\ 2 & 3 & 1 \end{pmatrix} \begin{pmatrix} 0 & 1 & 0 \\ 1 & 0 & 0 \\ 0 & 0 & -6 \end{pmatrix} \begin{pmatrix} 1 & 0 & 2 \\ 0 & 1 & 3 \\ 0 & 0 & 1 \end{pmatrix}.$$

Note that we have chosen the block diagonal factor so that the diagonal blocks of the triangular factors are identity matrices of order two and one.

Unfortunately, it is not sufficient simply to increase the order of the pivot from one to two whenever a pivot of order one is unsatisfactory. A pivot of order two can also be small or — just as bad — be near a singular matrix. This means we must search for a pair of diagonal elements to form the pivot. The search criterion cannot be very elaborate if large overheads are to be avoided. We will describe two strategies for choosing a satisfactory pivot of order one or two. By satisfactory, we mean that its use in the elimination algorithm will not introduce unduly large elements, as happened in Example 2.8. We will suppose that we are at the kth step of the reduction, so that α_{kk} is in the pivot position. We use a tolerance

$$\sigma = \frac{1 + \sqrt{17}}{8} \cong 0.64.$$

(The choice of the tolerance optimizes a bound on the growth of elements in the course of the reduction.)

The first strategy — *complete diagonal pivoting* — begins by locating the maximal off-diagonal element in the Schur complement, say α_{pq}, and the maximal element on the diagonal of the Schur complement, say α_{rr}. The choice of pivots is made as follows.

1. If $|\alpha_{rr}| \geq \sigma|\alpha_{pq}|$ use α_{rr} as a pivot.
2. Otherwise use the 2×2 principal matrix whose off-diagonal is α_{pq} as a pivot. (2.6)

The justification of this strategy is that in the first case the largest multiplier cannot be greater than σ^{-1}. On the other hand, in the second case the pivot block must have an inverse that is not too large.

This strategy involves an $O(n^3)$ overhead to find the largest element in each of the Schur complements. The effects of this overhead will depend on the implementation; but small or large it will remain proportionally the same as n increases. The following alternative — *partial diagonal pivoting* — has only an $O(n^2)$ overhead, which must wash out as n increases.

The strategy begins by finding an index $p > k$ such that

$$|\alpha_{pk}| \geq |\alpha_{ik}|, \qquad i = k+1, \ldots, n,$$

and an index $q \neq p$ such that

$$|\alpha_{pq}| \geq |\alpha_{pj}|, \qquad j = k, \ldots, p-1, p+1, \ldots, n.$$

In other words, α_{pk} is the largest element in the kth row and column of the current Schur complement, a_{kk} excluded. Likewise, α_{pq} is the largest element in the pth row and column, α_{pp} excluded. Note that the determination of these indices requires only $O(n)$ operations.

The final pivot is determined in four stages, the first three yielding a pivot of order one and the last a pivot of order two. We list the stages here, along with a brief justification of the first three.

Sec. 2. A Most Versatile Algorithm

1. If $|\alpha_{kk}| \geq \sigma|\alpha_{pk}|$, choose α_{kk} as a pivot. Since σ is near one, $|\alpha_{kk}|$ is near the pivot we would obtain by partial pivoting for size.
2. If $|\alpha_{pq}| \geq \sigma|\alpha_{pk}^2|/|\alpha_{kk}|$, choose α_{kk} as a pivot. This choice can increase elements in the Schur complement by no more than $|\alpha_{pk}^2|/|\alpha_{kk}|$. But by the criterion, α_{pq} is not much smaller than this size. Hence this choice of pivot at most increases the largest element of the matrix by a factor that is near two. (2.7)
3. If $|\alpha_{pp}| \geq \sigma|\alpha_{pq}|$, chose α_{pp} as a pivot. As in the first stage, this represents a close approximation to partial pivoting for size.
4. Otherwise, use the 2×2 submatrix whose diagonals are α_{11} and α_{pp} as a pivot.

It turns out that when we arrive at the fourth alternative, the 2×2 pivot must give bounded growth. The details are rather fussy (but straightforward), and we omit them.

Once a pivot has been determined, the elimination step must be performed. We will not give a detailed implementation, since it does not illustrate any new principles. However, when the pivot is a 2×2 block, the computation of the Schur complement can be done in various ways.

Suppose we have a 2×2 pivot and partition the active part of the matrix in the form

$$\begin{pmatrix} B & C^T \\ C & D \end{pmatrix},$$

where C is the pivot block. Then we must compute the Schur complement

$$S = D - CB^{-1}C^T.$$

One way to proceed is to set $\hat{C} = CB^{-1}$ and calculate S in the form

$$S = D - \hat{C}C^T.$$

Since C and $\hat{C}$ have but two columns, this is equivalent to subtracting two rank-one matrices from D, which can be done by level-two BLAS.

A disadvantage of this approach is that extra working storage must be supplied for $\hat{C}$. We can get around this difficulty by computing the spectral decomposition

$$B = V\Lambda V^T$$

of the pivot block (§4.4, Chapter 1). If we define $\hat{C} = CV$, the Schur complement can be written in the form

$$S = D - \hat{C}\Lambda^{-1}\hat{C}^T,$$

which again can be implemented in level-two BLAS. If we allow the matrix $\hat{C}$ to overwrite C we can recover C in the form $C = \hat{C}V^T$. Because V is orthogonal, the passage to $\hat{C}$ and back is stable.

The complexity of this algorithm can be determined by observing that the work done by pivoting on a block is essentially twice the work done by pivoting on a scalar. However, pivoting on a block advances the elimination by two steps. Hence the total operation count is the same as if we had only used scalar pivoting, i.e., the same as for the Cholesky algorithm. Hence:

The method of block diagonal pivoting takes

$\frac{1}{6}n^3$ *flam.*

This count omits the time to find the pivot. As we have mentioned, the first strategy (2.6) imposes an $O(n^3)$ overhead on the algorithm which may be negligible or significant, depending on the implementation details. The second strategy (2.7) requires only $O(n^2)$ work.

There are obvious variants of this algorithm for Hermitian matrices and complex symmetric matrices. For Hermitian matrices rounding errors can cause small imaginary components to appear on the diagonal, and they should be set to zero during each elimination step. (In general, it is not good computational practice to enforce reality in this manner. But for this particular algorithm no harm is done.)

2.3. HESSENBERG AND TRIDIAGONAL MATRICES

A matrix is upper Hessenberg if it is zero below the first subdiagonal — that is, if it has the form

$$\begin{pmatrix} X & X & X & X & X \\ X & X & X & X & X \\ 0 & X & X & X & X \\ 0 & 0 & X & X & X \\ 0 & 0 & 0 & X & X \end{pmatrix}.$$

A matrix is tridiagonal if it is zero below the first subdiagonal and above the first superdiagonal. Tridiagonal matrices have the form

$$\begin{pmatrix} X & X & 0 & 0 & 0 \\ X & X & X & 0 & 0 \\ 0 & X & X & X & 0 \\ 0 & 0 & X & X & X \\ 0 & 0 & 0 & X & X \end{pmatrix}.$$

Hessenberg and tridiagonal matrices are special cases of band matrices, which we will treat in the next subsection. But they arise in so many applications that they are worth a separate treatment.

SEC. 2. A MOST VERSATILE ALGORITHM

Structure and elimination

Up to now we have dealt exclusively with dense matrices having no special pattern of zero elements. For such matrices it is natural to think of Gaussian elimination as a factorization algorithm. For matrices with patterns of zero elements, however, it is more useful to regard Gaussian elimination as an algorithm that uses row operations to introduce zeros into the matrix (approach 2 in §1.1).

At the kth step of the algorithm, multiples of the kth row are subtracted from the other rows to annihilate nonzero elements in the kth column. This process produces an upper triangular matrix that is the U-factor of the original matrix. The multipliers, placed in the position of the elements they annihilate, constitute the nonzero elements of the L-factor.

Hessenberg matrices

Let us see how this version of the algorithm works for an upper Hessenberg matrix: say

$$H = \begin{pmatrix} h_{11} & h_{12} & h_{13} & h_{14} & h_{15} & h_{16} \\ h_{21} & h_{22} & h_{23} & h_{24} & h_{25} & h_{26} \\ 0 & h_{32} & h_{33} & h_{34} & h_{35} & h_{36} \\ 0 & 0 & h_{43} & h_{44} & h_{45} & h_{46} \\ 0 & 0 & 0 & h_{54} & h_{55} & h_{56} \\ 0 & 0 & 0 & 0 & h_{65} & h_{66} \end{pmatrix}.$$

To simplify the exposition, we will consider the problem of pivoting later.

In the first step, we generate a multiplier

$$\ell_{21} = \frac{h_{21}}{h_{11}}$$

and subtract ℓ_{21} times the first row from the second. This gives the matrix

$$\begin{pmatrix} h_{11} & h_{12} & h_{13} & h_{14} & h_{15} & h_{16} \\ 0 & h'_{22} & h'_{23} & h'_{24} & h'_{25} & h'_{26} \\ 0 & h_{32} & h_{33} & h_{34} & h_{35} & h_{36} \\ 0 & 0 & h_{43} & h_{44} & h_{45} & h_{46} \\ 0 & 0 & 0 & h_{54} & h_{55} & h_{56} \\ 0 & 0 & 0 & 0 & h_{65} & h_{66} \end{pmatrix},$$

where

$$h'_{2j} = h_{2j} - \ell_{21} h_{1j}, \qquad j = 2, 3, \ldots, 6.$$

The rows below the second are completely unaltered and hence contribute nothing to the work of the elimination.

The second step is to compute a multiplier

$$\ell_{32} = \frac{h_{32}}{h'_{22}}.$$

If ℓ_{32} times the second row is subtracted from the third row, the result is

$$\begin{pmatrix} h_{11} & h_{12} & h_{13} & h_{14} & h_{15} & h_{16} \\ 0 & h'_{22} & h'_{23} & h'_{24} & h'_{25} & h'_{26} \\ 0 & 0 & h'_{33} & h'_{34} & h'_{35} & h'_{36} \\ 0 & 0 & h_{43} & h_{44} & h_{45} & h_{46} \\ 0 & 0 & 0 & h_{54} & h_{55} & h_{56} \\ 0 & 0 & 0 & 0 & h_{65} & h_{66} \end{pmatrix},$$

where

$$h'_{3j} = h_{3j} - \ell_{32} h'_{2j}, \qquad j = 3, 4, \ldots, 6.$$

The general step should be clear at this point. At the end of the process we obtain an LU decomposition of A in the form

$$\begin{pmatrix} 1 & & & & & \\ \ell_{21} & 1 & & & & \\ 0 & \ell_{32} & 1 & & & \\ 0 & 0 & \ell_{43} & 1 & & \\ 0 & 0 & 0 & \ell_{54} & 1 & \\ 0 & 0 & 0 & 0 & \ell_{65} & 1 \end{pmatrix} \begin{pmatrix} h_{11} & h_{12} & h_{13} & h_{14} & h_{15} & h_{16} \\ 0 & h'_{22} & h'_{23} & h'_{24} & h'_{25} & h'_{26} \\ 0 & 0 & h'_{33} & h'_{34} & h'_{35} & h'_{36} \\ 0 & 0 & 0 & h'_{44} & h'_{45} & h'_{46} \\ 0 & 0 & 0 & 0 & h'_{55} & h'_{56} \\ 0 & 0 & 0 & 0 & 0 & h'_{66} \end{pmatrix}.$$

This reduction fails if any pivot element h'_{kk} is zero. It then becomes necessary to pivot. Complete pivoting destroys the Hessenberg structure. However, partial pivoting preserves it. In fact, there are only two candidates for a pivot at the kth stage: h'_{kk} and $h_{k+1,k}$. Since the rows containing these pivots have exactly the same structure of nonzero elements, interchanging them leaves the matrix upper Hessenberg. Algorithm 2.2 implements this scheme. Here are some comments.

- The reduction of a Hessenberg matrix is a comparatively inexpensive process. The bulk of the work is concentrated in statement 5, which requires about $n-k$ flam. Integrating this count from 0 to n, we find that:

Algorithm 2.2 requires

$\frac{1}{2}n^2$ flam.

This should be compared with the $O(n^3)$ flam required for Gaussian elimination on a full dense matrix.

SEC. 2. A MOST VERSATILE ALGORITHM

> Given an upper Hessenberg matrix H, this algorithm overwrites the upper part of the array H with the U-factor of H. The subdiagonal elements of the array contain the multipliers.
>
> 1. **for** $k = 1$ **to** $n-1$
> 2.   Choose a pivot index $p_k \in \{k, k+1\}$.
> 3. $H[k, k{:}n] \leftrightarrow H[p_k, k{:}n]$
> 4. $H[k+1, k] = H[k+1, k]/H[k, k]$
> 5. $H[k+1, k+1{:}n] = H[k+1, k+1{:}n] - H[k+1, k]*H[k, k+1{:}n]$
> 6. **end for** k

Algorithm 2.2: Reduction of an upper Hessenberg matrix

———————⋄———————

- The algorithm is decidedly row oriented. However, it passes only once over the entire matrix, so that the row orientation should make little difference unless the algorithm is used repeatedly on a large matrix. However, it can be recoded in column-oriented form. For the strategy see Algorithm 1.9, Chapter 4.

- The part of the array H below the first subdiagonal is not referenced and can be used to store other information. Alternatively, the matrix H can be represented in packed form.

- The treatment of the L-factor is quite different from its treatment in our previous algorithms. In the latter, whenever we pivoted we made the same interchanges in L, so that the final result was a factorization of a matrix A in which all the pivoting had been done initially. If we attempt to do the same thing in Algorithm 2.2, the $n-1$ elements of l would spread out through the lower part of the array H, destroying any information contained there. Consequently, we leave the elements of L in the place where they are generated. This treatment of L, incidentally, is typical of algorithms that take into account the zero structure of the matrix. It has algorithmic implications for the way the output is used to solve linear systems, which we will treat in a moment.

———

Most matrices with a structured arrangement of zero and nonzero elements can be reduced in more than one way. For example, an upper Hessenberg matrix can be reduced to upper triangular form by column operations beginning at the southeast corner. The result is a UL decomposition. Similarly, there are two strategies for reducing a lower Hessenberg matrix. Reducing the matrix by column operations beginning at the northwest corner gives an LU decomposition; reducing by row operations beginning at the southeast corner gives a UL decomposition.

The use of the output of Algorithm 2.2 to solve linear systems introduces something new. Heretofore we were able to apply all the interchanges from pivoting to the right-hand side at the very beginning. Since Algorithm 2.2 leaves the multipliers in

This algorithm uses the output of Algorithm 2.2 to solve the upper Hessenberg system $Hx = b$.

1. **for** $k = 1$ **to** $n-1$
2. $\quad b[k] \leftrightarrow b[p_k]$
3. $\quad b[k+1] = b[k+1] - H[k+1,k]*b[k]$
4. **end for** k
5. $xeuib(b, H, b)$

Algorithm 2.3: Solution of an upper Hessenberg system

———————◇———————

place, we must interleave the pivoting with the reduction of the right-hand side. Algorithm 2.3 shows how this is done with the output of Algorithm 2.2.

The bulk of the work in this algorithm is concentrated in *xeuib* (see Figure 2.2, Chapter 2), which requires $\frac{1}{2}n^2$ flam. Thus the a Hessenberg system requires a total of n^2 flam to solve—the same count as for matrix multiplication.

Tridiagonal matrices

The first thing to note about Gaussian elimination applied to tridiagonal matrices is that pivoting does not preserve the tridiagonal form. However, partial pivoting does preserve a band structure, so that the storage requirements for the algorithm are of the same order as the storage required for the matrix itself.

We begin with the matrix

$$T = \begin{pmatrix} a_1 & b_1 & 0 & 0 & 0 & 0 \\ c_1 & a_2 & b_2 & 0 & 0 & 0 \\ 0 & c_2 & a_3 & b_3 & 0 & 0 \\ 0 & 0 & c_3 & a_4 & b_4 & 0 \\ 0 & 0 & 0 & c_4 & a_5 & b_5 \\ 0 & 0 & 0 & 0 & c_5 & a_6 \end{pmatrix}. \tag{2.8}$$

As with Hessenberg matrices, complete pivoting is out of the question, and the only candidates for partial pivoting are the first and second elements in the first column. We can represent the pivoting step by writing

$$\begin{pmatrix} a'_1 & b'_1 & d_1 & 0 & 0 & 0 \\ c'_1 & a'_2 & b'_2 & 0 & 0 & 0 \\ 0 & c_2 & a_3 & b_3 & 0 & 0 \\ 0 & 0 & c_3 & a_4 & b_4 & 0 \\ 0 & 0 & 0 & c_4 & a_5 & b_5 \\ 0 & 0 & 0 & 0 & c_5 & a_6 \end{pmatrix},$$

SEC. 2. A MOST VERSATILE ALGORITHM

where the primes indicate a possible change in value. The element d_1 will be zero if there was no interchange; otherwise it will have the value of b_2 and b_2' will be zero.

For the elimination itself, we compute a multiplier

$$\ell_1 = \frac{c_1'}{a_1'}$$

and then subtract ℓ_1 times the first row from the second to get the matrix

$$\begin{pmatrix} a_1' & b_1' & d_1 & 0 & 0 & 0 \\ 0 & a_2'' & b_2'' & 0 & 0 & 0 \\ 0 & c_2 & a_3 & b_3 & 0 & 0 \\ 0 & 0 & c_3 & a_4 & b_4 & 0 \\ 0 & 0 & 0 & c_4 & a_5 & b_5 \\ 0 & 0 & 0 & 0 & c_5 & a_6 \end{pmatrix}.$$

For the second stage we pivot to get

$$\begin{pmatrix} a_1' & b_1' & d_1 & 0 & 0 & 0 \\ 0 & a_2''' & b_2''' & d_2 & 0 & 0 \\ 0 & c_2' & a_3' & b_3' & 0 & 0 \\ 0 & 0 & c_3 & a_4 & b_4 & 0 \\ 0 & 0 & 0 & c_4 & a_5 & b_5 \\ 0 & 0 & 0 & 0 & c_5 & a_6 \end{pmatrix}.$$

We then compute the multiplier

$$\ell_2 = \frac{c_2'}{a_2'''}$$

and eliminate to get

$$\begin{pmatrix} a_1' & b_1' & d_1 & 0 & 0 & 0 \\ 0 & a_2''' & b_2''' & d_2 & 0 & 0 \\ 0 & 0 & a_3'' & b_3'' & 0 & 0 \\ 0 & 0 & c_3 & a_4 & b_4 & 0 \\ 0 & 0 & 0 & c_4 & a_5 & b_5 \\ 0 & 0 & 0 & 0 & c_5 & a_6 \end{pmatrix}.$$

From this it is seen that each pivot step generates a new (possibly zero) element on the second superdiagonal, and the subsequent elimination step annihilates an element on the subdiagonal. The process continues until the matrix has been reduced to the triangular form

$$\begin{pmatrix} a_1' & b_1' & d_1 & 0 & 0 & 0 \\ 0 & a_2''' & b_2''' & d_2 & 0 & 0 \\ 0 & 0 & a_3''' & b_3''' & d_3 & 0 \\ 0 & 0 & 0 & a_4''' & b_4''' & d_4 \\ 0 & 0 & 0 & 0 & a_5''' & b_5''' \\ 0 & 0 & 0 & 0 & 0 & a_6'' \end{pmatrix}.$$

Let the tridiagonal matrix T be represented in the form (2.8). The following algorithm returns a pivoted LU decomposition of T. The three nonzero diagonals of the U-factor are returned in the arrays a, b, and d. The array c contains the multipliers.

1. $d(1{:}n{-}2) = 0$
2. **for** $k = 1$ **to** $n{-}1$
3.   Choose a pivot index $p_k \in \{k, k+1\}$
4.   **if** $(p_k \neq k)$
5. $a[k] \leftrightarrow c[k];\ b[k] \leftrightarrow a[k+1]$
6.   **if** $(k \neq n{-}1)\ d[k] \leftrightarrow b[k+1]$ **fi**
7.   **end if**
8. $c[k] = c[k]/a[k]$
9. $a[k+1] = a[k+1] - c[k]{*}b[k]$
10.   **if** $(k \neq n{-}1)\ b[k+1] = b[k+1] - c[k]{*}d[k]$ **fi**
11. **end for** k

Algorithm 2.4: Reduction of a tridiagonal matrix

―――――――◇―――――――

The notation for representing T has been chosen with an implementation in mind. Initially the matrix is contained in linear arrays a, b, and c. An additional array d is used to contain the extra superdiagonal generated by the pivoting. At the end, the arrays a, b, and d contain the U-factor. The multipliers can be stored in the array c as they are generated. Algorithm 2.4 gives an implementation.

Much of the code in this algorithm is devoted to pivoting. Arithmetically, the inner loop requires two additions and two multiplications for a total of $2n$ flam. However, there are also n divisions. Hence:

The operation count for Algorithm 2.4 is

$2n$ fladd $+ 2n$ flmlt $+ n$ fldiv.

On many machines the divisions will account for most of the work.

Even with the inclusion of divisions in the count, we are not really through. The pivoting carries an overhead that is proportional to the number of multiplications and divisions. Even as simple a thing as the if statement at the end of the loop on k slows down the algorithm. The algorithm is, in fact, a good example of why one should try to keep conditional statements out of inner loops. In this case we might let k run from 1 to $n-2$ and put the code for $k = n$ outside the loop.

Algorithm 2.5 for solving a linear system is analogous to the algorithm for Hessenberg systems, except that the call to *xeuib* is replaced by an explicit back substitution.

SEC. 2. A MOST VERSATILE ALGORITHM

> This algorithm uses the output of Algorithm 2.4 to solve the tridiagonal system $Tx = y$, overwriting y with the solution.
>
> 1. **for** $k = 1$ **to** $n-1$
> 2. $y[k] \leftrightarrow y[p_k]$
> 3. $y[k+1] = y[k+1] - c[k]*y[k]$
> 4. **end for** k
> 5. $y[n] = y[n]/a[n]$
> 6. $y[n-1] = (y[n-1] - b[n-1]*y[n])/a[n-1]$
> 7. **for** $k = n-2$ **to** 1 **by** -1
> 8. $y[k] = (y[k] - b[k]*y[k+1] - d[k]*y[k+2])/a[k]$
> 9. **end for** k

Algorithm 2.5: Solution of a tridiagonal system

———————⋄———————

> This algorithm takes a positive definite positive matrix whose diagonal is in the array a and superdiagonal is in b and overwrites a and b with the diagonal and superdiagonal of the Cholesky factor.
>
> 1. $a[1] = \sqrt{a[1]}$
> 2. **for** $k = 1$ **to** $n-1$
> 3. $b[k] = b[k]/a[k]$
> 4. $a[k+1] = \sqrt{a[k+1] - b[k]^2}$
> 5. **end for** i

Algorithm 2.6: Cholesky decomposition of a positive definite tridiagonal matrix

———————⋄———————

When T is positive definite, the subdiagonal is the same as the superdiagonal, and hence we can dispense with the array c. Moreover, pivoting is unnecessary, so that we can also dispense with the array d. It would appear that an additional array is needed to store the multipliers. However, if we compute the Cholesky decomposition $T = R^T R$, then R is bidiagonal, and its elements can overwrite the original elements in a and b. These considerations lead to Algorithm 2.6 for reducing a positive definite tridiagonal matrix.

An operation count for the algorithm is easy.

The operation count for Algorithm 2.6 is

$$n \text{ fladd} + n \text{ flmlt} + n \text{ fldiv} + n \text{ flsqrt}.$$

Depending on how square roots are implemented, this algorithm could be slower than simply performing Gaussian elimination on the matrix and storing the multipliers, especially since no pivoting is done.

2.4. BAND MATRICES

Recall (p. 12) that A is a *band matrix* with *lower band width* p and *upper band width* q if

$$i > j + p \implies a_{ij} = 0 \quad \text{and} \quad p < j - q \implies a_{ij} = 0. \tag{2.9}$$

The *band width* of A is $p+q+1$. In this subsection we will show how to factor band matrices.

The algorithm is analogous to the algorithm for factoring a tridiagonal matrix; however, there are more diagonals to deal with. In particular, since our algorithm must work for matrices having arbitrary band widths, we cannot store the diagonals in linear arrays with individual names — e.g., the arrays a, b, c, and d in Algorithm 2.8. However, before we turn to the problem of representing band matrices, it will be useful to consider the reduction of a band matrix in standard array storage.

We will use Wilkinson diagrams to describe the algorithm. The general algorithm is sufficiently well illustrated for the case $p = 2$ and $q' = 3$. In this case the leading part of A has the form

$$\begin{pmatrix} X & X & X & X & & & & \\ X & X & X & X & X & & & \\ X & X & X & X & X & X & & \\ & X & X & X & X & X & X & \\ & & X & X & X & X & X & X \\ & & & X & X & X & X & X & X \\ & & & & \ddots & \ddots & \ddots & \ddots & \ddots & \ddots \end{pmatrix}.$$

Since A is general, some form of pivoting will be necessary, and the only form of pivoting that results in a band matrix is partial pivoting. Consequently, at the first step we must select our pivots from the first three elements of the first columns. However, interchanging the rows may introduce new elements above the superdiagonal. These potential nonzero elements are indicated by the Y's in the following diagram. The 0 represents an element that cannot possibly become nonzero as a result of the interchanges.

$$\begin{pmatrix} X & X & X & X & Y & Y & & \\ X & X & X & X & X & 0 & & \\ X & X & X & X & X & X & & \\ & X & X & X & X & X & X & \\ & & X & X & X & X & X & X \\ & & & X & X & X & X & X & X \\ & & & & \ddots & \ddots & \ddots & \ddots & \ddots & \ddots \end{pmatrix}. \tag{2.10}$$

Sec. 2. A Most Versatile Algorithm

We next subtract multiples of the first row from the second and third to eliminate the $(2,1)$- and $(3,1)$-elements of A. The result is a matrix of the form

$$\begin{pmatrix} X & X & X & X & Y & Y & & & \\ 0 & X & X & X & X & Y & & & \\ 0 & X & X & X & X & X & & & \\ & X & X & X & X & X & X & & \\ & & X & X & X & X & X & X & \\ & & & X & X & X & X & X & X \\ & & & & \ddots & \ddots & \ddots & \ddots & \ddots & \ddots \end{pmatrix}.$$

Note how the 0 in (2.10) becomes a Y. This reflects the fact that the element could become nonzero as the two subdiagonal elements are eliminated.

The next step is analogous to the first. We choose a pivot from among the last three elements of the second column and interchange. This gives a matrix of the form

$$\begin{pmatrix} X & X & X & X & Y & Y & & & \\ 0 & X & X & X & X & Y & Y & & \\ 0 & X & X & X & X & X & 0 & & \\ & X & X & X & X & X & & & \\ & & X & X & X & X & X & & \\ & & & X & X & X & X & X & X \\ & & & & \ddots & \ddots & \ddots & \ddots & \ddots & \ddots \end{pmatrix}.$$

We then eliminate the $(3,2)$- and $(4,2)$-elements to get a matrix of the form

$$\begin{pmatrix} X & X & X & X & Y & Y & & & \\ 0 & X & X & X & X & Y & Y & & \\ 0 & 0 & X & X & X & X & Y & & \\ & 0 & X & X & X & X & X & & \\ & & X & X & X & X & X & X & \\ & & & X & X & X & X & X & X \\ & & & & \ddots & \ddots & \ddots & \ddots & \ddots & \ddots \end{pmatrix}.$$

Continuing one more step, we get

$$\begin{pmatrix} X & X & X & X & Y & Y & & & \\ 0 & X & X & X & X & Y & Y & & \\ 0 & 0 & X & X & X & X & Y & Y & \\ & 0 & 0 & X & X & X & X & Y & \\ & & 0 & X & X & X & X & X & \\ & & & X & X & X & X & X & X \\ & & & & \ddots & \ddots & \ddots & \ddots & \ddots & \ddots \end{pmatrix}.$$

The pattern is now obvious. As zeros are introduced into the lower band, the upper band expands by exactly the width of the lower band. The following code implements this algorithm, assuming that the matrix is stored in an array A of order n with elements outside the band explicitly set to zero.

1. **for** $k = 1$ **to** $n-1$
2. $iu = \min\{n, k+p\}$
3. $ju = \min\{n, k+p+q\}$
4.   Choose a pivot index $p_k \in \{k, \ldots, iu\}$.
5. $A[k, k{:}ju] \leftrightarrow A[p_k, k{:}ju]$ (2.11)
6. $A[k{+}1{:}iu, k] = A[k{+}1{:}iu, k]/A[k, k]$
7. $A[k{+}1{:}iu, k{+}1{:}ju] = A[k{+}1{:}iu, k{+}1{:}ju]$
       $- A[k{+}1{:}iu, k] * A[k, k{+}1{:}ju]$
8. **end for** k

We now turn to the implementation of this algorithm in a compact storage scheme. Here our notation for representing submatrices — for example, $A[k{+}1{:}iu, k{+}1{:}ju]$ — fails us. The reason is that any reasonable compact storage scheme will map rectangular submatrices onto parts of the array that are not rectangular subarrays. Anticipating this problem, we will recast (2.11) in terms of level-one and level-two BLAS (see §3.2, Chapter 2, for details about BLAS and strides).

Specifically, we will suppose we have BLAS subprograms to swap two vectors, to scale a vector, and to add a rank-one matrix to a general matrix. The following is the specification of these BLAS.

1. *swap*(n, x, *xstr*, y, *ystr*): This program swaps the n-vectors x and y having strides *xstr* and *ystr*.

2. *scale*(n, σ, x, *xstr*): This program overwrites the n-vector x having stride *xstr* with σx.

3. *apsxyt*(m, n, A, *astr*, σ, x, *xstr*, y, *ystr*): This program overwrites $A \in \mathbb{R}^{m \times n}$ having (row) stride *astr* with $A + \sigma x y^{\mathrm{T}}$, where x is an m-vector having stride *xstr* and y is an n-vector having stride *ystr*. (The name means "a plus s times x y transpose".)

To illustrate these BLAS we give an implementation of the third.

Sec. 2. A Most Versatile Algorithm

1. $apsxyt(m, n, A, astr, \sigma, x, xstr, y, ystr)$
2. $aj = 1$
3. $yj = 1$
4. **for** $j = 1$ **to** n
5. $xi = 1$; $aij = aj$
6.   **for** $i = 1$ **to** m
7. $A[aij] = A[aij] + \sigma*x[xi]*y[yj]$
8. $aij = aij+1$; $xi = xi+xstr$
9.   **end for** i
10. $aj = aj+astr$; $yj = yj+ystr$
11. **end for** j

In terms of these BLAS we can rewrite (2.11) as follows. Here we assume that the matrix A is contained in an array with stride *astr*

1. **for** $k = 1$ **to** $n-1$
2. $ni = \min\{p, n-k\}$
3. $nj = \min\{p+1, n-k\}$
4.   Choose a pivot index $p_k \in \{k, \ldots, k+ni\}$
5. $swap(nj+1, A[k,k], astr, A[p_k, k], astr)$
6. $scale(ni, 1/A[k,k], A[k+1,k], 1)$
7. $apsxyt(ni, nj, A[k+1, k+1], astr,$
           $A[k+1, k], 1, A[k, k+1], astr)$
8. **end for** k

Since the BLAS specify a vector by giving its origin, length, and stride, the variables *iu* and *ju* in (2.11) that specified the upper limits of i and j have been replaced by lengths *ni* and *nj* of the vectors involved in the computation.

We must now decide how we are going to represent A. Since we cannot store individual diagonals in separate arrays, a natural alternative is to store them along the rows (or columns) of a rectangular array. We will place them along the rows.

There are many ways we could store the diagonals, but the following scheme, illustrated below for $p = 2$ and $q = 3$, has the advantage that it lines rows of the matrix along diagonals of the array. In the illustration asterisks indicate unused elements. Since pivoting will add two additional super diagonal elements, we have included them in this structure. They are separated from the others by a horizontal line.

$$
\begin{array}{ccccccccc}
* & * & * & * & * & a_{16} & a_{27} & \cdots \\
* & * & * & * & a_{15} & a_{26} & a_{37} & \cdots \\
\hline
* & * & * & a_{14} & a_{25} & a_{36} & a_{47} & \cdots \\
* & * & a_{13} & a_{24} & a_{35} & a_{46} & a_{57} & \cdots \\
* & a_{12} & a_{23} & a_{34} & a_{45} & a_{56} & a_{67} & \cdots \\
a_{11} & a_{22} & a_{33} & a_{44} & a_{55} & a_{66} & a_{77} & \cdots \\
a_{21} & a_{32} & a_{43} & a_{54} & a_{65} & a_{76} & a_{87} & \cdots \\
a_{31} & a_{42} & a_{53} & a_{64} & a_{75} & a_{86} & a_{97} & \cdots \\
\end{array}
\qquad (2.12)
$$

The reader should verify that passing along a row of this array moves along a diagonal of the matrix, while passing down a column of the array passes down a column of the matrix. To move across a row of the matrix, however, one must move diagonally in the array toward the northeast.

It is precisely this diagonal storage of rows that makes the colon convention for representing submatrices unworkable. However, the BLAS we coded above can take diagonal storage in stride. For suppose the array in which A is stored has stride *astr*. If r is the address in memory of the first element of the first row, then the addresses of the elements of the entire row are

$$r,\ r + (astr{-}1),\ r + 2(astr{-}1),\ r + 3(astr{-}1),\ \ldots.$$

In other words a row in our representation is like a row in conventional storage but with its stride reduced by one. Thus we can convert our BLAS to this representation by reducing *astr* by one. There is one more difficulty. References like $A[k, k]$ must be translated to refer the position of a_{kk} in the new storage scheme. Fortunately, the correspondence is trivial. Let

$$m = p + q + 1.$$

Then

a_{ij} corresponds to $A[m{+}i{-}j, j]$.

These transformations are implemented in Algorithm 2.7. Here are some comments.

- The algorithm can be blocked in the usual way. But unless p and q are large, blocking will not improve the performance by much.

- The complexity of the algorithm is not easy to derive because it depends on three parameters: n, p, and q. The algorithm has three distinct stages depending on the value of the index k of the outer loop.

  1. For $k = 1, \ldots, n{-}p{-}q$, we have *ni* $= p$ and *nj* $= p{+}q$. Consequently the update of the Schur complement in statement 8 takes $p(p{+}q)$ flam for a total of $p(p{+}q)(n{-}p{-}q)$ flam.
  2. For $k = n{-}p{-}q{+}1, \ldots, n{-}p$, the length *ni* has the fixed value p, but *nj* decreases by one with each iteration of the loop. By standard integration techniques, we see that this part of the loop contributes $p^2 q + \frac{1}{2} p q^2$ flam.
  3. For the remaining values of k, the algorithm reduces the $p{\times}p$ matrix in the southeast corner, for an operation count of $\frac{1}{3} p^3$ flam.

Thus:

If $n > p + q$, the operation count for Algorithm 2.7 is

$$[p(p{+}q)(n{-}p{-}q) + p^2 q + \tfrac{1}{2} p q^2 + \tfrac{1}{3} p^3]\ \text{flam}.$$

SEC. 2. A MOST VERSATILE ALGORITHM

Let the matrix A with lower band width p and upper band width q be represented according to the scheme (2.12) in an array A having stride *astr*. The following algorithm overwrites the first $p+q+1$ rows of array A with the U-factor of the matrix A. The last p rows contain the multipliers.

1. $m = p+q+1$
2.   **for** $k = 1$ **to** $n-1$
3. $ni = \min\{p, n-k\}$
4. $nj = \min\{p+q, n-k\}$
5.   Choose a pivot index $p_k \in \{k, \ldots, k+ni\}$.
6.   *swap*($nj+1$, $A[m,k]$, *astr*-1, $A[m+p_k, k]$, *astr*-1)
7.   *scale*(ni, $1/A[m,k]$, $A[m+1,k]$, 1)
8.   *apsxyt*(ni, nj, $A[m, k+1]$, *astr*-1, -1,
                 $A[m+1, k]$, 1, $A[m-1, k+1]$, *astr*-1)
9.   **end for** k

Algorithm 2.7: Reduction of a band matrix

For p and q fixed and n large, the count is effectively $np(p+q)$.

- The algorithm to use the output of Algorithm 2.7 to solve banded systems consists of a forward application of the multipliers with interleaved pivoting followed by a back substitution phase. The forward and backward phases are straightforward, since our structure makes it easy to move down a column of the matrix. The implementation is left as an exercise.

2.5. NOTES AND REFERENCES

Positive definite matrices

The normal equations of least squares are positive definite, and it was to such systems that Gauss applied his elimination algorithm [130, 131, 1809, 1810]. Functions generally have a positive definite Hessian matrix at their minima, and in 1759 Lagrange [205] devised a test for positive definiteness that amounts to performing Gaussian elimination on the matrix and looking for negative diagonals in the Schur complements. Positive definite matrices also arise in the discretization of elliptic and parabolic partial differential equations.

The Cholesky decomposition, or rather the algorithm for computing it, was used by Cholesky to solve normal equations arising in geodesy. It was published on his behalf by Benoît [27, 1924] after his death.

In our implementation, we have assumed that A is stored in the upper half of the array. It could equally well be stored in the lower part, in which case the factorization would become $L^T L$, where L is lower triangular. The former option has been adopted here for consistency with the QR decomposition to be treated later, but the latter is also common.

Symmetric indefinite systems

Bunch and Parlett attribute the idea of using block elimination to stabilize the reduction of a symmetric indefinite system, and in [54, 51] they introduce and analyze the complete diagonal pivoting strategy (2.6). The elegant partial diagonal pivoting scheme (2.7) was proposed by Bunch and Kaufman [53]. In the literature the first scheme is usually called complete pivoting and the second partial pivoting; I have taken the liberty of inserting the words diagonal to avoid confusion with the standard usage of these terms. For a concise analysis of these schemes see [177, §10.4].

A curiosity of this algorithm is that if the upper half of A is stored then better column orientation can be obtained by computing the factorization from southeast to northwest — in effect computing a UL decomposition. Such are the implementations in LINPACK [99] and LAPACK [9].

The trick of using the spectral decomposition to save storage during the computation of the Schur complement is due to the authors of LAPACK. Its main drawback is that for small matrices it overwhelms the other computations.

Another approach, due to Parlett and Reid [254], is to reduce A to tridiagonal form by two-sided congruence transformations. Unfortunately, the extra work needed to preserve symmetry in a straightforward implementation gives the algorithm an operation count of $\frac{1}{3}n^3$ — no improvement over ordinary Gaussian elimination. Later Aasen [1] showed how to arrange the calculations to reduce the count to $\frac{1}{6}n^3$.

Band matrices

The storage scheme for band matrices presented here is due to the authors of LINPACK [99]. The ingenious use of level-two BLAS to move diagonally in a matrix is found in LAPACK [9].

3. THE SENSITIVITY OF LINEAR SYSTEMS

> *It is the City of Destruction, a populous place, but possessed with a very ill-conditioned and idle sort of people.*
>
> *The Pilgrims Progress, Second Part*
> John Bunyon

Linear systems of equations seldom come unadulterated. For example, the matrix A of the system may be measured, in which case the matrix at hand is not A itself but a perturbation $A + E$ of A. Or the elements of the matrix may be computed, in which case rounding errors insure that we will be working with a perturbed matrix. Even when A is known exactly, an algorithm like Gaussian elimination will effectively perturb A

SEC. 3. THE SENSITIVITY OF LINEAR SYSTEMS

(see §4). The question treated in this section is how do these perturbations affect the solution of a linear system.

In §3.1 we will present the classical perturbation theory, which bounds the norm of the error. The fact that norms are only a rough measure of the size of a vector or matrix can cause normwise bounds to be pessimistic. Consequently, in §3.2, we will treat componentwise bounds that to some extent alleviate this problem. In §3.3 we will be concerned with projecting a measure of the accuracy of the solution back on the original matrix. These *backward perturbation* bounds have many practical applications. Finally, in the last subsection, we will apply perturbation theory to analyze the method of *iterative refinement*, a technique for improving the accuracy of the solution of linear systems.

3.1. NORMWISE BOUNDS

In this subsection we will be concerned with the following problem.

> Let A be nonsingular and $Ax = b$. Let $\tilde{A} = A + E$. Determine conditions under which $\tilde{A}$ is nonsingular and bound the size of $\tilde{x} - x$, where $\tilde{A}\tilde{x} = b$. (3.1)

The solution of this problem depends on the sense in which we take the word size. We will begin with a *normwise perturbation analysis* in which we use norms to measure the size of the perturbations in A and x. The analysis expresses the bounds in terms of a *condition number*. A large condition number means that the problem is sensitive to at least some perturbations in A. However, if the class of perturbations is restricted, the condition number may overestimate the error — a situation called *artificial ill-conditioning*. We will discuss it at some length.

The basic perturbation theorem

We begin with a useful lemma — the vector analogue of Theorem 4.4, Chapter 2.

Lemma 3.1. *For any vector norm $\|\cdot\|$ suppose that*

$$\tilde{\rho} \equiv \frac{\|\tilde{x} - x\|}{\|\tilde{x}\|} < 1. \tag{3.2}$$

Then

$$\frac{\|\tilde{x} - x\|}{\|x\|} \leq \frac{\tilde{\rho}}{1 - \tilde{\rho}}. \tag{3.3}$$

Proof. From (3.2) we have

$$\tilde{\rho}\|\tilde{x}\| = \|\tilde{x} - x\| \geq \|\tilde{x}\| - \|x\|,$$

or

$$(1 - \tilde{\rho})\|\tilde{x}\| \leq \|x\|.$$

Hence

$$\frac{\|\tilde{x} - x\|}{\|x\|} \leq \frac{\|\tilde{x} - x\|}{(1-\tilde{\rho})\|\tilde{x}\|} = \frac{\tilde{\rho}}{1-\tilde{\rho}}. \blacksquare$$

The basic perturbation theorem is the following.

Theorem 3.2. *Let $\|\cdot\|$ denote a matrix norm and a consistent vector norm. If A is nonsingular and*

$$Ax = b \quad \text{and} \quad \tilde{A}\tilde{x} = b, \tag{3.4}$$

then

$$\frac{\|\tilde{x} - x\|}{\|\tilde{x}\|} \leq \|A^{-1}E\|. \tag{3.5}$$

If in addition

$$\|A^{-1}E\| < 1,$$

then $\tilde{A} = A + E$ is nonsingular and

$$\frac{\|\tilde{x} - x\|}{\|x\|} \leq \frac{\|A^{-1}E\|}{1 - \|A^{-1}E\|}. \tag{3.6}$$

Proof. From (3.4) it follows that $\tilde{x} - x = A^{-1}E\tilde{x}$, and (3.5) follows on taking norms.

Now assume that $\|A^{-1}E\| < 1$. The matrix $\tilde{A}$ is nonsingular if and only if the matrix $A^{-1}\tilde{A} = I + A^{-1}E$ is nonsingular. Hence by Theorem 4.18, Chapter 1, $\tilde{A}$ is nonsingular, and (3.6) follows immediately from (3.5) and Lemma 3.1. $\blacksquare$

Normwise relative error and the condition number

The left-hand side of (3.6) is called the *normwise relative error* in $\tilde{x}$ because it has the same form as the relative error in a scalar (see Definition 4.3, Chapter 2). The import of Lemma 3.1 is that if the relative error is even marginally less than one then it does not matter whether the denominator is $\|x\|$ or $\|\tilde{x}\|$. For example, if one of the bounds is 0.1, then the other can be no larger that 1.11 . . . .

For scalars there is a close relation between relative error and agreement of significant figures: if the relative error in α and β is ρ, then α and β agree to roughly $-\log \rho$ decimal digits. Unfortunately, the same thing is not true of the normwise relative error, as the following example shows.

Example 3.3. *Let*

$$x = \begin{pmatrix} 1.0000 \\ 0.0100 \\ 0.0001 \end{pmatrix} \quad \text{and} \quad \tilde{x} = \begin{pmatrix} 1.0002 \\ 0.0103 \\ 0.0002 \end{pmatrix}.$$

SEC. 3. THE SENSITIVITY OF LINEAR SYSTEMS

The relative error in $\tilde{x}$ as an approximation to x is $3 \cdot 10^{-4}$ in the ∞-norm. But the relative errors in the individual components are $2 \cdot 10^{-4}$, $3 \cdot 10^{-2}$, and 1. The large component is accurate, but the smaller components are inaccurate in proportion as they are small.

This example shows that the inequality (3.6) says a lot about the accuracy of the larger components of $\tilde{x}$ but less about the smaller components. Fortunately, the components of the solution of many linear systems are roughly equal in size. When they are not, they can sometimes be made so by rescaling the problem. Specifically, if $D_{\text{c}} = \text{diag}(\delta_1, \ldots, \delta_n)$ is diagonal, then the system $Ax = b$ can be written in the form

$$(AD_{\text{c}})(D_{\text{c}}^{-1}x) = b.$$

(The C in D_{c} stands for column, since D_{c} scales the columns of A.) If we have a rough idea of the sizes of the components of x, we can take $\delta_i \cong x_i$ ($i = 1, \ldots, n$), and the components of $D_{\text{c}}^{-1}x$ will all be nearly one. However, this approach has its own drawbacks — as we shall see when we consider artificial ill-conditioning.

The right-hand sides of (3.5) or (3.6) are not as easy to interpret as the left-hand sides. But if we weaken the bound, we can put them in a more revealing form. Specifically, from (3.5)

$$\frac{\|\tilde{x} - x\|}{\|\tilde{x}\|} \leq \|A^{-1}\|\|E\| = \kappa(A)\frac{\|E\|}{\|A\|}, \tag{3.7}$$

where

$$\kappa(A) = \|A\|\|A^{-1}\|.$$

If $\|A^{-1}\|\|E\| < 1$, then relative error in $\tilde{x}$ is bounded by

$$\frac{\|\tilde{x} - x\|}{\|x\|} \leq \frac{\kappa(A)\dfrac{\|E\|}{\|A\|}}{1 - \kappa(A)\dfrac{\|E\|}{\|A\|}}. \tag{3.8}$$

The number $\kappa(A)$ is called the *condition number of A with respect to inversion*, or, when the context is clear, simply the *condition number*. For all the commonly used norms it is greater than one, since

$$1 \leq \|I\| = \|AA^{-1}\| \leq \|A\|\|A^{-1}\| = \kappa(A).$$

We have already observed that the left-hand side of this bound can be regarded as a relative error in x. The factor

$$\frac{\|E\|}{\|A\|} = \frac{\|\tilde{A} - A\|}{\|A\|}$$

can likewise be regarded as a relative error in $\tilde{A}$. Thus the condition number $\kappa(A)$ tells us by how much the relative error in the matrix of the system $Ax = b$ is magnified in the solution.

There is a rule of thumb associated with the condition number. Let us suppose that the normwise relative errors reflect the relative errors in the elements of $\tilde{x}$ and $\tilde{A}$. Thus if $\frac{\|E\|}{\|A\|} = 10^{-t}$, then $\tilde{A}$ is accurate to about t decimal digits. If $\kappa(A) = 10^k$, then

$$\frac{\|\tilde{x} - x\|}{\|x\|} \leq 10^{-t+k};$$

that is, $\tilde{x}$ will be accurate to about $t-k$ digits. In other words:

If $\kappa(A) = 10^k$, expect $\tilde{x}$ to have k fewer correct digits than $\tilde{A}$.

When $\kappa(A)$ is large, the solution of the system $Ax = b$ is sensitive to perturbations in A, and the system is said to be *ill conditioned*. Generally speaking, ill-conditioned systems cannot be solved accurately, since they are sensitive to rounding errors made by the algorithm used to solve the system. Even if the algorithm is exact, rounding errors made when the matrix A is entered into the computer will perturb the solutions.

Example 3.4. *Let the elements of A be rounded on a machine with rounding unit ϵ_M to give a perturbed matrix $\tilde{A}$. Then*

$$\tilde{a}_{ij} = a_{ij}(1 + \epsilon_{ij}), \qquad |\epsilon_{ij}| \leq \epsilon_M.$$

In other words, $\tilde{A} = A + E$, where $e_{ij} = a_{ij}\epsilon_{ij}$. It follows that for any absolute norm

$$\|E\| \leq \|A\|\epsilon_M.$$

Hence by (3.7)

$$\frac{\|\tilde{x} - x\|}{\|\tilde{x}\|} \leq \kappa(A)\epsilon_M.$$

This says that the larger components of x may suffer a loss of $\log \kappa(A)$ decimal digits due to the rounding of A. The smaller components may lose even more digits.

The condition number in the two norm has a nice characterization in terms of singular values. It is the ratio of the largest singular value to the smallest:

$$\kappa_2(A) = \frac{\sigma_1}{\sigma_n}.$$

In many applications the elements of A are about one in magnitude, in which case a large condition number is equivalent to a small singular value.

Since the condition number involves $\|A^{-1}\|$, it would seem that to compute the condition number one would have to compute the inverse of A. However, once the matrix A has been factored — say by Gaussian elimination — there are $O(n^2)$ techniques to estimate its condition number. These *condition estimators* will be treated in §3, Chapter 5.

Perturbations of the right-hand side

It is sometimes important to assess perturbations in the right-hand side of the equation $Ax = b$. The following theorem gives the appropriate bounds.

Theorem 3.5. *Let A be nonsingular and $b \neq 0$. Let*

$$Ax = b \quad \text{and} \quad A\tilde{x} = b + e.$$

Then for any consistent norm

$$\frac{\|\tilde{x} - x\|}{\|x\|} \leq \frac{\kappa(A)}{\mu} \frac{\|e\|}{\|b\|}, \qquad (3.9)$$

where

$$\mu = \frac{\|A\|\|x\|}{\|b\|}.$$

Proof. We have $A(\tilde{x} - x) = e$ or $\tilde{x} - x = A^{-1}e$. Hence

$$\|\tilde{x} - x\| \leq \|A^{-1}\|\|e\|.$$

The result now follows on dividing this inequality by $\|x\|$ and applying the definitions of $\kappa(A)$ and μ. ∎

Once again, $\kappa(A)$ mediates the transfer of relative error from the data to the solution. However, the factor μ mitigates the effect of κ. In particular, it is easy to show that

$$1 \leq \mu \leq \kappa(A).$$

Hence the factor μ has the potential to reduce the factor of $\|e\|/\|b\|$ to one — i.e., to make the problem perfectly conditioned.

To see what μ means in terms of the original problem, suppose that $\|A\| = 1$. If $\mu = \kappa(A) = \|A^{-1}\|$, then $\|x\| = \|A^{-1}\|\|b\|$; that is, $\|x\|$ reflects the size of $\|A^{-1}\|$. On the other hand if $\mu = 1$, then $\|x\| = \|b\|$ and the size of $\|x\|$ tells us nothing about $\|A^{-1}\|$. In proportion as μ is near $\kappa(A)$ we say that x *reflects the ill-conditioning of A*. Problems that reflect the ill-conditioning of their matrix are insensitive to perturbations in the right-hand side.

It should be stressed that solutions of real-life problems usually do *not* reflect the ill-conditioning of their matrices. That is because the solutions have physical significance that makes it impossible for them to be large. And even when a right-hand side reflects the ill-conditioning of the matrix, the solution is still sensitive to errors in the matrix itself.

Artificial ill-conditioning

Unfortunately, Example 3.4 does not tell the whole story. Let us look at another example.

Example 3.6 (Artificial ill-conditioning). *Consider the system $Ax = b$ given below:*

$$\begin{pmatrix} 3.9885\text{e}{-}02 & -1.0263\text{e}{+}00 & 6.3481\text{e}{-}01 \\ -2.4828\text{e}{-}02 & 1.1535\text{e}{-}02 & 8.2041\text{e}{-}03 \\ 1.1587\text{e}{-}04 & -7.8646\text{e}{-}05 & -1.7603\text{e}{-}05 \end{pmatrix} \begin{pmatrix} 1 \\ 1 \\ 1 \end{pmatrix} = \begin{pmatrix} -3.5159\text{e}{-}01 \\ -5.0895\text{e}{-}03 \\ 1.9617\text{e}{-}05 \end{pmatrix}.$$

If we were to round the matrix A in the sixth digit, we might get an error matrix like

$$E_1 = \begin{pmatrix} 2.2434\text{e}{-}08 & -1.1262\text{e}{-}06 & -3.9372\text{e}{-}07 \\ -3.1642\text{e}{-}09 & -8.4355\text{e}{-}09 & 1.9456\text{e}{-}09 \\ 6.4209\text{e}{-}11 & 1.1048\text{e}{-}10 & -2.7933\text{e}{-}11 \end{pmatrix}.$$

Now the condition number of A is about $3.4\text{e}{+}5$, so that by Example 3.4 we should expect a relative accuracy of one or two figures in the solution $\tilde{x}_1$ obtained by perturbing A by E_1. In fact the bound on the normwise relative error is

$$\kappa(A)\frac{\|E_1\|_2}{\|A\|_2} = 0.3.$$

But when we solve the perturbed system $(A + E_1)\tilde{x}_1 = b$ we find that

$$\frac{\|\tilde{x}_1 - x\|_2}{\|x\|_2} = 2.3 \cdot 10^{-5}.$$

The error bound overestimates the actual error by almost four orders of magnitude.

The phenomenon exhibited in the above example is called *artificial ill-conditioning*. In trying to find out what is going on, the first thing to note is that there is nothing wrong with the bound in Theorem 3.2. On the contrary it is always realistic for at least one perturbation, as the following theorem shows.

Theorem 3.7. *Let A be nonsingular and let $Ax = b$. Let $\epsilon > 0$. If $\epsilon\|A^{-1}\|_2 < 1$, then there is a matrix E with $\|E\|_2 = \epsilon$ such that if $\tilde{x} = (A + E)^{-1}b$ then*

$$\frac{\|\tilde{x} - x\|_2}{\|x\|_2} \geq \frac{\epsilon\|A^{-1}\|_2}{1 + \epsilon\|A^{-1}\|_2}. \tag{3.10}$$

Proof. Let u be a vector of norm one such that $\|A^{-1}u\|_2 = \|A^{-1}\|_2$ and set

$$E = \epsilon \frac{ux^\mathrm{T}}{\|x\|_2}.$$

SEC. 3. THE SENSITIVITY OF LINEAR SYSTEMS

Then it is easily verified that $\|E\|_2 = \epsilon$,

$$\|A^{-1}Ex\|_2 = \epsilon \|A^{-1}\|_2 \|x\|_2, \tag{3.11}$$

and

$$\|A^{-1}E\|_2 = \epsilon \|A^{-1}\|_2. \tag{3.12}$$

Since $\epsilon\|A^{-1}\| < 1$, $A + E$ is nonsingular and $\tilde{x}$ is well defined. From the easily verified identity $\tilde{x} - x = (I - A^{-1}E)^{-1}A^{-1}Ex$, we have

$$\|\tilde{x} - x\|_2 = \|(I - A^{-1}E)^{-1}(AEx)\|_2 \equiv \|(I - A^{-1}E)y\|_2,$$

where by (3.11) $\|y\|_2 = \epsilon \|A^{-1}\|_2 \|x\|_2$. But by (3.12)

$$\begin{aligned}\|\tilde{x} - x\|_2 &= \|(I - A^{-1}E)y\|_2 \geq (1 + \epsilon\|A^{-1}\|_2)^{-1}\|y\|_2 \\ &= (1 + \epsilon\|A^{-1}\|_2)^{-1}\epsilon\|A^{-1}\|_2\|x\|_2,\end{aligned}$$

which is just (3.10). ∎

To see what (3.10) means, let us write it along with the corresponding upper bound from (3.8):

$$\frac{\epsilon\|A^{-1}\|_2}{1 - \epsilon\|A^{-1}\|_2} \geq \frac{\|\tilde{x} - x\|_2}{\|x\|_2} \geq \frac{\epsilon\|A^{-1}\|_2}{1 + \epsilon\|A^{-1}\|_2}.$$

The difference between the two bounds is in the sign of the denominator, which in both cases can be made as near one as we like by choosing ϵ small enough. Thus by making ϵ arbitrarily small, we can make the relative perturbation in x arbitrarily close to $\|E\|_2\|A^{-1}\|_2$. Although we have worked with the 2-norm for simplicity, analogous results hold for the other commonly used norms.

Having established that there are errors that makes the bounds (3.7) or (3.8) realistic, let us exhibit one such matrix of errors for Example 3.6.

Example 3.8 (Artificial ill-conditioning, continued). *Let*

$$E_2 = \begin{pmatrix} 2.2434\text{e}{-08} & -1.1262\text{e}{-06} & -3.9372\text{e}{-07} \\ -3.1642\text{e}{-07} & -8.4355\text{e}{-07} & 1.9456\text{e}{-07} \\ 6.4209\text{e}{-07} & 1.1048\text{e}{-06} & -2.7933\text{e}{-07} \end{pmatrix}.$$

The solution of the system $(A + E_2)x_2 = b$ *has error*

$$\frac{\|\tilde{x}_1 - x\|_2}{\|x\|_2} = 0.2,$$

while the error bound is

$$\kappa(A)\frac{\|E_2\|_2}{\|A\|_2} = 0.5.$$

Both numbers are now in the same ballpark.

We now have two errors — one for which the bound (3.7) works and one for which it does not. There is no question of one error being better or more realistic than the other. It depends on the application. If we are concerned with the effects of rounding the elements of A on the solution, then E_1 reflects the fact that we make only small relative errors in the components of A. On the other hand, if the elements of A were measured with an instrument that had an absolute accuracy of 10^{-6}, then E_2 more accurately reflects the error. Thus it is the structure of the error that creates the artificial ill-conditioning. The condition number has to be large enough to predict the results of perturbing by E_2. But it then overestimates the perturbations in the solution due to E_1.

Mathematically speaking, the overestimate results from the fact that for E_1 the right-hand side of the inequality

$$\|A^{-1}E_1\tilde{x}\|_2 \leq \|A^{-1}\|_2 \|E_1\|_2 \|\tilde{x}\|_2 \tag{3.13}$$

greatly overestimates the left-hand side. This suggests it may be possible to rescale the problem to strengthen the inequality. Specifically, we can replace the system $Ax = b$ with

$$(D_R \tilde{A} D_C)(D_C^{-1} x) = (D_R b),$$

where D_R and D_C are suitably chosen diagonal matrices. Since $D_R \tilde{A} D_C = D_R A D_C + D_R E D_C$, the matrix E inherits the scaling of A. Thus we wish to choose D_R and D_C so that the inequality

$$\|(D_C^{-1} A^{-1} D_R^{-1})(D_R E D_C)(D_C^{-1} \tilde{x})\|_2$$
$$\leq \|D_C^{-1} A^{-1} D_R^{-1}\|_2 \|D_R E D_C\|_2 \|D_C^{-1} \tilde{x}\|_2$$

is as sharp as possible.

The strategy recommended here is the following.

Choose D_R and D_C so that the elements of E are, as nearly as possible, equal.

There can be no completely rigorous justification of this recommendation, but the following theorem, which is stated without proof, is suggestive.

Theorem 3.9. *If the elements of E are uncorrelated random variables with variance σ^2, then*

$$\mathbf{E}\|A^{-1}Ex\|_2^2 = \sigma^2 \|A^{-1}\|_F^2 \|x\|_2^2.$$

Here $\mathbf{E}$ is the mathematical expectation — the average — and the number σ can be regarded as the size of a typical element of E. Thus the theorem says that if we regard σ as also representing $\|E\|_2$, then on the average the inequality (3.13) is sharp.

Example 3.10 (Artificial ill-conditioning, continued). *Rescaling works rather well for our running example. If we take*

$$D_R = \mathrm{diag}(1, 100, 10000),$$

SEC. 3. THE SENSITIVITY OF LINEAR SYSTEMS

then $D_R E_1 = E_2$, which is nicely balanced. Moreover,

$$\kappa(D_R A) \frac{\|D_R E\|_2}{\|D_R A\|_2} = 6.0\text{e}{-}5,$$

which is of a size with the observed error.

This error balancing strategy is not a panacea, and it is instructive to consider its limitations.

- The strategy requires some knowledge of the error. If the condition number is to be used to assess the sensitivity of solutions of systems whose matrices are already subject to error, then this knowledge must come from the application. If the error is the result of rounding the elements of the matrix, then balancing the elements of the matrix is equivalent to balancing the error (see Example 3.4). If the error is the result of an algorithm like Gaussian elimination, then an analysis of the algorithm may provide guidance on how to scale the matrix.

- Since there are only $2n$ free parameters in the matrices D_R and D_C (actually $2n-1$) and there are n^2 elements of the error matrix, we may not be able to achieve a balance. There are efficient algorithms, discussed in the notes and references, that achieve an optimal balance according to a least squares criterion.

- The matrix D_C also scales x and may have to be used to make the bound on $\|\tilde{x} - x\|$ meaningful. In this case only D_R can be used to balance the errors.

—

In spite of these limitations, error balancing can be a remarkably effective way of handling artificial ill-conditioning.

3.2. COMPONENTWISE BOUNDS

The difficulty with normwise perturbation analysis is that it attempts to summarize a complicated situation by the relation between three numbers: the normwise relative error in $\tilde{x}$, the condition number $\kappa(A)$, and the normwise relative error in $\tilde{A}$. If we are willing to compute the inverse of A, we can do better.

Theorem 3.11 (Bauer–Skeel). *Let A be nonsingular and let*

$$Ax = b \quad \text{and} \quad \tilde{A}\tilde{x} = b.$$

Then

$$|\tilde{x} - x| \leq |A^{-1}||E||\tilde{x}|. \tag{3.14}$$

Moreover, if for some consistent matrix norm

$$\||A^{-1}||E|\| < 1, \tag{3.15}$$

then $(I - |A^{-1}||E|)^{-1}$ is nonnegative and

$$|\tilde{x} - x| \leq (I - |A^{-1}||E|)^{-1}|A^{-1}||E||x|. \tag{3.16}$$

Proof. The bound (3.14) follows immediately on taking absolute values in the identity $\tilde{x} - x = A^{-1}E\tilde{x}$.

Turning now to (3.16), if (3.15) is satisfied then by Corollary 4.21, Chapter 1,

$$|(I + A^{-1}E)^{-1}| \leq (I - |A|^{-1}|E|)^{-1},$$

and the right-hand side is a fortiori nonnegative. The bound (3.16) now follows on taking absolute values in the identity

$$\tilde{x} - x = -(I + A^{-1}E)^{-1}A^{-1}Ex. \quad \blacksquare$$

Mathematically the bounds (3.14) and (3.16) differ, but as a practical matter they are essentially the same. For if $|A^{-1}||E|$ is reasonably small, then the factor $(I - |A^{-1}||E|)^{-1}$ will differ insignificantly from the identity, and either bound will give essentially the same result.

This is a good place to point out that there is a difference between a mathematician using perturbation bounds to prove a theorem and a person who wants some idea of how accurate the solution of a linear system is. The former must be punctilious; the latter can afford to be a little sloppy. For example, A^{-1} will generally be computed inaccurately. But as long as the problem is not so ill conditioned that A^{-1} has no figures of accuracy, it can be used with confidence in the bound. Similarly, if x and $\tilde{x}$ agree to one or two figures, it does not much matter which of the bounds (3.14) or (3.16) is used. Generally speaking, if the bounds say the solution is at all accurate, they are almost certainly overestimates.

The bounds can be quite an improvement over normwise bounds.

Example 3.12 (Artificial ill-conditioning, concluded). *In our continuing example, the error in $\tilde{x}_1$, component by component, is*

$$\begin{pmatrix} -2.0\text{e}-05 \\ -2.0\text{e}-05 \\ -2.9\text{e}-05 \end{pmatrix}.$$

On other hand if we estimate E_1 by $1.0\text{e}-6|A|$, then the componentwise bound is

$$1.0\text{e}-6|A^{-1}||A||x| = \begin{pmatrix} 6.0\text{e}-05 \\ 6.4\text{e}-05 \\ 9.7\text{e}-05 \end{pmatrix},$$

which is indeed a good bound.

The price to be paid for this improvement is the computation of A^{-1}, which more than doubles the work over what is required to solve a linear system (Algorithm 1.10, Chapter 3). However, if we can compute the ith row $a_i^{(-1)\text{T}}$ of A^{-1}, then it follows from (3.14) that

$$|\tilde{x}_i - x_i| \leq |a_i^{(-1)\text{T}}||E||\tilde{x}|. \tag{3.17}$$

SEC. 3. THE SENSITIVITY OF LINEAR SYSTEMS

Consequently, if we require bounds on only a few components of the solution, we can calculate the corresponding rows of the inverse by solving systems of the form

$$a_i^{(-1)\mathrm{T}} A = \mathbf{e}_i^\mathrm{T}.$$

An alternative is to weaken the Bauer–Skeel theorem so that it becomes a mixed bound with both normwise and componentwise features. Specifically, taking bounds in (3.14), we have the following corollary.

Corollary 3.13. *If $\|\cdot\|$ is an absolute norm, then*

$$\frac{\|\tilde{x} - x\|}{\|\tilde{x}\|} \leq \frac{\|\,|A^{-1}|\,|E|\,|\tilde{x}|\,\|}{\|\tilde{x}\|}. \tag{3.18}$$

This corollary converts the problem of computing a bound on the left-hand side of (3.18) to that of estimating $\|\,|A^{-1}|\,|E|\,|\tilde{x}|\,\|$. This can be done by the condition estimator described in §3.1, Chapter 5.

3.3. BACKWARD PERTURBATION THEORY

An important problem, closely related to the one we have been considering, is the following.

> *Given an ostensible solution $\tilde{x}$ of the nonsingular system $Ax = b$, find a perturbation E such that $(A + E)\tilde{x} = b$.* (3.19)

The solution of this *backward perturbation* problem requires that we have a computable measure of the quality of $\tilde{x}$ as an approximate solution of the system. We will measure the quality by the size of the residual vector

$$r = b - A\tilde{x}. \tag{3.20}$$

The problem (3.19) has the flavor of a backward rounding-error analysis (see §4.3, Chapter 2), in that it projects an error back on the original data — namely, the matrix A. However, the bound on this error is based on the residual, which can actually be computed. Consequently, backward perturbation results are used in many practical applications.

Normwise backward error bounds

The normwise solution to this backward perturbation problem is given in the following theorem.

Theorem 3.14 (Wilkinson). *Let $r = b - A\tilde{x}$. Then there is a matrix*

$$E = \frac{r\tilde{x}^\mathrm{T}}{\|\tilde{x}\|_2^2}$$

satisfying

$$\frac{\|E\|_2}{\|A\|_2} = \frac{\|r\|_2}{\|A\|_2\|\tilde{x}\|_2} \tag{3.21}$$

such that

$$(A + E)\tilde{x} = b. \tag{3.22}$$

Conversely if $\tilde{x}$ satisfies (3.22), then

$$\frac{\|r\|_2}{\|A\|_2\|\tilde{x}\|_2} \leq \frac{\|E\|_2}{\|A\|_2}.$$

Proof. The fact that E as defined above satisfies (3.21) and (3.22) is a matter of direct calculation. On the other hand, if $\tilde{x}$ satisfies (3.22), then

$$r = Ex,$$

and the bound follows on taking norms. ∎

We say that a quantity has been stably computed if it comes from a slight perturbation of the original data. What we have shown here is that for linear systems there is a one-one correspondence between stable solutions and solutions with small residuals. If we want to know when an ostensible solution has been computed stably, all we have to do is compute its residual.

Componentwise backward error bounds

There are many matrices E such that $(A + E)\tilde{x} = b$, of which the matrix exhibited in Theorem 3.14 is just one. The only thing required of E is that $E\tilde{x} = r$. For example, if one wishes to perturb only the ith column of A and $x_i \neq 0$ one can take

$$E = \frac{r\mathbf{e}_i^\mathrm{T}}{\tilde{x}_i}. \tag{3.23}$$

This is a special case of the following componentwise theorem.

Theorem 3.15 (Oettli–Prager). *Let $r = b - A\tilde{x}$. Let S and s be nonnegative and set*

$$\epsilon = \max_i \frac{|\rho_i|}{(S|\tilde{x}| + s)_i} \tag{3.24}$$

(here $0/0 = 0$ and otherwise $\rho/0 = \infty$). If $\epsilon \neq \infty$, there is a matrix E and a vector e with

$$|E| \leq \epsilon S \quad \text{and} \quad |e| \leq \epsilon s \tag{3.25}$$

such that

$$(A + E)\tilde{x} = b + e. \tag{3.26}$$

Moreover, ϵ is the smallest number for which such matrices exist.

SEC. 3. THE SENSITIVITY OF LINEAR SYSTEMS

Proof. From (3.24) we have

$$|\rho_i| \leq \epsilon(S|\tilde{x}| + s)_i.$$

This in turn implies that $r = D(S|\tilde{x}| + s)$, where $|D| \leq \epsilon I$. It is then easily verified that $E = DS \operatorname{diag}(\operatorname{sign}(\tilde{\xi}_1), \ldots, \operatorname{sign}(\tilde{\xi}_n))$ and $e = -Ds$ are the required backward perturbations.

On the other hand, given perturbations E and e satisfying (3.25) and (3.26) for some ϵ, we have

$$|r| = |b - A\tilde{x}| = |E\tilde{x} - e| \leq \epsilon(S|\tilde{x}| + s).$$

Hence $\epsilon \geq |\rho_i|/(S\tilde{x} + s)_i$, which shows that the ϵ defined by (3.24) is optimal. ∎

The proof of this theorem is constructive in that it contains a recipe for calculating E from S and s and r. It is an instructive exercise to verify that when $S = \mathbf{e}\mathbf{e}_i$ and $s = 0$, the resulting matrix E is precisely (3.23).

3.4. ITERATIVE REFINEMENT

Iterative refinement was introduced into numerical linear algebra as a method for improving the accuracy of approximate solutions of linear systems. But it is actually a general method of wide applicability in matrix computations—it can even be used when the equations involved are nonlinear. Here we give a general analysis of the method that can be specialized later.

Let $\mathbf{A}$ be a nonsingular linear operator on a finite-dimensional real or complex vector space, and consider the equation

$$\mathbf{r}(\mathbf{x}) \equiv \mathbf{b} - \mathbf{A}\mathbf{x} = 0, \qquad (3.27)$$

whose solution we will denote by $\mathbf{x}_*$. The notation used here emphasizes two points of generality. First, the use of boldface indicates that we are not assuming anything about the nature of the objects in the vector space. For example, the space could be the space of upper triangular matrices, and $\mathbf{A}$ could be the mapping $R \mapsto R^{\mathrm{T}} + R$. (This mapping arises in perturbation theory for Cholesky factors.) Second, we are attempting to find a point $\mathbf{x}_*$ that makes a function $\mathbf{r}(\mathbf{x})$ equal to zero. Although for the moment we assume that $\mathbf{r}$ is linear, we will see later that it can also be nonlinear.

Now let $\mathbf{x}_0$ be a putative solution of (3.27), and let $\mathbf{r}_0 = \mathbf{r}(\mathbf{x}_0)$. Then

$$\mathbf{x}_* = \mathbf{x}_0 + \mathbf{A}^{-1}\mathbf{r}_0.$$

Thus we can obtain $\mathbf{x}_*$ from $\mathbf{x}_0$ by the following algorithm.

1. $\mathbf{r}_0 = \mathbf{r}(\mathbf{x}_0)$
2. Solve $\mathbf{A}\mathbf{d}_0 = \mathbf{r}_0$
3. $\mathbf{x}_* = \mathbf{x}_0 + \mathbf{d}_0$

Since the above algorithm requires a solution of a linear equation involving $\mathbf{A}$, there seems to be no reason to prefer it over solving (3.27) directly. However, in many applications we cannot solve systems involving $\mathbf{A}$ exactly. For example, the results of the next section will show that if we are concerned with ordinary linear systems and use Gaussian elimination with a suitable pivoting strategy to solve (3.27) then the computed solution actually satisfies $\mathbf{A}_0 \mathbf{x} = \mathbf{b}$, where $\mathbf{A}_0$ is near $\mathbf{A}$. In other applications, the equations involving the operator $\mathbf{A}$ may be too difficult to solve directly, forcing us to use a simpler operator $\mathbf{A}_0$ in the place of $\mathbf{A}$.

There are other possible sources of error. For example, there may be errors in computing $\mathbf{r}$; i.e., its computed value may be $\mathbf{b} - \mathbf{A}\mathbf{x}_0 + \mathbf{g}_0$, where $\mathbf{g}_0$ has a known bound. There may also be errors in computing $\mathbf{x}_0 + \mathbf{d}_0$. Thus our actual calculation amounts to executing the following algorithm exactly for some $\mathbf{A}_0$, $\mathbf{g}_0$, and $\mathbf{h}_0$.

$$
\begin{aligned}
&1. \quad \mathbf{r}_0 = \mathbf{r}(\mathbf{x}_0) + \mathbf{g}_0 \\
&2. \quad \mathbf{d}_0 = \mathbf{A}_0^{-1} \mathbf{r}_0 \\
&3. \quad \mathbf{x}_1 = \mathbf{x}_0 + \mathbf{d}_0 + \mathbf{h}_0
\end{aligned}
\qquad (3.28)
$$

This algorithm can be applied iteratively to yield a sequence of approximate solutions $\mathbf{x}_0, \mathbf{x}_1, \ldots$. In general the sequence will not converge, but the following theorem shows that under appropriate circumstances it gets close to $\mathbf{x}_*$. Here we assume that $\|\cdot\|$ denotes a consistent operator and vector norm on the space in question.

Theorem 3.16. *Let (3.28) be applied iteratively to give the sequence* $\mathbf{x}_0, \mathbf{x}_1, \ldots$. *Let* $\mathbf{e}_k = \mathbf{x}_k - \mathbf{x}$ *and* $\mathbf{E}_k = \mathbf{A}_k - \mathbf{A}$. *If*

$$\frac{\|\mathbf{A}^{-1}\mathbf{E}_k\|}{1 - \|\mathbf{A}^{-1}\mathbf{E}_k\|} \leq \rho < 1, \quad \|\mathbf{g}_k\| \leq \gamma_k, \quad \text{and} \quad \|\mathbf{h}_k\| \leq \eta_k, \quad k = 0, 1, \ldots,$$

then

$$
\begin{aligned}
\|\mathbf{e}_k\| &\leq \rho^k \|\mathbf{e}_0\| + (1+\rho)\|\mathbf{A}^{-1}\|(\gamma_k + \rho\gamma_{k-1} + \cdots + \rho^k \gamma_0) \\
&\quad + (\eta_k + \rho\eta_{k-1} + \cdots + \rho^k \eta_0).
\end{aligned}
\qquad (3.29)
$$

In particular, if

$$\gamma_k \leq \gamma_+, \quad \text{and} \quad \eta_k \leq \eta_+, \quad k = 0, 1, \ldots,$$

then

$$\|\mathbf{e}_k\| \leq \rho^k \|\mathbf{e}_0\| + \frac{1+\rho}{1-\rho}\|\mathbf{A}^{-1}\|\gamma_+ + \frac{\eta_+}{(1-\rho)}. \qquad (3.30)$$

On the other hand if

$$\gamma = \limsup \gamma_k \quad \text{and} \quad \eta = \limsup \eta_k, \quad k = 0, 1, \ldots,$$

then

$$\limsup \|\mathbf{e}_k\| \leq \rho^k \|\mathbf{e}_0\| + \frac{1+\rho}{1-\rho}\|\mathbf{A}^{-1}\|\gamma + \frac{\eta}{(1-\rho)}. \qquad (3.31)$$

SEC. 3. THE SENSITIVITY OF LINEAR SYSTEMS

Proof. By Theorem 4.18, Chapter 1, $(I + A^{-1}E_k)^{-1} = I + F_k$, where

$$\|F_k\| \leq \rho.$$

Moreover,

$$\|(A + E_k)^{-1}\| \leq (1+\rho)\|A^{-1}\|.$$

Now

$$\begin{aligned}
x_1 &= x_0 + (A + E_0)(b - Ax_0 + g_0) + h_0 \\
&= x_0 + (I + A^{-1}E_0)^{-1}A^{-1}(b - Ax_0) + (A + E_0)^{-1}g_0 + h_0 \\
&= x_0 - (I + A^{-1}E_0)^{-1}e_0 + (A + E_0)^{-1}g_0 + h_0 \\
&= x_0 - (I + F_0)e_0 + (A + E_0)^{-1}g_0 + h_0 \\
&= x - F_0 e_0 + (A + E_0)^{-1}g_0 + h_0.
\end{aligned}$$

Hence

$$\begin{aligned}
\|e_1\| &\leq \|F_0\|\|e_0\| + \|(A + E_0)^{-1}\|\|g_0\| + \|h_0\| \\
&\leq \rho\|e_0\| + (1+\rho)\|A^{-1}\|\gamma_0 + \eta_0.
\end{aligned}$$

Similarly

$$\begin{aligned}
\|e_2\| &\leq \rho\|e_1\| + (1+\rho)\|A^{-1}\|\gamma_1 + \eta_1 \\
&\leq \rho^2\|e_0\| + (1+\rho)\|A^{-1}\|(\gamma_1 + \rho\gamma_0) + (\eta_1 + \rho\eta_0).
\end{aligned}$$

An obvious induction now gives (3.29). The bounds (3.30) and (3.31) follow directly from (3.29) and the definitions of $\gamma_\pm$ and $\eta_\pm$. ∎

The usual application of iterative refinement is to improve solutions of linear systems. In this case the error e_0 starts off larger than $\|A\|\gamma_+ + \eta_+$. The inequality (3.30) says that, assuming ρ is reasonably less than one, each iteration will decrease the error by a factor of ρ until the error is of a size with $\|A\|\gamma_+ + \eta_+$, at which point the iteration will stagnate. Thus the theorem provides both a convergence rate and a limit on the attainable accuracy.

The theorem also applies to nonlinear function r. For if r is differentiable at x_*, then

$$r(x_k) = b - Ax_k + o(\|e_k\|),$$

and we can incorporate the $o(\|e_k\|)$ term into the vector g_k. In this case the bound (3.31) says that the initial values of the $o(\|e_k\|)$ terms do not affect the limiting accuracy — though they may slow down convergence. Such an iteration is called self-correcting.

It should be stressed that the theorem does not tell us when a nonlinear iteration will converge — always a difficult matter. What it does says is that if the method does converge then nonlinearities have no effect on its asymptotic behavior.

3.5. NOTES AND REFERENCES

General references

Perturbation theory for linear systems is the staple of introductory numerical texts. For advanced treatments with further references see Stewart and Sun's *Matrix Perturbation Theory* [310] and Chapter 7 of Higham's *Accuracy and Stability of Numerical Algorithms* [177].

Normwise perturbation bounds

Perturbations bound for linear systems can be found as early as 1936 [350]. However, Turing [321, 1948] is responsible for the marriage of perturbation theory and rounding-error analysis. He introduced the condition number to quantify "the expression 'ill-condition' [which] is sometimes used merely as a term of abuse applicable to matrices or equations . . . ." An independent Russian tradition is suggested by the book of Faddeeva [116], which appeared in 1950 (see the bibliography in [115]).

Wilkinson [345, 1963, p. 103] pointed out the denominator in such bounds as (3.8) are harmless fudge factors. Forsythe and Moler [120, 1967] showed that they could be eliminated entirely by dividing by $\|\tilde{x}\|$ rather than by $\|x\|$.

Artificial ill-conditioning

Wilkinson [344, 1961] seems to have been the first to point out that artificial ill-conditioning could be laid to the unsharpness of bounds like $\|A^{-1}E\| \leq \|A^{-1}\|\|E\|$. His definitive words on the subject may be found on pages 192–193 of his *Algebraic Eigenvalue Problem* [346]. All other treatments of the subject, including the one here, are just elaborations of the good common sense contained in those two pages.

The authors of LINPACK [99, 1979] recommend equal error scaling to mitigate the effects of artificial ill-conditioning. Curtis and Reid [83] describe an algorithm for balancing the elements of a matrix in the least squares sense.

Componentwise bounds

Bauer [21, 1966] began the study of componentwise perturbation theory. His results, which chiefly concerned matrix inverses, never caught on (possibly because he wrote in German) until Skeel [281, 1979] established what is essentially (3.14). (We shall hear more of Skeel in the next subsection.) If we assume that $|E| \leq \epsilon|A|$, then on taking norms in (3.14), we get

$$\frac{\|\tilde{x} - x\|}{\|\tilde{x}\|} \leq \||A^{-1}\|A|\|\epsilon. \tag{3.32}$$

The number $\||A^{-1}\|A|\|$ is called the *Bauer–Skeel condition number*. It is invariant under row scaling of A but not under column scaling. In fact column scaling is the *bête noire* of all mixed bounds like (3.32) or (3.18). Such scaling changes the norm of the solution, and the bound must therefore also change. The only practical cure for

SEC. 4. THE EFFECTS OF ROUNDING ERROR 225

this problem is to work with formulas for the individual components of the solution [e.g., (3.17)].

For a survey of componentwise perturbation theory, see [176].

Backward perturbation theory

Theorem 3.14 is due to Wilkinson [345, 1963, p. 141]. Rigal and Gaches [269, 1967] established it for arbitrary operator norms. Theorem 3.15 is due to Oettli and Prager [245, 1964]. The statement and proof given here are taken verbatim from [310].

Iterative refinement

According to Higham [177, §9.10], Wilkinson gave a program for iterative refinement in 1948. The essence of the method is making do with the inverse of an approximation to the operator in question. Its prototype is the practice of replacing the derivative in Newton's method with a suitable, easily computable approximation. Some other applications of the method, notably to eigenvalue problems, will be found in [36, 45, 94, 98, 106, 162, 287].

The analysis given here parallels what may be found in the literature, e.g., [230, 288, 177] but with some variations that make it easier to apply to nonlinear problems.

4. THE EFFECTS OF ROUNDING ERROR

We have seen in Example 3.4 that if a matrix A with a condition number of 10^k is rounded to t decimal digits, we cannot guarantee more than $t-k$ digits of accuracy in the solution. Information has been lost, and no algorithm that knows only the rounded coefficients can be expected to recover the true solution. This situation is what J. H. Wilkinson calls "the fundamental limit of t-digit computation."

Although the accuracy of a computed solution is limited by the effects of rounding error, we would not be happy with an algorithm that introduced additional errors of its own. Such algorithms are said to be unstable. It was Wilkinson's great insight that stable algorithms coincide with the algorithms that have a backward rounding-error analysis. At this point you may want to review the material in §4.3, Chapter 2, where the ideas of stability, condition, and backward rounding error are treated in a general setting.

Since the focus of this work is algorithms, not rounding-error analysis, we will only presents the results of the analysis. In the next section we will treat triangular matrices. It turns out that the triangular systems that result from Gaussian elimination with pivoting tend to be solved accurately, even when the systems are ill conditioned. In §4.3 we will turn to Gaussian elimination itself. After stating the basic result, we will apply it to the solution of linear systems and to matrix inversion. In §4.4 we will consider the interrelated problems of scaling and pivoting. Finally, we consider iterative refinement applied to the solution of linear systems.

As usual ϵ_M stands for the rounding unit on the machine in question and $\epsilon'_M =$

$\epsilon_M/0.9$ is the adjusted rounding unit (see Theorem 4.10, Chapter 2).

4.1. Error Analysis of Triangular Systems

In this section we will treat the error analysis of triangular systems. The result is a backward error analysis with an extremely tight componentwise bound. The bound can be used as usual to predict the accuracy of the computed solution; however, as we shall see in the next subsection, the error in the computed solutions is often much less than the predicted error.

The results of the error analysis

The first problem one encounters in embarking on a rounding-error analysis of an algorithm is to decide which version of the algorithms to analyze. Fortunately, for triangular matrices all the algorithms considered up to now are essentially equivalent — in producing a component of the solution they perform the same operations in the same order, although they may interleave the operations for different components differently. For definiteness, we will treat the forward substitution for solving the lower triangular system $Lx = b$ (see §2, Chapter 2).

The basic result, which we state without proof, is the following.

Theorem 4.1. *Let L be a nonsingular lower triangular matrix of order n. Let $\tilde{x}$ denote the solution of the system $Lx = b$ computed by the forward substitution algorithm in floating-point arithmetic with rounding unit ϵ_M. Then there is a matrix E satisfying*

$$|E| \leq \epsilon'_M \begin{pmatrix} |\ell_{11}| & 0 & 0 & \cdots & 0 \\ |\ell_{21}| & 2|\ell_{22}| & 0 & \cdots & 0 \\ |\ell_{31}| & 2|\ell_{32}| & 3|\ell_{33}| & \cdots & 0 \\ \vdots & \vdots & \vdots & & \vdots \\ |\ell_{n1}| & 2|\ell_{n2}| & 3|\ell_{n3}| & \cdots & n|\ell_{nn}| \end{pmatrix} \quad (4.1)$$

such that

$$(L + E)\tilde{x} = b. \quad (4.2)$$

Here are some comments on this theorem.

- We could hardly expect a better analysis. The computed $\tilde{x}$ is the exact solution of a triangular system whose coefficients are small *relative* perturbations of the original — at most n times the size of the adjusted rounding unit. In most applications it is unlikely that these perturbations will be larger than the errors that are already present in the coefficients.

- Since the variations on the forward substitution algorithm in §2, Chapter 2, perform the same operations in the same order, the above theorem also applies to them. If the order of operations is varied, then the bound (4.1) can be replaced by

$$|\epsilon_{ij}| \leq n\epsilon'_M |\ell_{ij}|, \qquad i = 1, \ldots, n;\ j = 1, \ldots, i. \quad (4.3)$$

SEC. 4. THE EFFECTS OF ROUNDING ERROR

Note that in applying the bound we would probably make this simplification anyway.

- An obvious variant of the theorem applies to the solution of an upper triangular system $Ux = b$ by back substitution. The chief difference is that the error matrix assumes the form

$$E = \epsilon'_M \begin{pmatrix} n|u_{11}| & (n-1)|u_{12}| & (n-2)|u_{13}| & \cdots & |u_{1n}| \\ 0 & (n-1)|u_{22}| & (n-2)|u_{23}| & \cdots & |u_{2n}| \\ 0 & 0 & (n-2)|u_{33}| & \cdots & |u_{3n}| \\ \vdots & \vdots & \vdots & & \vdots \\ 0 & 0 & 0 & \cdots & |u_{nn}| \end{pmatrix}. \quad (4.4)$$

4.2. THE ACCURACY OF THE COMPUTED SOLUTIONS

The backward error bound (4.1) does not imply that the computed solution is accurate. If the system is ill conditioned, the small backward errors will be magnified in the solution. We can use the perturbation theory of the last subsection to bound their effect. Specifically, the bound (3.7) says that

$$\frac{\|\tilde{x} - x\|}{\|\tilde{x}\|} \leq \kappa(L)\frac{\|E\|}{\|L\|},$$

where $\kappa(L) = \|L\|\|L^{-1}\|$ is the condition number of L. From (4.3) we easily see that for any absolute norm

$$\|E\| \leq n\epsilon'_M \|L\|.$$

Consequently

$$\frac{\|\tilde{x} - x\|}{\|\tilde{x}\|} \leq n\kappa(L)\epsilon'_M. \quad (4.5)$$

Thus if $\kappa(L)$ is large, we may get inaccurate solutions. Let us look at an example.

Example 4.2. *Consider the matrix W_n illustrated below for $n = 6$:*

$$W_6 = \begin{pmatrix} 1 & 0 & 0 & 0 & 0 & 0 \\ -1 & 1 & 0 & 0 & 0 & 0 \\ -1 & -1 & 1 & 0 & 0 & 0 \\ -1 & -1 & -1 & 1 & 0 & 0 \\ -1 & -1 & -1 & -1 & 1 & 0 \\ -1 & -1 & -1 & -1 & -1 & 1 \end{pmatrix}.$$

The inverse of W_6 is

$$W_6^{-1} = \begin{pmatrix} 1 & 0 & 0 & 0 & 0 & 0 \\ 1 & 1 & 0 & 0 & 0 & 0 \\ 2 & 1 & 1 & 0 & 0 & 0 \\ 4 & 2 & 1 & 1 & 0 & 0 \\ 8 & 4 & 2 & 1 & 1 & 0 \\ 16 & 8 & 4 & 2 & 1 & 1 \end{pmatrix},$$

from which we see that

$$\|W_n^{-1}\|_\infty = 2^{n-1}.$$

Since $\|W_n\|_\infty = n$, the matrix W_n is ill conditioned.

The ill-conditioning of W_n is not artificial, as the following experiment shows. Let $b = W_n e$ and replace w_{11} by $1 + \epsilon_M$, where ϵ_M is the rounding unit for IEEE double precision arithmetic (about $2.2 \cdot 10^{-16}$). If we solve the perturbed system to get the vector $\tilde{x}$, we have the following results.

n	$\|W_n^{-1}\|_\infty \epsilon_M$	$\|\tilde{x} - e\|_\infty / \|e\|_\infty$
10	1.1e−13	4.6e−14
20	1.2e−10	4.7e−11
30	1.2e−07	4.8e−08
40	1.2e−04	4.9e−05
50	1.2e−01	5.0e−02

The second column of this table contains a ballpark estimate of the error. The third column contains the actual error. Although the former is pessimistic, it tracks the latter, which grows with the size of W_n.

In spite of this example, the ill-conditioning of most triangular systems is artificial, and they are solved to high accuracy. We will return to this point in the next subsection.

The residual vector

The error in the solution of a triangular system can also be thrown onto the right-hand side. For if $\tilde{x}$ denotes the computed solution and

$$r = b - L\tilde{x}$$

is the *residual vector*, then by definition $L\tilde{x} = b - r$. Since $(L + E)\tilde{x} = b$, we have $r = Ex$. On taking norms and applying (4.3), we get the following corollary,

Corollary 4.3. *The computed solution $\tilde{x}$ of the system $Lx = b$ satisfies*

$$L\tilde{x} = b - r,$$

SEC. 4. THE EFFECTS OF ROUNDING ERROR

where in any absolute norm

$$\|r\| \leq n\epsilon'_{\text{M}}\|L\|\|\tilde{x}\|. \tag{4.6}$$

If $n\epsilon_{\text{M}}\|L\|\|x\|$ is less than errors already present in b, then any inaccuracies of the solution can be regarded as coming from b rather than being introduced by the algorithm.

It is worth noting that the bound (4.6) is within a factor of n of what we could expect from the correctly rounded solution. For if $Lx^* = b$ and we round x^*, we get $\tilde{x} = x + g$, where

$$\gamma_i = \xi_i^* \epsilon_i, \qquad |\epsilon_k| \leq \epsilon_{\text{M}}.$$

Hence

$$\|b - L\tilde{x}\| = \|Lg\| \leq \epsilon_{\text{M}}\|L\|\|x^*\|.$$

4.3. ERROR ANALYSIS OF GAUSSIAN ELIMINATION

In this subsection we will show that the rounding errors made in the course of Gaussian elimination can be projected back on the original matrix — just as for the solution of triangular systems. The chief difference is that the errors are now absolute errors, and, depending on a quantity called the growth factor, they can be large. We will apply the error analysis to the solution of linear systems and to the invert-and-multiply algorithm.

The error analysis

As we have seen, Gaussian elimination comes in many variations. For a general dense matrix, these variations perform the same operations in the same order, and a general analysis will cover all of them. By Theorem 1.8, we can assume that any pivoting has been done beforehand, so that we can analyze the unpivoted algorithm.

The basic result is very easy to state.

Theorem 4.4. *Let the LU decomposition of the matrix A be computed by Gaussian elimination in floating-point arithmetic with rounding unit ϵ_{M}. Then the matrix*

$$E = A - LU$$

satisfies

$$|E| \leq n\epsilon'_{\text{M}}|L||U|. \tag{4.7}$$

Theorem 4.4 states that the computed LU decomposition of A is the exact decomposition of $A + E$, where E is proportional to n times the rounding unit. The factor n is usually an overestimate and applies only to the elements in the southeast. For matrices of special structure — e.g., Hessenberg matrices — the factor n can often be replaced by a small constant.

The sizes of the individual elements of E depend on the size of the computed factors L and U. To see the effects of large factors, let us return to the matrix of Example 2.8.

Example 4.5. *Let*

$$A = \begin{pmatrix} 10^{-6} & 1 & 2 \\ 1 & 0 & 3 \\ 2 & 3 & 0 \end{pmatrix}.$$

If we perform Gaussian elimination on this matrix in four-digit arithmetic, we get

$$L = \begin{pmatrix} 1 & 0 & 0 \\ 1 \cdot 10^6 & 1 & 0 \\ 2 \cdot 10^6 & 2 & 1 \end{pmatrix} \quad \text{and} \quad U = \begin{pmatrix} 1 \cdot 10^{-6} & 1 & 2 \\ 0 & -1 \cdot 10^6 & -2 \cdot 10^6 \\ 0 & 0 & 0 \end{pmatrix}.$$

Both these factors have large elements, and our analysis suggests that their product will not reproduce A well. In fact, the product is

$$\begin{pmatrix} 10^{-6} & 1 & 2 \\ 1 & 0 & 0 \\ 2 & 0 & 0 \end{pmatrix}.$$

The $(2,3)$- and $(3,2)$-elements have been obliterated.

The difficulties with the matrix A in the above example can be cured by partial pivoting for size. This amounts to interchanging the first and third rows of A, which yields the factors

$$L = \begin{pmatrix} 1 & 0 & 0 \\ 0.5 & 1 & 0 \\ 0.5 \cdot 10^{-6} & -0.6667 & 1 \end{pmatrix} \quad \text{and} \quad U = \begin{pmatrix} 2 & 3 & 0 \\ 0 & -1.5 & 3 \\ 0 & 0 & 4 \end{pmatrix}.$$

The product of these factors is

$$LU = \begin{pmatrix} 2 & 3 & 0 \\ 1 & 0 & 3 \\ 10^{-6} & 1.0000515 & 1.9999 \end{pmatrix}, \tag{4.8}$$

which is very close to A with its first and third row interchanged.

The product (4.8) illustrates a point that is easy to overlook — namely, the matrix $A + E$ from Theorem 4.4 need not be representable as an array of floating-point numbers. In fact, if (4.8) is rounded to four digits, the result is A itself.

We have given componentwise bounds, but it is also possible to give bounds in terms of norms. To do this, it will be convenient to introduce some notation.

SEC. 4. THE EFFECTS OF ROUNDING ERROR

Definition 4.6. *Let $A = LU$ be an LU decomposition of A. Then the* GROWTH FACTOR *with respect to the norm $\|\cdot\|$ is*

$$\gamma(A) \stackrel{\text{def}}{=} \frac{\|\,|L|\,|U|\,\|}{\|A\|}.$$

If L_{ϵ_M} and U_{ϵ_M} are L- and U-factors computed in floating-point arithmetic with rounding unit ϵ_M, then

$$\gamma_{\epsilon_M}(A) \stackrel{\text{def}}{=} \frac{\|\,|L_{\epsilon_M}|\,|U_{\epsilon_M}|\,\|}{\|A\|}.$$

With this definition we have the following corollary of Theorem 4.4.

Corollary 4.7. *For any absolute norm $\|\cdot\|$ the backward error E satisfies*

$$\frac{\|E\|}{\|A\|} \leq n\gamma_{\epsilon_M}(A)\epsilon_M. \tag{4.9}$$

Proof. Take norms in (4.7) and use the definition of $\gamma_{\epsilon_M}(A)$.

For any absolute norm we have

$$\|A\| = \|\,|A|\,\| \leq \|\,|L|\,|U|\,\|,$$

so that $\gamma(A) \geq 1$. In general, $\gamma_{\epsilon_M}(A)$ will be a reasonable approximation to $\gamma(A)$ and hence will also be greater than one. Consequently, $\gamma_{\epsilon_M}(A)$ serves as a magnification factor mediating how the rounding error in the arithmetic shows up in the relative backward error.

The condition of the triangular factors

We are going to apply Theorem 4.4 to analyze what happens when Gaussian elimination is used to solve linear systems and invert matrices. But these analyses are misleading without the addition of an important empirical observation.

> When a matrix is decomposed by Gaussian elimination with partial or complete pivoting for size, the resulting L-factor tends to be well conditioned while any ill-conditioning in the U-factor tends to be artificial. (4.10)

Why this should be so is imperfectly understood, but the effect is striking.

Example 4.8. *The matrix*

$$A = \begin{pmatrix} 4.1209\text{e}{-01} & -3.6937\text{e}{-01} & 2.9024\text{e}{-02} & -1.3198\text{e}{-01} \\ 8.1818\text{e}{-02} & -6.1902\text{e}{-02} & 1.2648\text{e}{-02} & -2.2603\text{e}{-02} \\ -4.4000\text{e}{-01} & 3.7436\text{e}{-01} & -4.1898\text{e}{-02} & 1.3461\text{e}{-01} \\ 4.1725\text{e}{-01} & -3.4564\text{e}{-01} & 4.4889\text{e}{-02} & -1.2468\text{e}{-01} \end{pmatrix}$$

was generated in the form

$$A = U \text{diag}(1.0000\text{e}+00, 2.1544\text{e}-02, 4.6416\text{e}-04, 1.0000\text{e}-05)V^\text{T},$$

where U and V are random orthogonal matrices. Thus $\kappa_2(A) = 10^5$. The L-factor resulting from Gaussian elimination with partial pivoting for size is

$$L = \begin{pmatrix} 1.0000\text{e}+00 & & & \\ -9.3657\text{e}-01 & 1.0000\text{e}+00 & & \\ -1.8595\text{e}-01 & -4.1108\text{e}-01 & 1.0000\text{e}+00 & \\ -9.4829\text{e}-01 & -4.9906\text{e}-01 & 9.0041\text{e}-02 & 1.0000\text{e}+00 \end{pmatrix}$$

and its condition number in the 2-norm is $4.2\text{e}+00$. The corresponding U-factor is

$$U = \begin{pmatrix} -4.4000\text{e}-01 & 3.7436\text{e}-01 & -4.1898\text{e}-02 & 1.3461\text{e}-01 \\ & -1.8753\text{e}-02 & -1.0216\text{e}-02 & -5.9071\text{e}-03 \\ & & 6.5700\text{e}-04 & -4.4270\text{e}-07 \\ & & & 1.8446\text{e}-05 \end{pmatrix}$$

and its condition number is $\kappa_2(U) = 3.4\text{e}+04$. But if we row-scale U so that its diagonal elements are one, we obtain a matrix whose condition number is $2.9\text{e}+00$.

Since row-scaling a triangular system has no essential effect on the accuracy of the computed solution, systems involving the U-factor will be solved accurately. Systems involving the L-factor will also be solved accurately because L is well conditioned. As we shall see, these facts have important consequences.

It should be stressed that this phenomenon is not a mathematical necessity. It is easy to construct matrices for which the L-factor is ill conditioned and for which the ill-conditioning in the U-factor is genuine. Moreover, the strength of the phenomenon depends on the pivoting strategy, being weakest for no pivoting and strongest for complete pivoting. For more see the notes and references.

The solution of linear systems

Turning now to the solution of linear systems, suppose that the computed LU decomposition is used to solve the system $Ax = b$. The usual algorithm is

1. Solve $Ly = b$
2. Solve $Ux = y$ (4.11)

The result of the error analysis of this algorithm is contained in the following theorem.

Theorem 4.9. *Let $A + E = LU$, where LU is the LU decomposition of A computed by Gaussian elimination with rounding unit ϵ_M. Let the solution of the system $Ax = b$ be computed from L and U by (4.11). Then the computed solution $\tilde{x}$ satisfies*

$$(A + H)\tilde{x} = b,$$

SEC. 4. THE EFFECTS OF ROUNDING ERROR

where

$$|H| \leq (3 + n\epsilon'_M)n\epsilon'_M|L||U|.$$

Hence if $\|\cdot\|$ is an absolute norm,

$$\frac{\|H\|}{\|A\|} \leq (3 + n\epsilon'_M)n\gamma_{\epsilon_M}(A)\epsilon'_M.$$

Proof. By Theorem 4.1, the computed y satisfies

$$(L+F)y = b, \qquad |F| \leq n\epsilon'_M|L|,$$

and the compute x satisfies

$$(U+G)\tilde{x} = y, \qquad |G| \leq n\epsilon'_M|U|.$$

Combining these results, we have

$$b = (L+F)(U+G)\tilde{x} = (LU + LG + FU + FG)\tilde{x}$$
$$= (A + E + LG + FU + FG)\tilde{x} \equiv (A + H)\tilde{x}.$$

But

$$|H| \leq |E| + |L||G| + |F||U| + |F||G| \leq (3 + n\epsilon_M)n\epsilon'_M|L||U|.$$

The norm bound on H follows on taking norms in the bound on $|H|$ and applying the definition of $\gamma_{\epsilon_M}(A)$. ∎

Thus the computed solution is the exact solution of a slightly perturbed system. The bound on the perturbation is greater by a factor of essentially three than the bound on the perturbation E produced by the computation of the LU decomposition. However, this bound does not take into account the observation (4.10) that the L-factor produced by Gaussian elimination with pivoting tends to be well conditioned while any ill-conditioning in the U-factor tends to be artificial. Consequently, the first triangular system in (4.11) will be solved accurately, and the solution of the second will not magnify the error. This shows that:

> *If Gaussian elimination with pivoting is used to solve the system $Ax = b$, the result is usually a vector $\hat{x}$ that is near the solution of the system $(A + E)\tilde{x} = b$, where E is the backward error from the elimination procedure.* (4.12)

Since this observation is not a theorem, we cannot quantify the nearness of $\hat{x}$ to $\tilde{x}$. But in many matrix algorithms one must take the phenomenon into account to really understand what is going on.

The stability result for linear systems implies that if the growth factor is small, the computed solution almost satisfies the original equations.

Corollary 4.10. *Let* $r = b - A\tilde{x}$. *Then*

$$\frac{\|r\|}{\|A\|\|\tilde{x}\|} \leq (3 + n\epsilon'_M) n\gamma_{\epsilon_M}(A)\epsilon'_M.$$

Proof. Since $b - (A + H)x = 0$, we have $r = H\tilde{x}$. Hence $\|r\| \leq \|H\|\|\tilde{x}\| \leq (3 + n\epsilon_M)n\gamma_{\epsilon_M}(A)\epsilon_M \|A\|\|\tilde{x}\|$. ∎

We have seen in Theorem 3.14 that the converse of this corollary is also true. If a purported solution has a small residual, it comes from a slightly perturbed problem. This converse has an important practical implication. If we want to know if an algorithm has solved a linear system stably, all we have to do is compare its residual, suitably scaled, to the rounding unit.

Matrix inversion

We conclude our treatment of backward stability with a discussion of matrix inversion. For definiteness, let us suppose that we compute the inverse X of A by solving the systems

$$Ax_i = \mathbf{e}_i, \quad i = 1, \ldots, n.$$

Then each column of the inverse satisfies

$$(A + H_i)x_i = b_i.$$

However, it does not follow from this that there is a single matrix H such that $X = (A+H)^{-1}$, and in general there will not be — matrix inversion is not backward stable.

However, it is almost stable. By the observation (4.12) each x_i will tend to be near the solution of the system $(A + E)\tilde{x}_i = \mathbf{e}_i$, where E is the backward error from Gaussian elimination. Thus the computed inverse will tend to be near the inverse of matrix $A + E$, where E is small.

Unfortunately in some applications nearly stable is not good enough. The following example shows that this is true of the invert-and-multiply algorithm for solving a linear system.

Example 4.11. *The following equation displays the first five digits of a matrix having singular values* $1, 10^{-7}$, *and* 10^{-14}:

$$A \cong \begin{pmatrix} 9.8127\text{e}{-}02 & -5.3667\text{e}{-}02 & -2.6485\text{e}{-}02 \\ -7.6139\text{e}{-}01 & 4.1641\text{e}{-}01 & 2.0550\text{e}{-}01 \\ 3.7355\text{e}{-}01 & -2.0430\text{e}{-}01 & -1.0083\text{e}{-}01 \end{pmatrix}.$$

With $x_* = \mathbf{e}$, *I computed* $b = Ax_*$ *and solved the equation by Gaussian elimination with partial pivoting to get a solution* x_g *and by invert-and-multiply to get a solution*

SEC. 4. THE EFFECTS OF ROUNDING ERROR

x_i. *The following is a table of the relative errors and the relative residuals for the two solutions.*

	g	i
$\frac{\|x-x_*\|}{\|x_*\|}$	1.5e−03	8.5e−04
$\frac{\|b-Ax\|}{\|b\|}$	5.0e−16	1.6e−03

The invert-and-multiply solution is slightly more accurate than the solution by Gaussian elimination. But its residual is more than 12 orders of magnitude larger.

A simplified analysis will show what is going on here. Suppose that we compute the correctly rounded inverse — that is, the computed matrix X satisfies $X = A^{-1} + F$, where $\|F\| \leq \alpha \|A^{-1}\| \epsilon_M$, for some constant α. If no further rounding errors are made, the solution computed by the invert-and-multiply algorithm is $A^{-1}b + Fb$, and the residual is $r = -AFb$. Hence

$$\frac{\|r\|}{\|b\|} \leq \alpha \kappa(A) \epsilon_M.$$

Thus the residual (compared to $\|b\|$) can be larger than the rounding unit by a factor of $\kappa(A)$. In other words, if A is ill conditioned expect large residuals from the invert-and-multiply algorithm.

4.4. PIVOTING AND SCALING

We have seen that the backward componentwise errors induced by Gaussian elimination are bounded by $n\epsilon'_M |L||U|$. For a stable reduction it is imperative that the components of $|L||U|$ not be grossly larger than the corresponding components of A. Pivoting for size is a means of controlling the magnitudes of the elements of $|L||U|$. In this subsection, we will treat the two most common pivoting strategies: partial pivoting and complete pivoting. We will also consider classes of matrices for which pivoting is unnecessary.

In all pivoting strategies, scaling the rows and columns of a matrix can change the pivots selected. This leaves the knotty and poorly understood problem of how a matrix should be scaled before starting the elimination. We will discuss this problem at the end of this subsection.

On scaling and growth factors

First some general observations on scaling. Most pivot strategies are not invariant under scaling — the choice of pivots changes when A is replaced by $D_R A D_C$, where D_R and D_C are nonsingular diagonal matrices. It is important to realize that when the order of pivots is fixed the bound on the backward rounding error E from Gaussian elimination inherits the scaling; that is,

$$|E| \leq n\epsilon'_M |D_R||L||U||D_C| \tag{4.13}$$

[cf. (4.7)]. In particular, to the extent that the bounds are valid, the *relative* backward error in any component of A remains unchanged by the scaling. This means that if a fixed pivoting sequence is found to be good for a particular matrix, the same sequence will be good for all scaled versions of the matrix.

The normwise growth factor

$$\gamma(A) = \frac{\||L||U|\|}{\|A\|}$$

is not easy to work with. However, if partial or complete pivoting for size is used in computing the LU decomposition, the components of $|L|$ are not greater than one, and any substantial growth will be found in the elements of U. For this reason we will analyze the growth in terms of the number

$$\bar{\gamma}_n = \frac{\max |u_{ij}|}{\max |a_{ij}|}. \qquad (4.14)$$

The absence of the matrix A and the presence of the subscript n indicates that we will be concerned with the behavior of $\bar{\gamma}_n$ for arbitrary matrices as a function of the order of the matrix.

The backward error bound for Gaussian elimination is cast in terms of the *computed* L- and U-factors, and strictly speaking we should include the effects of rounding errors in our analysis of $\bar{\gamma}_n$. However, this is a tedious business that does not change the results in any essential way. Hence we will analyze the growth for the exact elimination procedure.

Partial and complete pivoting

The three most common pivoting strategies for dense matrices are partial pivoting, complete pivoting, and diagonal pivoting. Diagonal pivoting is not generally used to control the effects of rounding error. Hence we treat only partial and complete pivoting here.

- **Partial pivoting.** The best general result on partial pivoting is discouraging.

Theorem 4.12. *Let the LU decomposition $PA = LU$ be computed by partial pivoting for size. Then*

$$\bar{\gamma}_n \leq 2^{n-1}.$$

Proof. Assume that pivoting has been done initially. Let $a_{ij}^{(k)}$ be the elements of the kth Schur complement, and let $\alpha_k = \max_{ij} |a_{ij}^{(k)}|$. Now

$$a_{ij}^{(k+1)} = a_{ij}^{(k)} - \ell_{ik} a_{ki}^{(k)}.$$

Since $|\ell_{ik}| \leq 1$,

$$|a_{ij}^{(k+1)}| = |a_{ij}^{(k)}| + |\ell_{ik}||a_{ki}^{(k)}| \leq 2\alpha_k.$$

SEC. 4. THE EFFECTS OF ROUNDING ERROR

Hence $\alpha_{k+1} \leq 2\alpha_k \leq 2^{k-1}\alpha_1$. Since the kth row of U consists of the elements $a_{ki}^{(k)}$, the result follows. ∎

The discouraging aspect of this bound is that it suggests that we cannot use Gaussian elimination with partial pivoting on matrices larger than roughly $-\log_2 \epsilon_M$. For at that size and beyond, the backward error could overwhelm the elements of the matrix. For IEEE standard arithmetic, this would confine us to matrices of order, say, 50 or less.

Moreover, the bound can be attained, as the following example shows.

Example 4.13. *Let*

$$A = \begin{pmatrix} 1 & 0 & 0 & \cdots & 1 \\ -1 & 1 & 0 & \cdots & 1 \\ -1 & -1 & 1 & \cdots & 1 \\ \vdots & \vdots & \vdots & & \vdots \\ -1 & -1 & -1 & \cdots & 1 \end{pmatrix}.$$

Then if we break ties in the choice of pivot in favor of the diagonal element, it is easily seen that each step of Gaussian elimination doubles the components of the last column, so that the final U has the form

$$U = \begin{pmatrix} 1 & 0 & 0 & \cdots & 1 \\ 0 & 1 & 0 & \cdots & 2 \\ 0 & 0 & 1 & \cdots & 4 \\ \vdots & \vdots & \vdots & & \vdots \\ 0 & 0 & 0 & \cdots & 2^{n-1} \end{pmatrix}.$$

In spite of this unhappy example, Gaussian elimination with partial pivoting is the method of choice for the solution of dense linear systems. The reason is that the growth suggested by the bound rarely occurs in practice. The reasons are not well understood, but here is the bill of particulars.

1. Proof by authority. Writing in 1965, Wilkinson stated, "No example which has arisen naturally has in my experience given an increase by a factor as large as 16."
2. Matrices with small singular values (i.e., ill-conditioned matrices) often exhibit a systematic decrease in the elements of U.
3. When A has special structure, the bound on $\bar{\gamma}_n$ may be quite a bit smaller than 2^{n-1}. For Hessenberg matrices $\bar{\gamma}_n \leq n$. For tridiagonal matrices $\bar{\gamma}_n \leq 2$.
4. Example 4.13 is highly contrived. Moreover, all examples that exhibit the same growth are closely related to it.

5. Attempts to find uncontrived examples give much lower growths. Matrices of standard normal deviates exhibit a growth of order $\sqrt{n}$. Certain orthogonal matrices exhibit very slow growth.

Against all this must be set the fact that two examples have recently surfaced in which partial pivoting gives large growth. Both bear a family resemblance to the matrix in Example 4.13. The existence of these examples suggests that Gaussian elimination with partial pivoting cannot be used uncritically—when in doubt one should monitor the growth. But the general approbation of partial pivoting stands.

- **Complete pivoting.** Complete pivoting can be recommended with little reservation. It can be shown that

$$\bar{\gamma}_n \leq n^{\frac{1}{2}} \left(2 \cdot 3^{\frac{1}{2}} \cdots n^{\frac{1}{n-1}}\right)^{\frac{1}{2}}. \tag{4.15}$$

The bound is not exactly small—for $n = 1000$ it is about seven million—but for many problems it would be satisfactory. However, the bound is rendered largely irrelevant by the fact that until recently no one has been able to devise an example for which $\bar{\gamma}_n$ is greater than n. For many years it was conjectured that n was an upper bound on the growth, but a matrix of order 25 has been constructed for which the $\bar{\gamma}_n$ is about 33.

—

Given the general security of complete pivoting and the potential insecurity of partial pivoting, it is reasonable to ask why not use complete pivoting at all times. There are three answers.

1. Complete pivoting adds an $O(n^3)$ overhead to the algorithm—the time required to find the maximum elements in the Schur complements. This overhead is small on ordinary computers, but may be large on supercomputers.
2. Complete pivoting can be used only with unblocked classical Gaussian elimination. Partial pivoting can be use with blocked versions of all the variants of Gaussian elimination except for Sherman's march. Thus partial pivoting gives us more flexibility to adapt the algorithm to the machine in question.
3. Complete pivoting frequently destroys the structure of a matrix. For example, complete pivoting can turn a banded matrix into one that is not banded. Partial pivoting, as we have seen, merely increases the band width.

Matrices that do not require pivoting

Unless something is known in advance about a matrix, we must assume that some form of pivoting is necessary for its stable reduction by Gaussian elimination. However, there are at least three classes of matrices that do not require pivoting—in fact, for which pivoting can be deleterious. Here we will consider positive definite matrices, diagonally dominant matrices, and totally positive matrices.

Sec. 4. The Effects of Rounding Error

- **Positive definite matrices.** The reason no pivoting is required for positive definite matrices is contained in the following result.

Theorem 4.14. *Let A be positive definite, and let a_{kk} be a maximal diagonal element of A. Then*

$$a_{kk} > \max_{i \neq j} |a_{ij}|.$$

Proof. Suppose that for some a_{ij} we have $|a_{ij}| \geq a_{kk}$. Since A is positive definite, the matrix

$$B = \begin{pmatrix} a_{ii} & a_{ij} \\ a_{ij} & a_{jj} \end{pmatrix}$$

must be positive definite and hence have positive determinant. But

$$\det(B) = a_{ii}a_{jj} - a_{ij}^2 \leq a_{kk}^2 - a_{ij}^2 \leq 0. \quad \blacksquare$$

In other words, the element of a positive definite matrix A that is largest in magnitude will be found on the diagonal. By Theorem 2.6, the Schur complements generated by Gaussian elimination are positive definite. When we perform one step of Gaussian elimination on A, the diagonal elements of the Schur complement are given by

$$a_{ii} - \frac{a_{1i}^2}{a_{11}} \leq a_{ii},$$

the inequality following from the fact that $a_{ii}^2 \geq 0$ and $a_{ii} > 0$. Hence the diagonal elements of a positive definite matrix are not increased by Gaussian elimination. Since we have only to look at the diagonal elements to determining the growth factor, we have the following result.

For Gaussian elimination applied to a positive definite matrix $\bar{\gamma}_n = 1$.

Several times we have observed that pivoting can destroy structure. For example, partial pivoting increases the bandwidth of a band matrix. Since positive definite matrices do not require pivoting, we can avoid the increase in band width with a corresponding savings in work and memory.

Partial pivoting is not an option with positive definite matrices, for the row interchanges destroy symmetry and hence positive-definiteness. Moreover, Theorem 4.14 does not imply that a partial pivoting strategy will automatically select a diagonal element — e.g., consider the matrix

$$A = \begin{pmatrix} 0.5 & 1.0 \\ 1.0 & 4.0 \end{pmatrix}.$$

For this reason positive definite systems should not be trusted to a general elimination algorithm, since most such algorithms perform partial pivoting.

- **Diagonally dominant matrices.** Diagonally dominant matrices occur so frequently that they are worthy of a formal definition.

Definition 4.15. *A matrix A of order n is* DIAGONALLY DOMINANT BY ROWS *if*

$$|a_{ii}| \geq \sum_{j \neq i} |a_{ij}|, \qquad i = 1, \ldots, n. \tag{4.16}$$

It is STRICTLY DIAGONALLY DOMINANT *if strict inequality holds in* (4.16). *The matrix A is* (STRICTLY) DIAGONALLY DOMINANT BY COLUMNS *if A^T is (strictly) diagonally dominant by rows.*

The following theorem lists the facts we need about diagonal dominance. Its proof is quite involved, and we omit it here.

Theorem 4.16. *Let A be strictly diagonally dominant by rows, and let A be partition in the form*

$$A = \begin{pmatrix} A_{11} & A_{12} \\ A_{21} & A_{22} \end{pmatrix},$$

where A_{11} is square. Then A_{11} is nonsingular. Moreover, the Schur complement

$$S = A_{22} - A_{21} A_{11}^{-1} A_{12}$$

is strictly diagonally dominant by rows, and

$$|S|\mathbf{e} < |A_{21}|\mathbf{e} + |A_{22}|\mathbf{e}. \tag{4.17}$$

If A is diagonally dominant by rows and A_{11} is nonsingular so that S exists, then S is diagonally dominant by rows, and

$$|S|\mathbf{e} \leq |A_{21}|\mathbf{e} + |A_{22}|\mathbf{e}. \tag{4.18}$$

An analogous theorem holds for matrices that are diagonally dominant by columns. Note that the statement that A_{11} is nonsingular is essentially a statement that any strictly diagonally dominant matrix is nonsingular.

To apply these results to Gaussian elimination, let A be diagonally dominant by columns, and assume that all the leading principal submatrices of A are nonsingular, so that A has an LU decomposition. Since the Schur complements of the leading principal minors are diagonally dominant by columns, Gaussian elimination with partial pivoting is the same as Gaussian elimination without pivoting, provided ties are broken in favor of the diagonal element. Moreover, by (4.18) the sum of the magnitudes of elements in any column of the Schur complement are less than or equal to the sum of the magnitudes of elements in the corresponding column of the original matrix and hence are less than or equal to twice the magnitude of the corresponding diagonal element of the original matrix. Since the largest element of a diagonally dominant matrix may be found on its diagonal, it follows that the growth factor $\bar{\gamma}_n$ cannot be greater than two. We have thus proved the following theorem.

SEC. 4. THE EFFECTS OF ROUNDING ERROR

Theorem 4.17. *Let A be diagonally dominant by columns with nonsingular leading principal submatrices. Then Gaussian elimination on A with no pivoting is the same as Gaussian elimination with partial pivoting, provided ties are broken in favor of the diagonal. Moreover, $\bar{\gamma}_n \leq 2$.*

The theorem shows that Gaussian elimination on a matrix that is diagonally dominant by columns is unconditionally stable. Since up to diagonal scaling factors Gaussian elimination by columns produces the same LU decomposition as Gaussian elimination by rows, Gaussian elimination without pivoting on a matrix that is diagonally dominant by rows is also stable.

The comments on pivoting and structure made about positive definite matrices apply also to diagonally dominant matrices. In particular, partial pivoting on a matrix that is diagonally dominant by rows can destroy the diagonal dominance.

- **Totally positive matrices.** A matrix A of order n is *totally positive* if every square submatrix of A has nonnegative determinant. It is not difficult to show that if A is totally positive with nonsingular leading principal submatrices and $A = LU$ is an LU-decomposition of A then $L \geq 0$ and $U \geq 0$. It follows that $\|A\| = \|\,|L|\,|U|\,\|$, and hence that the growth factor $\gamma(A)$ is one. Consequently totally positive matrices can be reduced without pivoting. As with the preceding classes of matrices, pivoting can destroy total positivity.

Scaling

By scaling we mean the replacement of the matrix A by $D_R A D_C$, where D_R and D_C are diagonal matrices. There are two preliminary observations to be made. We have already observed that the backward error committed in Gaussian elimination inherits any scaling. Specifically, the error bound (4.7) becomes

$$|E| \leq n\epsilon'_M D_R |L||U| D_C.$$

Thus if the pivot order is fixed, the elements suffer essentially the same relative errors (exactly the same if the computation is in binary and the diagonals of D_R and D_C are powers of two).

A second observation is that by scaling we can force partial or complete pivoting to choose any permissible sequence of nonzero elements. For example, if for the matrix

$$A = \begin{pmatrix} 10^{-6} & 1 & 2 \\ 1 & 0 & 3 \\ 2 & 3 & 0 \end{pmatrix}$$

of Example 4.5 we take $D_R = D_C = \mathrm{diag}(1, 10^{-7}, 10^{-7})$, then

$$D_R A D_C = \begin{pmatrix} 10^{-6} & 10^{-7} & 2\cdot 10^{-7} \\ 10^{-7} & 0 & 3\cdot 10^{-14} \\ 2\cdot 10^{-2} & 3\cdot 10^{-2} & 0 \end{pmatrix}. \tag{4.19}$$

Thus partial or complete pivoting chooses the $(1,1)$-element as a pivot. This scaling procedure can be repeated on the Schur complements.

The two observations put us in a predicament. The second observation and the accompanying example show that we can use scaling to force partial or complete pivoting to choose a bad pivot. The first observation says that the bad pivot continues to have the same ill effect on the elimination. For example, Gaussian elimination applied to (4.19) still obliterates the $(2,3)$- and $(3,2)$-elements. What scaling strategy, then, will give a good pivot sequence?

There is no easy answer to this question. Here are three scaling strategies that are sometimes suggested.

1. If A is contaminated with errors, say represented by a matrix G, scale A so that the elements of $D_C G D_R$ are as nearly equal as possible. Under this scaling, the more accurate elements will tend to be chosen as pivots.

2. For complete pivoting, if A contains no errors or relative errors on the order of the rounding unit, scale A so that its elements are as nearly equal as possible. This is really the first strategy with the elements G regarded as a small multiple of the elements of A.

3. For partial pivoting (in which column scaling does not affect the choice of pivots), scale the rows of A so that they have norm one in, say, the 1-norm. This process is called *row equilibration* and is probably the most commonly used strategy.

None of these strategies is foolproof. However, in practice they seem to work well — perhaps because computations in 64-bit floating-point arithmetic tend to be very forgiving. Moreover, iterative refinement, to which we now turn, can often be used to restore lost accuracy.

4.5. Iterative refinement

The iterative refinement algorithm of §3.4 can be applied to improve the quality of the solution x_0 of a linear system $Ax = b$ computed by Gaussian elimination. As such it takes the following form.

1. $r_0 = b - Ax_0$
2. Solve the system $Ad_0 = r_0$ (4.20)
3. $x_1 = x_0 + d_0$

In exact arithmetic this x_1 will be the solution of the system. In the presence of rounding error, x_1 will not be exact, but under some circumstances it will be nearer the solution of x_0. If x_1 is not satisfactory, the algorithm can be repeated iteratively with x_1 replacing x_0.

The advantage of this algorithm is that we can reuse the decomposition of A to compute d_k. Thus the cost of an iteration is only $O(n^2)$. However, what we get out of an iteration depends critically on how r_k is computed.

SEC. 4. THE EFFECTS OF ROUNDING ERROR

A general analysis

We are going to use Theorem 3.16 to analyze the algorithm when it is carried out in floating-point arithmetic. To apply the theorem, we must compute the bounds ρ, γ_+, and η_+ in (3.30). In doing so we will make some reasonable simplifying assumptions.

1. The vectors x_k are bounded in norm by a constant ξ.
2. The vectors Ax_k approximate b.
3. The correction d_k is smaller in norm than x_k.
4. The residual r_k may be computed with a rounding unit $\hat{\epsilon}_M$ that is different from the other computations.

The first three conditions are what one would expect from a converging iteration. The last represents a degree of freedom in the computation of the residual.

Using these assumptions we can derive the following bounds, which for brevity we state without proof. First, from the error analysis of Gaussian elimination we get

$$\rho = \frac{c_E \kappa(A)\epsilon_M}{1 - c_E \kappa(A)\epsilon_M},$$

where c_E is a slowly growing function of n. In what follows, we will assume that

$$\rho < \frac{1}{4}.$$

For the residual it can be shown that the computed vector satisfies

$$r = b - (A + G)x,$$

where

$$|G| \leq c_r |A| \hat{\epsilon}_M$$

with c_r a slowly growing function of n. It follows that

$$\|g\| = \|Gx\| \leq c_r \|A\| \xi \hat{\epsilon}_M \equiv \gamma_+.$$

Finally we can bound the error in the correction by

$$\|h\| \leq 2\xi \epsilon_M \equiv \eta_+.$$

If we combine these bounds with (3.30) we get the following result:

$$\frac{\|e_k\|}{\xi} \leq \rho^k \frac{\|e_0\|}{\xi} + 2 c_r \kappa(A) \hat{\epsilon}_M + 3\epsilon_M. \qquad (4.21)$$

The first term on the right-hand side of (4.21) says that the initial error is decreased by a factor of ρ at each iteration. This decrease continues until the other two terms dominate, at which point convergence ceases. The point at which this happens will depend on $\hat{\epsilon}_M$ — the precision to which the residual is computed. We will consider two cases: double and single precision.

Double-precision computation of the residual

If the residual is computed in double precision, we have

$$\hat{\epsilon}_M \leq \epsilon_M^2.$$

If $2c_r\kappa(A)\epsilon_M$ is less than one, the attainable accuracy is limited by the term $3\epsilon_M$. Thus with double-precision computation of the residual iterative refinement produces a result that is effectively accurate to working precision. If A is ill conditioned the convergence will be slow, but ultimately the solution will attain almost full accuracy.

Single-precision computation of the residual

The bound (4.21) suggests that there is no point in performing iterative refinement with single-precision computation of the residual. For when $\hat{\epsilon}_M = \epsilon_M$, the term $2c_r\kappa(A)\hat{\epsilon}_M$, which limits the attainable accuracy of the method, is essentially the accuracy we could expect from a backward stable method. However, this line of reasoning overlooks other possible benefits of iterative refinement such as the following.

> *Iterative refinement in fixed precision tends to produce solutions that have small componentwise backward error.*

The formal derivation of this result is quite detailed. But it is easy to understand why it should be true. The computed residual is $r = b - (A + G)x$, where $|G| \leq c_r|A|\epsilon_M$ — i.e., $A + G$ is a componentwise small relative perturbation of A. Now let's shift our focus a bit and pretend that we were really trying to solve the system $(A + G)x = b$. Then our residual calculation gives a nonzero vector r that considered as the residual of the system $(A + G)x = b$ is completely accurate. Consequently, one step of iterative refinement will move us nearer the solution of $(A + G)x = b$ — a solution that by definition has a small relative componentwise backward error with respect to the original system $Ax = b$.

Assessment of iterative refinement

The major drawback of iterative refinement is that one has to retain the original matrix. In the days of limited storage, this weighed heavily against the procedure. Today our ideas of what is a small matrix have changed. For example, eight megabytes of memory can easily accommodate a matrix of order 500 and its LU decomposition — both in double precision. Thus in most applications the benefits of fixed-precision iterative refinement can be had for little cost.

Iterative refinement with the residual calculated in double precision has the added disadvantage of being a mixed-precision computation. As the working precision increases, eventually a point will be reached where special software must be used to calculate the residual in double precision. The positive side of this is that given such software, the solution can be calculated to any degree of accuracy for the cost of computing a single, low-precision LU decomposition.

4.6. NOTES AND REFERENCES

General references

Elementary rounding-error analysis is treated in most textbooks on matrix computations. For more detailed treatment see Wilkinson's two books, *Rounding Errors in Algebraic Processes* [345, 1963] and *The Algebraic Eigenvalue Problem* [346, 1965], which contain many samples of rounding-error analysis interpreted with great good sense. Special mention must be made of Higham's book *Accuracy and Stability of Numerical Algorithms* [177], to which the reader is referred for more details and references.

Historical

It is a commonplace that rounding-error analysis and the digital computer grew up together. In the days of hand computation, the person performing the computations could monitor the numbers and tell when a disaster occurred. In fact the principal source of errors were simple blunders on the part of the computer, and the computational tableaus of the time contained elaborate checks to guard against them (e.g., see [112, 1951]). With the advent of the digital computer intermediate quantities were not visible, and people felt the need of mathematical reassurance.

Nonetheless, the first rounding-error analysis of Gaussian elimination predated the digital computer. The statistician Hotelling [186, 1943] gave a forward error analysis that predicted an exponential growth of errors and ushered in a brief period of pessimism about the use of direct methods for the solution of linear systems. This pessimism was dispelled in 1947 by von Neumann and Goldstine [331], who showed that a positive definite system would be solved to the accuracy warranted by its condition. This was essentially a weak stability result, but, as Wilkinson [348] points out, backward error analysis was implicit in their approach. In an insightful paper [321, 1948], Turing also came close to giving a backward rounding-error analysis.

The first formal backward error analysis was due to Givens [145, 1954]. He showed that the result of computing a Sturm sequence for a symmetric tridiagonal matrix is the same as exact computations on a nearby system. However, this work appeared only as a technical report, and the idea languished until Wilkinson's definitive paper on the error analysis of direct methods for solving linear systems [344, 1961]. Wilkinson went on to exploit the technique in a variety of situations.

The error analyses

The basic error analyses of this section go back to Wilkinson [344, 1961]. A proof of Theorem 4.1 may be found in [177]. The theorem, as nice as it is, does not insure that the solution will be accurate. However, the triangular systems that one obtains from Gaussian elimination tend to be artificially ill conditioned and hence are solved accurately.

Wilkinson's original analysis of Gaussian elimination differs from the one given here in two respects. First, he assumes pivoting for size has been performed so that the multipliers are bounded by one. Second, the multiplier of the backward error involves the maximum element of all the intermediate Schur complements, not just the elements of L and U. Although his analysis is componentwise, he is quick to take norms. The first componentwise bound in the style of Theorem 4.4 is due to Chartres and Geuder [65, 1967], and their bound is essentially the same, though not expressed in matrix form.

Condition of the L- and U-factors

In his original paper on error analysis, Wilkinson [344] noted that the solutions of triangular systems are often more accurate than expected. In this and subsequent publications [345, 346] he treats the phenomenon as a property of triangular matrices in general. Recent work by the author [309] suggests that is more likely a property of the triangular factors that resulted from pivoted Gaussian elimination. Specifically, the pivoting strategy tends to make the diagonal elements of the factors reveal small singular values, and any ill-conditioning in such a triangular matrix is necessarily artificial. But this subject needs more work.

Inverses

The fact that there is no backward error analysis of matrix inversion was first noted by Wilkinson [344, 1961]. But because triangular systems from Gaussian elimination tend to be solved accurately, the computed inverse will generally be near the inverse of a slightly perturbed matrix. Unfortunately, as we have seen (Example 4.11), near is not good enough for the invert-and-multiply method for solving linear systems. For this reason, the invert-and-multiply algorithm has rightly been deprecated. However, if the matrix in question is known to be well conditioned, there is no reason not to use it. A trivial example is the solution of orthogonal systems via multiplication by the transpose matrix.

The backward error analysis of the LU decomposition does not imply that the computed L- and U-factors are accurate. In most applications the fact that the product LU reproduces the original matrix to working accuracy is enough. However, there is a considerable body of literature on the sensitivity of the decomposition [18, 315, 304, 308]. For a summary of these results and further references see [177].

Growth factors

Definition 4.6, in which growth factors are defined in terms of norms, is somewhat unconventional and has the drawback that one needs to know something special about $|L|$ and $|U|$ to compute them. Usually the growth factors are defined by something like (4.14), with the assumption that a pivoting strategy has kept the elements of L under control. Whatever the definition, one must choose whether to work with the exact factors or the computed factors.

The matrix in Example 4.13, which shows maximal growth under partial pivoting, is due to Wilkinson [344]. N. J. and D. J. Higham [178] show that any matrix that attains that growth must be closely related. The observation that Hessenberg and tridiagonal matrices have reasonable bounds for their growth factors is also due to Wilkinson [344]. Trefethen and Schreiber [320] have made an extensive investigation of pivot growth in random matrices. Higham and Higham [178] have exhibited orthogonal matrices that exhibit modest growth. For a practical example in which partial pivoting fails see [121].

The bound (4.15) for Gaussian elimination with complete pivoting is due to Wilkinson, who observed that it could not be attained. For further references on complete pivoting, see [177].

Wilkinson [344] showed that pivoting was unnecessary for positive definite matrices and matrices that are diagonally dominant by columns. That the same is true of matrices that are diagonally dominant by rows is obvious from the fact that Gaussian elimination by rows or columns gives the same sequence of Schur complements. Cryer [81] established the nonnegativity of the L- and U-factors of totally positive matrices; the connection with the stability of Gaussian elimination was made by de Boor and Pinkus [90].

Scaling

Bauer [19, 1963] was the first to observe that scaling affects Gaussian elimination only by changing the choice of pivots. Equal-error scaling is recommended by the authors of LINPACK [99]. For another justification see [292].

A strategy that was once in vogue was to scale to minimize the condition number of the matrix of the system (e.g., see [20]). Given the phenomenon of artificial ill-conditioning, the theoretical underpinnings of this strategy are at best weak. It should be noted, however, that balancing the elements of a matrix tends to keep the condition number in the usual norms from getting out of hand [20, 324, 323, 310].

Since row and column scaling use $2n-1$ free parameters to adjust the sizes of the n^2 elements of a matrix, any balancing strategy must be a compromise. Curtis and Reid [83] describe an algorithm for balancing according to a least squares criterion.

Iterative refinement

Iterative refinement is particularly attractive on machines that can accumulate inner products in double precision at little additional cost. But the double-precision calculation of the residual is difficult to implement in general software packages. The authors of LINPACK, who were not happy with mixed-precision computation, did not include it in their package. They noted [99, p. 1.8], "Most problems involve inexact input data and obtaining a highly accurate solution to an imprecise problem may not be justified." This is still sound advice.

The fact that iterative refinement with single-precision computation of the residual could yield componentwise stable solutions was first noted by Skeel [282, 1980]. For a complete analysis of this form of the method see [177]. For implementation details see the LAPACK code [9].

4

THE QR DECOMPOSITION AND LEAST SQUARES

A useful approach to solving matrix problems is to ask what kinds of transformations preserve the solution and then use the transformations to simplify the problem. For example, the solution of the system $Ax = b$ is not changed when the system is premultiplied by a nonsingular matrix. If we premultiply by elementary lower triangular matrices to reduce A to upper triangular form, the result is Gaussian elimination — as we saw in §1.1, Chapter 3.

A problem that finds application in many fields is the least squares problem finding a vector b that minimizes $\|y - Xb\|_2$. The natural transformations for this problem are orthogonal matrices, since $\|y - Xb\|_2 = \|Q^T(y - Xb)\|_2$ whenever Q is orthogonal. The result of triangularizing X by orthogonal transformations is a decomposition called the QR decomposition, which is the focus of this chapter.

The first section of this chapter is devoted to the QR decomposition itself. The second section treats its application to least squares, with particular attention being paid to the tricky matter of whether to use the QR decomposition or the normal equations, which is the traditional way of solving least squares problems. The last section is devoted to updating and downdating — the art of solving least squares problems that have been modified by the addition or deletion of an observation or a parameter.

Statisticians are one of the chief consumers of least squares, and there is a notational divide between them and numerical analysts. The latter take their cue from linear systems and write an overdetermined system in the form $Ax \cong b$, where A is $m \times n$. A statistician, confined to Latin letters, might write $Xb \cong y$, where X is $n \times p$. Since the analogy between least squares and linear systems is at best slippery, I have chosen to adopt the statistical notation. Hence throughout this chapter

> X *will be a real $n \times p$ matrix of rank p.*

The extension of the results of this chapter to complex matrices is not difficult. The case where $\text{rank}(X) < p$ will be treated in the next chapter.

1. THE QR DECOMPOSITION

The QR decomposition of X is an orthogonal reduction to triangular form — that is, a decomposition of the form

$$Q^T X = \begin{pmatrix} R \\ 0 \end{pmatrix},$$

where Q is orthogonal and R is upper triangular. We will begin this section by establishing the existence of the QR decomposition and describing its properties. In the next subsection we will show how to compute it by premultiplying X by a sequence of simple orthogonal matrices called Householder transformations. In the following section we will introduce another class of orthogonal matrices — the plane rotations, which are widely used to introduce zeros piecemeal into a matrix. We will conclude with an alternative algorithm — the Gram–Schmidt algorithm.

1.1. BASICS

In this subsection we will establish the existence of the QR decomposition and give some of its basic properties.

Existence and uniqueness

We have already established (Theorem 4.24, Chapter 1) the existence of a QR factorization of a matrix X into the product $Q_X R$ of an orthonormal matrix Q_X and an upper triangular matrix R with positive diagonal elements. It is easy to parlay this factorization into a full blown decomposition. By (4.2), Chapter 1, we know that there is an orthogonal matrix $Q_\perp$ such that

$$Q = (Q_X \ Q_\perp)$$

is orthogonal. Since the column space of Q_X forms an orthonormal basis for the column space of X, we have $Q_\perp^T X = 0$. It follows that

$$Q^T X = \begin{pmatrix} Q_X^T X \\ Q_\perp^T X \end{pmatrix} = \begin{pmatrix} R \\ 0 \end{pmatrix},$$

which establishes the existence of the decomposition. We summarize in the following theorem.

Theorem 1.1. *Let $X \in \mathbb{R}^{n \times p}$ be of rank p. Then there is an orthogonal matrix Q such that*

$$Q^T X = \begin{pmatrix} R \\ 0 \end{pmatrix}, \tag{1.1}$$

where R is upper triangular with positive diagonal elements. The matrix R is unique, as are the first p columns of Q.

Sec. 1. The QR Decomposition

There are three comments to be made on this result.

- The matrix R in the QR factorization is called the R-*factor*. The matrix Q_X is usually called the Q-*factor*, since it is the matrix in the QR factorization of Theorem 4.24, Chapter 1. But sometimes the term refers to the entire matrix Q.

- From (1.1) we have

$$X^{\mathrm{T}} X = R^{\mathrm{T}} R.$$

Thus:

> The R-factor of X is the Cholesky factor of $X^{\mathrm{T}} X$. The Q factor is $X R^{-1}$.

This result suggests a computational algorithm: namely,

1. Form the cross-product matrix $C = X^{\mathrm{T}} X$
2. Calculate the Cholesky decomposition $C = R^{\mathrm{T}} R$
3. $Q_X = X R^{-1}$

Unfortunately, this algorithm can be quite unstable and is to be avoided unless we know a priori that R is well conditioned. For suppose that $\|X\|_2 = 1$. Then $\|R\|_2 = \|X\|_2 = 1$. Hence if R is ill conditioned, R^{-1} must be large. But $\|Q\|_2 = 1$. Hence if we compute Q in the form $X R^{-1}$ we will get cancellation and the columns of Q will deviate from orthogonality.

- From the relation $X = Q_X R$ and the nonsingularity of R, we see that:

> The columns of Q_X form an orthonormal basis for $\mathcal{R}(X)$.

Also

> The columns of $Q_\perp$ form an orthonormal basis for the orthogonal complement of $\mathcal{R}(X)$.

Unlike Q_X, the matrix $Q_\perp$ is not unique. It can be any orthonormal basis for the orthogonal complement of $\mathcal{R}(X)$.

Projections and the pseudoinverse

Since Q_X is an orthonormal basis for $\mathcal{R}(X)$, the matrix

$$P_X = Q_X Q_X^{\mathrm{T}} \tag{1.2}$$

is the orthogonal projection onto $\mathcal{R}(X)$. Similarly

$$P_\perp = Q_\perp Q_\perp^{\mathrm{T}} = I - Q_X Q_X^{\mathrm{T}} \tag{1.3}$$

is the projection onto the orthogonal complement of $\mathcal{R}(X)$.

It is worth noting that (1.3) gives us two distinct representations of $P_\perp$. Although they are mathematically equivalent, their numerical properties differ. Specifically, if we have only a QR factorization of X, we must compute $P_\perp y$ in the form

$$P_\perp y = y - Q_X(Q_X^T y).$$

If there is cancellation, the resulting vector may not be orthogonal to $\mathcal{X}$. On the other hand, if we have a full QR decomposition, we can compute

$$P_\perp y = Q_\perp(Q_\perp^T y).$$

This expresses $P_\perp y$ explicitly as a linear combination of the columns of $Q_\perp$, and hence it will be orthogonal to $\mathcal{R}(X)$ to working accuracy. We will return to this point when we discuss the Gram–Schmidt algorithm (§1.4).

The above formulas for projections are the ones customarily used by numerical analysts. People in other fields tend to write the projection in terms of the original matrix X. The formula, which we have already given in §4.2, Chapter 1, can be easily derived from (1.2). If we write $Q = XR^{-1}$, then

$$P_X = (XR^{-1})(XR^{-1})^T = X(R^T R)^{-1} X^T = X(X^T X)^{-1} X^T.$$

This formula can be written more succinctly in terms of the *pseudoinverse* of X, which is defined by

$$X^\dagger = (X^T X)^{-1} X^T. \qquad (1.4)$$

It is easy to verify that

$$X^\dagger X = I$$

and

$$X X^\dagger = P_X.$$

There are alternative expressions for $X^\dagger$ in terms of the QR and singular value factorizations of X. Specifically,

$$X^\dagger = R^{-1} Q_X^T. \qquad (1.5)$$

If $X = U_X \Sigma V^T$ is the singular value factorization of X, then

$$X^\dagger = V \Sigma^{-1} U_X^T. \qquad (1.6)$$

The verification of these formulas is purely computational.

The pseudoinverse is a useful notational device, but like the inverse of a square matrix there is seldom any reason to compute it.

SEC. 1. THE QR DECOMPOSITION

The partitioned factorization

One nice feature of the QR factorization of X is that it gives us the QR factorization of any leading set of columns of X. To see this, partition the factorization $X = Q_X R$ in the form

$$(X_1 \; X_2) = (Q_1 \; Q_2) \begin{pmatrix} R_{11} & R_{12} \\ 0 & R_{22} \end{pmatrix}.$$

Then:

> The QR factorization of X_1 is $X_1 = Q_1 R_{11}$.

If we compute the second column of the partition we get

$$X_2 = Q_1 R_{12} + Q_2 R_{22}.$$

Let $P_1^\perp$ be the projection onto the orthogonal complement of $\mathcal{R}(X_1)$. Then from the above equation

$$P_1^\perp X_2 = Q_2 R_{22}.$$

In other words:

> The matrix $Q_2 R_{22}$ is the projection of X_2 onto the orthogonal complement of $\mathcal{R}(X_1)$.

One final result. Consider the partitioned cross-product matrix

$$\begin{pmatrix} X_1^T X_1 & X_1^T X_2 \\ X_2^T X_1 & X_2^T X_2 \end{pmatrix} = \begin{pmatrix} R_{11} & R_{12} \\ 0 & R_{22} \end{pmatrix}^T \begin{pmatrix} R_{11} & R_{12} \\ 0 & R_{22} \end{pmatrix}.$$

The right-hand side of this equation is a Cholesky factorization of the left-hand side. By Theorem 1.6, Chapter 3, the matrix $R_{22}^T R_{22}$ is the Schur complement of $X_1^T X_1$. Hence:

> The matrix R_{22} is the Cholesky factor of the Schur complement of $X_1^T X_1$ in $X^T X$.

Relation to the singular value decomposition

We conclude this tour of the QR decomposition by showing its relation to the singular value decomposition. Specifically, we have the following theorem. The proof is purely computational.

Theorem 1.2. *Let*

$$(Q_X \; Q_\perp)^T X = \begin{pmatrix} R \\ 0 \end{pmatrix}$$

be a QR decomposition of X and let

$$W^\mathrm{T} R V = \Sigma$$

be the singular value decomposition of R. Set

$$U = (Q_X W \; Q_\perp).$$

Then

$$U^\mathrm{T} X V = \begin{pmatrix} \Sigma \\ 0 \end{pmatrix}$$

is a singular value decomposition of X.

Thus the singular values of X and R are the same, as are their left singular vectors.

1.2. HOUSEHOLDER TRIANGULARIZATION

A compelling reason for preferring a QR decomposition to a QR factorization is that the former provides an orthonormal basis for the orthogonal complement of $\mathcal{R}(X)$. However, this asset can also be a liability. In many applications, the number n of rows of X greatly exceeds the number p of columns. For example, n might be a thousand while p is twenty. It would then require a million words to store Q but only twenty thousand to store Q_X.

A cure for this problem can be found by considering Gaussian elimination. In §1.1, Chapter 3, we showed how Gaussian elimination could be regarded as premultiplying a matrix by elementary lower triangular matrices to reduce it to triangular form. If we were to apply this procedure to an $n \times p$ matrix, it would be equivalent to multiplying the matrix by an $n \times n$ lower triangular matrix, whose explicit storage would require $\frac{1}{2}n^2$ locations. But the transformations themselves, whose nonzero elements are the multipliers, can be stored in less than np locations.

In this subsection we are going to show how to find a sequence $H_1, \ldots, H_m$ of orthogonal matrices such that

$$H_m H_{m-1} \cdots H_1 X = \begin{pmatrix} R \\ 0 \end{pmatrix},$$

where R is upper triangular. Since the product $H_m H_{m-1} \cdots H_1$ is orthogonal, it is equal to Q^T in a QR decomposition of A. However, if our transformations can be stored and manipulated economically, we can in effect have our entire Q at the cost of storing only the transformation.

Householder transformations

Before we introduce Householder transformations, let us look ahead and see how we are going to use them to triangularize a matrix. Partition X in the form $(x_1 \; X_2)$ and

Sec. 1. The QR Decomposition

let H be an orthogonal matrix whose first row is $x_1/\|x_1\|_2$. Then $Hx_1 = \|x_1\|e_1$. It follows that

$$HX = (Hx_1 \ HX_2) = (\|x_1\|e_1 \ HX_2) \equiv \begin{pmatrix} \rho_{11} & r_{12}^T \\ 0 & X_{22} \end{pmatrix}.$$

If this process is applied recursively to X_{22}, the result is an upper triangular matrix R (whose first row is already sitting in the first row of HX).

The key to this algorithm lies in constructing an orthogonal transformation H such that Hx_1 is a multiple of e_1. For the method to be useful, the transformation must be cheap to store. In addition, we must be able form the product HX_2 cheaply. The following class of transformations fills the bill.

Definition 1.3. *A* Householder transformation *(also known as an* elementary reflector*) is a matrix of the form*

$$H = I - uu^T,$$

where

$$\|u\|_2 = \sqrt{2}.$$

There is a certain arbitrariness in this definition. We could equally well define a Householder transformation to be a matrix of the form $I - \rho uu^T$, where $\rho\|u\|_2^2 = 2$. For example, Householder's original definition was $H = I - 2uu^T$, where $\|u\|_2 = 1$.

Householder transformations are symmetric. They are orthogonal, since

$$\begin{aligned} H^T H &= H^2 \\ &= (I - uu^T)(I - uu^T) \\ &= I - 2uu^T + uu^T uu^T \\ &= I - 2uu^T + 2uu^T \\ &= I. \end{aligned}$$

A Householder transformation of order n can be stored by storing the vector u, which requires n locations. The product HX can be computed in the form

$$HX = (I - uu^T)X = X - u(u^T X).$$

This leads to the following simple algorithm to overwrite X by HX

$$\begin{array}{ll} 1. & v^T = u^T X \\ 2. & X = X - uv^T \end{array} \tag{1.7}$$

It is easy to see that the operation count for this algorithm is $2np$ flam, which is satisfactorily small.

We must now show that Householder transformations can be used like elementary lower triangular matrices to introduce zeros into a vector. The basic construction is contained in the following theorem.

Theorem 1.4. *Suppose* $\|x\|_2 = 1$, *and let*

$$u = \frac{x \pm \mathbf{e}_1}{\sqrt{1 \pm \xi_1}}. \tag{1.8}$$

If u is well defined, then

$$Hx \equiv (I - uu^{\mathrm{T}})x = \mp \mathbf{e}_1.$$

Proof. We first show that H is a Householder transformation by showing that $\|u\|_2 = \sqrt{2}$. Specifically,

$$\begin{aligned}
\|u\|_2^2 &= \frac{(x \pm \mathbf{e}_1)^{\mathrm{T}}(x \pm \mathbf{e}_1)}{1 \pm \xi_1} \\
&= \frac{x^{\mathrm{T}}x \pm 2\mathbf{e}_1^{\mathrm{T}}x + \mathbf{e}_1^{\mathrm{T}}\mathbf{e}_1}{1 \pm \xi_1} \\
&= \frac{1 \pm 2\xi_1 + 1}{1 \pm \xi_1} \\
&= 2.
\end{aligned}$$

Now $Hx = x - (u^{\mathrm{T}}x)u$. But

$$\begin{aligned}
u^{\mathrm{T}}x &= \frac{(x \pm \mathbf{e}_1)^{\mathrm{T}}x}{\sqrt{1 \pm \xi_1}} \\
&= \frac{x^{\mathrm{T}}x \pm \mathbf{e}_1^{\mathrm{T}}x}{\sqrt{1 \pm \xi_1}} \\
&= \frac{1 \pm \xi_1}{\sqrt{1 \pm \xi_1}} \\
&= \sqrt{1 \pm \xi_1}.
\end{aligned}$$

Hence

$$\begin{aligned}
Hx &= x - (u^{\mathrm{T}}x)u \\
&= x - \sqrt{1 \pm \xi_1}\,\frac{x \pm \mathbf{e}_1}{\sqrt{1 \pm \xi_1}} \\
&= x - (x \pm \mathbf{e}_1) \\
&= \mp \mathbf{e}_1. \quad \blacksquare
\end{aligned}$$

There are two comments to be made about this construction.

- Equation (1.8) shows that in general there are two Householder transformations that will reduce x. However, for numerical stability we take the sign $\pm$ to be the same as the sign of the first component of x. If we take the opposite sign, cancellation can occur in the computation of the first component of u, with potentially disastrous results.

SEC. 1. THE QR DECOMPOSITION

This algorithm takes a vector x and produces a vector u that generates a Householder transformation $H = I - uu^T$ such that $Hx = \mp\|x\|_2 \mathbf{e}_1$. The quantity $\mp\|x\|_2$ is returned in ν.

```
1.   housegen(x, u, ν)
2.      u = x
3.      ν = ‖u‖₂
4.      if ν = 0; u[1] = √2; return ; fi
5.      u = x/ν
6.      if (u[1] ≥ 0)
7.         u[1] = u[1] + 1
8.         ν = -ν
9.      else
10.        u[1] = u[1] - 1
11.     end if
12.     u = u/√|u[1]|
13.  end housegen
```

Algorithm 1.1: Generation of Householder transformations

───────◊───────

(However, the alternate transformation can be computed stably. See the notes and references.)

- If $\|x\|_2 \neq 1$, we can generate u from $x/\|x\|_2$, in which case

$$Hx = \mp\|x\|_2 \mathbf{e}_1.$$

───

Combining these two observations we get Algorithm 1.1 — a program to generate a Householder transformation. Note that when $x = 0$, any u will do. In this case the program *housegen* returns $u = \sqrt{2}\mathbf{e}_1$. This choice does not make H the identity, but the identity with its $(1,1)$-element changed to -1. (In fact, it is easy to see that a Householder transformation can never be the identity matrix, since it transforms u into $-u$.)

We have observed that it is straightforward to extend the algorithms in this chapter to complex matrices. However, the generation of complex Householder transformations is a little tricky. What we do is to start from

$$u = \rho \frac{x}{\|x\|_2},$$

where ρ is a scalar of absolute value one chosen to make the first component of u non-negative. We then proceed as usual. The resulting Householder transformation satis-

fies
$$Hx = -\bar{\rho}\|x\|_2 \mathbf{e}_1.$$

It is a good exercise to work out the details.

Householder triangularization

Let us now return the orthogonal triangularization of X. A little northwest indexing will help us derive the algorithm. Suppose that we have determined Householder transformations $H_1, \ldots, \hat{H}_{k-1}$ so that

$$H_{k-1} \cdots H_1 X = \begin{pmatrix} R_{11} & r_{1k} & R_{1,k+1} \\ 0 & x_{kk} & X_{k,k+1} \end{pmatrix},$$

where R_{11} is upper triangular. Let $\hat{H}_k$ be a Householder transformation such that

$$\hat{H}_k x_{kk} = \rho_{kk} \mathbf{e}_1.$$

Set $H_k = \text{diag}(I_{k-1}, \hat{H}_k)$. Then

$$H_k H_{k-1} \cdots H_1 X = \begin{pmatrix} R_{11} & r_{1k} & R_{1,k+1} \\ 0 & \hat{H}_k x_{kk} & \hat{H}_k X_{k,k+1} \end{pmatrix}$$

$$\equiv \begin{pmatrix} R_{11} & r_{1k} & R_{1,k+1} \\ 0 & \rho_{kk} & r_{k,k+1}^{\mathrm{T}} \\ 0 & 0 & X_{k+1,k+1} \end{pmatrix}.$$

This process clearly advances the triangularization by one step.

Algorithm 1.2 is an implementation of this process. There are many things to be said about it.

- Technically, the algorithm does not compute a QR decomposition, since the diagonal elements of R can be negative. In practice, this deviation makes no difference, and most programs do not attempt to clean up the signs.

- We have been very free with storage, throwing away X and placing U and R in separate arrays. In fact, U is lower trapezoidal and R is upper triangular, so that they both can be stored in X. The only problem with this arrangement is that they both compete for the diagonal. Most programs give the diagonal to R and store the diagonal of U separately.

Whatever the storage scheme, if $n \gg p$, as often happens in least squares problems, the storage is $pO(n)$, much less that the n^2 locations required to store an explicit Q.

- The bulk of the work in the algorithm is done in statements 4 and 5. Each of these statements requires about $(n-k)(p-k)$ flam. Integrating, we get a count of

$$2 \int_0^p (n-k)(p-k)\, dk \cong np^2 - \frac{1}{3}p^3.$$

Sec. 1. The QR Decomposition

Given an $n \times p$ matrix X, let $m = \min\{n, p\}$. This algorithm computes a sequence $H_1, \ldots, H_m$ of Householder transformations such that

$$H_m \cdots H_1 X = \begin{pmatrix} R \\ 0 \end{pmatrix},$$

where R is upper triangular. The generators of the Householder transformation are stored in the array U.

1. hqrd(X, U, R)
2.   for $k = 1$ to $\min\{p, n\}$
3.   housegen($X[k{:}n, k]$, $U[k{:}n, k]$, $R[k, k]$)
4. $v^T = U[k{:}n, k]^T * X[k{:}n, k+1{:}p]$
5. $X[k{:}n, k+1{:}p] = X[k{:}n, k+1{:}p] - U[k{:}n, k] * v^T$
6. $R[k, k+1{:}p] = X[k, k+1{:}p]$
7.   end for k
8. end hqrd

Algorithm 1.2: Householder triangularization

———◇———

Hence:

Algorithm 1.2 requires $(np^2 - \frac{1}{3}p^3)$ flam.

When $n \gg p$, the np^2 term dominates. On the other hand, when $n = p$, the count reduces to $\frac{2}{3}n^3$ flam, which is twice the count for Gaussian elimination.

• If we partition $X = (X_1 \; X_2)$, where X_1 has q columns, then $H_1 \cdots H_q$ is the orthogonal part of the QR decomposition of Q. Thus, having computed the factored decomposition of X, we have a factored decomposition of every initial set of columns.

• The algorithm is backward stable, as we shall see in Theorem 1.5.

• The algorithm can be blocked, but the process is more complicated than with the variants of Gaussian elimination. The reason is that the transformations must be further massaged so that their effect can be expressed in terms of matrix-matrix operations. This topic is treated at the end of this subsection.

• The algorithm works when $n < p$. In this case the final matrix has the form

$(R_{11} \; R_{1,n+1})$,

where R_{11} is upper triangular. The operation count changes to $(pn^2 - \frac{1}{3}n^3)$ flam.

• The algorithm can be applied to a matrix that is not of full rank. Thus it gives a constructive proof that matrices of any rank have a QR decomposition. However, R

is no longer unique, and it is necessarily singular. In particular it must have at least one zero on the diagonal — at least mathematically.

- Both row and column exchanges can be incorporated into the algorithm. Column pivoting for size is widely used, chiefly as a method for revealing rank but also as a technique for avoiding instabilities in graded matrices [see (1.14) and §2.4]. We will discuss column pivoting more fully in the next chapter (2.1, Chapter 5). Row pivoting is rarely used, although it is often desirable to exchange rows before the reduction to bring the larger rows to the top [see (1.14)].

Computation of projections

After a Householder reduction of X, the orthogonal part of its QR decomposition is given by

$$Q = H_1 H_2 \cdots H_m,$$

where $m = \min\{n-1, p\}$. We will now show how to use this factored form to compute projections of a vector y onto $\mathcal{R}(X)$ and its orthogonal complement.

Let the orthogonal part of the QR decomposition be partitioned as usual in the form

$$Q = (Q_X \quad Q_\perp).$$

Let

$$Q^\mathrm{T} y = \begin{pmatrix} Q_X^\mathrm{T} y \\ Q_\perp^\mathrm{T} y \end{pmatrix} \equiv \begin{pmatrix} z_X \\ z_\perp \end{pmatrix}.$$

Since $P_X y = Q_X Q_X^\mathrm{T} y = Q_X z_X$, we have

$$P_X y = (Q_X \quad Q_\perp) \begin{pmatrix} z_X \\ 0 \end{pmatrix}.$$

Thus to compute $P_X y$ all we have to do is to compute $z = Q^\mathrm{T} y$, zero out the last $n-p$ components of z to get an new vector $\hat{z}$, and then compute $P_X y = Q\hat{z}$. Similarly, to compute $P_\perp y$ we zero out the first p components of z and multiply by Q.

Algorithm 1.3 is an implementation of this procedure. It is easily seen that it requires $(2np - \frac{1}{2}p^2)$ flam to perform the forward multiplication and the same for each of the back multiplications. If $n \gg p$, the total is essentially $6np$ flam, which compares favorably with multiplication by an $n \times p$ matrix.

As another illustration of the manipulation of the Householder QR decomposition, suppose we wish to compute Q_X of the QR factorization. We can write this matrix in the form

$$Q_X = (Q_X \quad Q_\perp) \begin{pmatrix} I_p \\ 0 \end{pmatrix}.$$

SEC. 1. THE QR DECOMPOSITION

> This algorithm takes a vector y and the output of Algorithm 1.2 and computes $y_X = P_X y$ and $y_\perp = P_\perp y$.
>
> 1. $hproj(n, p, U, y, y_X, y_\perp)$
> 2. $y_X = y$
> 3.   **for** $k = 1$ **to** p
> 4. $\nu = U[k{:}n, k]^\mathrm{T} * y_X[k{:}n]$
> 5. $y_X[k{:}n] = y_X[k{:}n] - \nu * U[k{:}n, k]$
> 6.   **end for** k
> 7. $y_\perp = y_X$
> 8. $y_\perp[1{:}p] = 0$
> 9. $y_X[p+1{:}n] = 0$
> 10.   **for** $k = p$ **to** 1 **by** -1
> 11. $\nu = U[k{:}n, k]^\mathrm{T} * y_X[k{:}n]$
> 12. $y_X[k{:}n] = y_X[k{:}n] - \nu * U[k{:}n, k]$
> 13. $\nu = U[k{:}n, k]^\mathrm{T} * y_\perp[k{:}n]$
> 14. $y_\perp[k{:}n] = y_\perp[k{:}n] - \nu * U[k{:}n, k]$
> 15.   **end for** k
> 16. **end** $hproj$

Algorithm 1.3: Projections via the Householder decomposition

———◇———

Consequently, we can generate Q_X by computing the product of Q and $(I_p\ 0)^\mathrm{T}$. The algorithm is simplicity itself.

1. $QX[1{:}p, :] = I_p$
2. $QX[p+1{:}n, :] = 0$
3.   **for** $k = p$ **to** 1 **by** -1
4. $v^\mathrm{T} = U[k{:}n, k]^\mathrm{T} * QX[k{:}n, k{:}p]$
5. $QX[k{:}n, k{:}p] = QX[k{:}n, k{:}p] - U[k{:}n, k] * v^\mathrm{T}$
6.   **end for** k

(1.9)

Note that at the kth step of the algorithm it is only necessary to work with $QX[k{:}n, k{:}p]$, the rest of the array being unaffected by the transformation H_k. The operation count for the algorithm is $(np^2 - \frac{1}{3}p^3)$ flam — the same as for Householder triangularization.

Numerical stability

The hard part about the error analysis of Householder transformations is to decide what to prove. There are three problems.

The first problem is that Householder transformations are used both to triangularize matrices and then later to compute such things as projections. We will sidestep this problem by giving a general analysis of what it means to multiply a vector by a sequence of Householder transformations and then apply the general analysis to specific cases.

The second problem is that when a transformation is used to introduce zeros into a vector, we do not actually transform the vector but set the components to zero. Fortunately, the error analysis can be extended to this case.

The third problem is that we must deal with three different kinds of transformations.

1. The transformations we would have computed if we had done exact computations. We have been denoting these transformations generically by H.

2. The transformations we would have computed by exact computation in the course of the inexact reduction. We will denote these by $\hat{H}$. (1.10)

3. The transformations we actually apply. This includes the errors made in generating the transformation (Algorithm 1.1) and those made in applying the transformation via (1.7). We will use the fl notation to describe the effects of these transformations.

The key to solving the problem of multiple classes of transformations is to forget about the first kind of transformation, which is unknowable, and focus on the relation between the second and the third.

With these preliminaries, the basic result can be stated as follows.

Theorem 1.5 (Wilkinson). *Let $\hat{Q} = \hat{H}_1 \cdots \hat{H}_m$ be a product of Householder transformations, and let $b = \text{fl}(\hat{H}_m \cdots \hat{H}_1 a)$. Then*

$$b = \hat{Q}^{\mathrm{T}}(a+e),$$

where

$$\frac{\|e\|_2}{\|a\|_2} \leq \varphi(n,m)\epsilon_{\mathrm{M}}. \qquad (1.11)$$

Here φ is a slowly growing function of n and m.

For a proof consult the references cited at the end of this section.

The theorem is a classical backward error analysis. The computed value of b is the result of performing exact computations on a slightly perturbed a. It is worth restating that the errors in computing b include the errors in generating the transformations and that the transformations can either be applied normally to b or introduce explicit zeros into b.

Let us now apply Theorem 1.5 to the reduction of X to triangular form. Let x_k be the kth column of X, and let $\hat{H}_i$ be the exact transformations computed during the course of the reduction [item 2 in (1.10)]. Then x_k is transformed into

$$\text{fl}(\hat{H}_k \cdots \hat{H}_1)x_k = \text{fl}(\hat{H}_p \cdots \hat{H}_1)x_k$$

Sec. 1. The QR Decomposition

(the right-hand side follows from the fact that $\hat{H}_{p+1}, \ldots, \hat{H}_p$ operate only on the zero part of $\mathrm{fl}(\hat{H}_k \cdots \hat{H}_1)x_k)$. If, as above, we set $\hat{Q}_k^\mathrm{T} = \hat{H}_p \cdots \hat{H}_1$, then by Theorem 1.5 there is a vector e_k such that

$$\hat{Q}^\mathrm{T}(x_i + e_k) = \begin{pmatrix} \hat{r}_k \\ 0 \end{pmatrix},$$

where $\hat{r}$ is the computed value of the kth column of R. From (1.11) and the fact that x_k is multiplied by only k transformations we see that the kth column of E satisfies

$$\frac{\|e_k\|_2}{\|x_k\|_2} \leq \varphi(n,k)\epsilon_\mathrm{M}, \qquad k = 1, 2, \ldots, p. \tag{1.12}$$

If we combine these bounds, we get

$$\frac{\|E\|_\mathrm{F}}{\|X\|_\mathrm{F}} \leq \varphi(n,m)\epsilon_\mathrm{M}. \tag{1.13}$$

This is the usual bound reported in the literature, but it should be kept in mind that it is derived from the more flexible columnwise bound (1.12).

In assessing these bounds it is important to understand that it does not say that $\hat{Q}$ and $\hat{R}$ are near the matrices that would be obtained by exact computation with X. For example, the column spaces of X and $X + E$ may differ greatly, in which case the compute $\hat{Q}_X$ will differ greatly from its exact counterpart. This phenomenon is worth pursuing.

Example 1.6. *Let*

$$X = \begin{pmatrix} 1 & 1+\epsilon \\ 1 & 1 \\ 1 & 1 \end{pmatrix} \quad \text{and} \quad \tilde{X} = \begin{pmatrix} 1 & 1+\epsilon \\ 1 & 1+\epsilon \\ 1 & 1 \end{pmatrix}.$$

Then no matter how small ϵ is,

$$\mathcal{R}(X) \text{ is spanned by } \begin{pmatrix} 1 \\ 1 \\ 1 \end{pmatrix} \text{ and } \begin{pmatrix} 1 \\ 0 \\ 0 \end{pmatrix},$$

while

$$\mathcal{R}(\tilde{X}) \text{ is spanned by } \begin{pmatrix} 1 \\ 1 \\ 1 \end{pmatrix} \text{ and } \begin{pmatrix} 1 \\ 1 \\ 0 \end{pmatrix}.$$

These are clearly different spaces. And in fact the Q factors of X and $\tilde{X}$ are

$$Q = \begin{pmatrix} -0.5774 & 0.8165 & 0.0000 \\ -0.5774 & -0.4082 & -0.7071 \\ -0.5774 & -0.4082 & 0.7071 \end{pmatrix}$$

and
$$\tilde{Q} = \begin{pmatrix} -0.5774 & -0.4082 & -0.7071 \\ -0.5774 & -0.4082 & 0.7071 \\ -0.5774 & 0.8165 & 0.0000 \end{pmatrix}.$$

A consequence of this example is that when we use (1.9) to compute $\hat{Q}_X$, the columns of the resulting matrix may not span $\mathcal{R}(X)$. However, we can use our error analysis to show that the columns of the computed matrix — call it $\bar{Q}_X$ — are orthogonal to working accuracy. Specifically,

$$\bar{Q}_X = \text{fl}\left[\hat{H}_1 \cdots \hat{H}_p \begin{pmatrix} I_p \\ 0 \end{pmatrix}\right].$$

Hence by Theorem 1.5

$$\bar{Q}_X = \hat{H}_1 \cdots \hat{H}_p \left[\begin{pmatrix} I_p \\ 0 \end{pmatrix} + E\right],$$

where the columns of e satisfy

$$\|e_j\|_2 \leq \varphi(n,p)\epsilon_{\text{M}}.$$

It follows from the exact orthogonality of the product $\hat{H}_1 \cdots \hat{H}_p$ that

$$\bar{Q}_X^{\text{T}} \bar{Q}_X = I + E_1 + E_1^{\text{T}} + E^{\text{T}} E,$$

where E_1 consists of the first p rows of E. Ignoring the second-order term, we have

$$\|I - \bar{Q}_X^{\text{T}} \bar{Q}_X\|_{\text{F}} \lesssim 2\sqrt{p}\varphi(n,p)\epsilon_{\text{M}}.$$

This ability to produce almost exactly orthogonal bases is one of the strong points of orthogonal triangularization.

Graded matrices

An important feature of the backward error bound (1.12) is that it is independent of the scaling of the columns of X. Unfortunately, the backward error is not independent of row scaling, as the following example shows.

Example 1.7. *Consider the linear system*

$$\begin{pmatrix} 9.3040\text{e}{-01} & 9.1960\text{e}{-02} & 7.0120\text{e}{-01} \\ 8.4620\text{e}{-08} & 6.5390\text{e}{-08} & 9.1030\text{e}{-08} \\ 5.2690\text{e}{-15} & 4.1600\text{e}{-15} & 7.6220\text{e}{-15} \end{pmatrix} \begin{pmatrix} 1 \\ 1 \\ 1 \end{pmatrix} = \begin{pmatrix} 1.7236\text{e}{+00} \\ 2.4104\text{e}{-07} \\ 1.7051\text{e}{-14} \end{pmatrix}.$$

Denoting this system by $Ax = b$, we can solve it by the following algorithm.

SEC. 1. THE QR DECOMPOSITION

1. Compute the QR decomposition of A
2. $z = Q^T b$
3. Solve the system $Rx = z$

If we do this in double precision arithmetic we get

$$x = \begin{pmatrix} 1.00000000000000 \\ 1.00000000000000 \\ 1.00000000000000 \end{pmatrix},$$

which is fully accurate.

Now consider the system

$$\begin{pmatrix} 5.2690\text{e}-15 & 4.1600\text{e}-15 & 7.6220\text{e}-15 \\ 8.4620\text{e}-08 & 6.5390\text{e}-08 & 9.1030\text{e}-08 \\ 9.3040\text{e}-01 & 9.1960\text{e}-02 & 7.0120\text{e}-01 \end{pmatrix} \begin{pmatrix} 1 \\ 1 \\ 1 \end{pmatrix} = \begin{pmatrix} 1.7051\text{e}-14 \\ 2.4104\text{e}-07 \\ 1.7236\text{e}+00 \end{pmatrix},$$

which differs from the first in having its first and third rows interchanged. If we try our procedure on this system, we get

$$x = \begin{pmatrix} 1.11689349448549 \\ 1.07908877773654 \\ 0.83452551159448 \end{pmatrix},$$

which is almost completely inaccurate.

In trying to find out what is going on it is important to keep in mind where the mystery is. It is not mysterious that one can get inaccurate results. The error analysis says that the backward relative normwise error $\|E\|_F / \|A\|_F$ is small. But each system has a very small row, which can be overwhelmed by that error. In fact this is just what has happened in the second system. The backward error is

$$A - QR = \begin{pmatrix} -1.0\text{e}-16 & 1.7\text{e}-17 & 1.6\text{e}-16 \\ 0 & 1.3\text{e}-23 & 3.9\text{e}-23 \\ 0 & 2.7\text{e}-17 & 2.2\text{e}-16 \end{pmatrix}.$$

The backward error in the first row is almost as large as the row itself.

The mystery comes when we compute the backward error in the first system:

$$\begin{pmatrix} 0 & 2.7\text{e}-17 & 1.1\text{e}-16 \\ 0 & 0 & 0 \\ 0 & 7.8\text{e}-31 & 0 \end{pmatrix}.$$

This represents a very small relative error in each of the elements of the matrix A, which accounts for the accuracy of the solution. But what accounts for the low relative backward error?

There is no truly rigorous answer to this question. The matrices of these two systems are said to be *graded*, meaning their elements show an upward or downward trend as we pass from the top to the bottom of the matrix. The second system grades up, and it is easy to see why it is a disaster. When we normalize its first column, preparatory to computing the first Householder transformation, we get

$$\begin{pmatrix} 5.663155631986219\mathrm{e}{-}15 \\ 9.095012897678379\mathrm{e}{-}08 \\ 9.999999999999958\mathrm{e}{-}01 \end{pmatrix}.$$

We now add one to the first component to get

$$1.000000000000006\mathrm{e}{+}00.$$

Only the rounded first digit of the first component is preserved. The loss of information in that first component is sufficient to account for the inaccuracy. (Actually, all the elements in the first row are affected, and it is an instructive exercise to see how this comes about.)

On the other hand if the matrix is graded downward, the results of Householder reduction are often quite satisfactory. The reason is that the vectors generating the Householder transformations tend to share the grading of the matrix. In this case when we apply the transformation to a column of A in the form

$$a - (u^\mathrm{T} a)u,$$

the corresponding components of the terms a and $(u^\mathrm{T} a)u$ are roughly the same size so that large components cannot wash out small ones. However, we cannot rule out the possibility that an unfortunate cancellation of large elements will produce a u that is not properly graded.

Example 1.8. *Consider the matrix*

$$A = \begin{pmatrix} -9.1600\mathrm{e}{-}01 & -9.1600\mathrm{e}{-}01 & 3.5730\mathrm{e}{-}01 \\ 1.4900\mathrm{e}{+}00 & 1.4900\mathrm{e}{+}00 & -5.1380\mathrm{e}{-}01 \\ 2.1490\mathrm{e}{-}15 & 4.2980\mathrm{e}{-}15 & -8.6700\mathrm{e}{-}17 \end{pmatrix},$$

whose leading 2×2 matrix is exactly singular. If we perform one step of Householder triangularization, we get the matrix

$$\begin{pmatrix} 1.7490\mathrm{e}{+}00 & 1.7490\mathrm{e}{+}00 & -6.2483\mathrm{e}{-}01 \\ 0.0000\mathrm{e}{-}00 & 3.3307\mathrm{e}{-}16 & 3.5297\mathrm{e}{-}02 \\ 0.0000\mathrm{e}{+}00 & 2.1490\mathrm{e}{-}15 & 7.0525\mathrm{e}{-}16 \end{pmatrix}.$$

Note that owing to rounding error the $(2,2)$-element, which should be exactly zero, is at the level of the rounding unit and only an order of magnitude different from the $(3,2)$-element. Consequently the next Householder transformation will be reasonably

Sec. 1. The QR Decomposition

balanced and will mix the $(2,3)$- and $(3,3)$-elements, largely destroying the latter. In fact the backward error for the full reduction is

$$\begin{pmatrix} 3.3307\text{e}-16 & 1.1102\text{e}-16 & -1.1102\text{e}-16 \\ 0.0000\text{e}+00 & -2.2204\text{e}-16 & 1.1102\text{e}-16 \\ -3.9443\text{e}-31 & -7.8886\text{e}-31 & -4.7343\text{e}-18 \end{pmatrix}.$$

The relative backward error in the $(3,3)$-element is $5.4 \cdot 10^{-2}$ — i.e., only two figures are accurate.

It is worth observing that if we interchange the second and third columns of A, the leading 2×2 matrix is well conditioned and the problem goes away. In this case the backward error is

$$\begin{pmatrix} 3.3307\text{e}-16 & -1.1102\text{e}-16 & 1.1102\text{e}-16 \\ 0.0000\text{e}+00 & 1.1102\text{e}-16 & -2.2204\text{e}-16 \\ -3.9443\text{e}-31 & 2.2187\text{e}-31 & -7.8886\text{e}-31 \end{pmatrix},$$

which is entirely satisfactory.

To summarize:

> *When using Householder transformations to triangularize a graded matrix, permute the rows so that the grading is downward and pivot for size on the columns.* (1.14)

Blocked reduction

In §3.3, Chapter 2, we described a technique, called blocking, that could potentially enhance the performance of algorithms on machines with hierarchical memories. For Householder triangularization, the analogue of the algorithm in Figure 3.3, Chapter 2, is the following. In Algorithm 1.2, having chosen a block size m, we generate the Householder transformations $(I - u_1 u_1^T), \ldots, (I - u_m u_m^T)$ from $X[1{:}n, 1{:}m]$ but defer applying them to the rest of the matrix until they have all been generated. At that point we are faced with the problem of computing

$$[(I - u_1 u_1^T) \cdots (I - u_m u_m^T)]^T X[1{:}n, m+1{:}p]. \tag{1.15}$$

Blocking requires that we apply the transformations simultaneously. Unfortunately, (1.15) is in the form of a sequence of individual transformations.

Fortunately, there is an ingenious cure for this problem. It is possible to write the product

$$(I - u_1 u_1^T)(I - u_2 u_2^T) \cdots (I - u_m u_m^T) \tag{1.16}$$

in the form

$$I - UTU^T,$$

where T is upper triangular. Specifically, we have the following theorem, which applies not just to Householder transformations but to any product of the form (1.16).

This algorithm takes a sequence of m vectors contained in the array U and returns an upper triangular matrix T such that $(I - u_1 u_1^{\mathrm{T}})(I - u_2 u_2^{\mathrm{T}}) \cdots (I - u_m u_m^{\mathrm{T}}) = I - UTU^{\mathrm{T}}$.

1. $\text{utu}(m, U, T)$
2.   **for** $j = 1$ **to** m
3. $T[j, j] = 1$
4. $T[1{:}j{-}1, j] = U[:, 1{:}j{-}1]^{\mathrm{T}} * U[:, j]$
5. $T[1{:}j{-}1, j] = -T[1{:}j{-}1, 1{:}j{-}1] * T[1{:}j{-}1, j]$
6.   **end for** j
7. **end** utu

Algorithm 1.4: UTU representation of $\prod_i (I - u_i u_i^{\mathrm{T}})$

Theorem 1.9. *The product* $(I - UTU^{\mathrm{T}})(I - u)$ *can be written in the form*

$$(I - UTU^{\mathrm{T}})(I - u) = I - \begin{pmatrix} U & u \end{pmatrix} \begin{pmatrix} T & -TU^{\mathrm{T}} u \\ 0 & 1 \end{pmatrix} \begin{pmatrix} U^{\mathrm{T}} \\ u^{\mathrm{T}} \end{pmatrix}. \qquad (1.17)$$

Proof. Direct verification. ∎

Starting with $I - u_1 u_1^{\mathrm{T}}$, we can apply (1.17) successively to express the product $(I - u_1 u_1^{\mathrm{T}})(I - u_2 u_2^{\mathrm{T}}) \cdots (I - u_m u_m^{\mathrm{T}})$ in the form

$$(I - u_1 u_1^{\mathrm{T}})(I - u_2 u_2^{\mathrm{T}}) \cdots (I - u_m u_m^{\mathrm{T}}) = UTU^{\mathrm{T}},$$

where T is unit upper triangular. We will call this the *UTU form of the product*. Note that the vectors u_j appear unchanged in U. The only new item is the matrix T. The procedure for generating T is implemented in Algorithm 1.4. Two comments.

- If the vectors are of length n, then the algorithm takes

$$\left(\tfrac{1}{2} m^2 n + \tfrac{1}{3} m^3 \right) \text{ flam.}$$

For large n the first term dominates.

- The UTU representation involves a curious reversal of the order of multiplication and the order of storage. The vectors $u_1, u_2, \ldots, u_m$ will naturally be stored in U beginning with u_1. On the other hand, if the transformations $I - u_i u_i^{\mathrm{T}}$ are premultiplied in the natural order, the result is a premultiplication by the matrix $(I - u_m u_m^{\mathrm{T}})(I - u_{m-1} u_{m-1}^{\mathrm{T}}) \cdots (I - u_1 u_1^{\mathrm{T}}) = I - UT^{\mathrm{T}} U^{\mathrm{T}}$.

We can use this algorithm in a blocked Householder triangularization of a matrix X. We choose a block size m and perform an ordinary Householder triangularization

Sec. 1. The QR Decomposition

This algorithm takes an $n \times p$ matrix X and a block size m and produces $q = \lceil p/m \rceil$ orthogonal transformations $I - U_k T_k U_k^T$ in UTU form such that

$$(I - U_q T_p U_q)^T \cdots (I - U_1 T_1 U_1)^T X = \begin{pmatrix} R \\ 0 \end{pmatrix},$$

where R is upper triangular. The algorithm uses Algorithms 1.2 (*hqrd*) and 1.4 (*utu*).

1. **bhqrd**(m, X, U, T, R)
2. $q = 0$
3. **for** $k = 1$ **to** p **by** m
4. $\quad q = q+1$
5. $\quad l = \min\{p, k+m-1\}$
6. $\quad$ hqrd($X[k{:}n, k{:}l]$, $U[k{:}n, k{:}l]$, $R[k{:}l, k{:}l]$)
7. $\quad$ utu($l-k+1$, $U[k{:}n, k{:}l]$, T_q)
8. $\quad V = U[k{:}n, k{:}l]^T * X[k{:}n, l+1{:}p]$
9. $\quad V = T_q^T * V$
10. $\quad X[k{:}n, l+1{:}p] = X[k{:}n, l+1{:}p] - U[k{:}n, k{:}l] * V$
11. $\quad R[k{:}l, l+1{:}p] = X[k{:}l, l+1{:}p]$
12. **end for** k
13. **end bhqrd**

Algorithm 1.5: Blocked Householder triangularization

─────────◇─────────

on $X[:, 1{:}m]$. The transformations are then put in UTU form, after which they can be applied to the rest of the matrix as

$$X[1{:}n, m+1{:}p] = X[1{:}n, m+1{:}p] - U*(T^T*(U^T*X[1{:}n, m+1{:}p]))$$

The process is then repeated with the next set of m columns. We can use Algorithm 1.2 to reduce the blocks.

Algorithm 1.5 implements this procedure. Here are some observations.

• The transpose in statement 9 reflects the inconsistency between order of storage and order of application in the UTU representation.

• If m is not large compared with p, the blocked algorithm requires about np^2 flam, the same as for the unblocked algorithm. In this case, the overhead to form the UTU representations is negligible.

• We have not tried to economize on storage. In practice, the vectors in U and the matrix R would share the storage originally occupied by X. The matrices T_j could occupy an $m \times p$ array (or $\frac{1}{2} m \times p$ array, if packed storage is used).

- The UTU form of the transformations enjoy the same numerical properties as the original transformations. In particular the natural analogue of Theorem 1.5 holds.

- Because the application of the transformations in a block are deferred, one cannot column pivot for size as a block of transformations are accumulated. This is a serious drawback to the algorithm in some applications.

- In the unblocked form of the algorithm it is possible to recover the QR decomposition of any initial set of columns of X. Because the blocked algorithm recasts each block of Householder transformations as a single UTU transformation, we can only recover initial decompositions that are conformal with the block structure.

—

With Gaussian elimination, blocking is unlikely to hurt and may help a great deal. For triangularization by Householder transformations the situation is mixed. If one needs to pivot or get at initial partitions of the decomposition — as is true of many applications in statistics — then the blocked algorithm is at a disadvantage. On the other hand, if one just needs the full decomposition, blocking is a reasonable thing to do. This is invariably true when Householder transformations are used to compute an intermediate decomposition — as often happens in the solution of eigenvalue problems.

1.3. TRIANGULARIZATION BY PLANE ROTATIONS

In some applications the matrix we want to triangularize has a special structure that reduces the size of the Householder transformations. For example, suppose that H is upper Hessenberg, i.e., that H has a Wilkinson diagram of the form

$$H = \begin{pmatrix} X & X & X & X & X \\ X & X & X & X & X \\ 0 & X & X & X & X \\ 0 & 0 & X & X & X \\ 0 & 0 & 0 & X & X \end{pmatrix}.$$

Then only the subdiagonal of H has to be annihilated. In this case it would be inefficient to apply the full Householder triangularization to H. Instead we should apply 2×2 transformations to the rows of H to put zeros on the subdiagonal (details later). Now applying a 2×2 Householder transformation to a vector requires 3 fladd + 4 flmlt. On the other hand to multiply the same vector by a 2×2 matrix requires 2 fladd + 4 flmlt. If the order of X is large enough, it will pay us to reconstitute the Householder transformation as a matrix before we apply it.

An alternative is to generate a 2×2 orthogonal matrix directly. The matrices that are conventionally used are called plane rotations. This subsection is devoted to the basics properties of these transformations.

Plane rotations

We begin with a definition.

Sec. 1. The QR Decomposition

Definition 1.10. *A* PLANE ROTATION *(also called a* GIVENS ROTATION*) is a matrix of the form*

$$P = \begin{pmatrix} c & s \\ -s & c \end{pmatrix},$$

where

$$c^2 + s^2 = 1.$$

It is easy to verify that a plane rotation is orthogonal. Moreover, since $c^2 + s^2 = 1$ there is a unique angle $\theta \in [0, 2\pi)$ such that

$$c = \cos\theta \quad \text{and} \quad s = \sin\theta.$$

The vector

$$\begin{pmatrix} c & s \\ -s & c \end{pmatrix} \begin{pmatrix} a \\ b \end{pmatrix} = \begin{pmatrix} ca + sb \\ cb - sa \end{pmatrix} \tag{1.18}$$

is obtained by rotating the vector $(a \; b)^{\mathrm{T}}$ clockwise through the angle θ.

Rotations can be used to introduce zeros in the following manner. Let $(a \; b)^{\mathrm{T}} \neq 0$ be given, and set

$$c = \frac{a}{\sqrt{a^2 + b^2}} \quad \text{and} \quad s = \frac{b}{\sqrt{a^2 + b^2}}.$$

Then from (1.18) we see that

$$\begin{pmatrix} c & s \\ -s & c \end{pmatrix} \begin{pmatrix} a \\ b \end{pmatrix} \begin{pmatrix} \sqrt{a^2 + b^2} \\ 0 \end{pmatrix}.$$

Rotations would not be of much use if we could only apply them to 2-vectors. However, we can apply them to rows and columns of matrices. Specifically, define a *rotation in the (i,j)-plane* as a matrix of the form

$$P_{ij} = \begin{array}{c} \\ \\ \\ i \\ \\ j \\ \\ \\ \end{array} \begin{pmatrix} 1 & \cdots & 0 & 0 & \cdots & 0 & 0 & \cdots & 0 \\ \vdots & \ddots & \vdots & \vdots & & \vdots & \vdots & & \vdots \\ 0 & \cdots & 1 & 0 & \cdots & 0 & 0 & \cdots & 0 \\ 0 & \cdots & 0 & c & \cdots & s & 0 & \cdots & 0 \\ \vdots & & \vdots & \vdots & \ddots & \vdots & \vdots & & \vdots \\ 0 & \cdots & 0 & -s & \cdots & c & 0 & \cdots & 0 \\ 0 & \cdots & 0 & 0 & \cdots & 0 & 1 & \cdots & 0 \\ \vdots & & \vdots & \vdots & & \vdots & \vdots & \ddots & \vdots \\ 0 & \cdots & 0 & 0 & \cdots & 0 & 0 & \cdots & 1 \end{pmatrix}. \tag{1.19}$$

> The following algorithm generates a plane rotation from the quantities a and b. It overwrites a with $\sqrt{a^2+b^2}$ and b with 0.
>
> 1. $\text{rotgen}(a, b, c, s)$
> 2. $\quad \tau = |a| + |b|$
> 3. $\quad \textbf{if } (\tau = 0)$
> 4. $\quad\quad c = 1; s = 0; \textbf{return}$
> 5. $\quad \textbf{end if}$
> 6. $\quad \nu = \tau * \sqrt{(a/\tau)^2 + (b/\tau)^2}$
> 7. $\quad c = a/\nu; s = b/\nu$
> 8. $\quad a = \nu; b = 0$
> 9. $\textbf{end } \text{rotgen}$

Algorithm 1.6: Generation of a plane rotation

———⋄———

In other words, a rotation in the (i,j)-plane is an identity matrix in which a plane rotation has been embedded in the submatrix corresponding to rows and columns i and j.

To see the effect of a rotation in the (i,j)-plane on a matrix, let X be a matrix and let

$$Y = P_{ij}X.$$

If X and Y are partitioned by rows, then

1. $y_i^{\mathrm{T}} = cx_i^{\mathrm{T}} + sx_j^{\mathrm{T}}$,
2. $y_j^{\mathrm{T}} = cx_j^{\mathrm{T}} - sx_i^{\mathrm{T}}$,
3. $y_k^{\mathrm{T}} = x_k^{\mathrm{T}}$, whenever $k \neq i, j$.

Thus premultiplication by a rotation in the (i,j)-plane combines rows i and j and leaves the others undisturbed. By an appropriate choice of c and s we can introduce a zero anywhere we want in the ith or jth row.

Similarly, postmultiplication by P_{ij}^{T} affects the columns in the same way, and we can introduce a zero anywhere in the ith or jth column.

Algorithms 1.6 and 1.7 display utility routines for generating and applying rotations. Here are some observations on these programs.

- The program *rotgen* overwrites a and b with the values they would get if the rotation were actually applied to them. This is usually what we want, since a and b will be paired elements in the two rows that are being transformed. The overwriting has the advantage that b becomes exactly zero, something that otherwise might not occur owing to rounding error.

Sec. 1. The QR Decomposition

The following function applies a rotation to two vectors x and y, overwriting the vectors.

1. $rotapp(c, s, x, y)$
2. $\quad t = c*x + s*y$
3. $\quad y = c*y - s*x$
4. $\quad x = t$
5. **end** *rotapp*

Algorithm 1.7: Application of a plane rotation

———⋄———

- The scaling factor τ is introduced to avoid overflows and make underflows harmless (see Algorithm 4.1, Chapter 2, for more details).

- Since the vectors in *rotapp* overwrite themselves, it is necessary to create a third vector to contain intermediate values. In a real-life implementation one must take care that the program does not call a storage allocator each time it is invoked.

- As a sort of shorthand we will write

$$rotapp(c, s, x, y)$$

even when x and y are scalars. In applications the operations should be written out in scalar form to avoid the overhead of invoking *rotapp*.

- In a BLAS implementation the vectors x and y would be accompanied by strides telling how they are allocated in memory.

- If we are computing in complex arithmetic, considerable savings can be effected by scaling the rotation so that the cosine c is real. To multiply a complex 2-vector by a complex plane rotation requires 16 flmlt + 4 fladd. If the cosine is real, this count becomes 12 flmlt + 2 fladd. The price to be paid is that ν becomes complex.

———

Transformations that can introduce a zero into a matrix also have the power to destroy zeros that are already there. In particular, plane rotations are frequently used to move a nonzero element around in a matrix by successively annihilating it in one position and letting it pop up elsewhere. In designing algorithms like this, it is useful to think of the transformations as a game played on a Wilkinson diagram. Here are the rules for one step of the game.

Begin by selecting two rows of the matrix (or two columns):

1	2	3	4	5
X	$\hat{\text{X}}$	0	X	0
X	X	0	0	X

Then after the transformation.

1. The element of your choice becomes zero ($\widehat{\text{X}}$ in column 2).
2. A pair of X's remains a pair of X's (column 1).
3. A pair of 0's remains a pair of 0's (column 3).
4. A mixed pair becomes a pair of X's (columns 4 and 5).

Thus our transformed pair of rows is

1	2	3	4	5
X	0	0	X	X
X	X	0	X	X

In this example we have actually lost a zero element.

Reduction of a Hessenberg matrix

Plane rotations can be used to triangularize a general matrix. However, they are most useful with structured matrices — like the Hessenberg matrix we used to motivate the introduction of plane rotations. Since we will see many examples of the use of plane rotations later, here we will just show how to triangularize a Hessenberg matrix. Actually, with a later example in mind (Algorithm 2.2), we will triangularize an augmented Hessenberg matrix with an extra row.

The algorithm is best derived by considering the following sequence of Wilkinson diagrams.

$$
\begin{array}{l}
\rightarrow \\
\rightarrow
\end{array}
\begin{array}{ccccc}
X & X & X & X & X \\
\widehat{X} & X & X & X & X \\
0 & X & X & X & X \\
0 & 0 & X & X & X \\
0 & 0 & 0 & X & X \\
0 & 0 & 0 & 0 & X
\end{array}
\stackrel{P_{12}}{\Longrightarrow}
\rightarrow
\begin{array}{ccccc}
X & X & X & X & X \\
0 & X & X & X & X \\
0 & \widehat{X} & X & X & X \\
0 & 0 & X & X & X \\
0 & 0 & 0 & X & X \\
0 & 0 & 0 & 0 & X
\end{array}
\stackrel{P_{23}}{\Longrightarrow}
$$

$$
\begin{array}{l}
\\
\\
\rightarrow \\
\rightarrow
\end{array}
\begin{array}{ccccc}
X & X & X & X & X \\
0 & X & X & X & X \\
0 & 0 & X & X & X \\
0 & 0 & \widehat{X} & X & X \\
0 & 0 & 0 & X & X \\
0 & 0 & 0 & 0 & X
\end{array}
\stackrel{P_{34}}{\Longrightarrow}
\rightarrow
\begin{array}{ccccc}
X & X & X & X & X \\
0 & X & X & X & X \\
0 & 0 & X & X & X \\
0 & 0 & 0 & X & X \\
0 & 0 & 0 & \widehat{X} & X \\
0 & 0 & 0 & 0 & X
\end{array}
\stackrel{P_{45}}{\Longrightarrow}
$$

$$
\begin{array}{l}
\\
\\
\\
\\
\rightarrow \\
\rightarrow
\end{array}
\begin{array}{ccccc}
X & X & X & X & X \\
0 & X & X & X & X \\
0 & 0 & X & X & X \\
0 & 0 & 0 & X & X \\
0 & 0 & 0 & X & X \\
0 & 0 & 0 & 0 & \widehat{X}
\end{array}
\stackrel{P_{56}}{\Longrightarrow}
\begin{array}{ccccc}
X & X & X & X & X \\
0 & X & X & X & X \\
0 & 0 & X & X & X \\
0 & 0 & 0 & X & X \\
0 & 0 & 0 & 0 & X \\
0 & 0 & 0 & 0 & 0
\end{array}
$$

Sec. 1. The QR Decomposition

This algorithm takes an augmented upper Hessenberg matrix $H \in \mathbb{R}^{(n+1) \times n}$ and reduces it to triangular form by a sequence of plane rotations.

1. **for** $k = 1$ **to** n
2.   *rotgen*($H[k,k]$, $H[k+1,k]$, c, s)
3.   *rotapp*(c, s, $H[k, k+1{:}n]$, $H[k+1, k+1{:}n]$)
4. **end for** k

Algorithm 1.8: Reduction of an augmented Hessenberg matrix by plane rotations

---◇---

The meaning of this sequence is the following. The arrows in a particular diagram point to the rows on which the plane rotation will operate. The X with a hat is the element that will be annihilated. On the double arrow following the diagram is the name of the rotation that effects the transformation. Thus the above sequence describes a transformation $P_{45} P_{34} P_{23} P_{12} X$ by a sequence of rotations in the $(i, i+1)$-plane that successively annihilates the elements x_{21}, x_{32}, x_{43}, and x_{54}.

Algorithm 1.8 implements this procedure. The very simple code is typical of algorithms involving plane rotations. An operation count is easily derived. The application of a rotation to a pair of scalars requires 2 fladd and 4 flmlt. Since statement 3 performs this operation about $n-k$ times, we find on integrating from $k = 0$ to $k = n$ that

Algorithm 1.8 *requires*

$$n^2 \text{ fladd} + 2n^2 \text{ flmlt} = \tfrac{1}{2} n^2 \text{ flrot}.$$

Here we have introduced the notation "flrot" as an abbreviation for 2 fladd + 4 flmlt (see Figure 2.1, Chapter 2).

Algorithm 1.8 has the disadvantage that it is row oriented. Now in many applications involving plane rotations the matrices are not very large, and the difference between column and row orientation is moot. However, if we are willing to store our rotations, we can apply them to each column until we reach the diagonal and then generate the next rotation. Algorithm 1.9 is an implementation of this idea. It should be stressed that this algorithm is numerically the exact equivalent of Algorithm 1.8. The only difference is the way the calculations are interleaved. Note the inefficient use of *rotapp* with the scalars $H[i, k]$ and $H[i, k+1]$.

Numerical properties

Plane rotations enjoy the same stability properties as Householder transformations. Specifically, Theorem 1.5 continues to hold when the Householder transformations are replaced by plane rotations. However, in many algorithms some of the plane rotations are nonoverlapping. For example, in Algorithm 1.8 each row is touched by at most

This algorithm is a column-oriented reduction by plane rotations of an augmented upper Hessenberg matrix $H \in \mathbb{R}^{(n+1)\times}$ to triangular form.

1. **for** $k = 1$ **to** n
2.   **for** $i = 1$ **to** $k-1$
3.   *rotapp*($c[i]$, $s[i]$, $H[i,k]$, $H[i+1,k]$)
4.   **end for** i
5.   *rotgen*($H[k,k]$, $H[k+1,k]$, $c[k]$, $s[k]$)
6. **end for** k

Algorithm 1.9: Column-oriented reduction of an augmented Hessenberg matrix

———⋄———

two rotations. This sometimes makes it possible to reduce the constant multiplying the rounding unit in the error bounds.

Plane rotations tend to perform better than Householder transformations on graded matrices. For example, if a plane rotation is generated from a vector whose grading is downward, say

$$\begin{pmatrix} 1 \\ \epsilon \end{pmatrix},$$

then up to second-order terms in ϵ it has the form

$$\begin{pmatrix} 1 & \epsilon \\ -\epsilon & 1 \end{pmatrix}.$$

Thus it is a perturbation of the identity and will not combine small and large elements. On the other hand, if the grading is upward, say

$$\begin{pmatrix} \epsilon \\ 1 \end{pmatrix},$$

then up to second-order terms in ϵ the rotation has the form

$$\begin{pmatrix} \epsilon & 1 \\ -1 & \epsilon \end{pmatrix}.$$

Thus the rotation is effectively an exchange matrix (with a sign change) and once again does not combine large and small elements.

However, as with Householder transformations, we can prove nothing in general. It is possible for an unfortunate cancellation to produce a balanced transformation that combines large and small elements. In fact, the matrix A of Example 1.8 serves as a counterexample for plane rotations as well as Householder transformations.

1.4. THE GRAM–SCHMIDT ALGORITHM

Although the Householder form of the QR decomposition contains the full decomposition in factored form, there are occasions when we need an explicit QR factorization. We can compute Q_X using the algorithm (1.9) but at the cost of doubling our work. It is therefore reasonable to search for cheaper alternatives. In fact, we have one at hand — the Gram–Schmidt algorithm used in the proof of Theorem 4.24, Chapter 1, where we established the existence of the QR factorization. The purpose of this subsection is to examine this algorithm in more detail.

There are two versions of the Gram–Schmidt algorithm: the classical algorithm and the modified algorithm. From a numerical point of view, both represent compromises. Specifically, if R is ill conditioned, they are both guaranteed to produce nonorthogonal vectors. There is a fix-up — reorthogonalization — but it requires additional work. However, if orthogonality is not the prime concern (and sometimes it is not), the modified Gram–Schmidt algorithm has superior numerical properties.

We will begin by rederiving the classical Gram–Schmidt algorithm and then turn to the modified algorithm. We will then discuss their numerical properties and conclude with a treatment of reorthogonalization.

The classical and modified Gram–Schmidt algorithms

The classical Gram–Schmidt algorithm can be regarded as a method of projection. Suppose that X has linearly independent columns and partition X in the form

$$X = (X_1 \ X_2),$$

where X_1 has $k-1$ columns. Suppose we have computed the QR factorization

$$X_1 = Q_1 R_{11}$$

and we want to compute the QR factorization of $(X_1 \ x_k)$, where as usual x_k is the kth column of X.

The projection of x_k onto the orthogonal complement of $\mathcal{R}(X_1) = \mathcal{R}(Q_1)$ is

$$x_k^\perp = (I - Q_1 Q_1^T) x_k = x_k - Q_1(Q_1^T x_k) \equiv x_k - Q_1 r_{1k}.$$

Now $x_k^\perp$ cannot be zero, for that would mean that it lies in $\mathcal{R}(X_1)$. Hence if we define

$$\rho_{kk} = \|x_k^\perp\| \quad \text{and} \quad q_k = x_k^\perp / \rho_{kk},$$

then $q_k \perp \mathcal{R}(Q_1)$ and

$$x_k = (Q_1 \ q_k) \begin{pmatrix} r_{1k} \\ \rho_{kk} \end{pmatrix}.$$

It follows that

$$(X_1 \ x_k) = (Q_1 \ q_k) \begin{pmatrix} R_{11} & r_{1k} \\ 0 & \rho_{kk} \end{pmatrix}$$

Given an $n \times p$ matrix X with linearly independent columns, this algorithm computes the QR factorization of X.

1. **for** $k = 1$ **to** p
2. $Q[:,k] = X[:,k]$
3.   **if** $(k \neq 1)$
4. $R[1{:}k{-}1, k] = Q[:, 1{:}k{-}1]^{\mathrm{T}} * Q[:, k]$
5. $Q[:, k] = Q[:, k] - Q[:, 1{:}k{-}1] * R[1{:}k{-}1, k]$
6.   **end if**
7. $R[k, k] = \|Q[:, k]\|_2$
8. $Q[:, k] = Q[:, k]/R[k, k]$
9. **end for** k

Algorithm 1.10: The classical Gram–Schmidt algorithm

———◇———

is the QR factorization of $(X_1 \ x_k)$.

Thus we can compute a QR factorization of X by successive projections and normalizations, as in Algorithm 1.10. Two comments.

- The algorithm requires $(np^2 - \frac{1}{3}p^3)$ flam, the same as for the Householder reduction. However, it gives the QR factorization immediately.

- If we replace Q by X everywhere in the algorithm, the algorithm overwrites X with Q_X.

———

We can derive a different version of the Gram–Schmidt algorithm by writing the kth column of X in the form

$$x_k = q_1 r_{1k} + q_2 r_{2k} + \cdots + q_k r_{kk}. \tag{1.20}$$

On multiplying this relation by q_1^{T} we get

$$r_{1k} = q_1^{\mathrm{T}} x_k.$$

Subtracting $q_1 r_{1k} = q_1 q_1^{\mathrm{T}} x_k$ from (1.20) we get

$$(I - q_1 q_1^{\mathrm{T}}) x_k = q_2 r_{2k} + \cdots + q_k r_{kk}. \tag{1.21}$$

To continue the process multiply (1.21) by q_2^{T} to get

$$r_{2k} = q_2^{\mathrm{T}}(I - q_1 q_1^{\mathrm{T}}) x_k,$$

and then subtract $q_2 r_{2k}$ from (1.21) to get

$$(I - q_1 q_1^{\mathrm{T}})(I - q_1 q_1^{\mathrm{T}}) x_k = q_3 r_{3k} + \cdots + q_k r_{kk}.$$

Sec. 1. The QR Decomposition

Given an $n \times p$ matrix X with linearly independent columns, this algorithm computes the QR factorization of X by the modified Gram–Schmidt method in a version that constructs R column by column.

1. **for** $k = 1$ **to** p
2. $Q[:,k] = X[:,k]$
3.   **for** $i = 1$ **to** $k-1$
4. $R[i,k] = Q[:,i]^\mathrm{T} * Q[:,k]$
5. $Q[:,k] = Q[:,k] - R[i,k] * Q[:,i]$
6.   **end for** i
7. $R[k,k] = \|Q[:,k]\|_2$
8. $Q[:,k] = Q[:,k]/R[k,k]$
9. **end for** k

Algorithm 1.11: The modified Gram–Schmidt algorithm: column version

---

Given an $n \times p$ matrix X with linearly independent columns, this algorithm computes the QR factorization of X by the modified Gram–Schmidt method, in a version that constructs R row by row.

1. $Q = X$
2. **for** $k = 1$ **to** p
3. $R[k,k] = \|Q(:,k)\|_2$
4. $Q[:,k] = Q[:,k]/R[k,k]$
5. $R[k,k+1{:}p] = Q[:,k]^\mathrm{T} * Q[:,k+1{:}p]$
6. $Q[:,k+1{:}p] = Q[:,k+1{:}p] - Q[:,k] * R[k,k+1{:}p]$
7. **end for** k

Algorithm 1.12: The modified Gram–Schmidt algorith: row version

---

This process can be continued until all that is left is $q_k r_{kk}$, from which r_{kk} can be obtained by normalization.

Algorithm 1.11 implements this scheme. It is called the *modified Gram–Schmidt algorithm*—a slightly misleading name, since it is no mere rearrangement of the classical Gram–Schmidt algorithm but a new algorithm with, as we shall see, greatly different numerical properties.

Algorithm 1.11 builds up R column by column. A different interleaving of the computations, shown in Algorithm 1.12, builds up R row by row. It should be stressed that this algorithm is numerically the exact equivalent of Algorithm 1.11 in the sense that it will produce exactly the same results in computer arithmetic. However, it is richer in matrix-vector operations, which makes it a better candidate for optimization.

Modified Gram–Schmidt and Householder triangularization

The superior properties of the modified Gram–Schmidt algorithm are a consequence of a remarkable relation between the algorithm and Householder triangularization.

Theorem 1.11. *Let $\bar{Q}$ and $\bar{R}$ denote the matrices computed by the modified Gram–Schmidt algorithm in floating-point arithmetic. If Householder triangularization is applied to the matrix*

$$\begin{pmatrix} 0_p \\ X \end{pmatrix}$$

in the same arithmetic, the result is

$$\begin{pmatrix} \bar{R} \\ 0 \end{pmatrix}.$$

Moreover, the vectors defining the Householder transformations H_k are

$$u_k = \begin{pmatrix} \mathbf{e}_k \\ \bar{q}_k \end{pmatrix}.$$

Proof. The identities can be verified by comparing the Householder triangularization with the version of the modified Gram–Schmidt algorithm given in Algorithm 1.12. One step is sufficient to see the pattern. ∎

Theorem 1.11 is useful in deriving new algorithms. However, its greatest value is that it explains the numerical properties of the modified Gram–Schmidt algorithm.

Error analysis of the modified Gram–Schmidt algorithm

In discussing the error analysis of Householder triangularization we made a distinction between the Householder transformations $\hat{H}_k$ that would have been generated by exact computations in the course of the reduction and the transformations that we actually generate and apply [see (1.10)]. If we apply Theorem 1.5 to the reduction of Theorem 1.11, we find that there are small matrices E_1 and E_2 such that

$$\begin{pmatrix} E_1 \\ A + E_2 \end{pmatrix} = (\hat{H}_1 \cdots \hat{H}_p) \begin{pmatrix} \bar{R} \\ 0 \end{pmatrix} \equiv \begin{pmatrix} \hat{Q}_1 \\ \hat{Q}_2 \end{pmatrix} \bar{R}. \tag{1.22}$$

Equivalently,

$$A + E_2 = \hat{Q}_2 R.$$

Now $\hat{Q}_2$ is not orthonormal. But from the relation (1.22), we can show that there is an orthonormal matrix $\check{Q}$, satisfying $A + E = \check{Q}R$, where $\|e_j\|_F \leq \|e_j^{(1)}\|_F + \|e_j^{(2)}\|_F$ (the proof of this fact is not trivial). From this and the bound (1.12) we get the following theorem.

Sec. 1. The QR Decomposition

Theorem 1.12. *Let $\bar{Q}$ and $\bar{R}$ denote the matrices computed by the modified Gram–Schmidt algorithm in floating-point arithmetic with rounding unit ϵ_M. Then there is an orthonormal matrix $\check{Q}$ such that*

$$X + E = \check{Q}\bar{R}, \tag{1.23}$$

where the columns of E satisfy

$$\frac{\|e_j\|_F}{\|x_j\|_F} \leq \varphi(n,p)\epsilon_M. \tag{1.24}$$

Here φ is a slowly growing function of n and p. Moreover, there is a matrix F such that

$$X + F = \bar{Q}\bar{R}, \tag{1.25}$$

where the columns of F satisfy a bound of the form (1.24).

Equation (1.23) says that the factor $\bar{R}$ computed by the modified Gram–Schmidt algorithm is the exact R-factor of a slightly perturbed X. The bound is columnwise — as might be expected, since scaling a column of X does not materially affect the course of the algorithm. Unfortunately, $\check{Q}$ can have little to do with the computed $\bar{Q}$.

Equation (1.25), on the other hand, says that the product of the factors we actually compute reproduces X accurately. Unfortunately, there is nothing to insure that the columns of $\bar{Q}$ are orthogonal. Let us look at this problem more carefully.

Loss of orthogonality

The classical and modified Gram–Schmidt algorithms are identical when they are applied to two vectors. Even in this simple case the resulting Q can be far from orthonormal. Suppose, for example, that x_1 and x_2 are nearly proportional. If $P_1^\perp$ denotes the projection onto the orthogonal complement of x_1, then $P_1^\perp x_2$ will be small compared to x_2. Now the Gram–Schmidt algorithm computes this projection in the form

$$P_1^\perp x_2 = x_2 - (q_1^T x_2)q_1,$$

where $q_1 = x_1/\|x_1\|_2$. The only way we can get a small vector out of this difference is for there to be cancellation, which will magnify the inevitable rounding errors in x_2 and q_1. Rounding errors are seldom orthogonal to anything useful.

A numerical example will make this point clear.

Example 1.13. *The matrix*

$$U = \begin{pmatrix} -0.88061 & -0.47384 \\ -0.47384 & 0.88061 \end{pmatrix}$$

is exactly orthogonal. Let us take the first column u_1 of U as x_1. For x_2 we round u_1 to three digits:

$$x_2 = \begin{pmatrix} -0.88100 \\ -0.47400 \end{pmatrix}.$$

When the Gram–Schmidt algorithm is applied to these vectors in five-digit decimal arithmetic, we get

$$\mathrm{fl}(P_1^\perp x_2) = \begin{pmatrix} -0.88100 \\ -0.47400 \end{pmatrix} - \begin{pmatrix} -0.88096 \\ -0.47403 \end{pmatrix} = \begin{pmatrix} -0.00004 \\ 0.00003 \end{pmatrix}.$$

The angle between this vector and x_1 is approximately 65 degrees!

It is worth noting that if we write $P_1^\perp$ in the form $u_2 u_2^T$ then we can compute the projection in the form

$$\mathrm{fl}[(u_2^T x_2) u_2] = \begin{pmatrix} -0.000020802 \\ 0.000038659 \end{pmatrix}.$$

This vector is almost exactly orthogonal to x_1 (though it is not accurate, since there is cancellation in the computation of the inner product $q_2^T x_2$). This is the kind of result we would get if we used the basis $Q_\perp$ from Householder's triangularization to compute the projection. Thus when it comes to computing projections, orthogonal triangularization is superior to both versions of Gram–Schmidt.

When $p > 2$, the classical and modified Gram–Schmidt algorithms go their separate ways. The columns of Q produced by the classical Gram–Schmidt can quickly lose all semblance of orthogonality. On the other hand, the loss of orthogonality in the Q produced by the modified Gram–Schmidt algorithm is proportional to the condition number of R. Specifically, we have the following theorem.

Theorem 1.14. *Let $X = QR$ be the QR factorization of X. Let $\bar{Q}$ and $\bar{R}$ be the QR factorization computed by the modified Gram–Schmidt algorithm in floating-point arithmetic with rounding unit ϵ_M. Let $\check{Q}$ be the orthogonal matrix whose existence is guaranteed by Theorem 1.12. Then there is a constant γ such that*

$$\|\check{Q} - \bar{Q}\|_F \leq \frac{\gamma \kappa_F(R) \epsilon_M}{1 - \gamma \kappa_F(R) \epsilon_M}. \tag{1.26}$$

Proof. From the bounds of Theorem 1.12, we can conclude that there is a constant γ such that

$$\|E\|_F + \|F\|_F \leq \gamma \|X\|_F \epsilon_M = \gamma \|R\|_F \epsilon_M.$$

From (1.23) and (1.25), we have $\check{Q} - \bar{Q} = (E - F)\bar{R}^{-1}$. Hence

$$\|\check{Q} - \bar{Q}\|_F \leq \gamma \|X\|_F \|\bar{R}^{-1}\|_2 \epsilon_M \leq \gamma \|R\|_F \|\bar{R}^{-1}\|_2 \epsilon_M. \tag{1.27}$$

SEC. 1. THE QR DECOMPOSITION

Since $X+E = \check{Q}\bar{R}$ and $\check{Q}$ is exactly orthogonal, the smallest singular value of $\bar{R}$ is the same as the smallest singular value of $X + E$, which is bounded below by $\sigma - \|E\|_F$, where σ is the smallest singular value of X and R (they are the same). Thus

$$\begin{aligned}\|\bar{R}^{-1}\|_2 &\leq \frac{1}{\sigma - \|E\|_F} \\ &= \frac{\sigma^{-1}}{1 - \sigma^{-1}\|E\|_F} \\ &= \frac{\|R^{-1}\|_2}{1 - \|R^{-1}\|_2\|E\|_F} \\ &\leq \frac{\|R^{-1}\|_F}{1 - \gamma\|R\|_F\|R^{-1}\|_F \epsilon_M}.\end{aligned}$$

The result follows on substituting this upper bound in (1.27). ∎

Two comments on this theorem.

- If $X = QR$ is a QR factorization of X, then the pseudoinverse of X is given by $X^\dagger = R^{-1}Q^T$ [see (1.5)]. Since the columns of Q are orthonormal, we have $\|R^{-1}\| = \|X^\dagger\|$, where $\|\cdot\|$ is either the spectral or Frobenius norm. Consequently, if we define the *condition number* of X by

$$\kappa(X) = \|X\|\|X^\dagger\|, \tag{1.28}$$

we have

$$\kappa_p(X) = \kappa_p(R), \qquad p = 2, F.$$

We will be seeing more of $\kappa(X)$.

- If R_k is the leading principal submatrix of R of order k, then

$$\kappa_F(R_k) \leq \kappa_F(R_{k+1}), \qquad k = 1, \ldots, p - 1.$$

Since R_k is the R-factor of the matrix X_k consisting of the first k columns of X, we should see a progressive deterioration in the orthogonality of the vectors q_k as k increases. The following example demonstrates this effect and also shows the dramatic failure of the classical Gram–Schmidt algorithm.

Example 1.15. *A 50×10 matrix X was generated by computing*

$$U\mathrm{diag}(1, 10^{-1}, \ldots, 10^{-9})V^T,$$

where U and V are random orthonormal matrices. Thus the singular values of X are $1, 10^{-1}, \ldots, 10^{-9}$ and $\kappa_F(X) \cong 10^9$. Both the classical Gram–Schmidt and modified

Gram–Schmidt algorithms were applied to this matrix with the following results.

k	$\kappa(X_k)$	$\|I - Q_{\text{CGS}}^{\text{T}} Q_{\text{CGS}}\|_2$	$\|I - Q_{\text{MGS}}^{\text{T}} Q_{\text{MGS}}\|_2$
1	1.0e+00	1.4e−15	1.4e−15
2	1.2e+02	7.1e−14	7.1e−14
3	3.4e+02	1.9e−12	1.2e−13
4	2.7e+03	2.0e−11	1.1e−12
5	4.4e+04	2.3e−09	2.1e−11
6	3.7e+05	9.8e−07	1.7e−10
7	5.6e+06	7.5e−04	1.9e−09
8	2.3e+07	4.0e−02	4.7e−09
9	9.7e+07	9.3e−01	3.3e−08
10	1.0e+09	1.9e+00	3.2e−07

The kth row concerns the factorization of the matrix X_k consisting of the first k columns of X. It lists the condition of X_k and the departure from orthogonality of the vectors generated by the classical and modified Gram–Schmidt algorithms. The condition number increases, as mentioned above. The vectors generated by the classical Gram–Schmidt algorithm quickly lose orthogonality. The loss of orthogonality for the modified Gram–Schmidt algorithm is roughly the rounding unit $(2 \cdot 10^{-16})$ divided by the condition number.

Reorthogonalization

The loss of orthogonality generated by the Gram–Schmidt algorithm is acceptable in some applications. However, in others — updating, for example — we demand more. Specifically, given a vector a vector x and an orthonormal matrix Q we need to compute quantities $x_\perp$, r, and ρ such that

1. $\rho^{-1}\|x_\perp\| = 1$,
2. $x = Qr + x_\perp$ to working accuracy,
3. $\mathcal{R}(Q) \perp \rho^{-1} x_\perp$ to working accuracy.

The first item says that $x_\perp \neq 0$ so that it can be normalized. The second item says that if we set $q = \rho^{-1} x_\perp$, then $x = Qr + \rho q$; i.e., r and ρ can be regarded as forming the last column of a QR factorization. The third says that to working accuracy q is orthogonal to $\mathcal{R}(Q)$. It is the last item that gives the Gram–Schmidt algorithms trouble. The cure is reorthogonalization.

To motivate the reorthogonalization procedure, suppose that we have computed a nonzero vector $\hat{x}_\perp$ that satisfies

$$\hat{x}_\perp = x - Qr$$

but that $\hat{x}_\perp$ is not sufficiently orthogonal to $\mathcal{R}(Q)$. If we ignore rounding error and define

$$s = Q^{\text{T}} \hat{x}_\perp \quad \text{and} \quad \bar{x}_\perp = \hat{x}_\perp - Qs,$$

SEC. 1. THE QR DECOMPOSITION

then
$$x = \bar{x}_\perp + Q(r + s).$$

By construction $\bar{x}_\perp$ is exactly orthogonal to $\mathcal{R}(Q)$. It is not unreasonable to expect that in the presence of rounding error the orthogonality of $\bar{x}_\perp$ will be improved. All this suggests the following iterative algorithm for orthogonalizing x against the columns of Q.

1. $x_\perp = x$
2. $r = 0$
3.   **while (true)**
4. $s = Q^T x_\perp$
5. $x_\perp = x_\perp - Qs$
6. $r = r + s$
7.   **if** ($x_\perp$ is satisfactory) **leave** the loop; **fi**
8.   **end while**
9. $\rho = \|x_\perp\|_2$
10. $q = x_\perp/\rho$

(1.29)

Let us see what happens when this algorithm is applied to the results of Example 1.13.

Example 1.16. *In attempting to orthogonalize*

$$x = \begin{pmatrix} -8.8100\text{e}{-01} \\ -4.7400\text{e}{-01} \end{pmatrix} \quad \text{against} \quad q = \begin{pmatrix} -8.8061\text{e}{-01} \\ -4.7384\text{e}{-01} \end{pmatrix}$$

we obtained the vector

$$\hat{x} = \begin{pmatrix} -4.0000\text{e}{-05} \\ 3.0000\text{e}{-05} \end{pmatrix},$$

which is decidedly not orthogonal to q. If we reorthogonalize $\hat{x}$ in five-digit arithmetic, we get

$$\bar{x} = \begin{pmatrix} 2.1499\text{e}{-05} \\ -3.9955\text{e}{-05} \end{pmatrix},$$

for which

$$\frac{|q^T \bar{x}|}{\|\bar{x}\|_2} = 9.4 \cdot 10^{-7}.$$

Thus $\bar{x}$ is orthogonal to q to working accuracy.

In order to turn (1.29) into a working program we must deal with the possibility that the iteration does not terminate. We begin by showing that if $\hat{x}_\perp$ is sufficiently large compared to x then it is orthogonal to x. To do so we will work with a simplified model of the computation. Specifically, we will assume that the columns of Q are exactly orthonormal and that

$$\hat{x}_\perp \equiv \text{fl}[x - Q(Q^T x)] = x_\perp + e, \qquad (1.30)$$

where

$$\frac{\|e\|_2}{\|x\|_2} \leq \gamma \epsilon_M \qquad (1.31)$$

for some modest constant γ.

These assumptions allow us to determine when $\hat{x}_\perp$ is reasonably orthogonal to $\mathcal{R}(Q)$ without having to compute $Q^T \hat{x}_\perp$. From Theorem 4.37, Chapter 1, we know that $\|Q^T \hat{x}_\perp\|_2 / \|\hat{x}_\perp\|_2$ is the sine of the angle between $\hat{x}_\perp$ and $\mathcal{R}(Q)^\perp$. By (1.30), $Q^T \hat{x}_\perp = Q^T e$. Thus from (1.31)

$$\frac{\|Q^T \hat{x}_\perp\|_2}{\|x_\perp\|_2} \leq \gamma \frac{\|x\|_2}{\|\hat{x}_\perp\|_2} \epsilon_M.$$

Hence if

$$\frac{\|\hat{x}_\perp\|_2}{\|x\|_2} \geq \alpha \qquad (1.32)$$

for some constant α near one, then

$$\frac{\|Q^T \hat{x}_\perp\|_2}{\|x_\perp\|_2} \leq \alpha^{-1} \gamma \epsilon_M \qquad (1.33)$$

is small and $\hat{x}_\perp$ lies almost exactly in $\mathcal{R}(Q)^\perp$. This means that we can tell if the current $x_\perp$ is satisfactory by choosing a tolerance α — e.g., $\alpha = \frac{1}{2}$ — and demanding that x and $x_\perp$ satisfy (1.32).

This analysis also suggests that the loop in (1.29) is unlikely be executed more than twice. The reason is that loss of orthogonality can occur only when x is very near $\mathcal{R}(Q)$. If the loss of orthogonality is not catastrophic, the vector $\hat{x}_\perp$ will not be near $\mathcal{R}(Q)$, and the next iteration will produce a vector that is almost exactly orthogonal. On the other had if there is a catastrophic loss of orthogonality, the vector $\hat{x}_\perp$ will be dominated by the vector e of rounding errors. This vector is unlikely to be near $\mathcal{R}(Q)$, and once again the next iterate will give an orthogonal vector.

There still remains the unlikely possibility that the vector e and its successors are all very near $\mathcal{R}(Q)$, so that the vectors $x_\perp$ keep getting smaller and smaller without becoming orthogonal. Or it may happen that one of the iterates becomes exactly zero. In either case, once the current $x_\perp$ is below the rounding unit times the norm of the original x, we may replace $x_\perp$ with an arbitrary vector of the same norm, and the relation $x = Qr + x_\perp$ will remain valid to working accuracy. In particular, if we choose

SEC. 1. THE QR DECOMPOSITION

This algorithm takes an orthonormal matrix Q and a nonzero vector x and returns a vector q of norm one, a vector r, and a scalar ρ such that $x = Qr + \rho q$ to working accuracy. Moreover, $Q^T q \cong 0$ in proportion as the parameter α is near one.

```
1.    gsreorthog(Q, x, q, r, ρ)
2.        ν = σ = ||x||₂
3.        x⊥ = x
4.        r = 0
5.        while (true)
6.            s = Qᵀx⊥
7.            r = r + s
8.            x⊥ = x⊥ − Qs
9.            τ = ||x⊥||₂
10.           if (τ/σ ≥ α) leave the loop; fi
11.           if (τ > 0.1*ν*ε_M)
12.               σ = τ
13.           else
14.               ν = σ = 0.1*σ*ε_M
15.               i = index of the row of minimal 1-norm in Q
16.               x⊥ = σ*eᵢ
17.           end if
18.       end while
19.       ρ = ||x⊥||₂
20.       q = x⊥/ρ
21.   end for gsreorthog
```

Algorithm 1.13: Classical Gram–Schmidt orthogonalization with reorthogonalization

———◇———

$x_\perp$ to have a significant component in $\mathcal{R}(Q)^\perp$, then the iteration will terminate after one or two further iterations. A good choice is the vector $\mathbf{e}_i$, where i is the index of the row of least 1-norm in Q.

Although the analysis we have given here is informal in that it assumes the exact orthonormality of Q, it can be extended to the case where the columns of Q are nearly orthonormal.

Algorithm 1.13 implements this reorthogonalization scheme. The value of $||x||_2$ is held in ν. The current $||x_\perp||_2$ is held in σ and the value of $||\hat{x}_\perp||_2$ is held in τ. Thus statement 10 tests for orthogonality by comparing the reduction in $||\hat{x}_\perp||_2$. If that test fails, the algorithm goes on to ask if the current $x_\perp$ is negligible compared to the original x (statement 11). If it is not, another step of reorthogonalization is performed. If it is, the original vector x is replaced by a suitable, small vector, after which the algorithm will terminate after no more than two iterations.

Here are some comments on the algorithm.

- We have focused on the classical Gram–Schmidt algorithm, which needs the most protection against loss of orthogonality. However, the procedure works equally well with the modified algorithm.

- If $\alpha = \frac{1}{2}$, both the classical and modified algorithms give matrices that are orthonormal to working accuracy. In this case, the algorithm will usually perform two orthogonalizations for each vector orthogonalized. Thus the price for guaranteed orthogonality is a doubling of the work. This makes Gram–Schmidt comparable to Householder triangularization when it comes to producing an explicit orthonormal basis. Which algorithm to prefer will depend on the application.

- The constant 0.1 in statement 11 is a somewhat arbitrary shrinking factor that insures that the current $x_\perp$ is truly negligible compared to the original x. In particular, further corrections to r will be below the level of rounding error and will not register.

- If x is very small, the subtraction $x_\perp = x_\perp - Qs$ may underflow, causing the algorithm to fail. A better implementation would first normalize x and then readjust r and ρ at the end.

- If Q is $n \times k$, the computation of the 1-norms of the rows of Q requires $n \times k$ compares to form the absolute values and the same number of additions to compute the norm. Since the algorithm usually terminates after two steps, this is an acceptable overhead. However, if the algorithm is used to orthogonalize a sequence of vectors x_k, the 1-norms of the rows of Q can be updated as columns are added at a cost of n compares and additions.

- When $x \in \mathcal{R}(Q)$, the vector q generated by the algorithm is essentially a space filler, contrived to hold a spot in a QR factorization. In some applications, it is sufficient to return a zero vector as an indication that $x \in \mathcal{R}(Q)$. In this case all one needs to do is at most two orthogonalizations, and return zero if the α-test fails both times. This algorithm is sometimes called the *twice-is-enough algorithm*.

1.5. NOTES AND REFERENCES

General references

Anyone concerned with least squares problems should obtain a copy of Åke Björck's *Numerical Methods for Least Squares Problems* [41]. This definitive survey touches all aspects of the subject and has a bibliography of 860 entries. Lawson and Hanson's classic, *Solving Least Squares Problems* [213], has recently been reprinted by SIAM with a survey of recent developments. Most texts on matrix computations (e.g., [86, 153, 319, 288, 333]) contain sections devoted to the QR decomposition and least squares.

The QR decomposition

For historical comments on the QR factorization see §4.6, Chapter 1.

SEC. 1. THE QR DECOMPOSITION

The distinction between a full decomposition and an abbreviated factorization is not standard; but it is useful, and we have applied it to both the QR and singular value decompositions. What we have called the QR factorization is sometimes called the Gram–Schmidt decomposition (e.g., in [41, 84]); however, neither Gram nor Schmidt saw their processes as computing a factorization of a matrix.

As we noted for the LU decomposition, the sensitivity of the QR decomposition to perturbations in the original matrix is largely a matter for specialists. Results and further references may be found in [177, 291, 304, 314, 354].

The pseudoinverse

In Theorem 3.23, Chapter 1, we showed that any matrix X of full column rank has a left inverse X^I satisfying $X^I X = I$. The pseudoinverse $X^\dagger = X(X^T X)^{-1} X^T$ is one of many possible choices — but an important one. It has the useful property that, of all left-inverses, it has minimal Frobenius norm. This result (though not phrased in terms of matrices) was essentially established by Gauss [133, 1823] to support his second justification of least squares. The modern formulation of the pseudoinverse is due to Moore [236, 1920], Bjerhammer [34, 1951], and Penrose [259, 1955], all of whom considered the case where X is not of full rank.

For full-rank matrices, the pseudoinverse is a useful notational device, whose formulas can be effectively implemented by numerical algorithms. As with the matrix inverse, however, one seldom has to compute the pseudoinverse itself. For matrices that are not of full rank, one is faced with the difficult problem of determining rank — usually in the presence of error. We will treat this important problem in the next chapter.

The pseudoinverse is only one of many *generalized inverses* which have been proposed over the years (Penrose's paper seems to have triggered the vogue). For a brief introduction to the subject via the singular value decomposition see [310, §III.1.1]. For an annotated bibliography containing 1776 entries see the collection edited by Nashed and Rall [241].

Householder triangularization

Householder transformations seem first to have appeared in a text by Turnbull and Aitken [322, 1932], where they were used to establish Schur's result [274, 1909] that any square matrix can be triangularized by a unitary similarity transformation. They also appear as a special case of a class of transformations in [117, 1951]. Householder [188, 1958], who discovered the transformations independently, was the first to realize their computational significance.

Householder called his transformations *elementary Hermitian matrices* in his *Theory of Matrices in Numerical Analysis* [189], a usage which has gone out fashion. Since the Householder transformation $I - uu^T$ reflects the vector u through its orthogonal complement (which remains invariant), these transformations have also been called *elementary reflectors*.

Householder seems to have missed the fact that there are two transformations that will reduce a vector to a multiple of e_1 and that the natural construction of one of them is unstable. This oversight was corrected by Wilkinson [343]. Parlett [252, 253] has shown how to generate the alternative transformation in a stable manner.

Although Householder derived his triangularization algorithm for a square matrix, he pointed out that it could be applied to rectangular matrices. We will return to this point in the next section, where we treat algorithms for least squares problems.

Rounding-error analysis

The rounding-error analysis of Householder transformations is due to Wilkinson [346, 347]. Higham gives a proof of Theorem 1.5 in [177, §18.3].

Martin, Reinsch, and Wilkinson [225, 1968] noted that graded matrices must be oriented as suggested in (1.14) to be successfully reduced by Householder transformations. Simultaneously, Powell and Reid [264, 1968] showed that under a combination of column pivoting on the norms of the columns and row pivoting for size the reduction is rowwise stable. Cox and Higham [77] give an improved analysis, in which they show that the row pivoting for size can be replaced by presorting the rows. Unfortunately, these results contain a growth factor which can be large if an initial set of rows is intrinsically ill conditioned — something that can easily occur in the weighting method for constrained least squares (§2.4).

Blocked reduction

The first blocked triangularization by orthogonal transformations is due to Bischof and Van Loan [33]. They expressed the product of k Householder transformations in the form WY^T where W and Y are $n \times k$ matrices. The UTU representation (Theorem 1.9), which requires only half the storage, is due to Schreiber and Van Loan [273]. For an error analysis see [177, §18.4].

Plane rotations

Rotations of the form (1.19) were used by Jacobi [190, 1846] in his celebrated algorithm for the symmetric eigenvalue problem. They are usually distinguished from plane rotations because Jacobi chose his angle to diagonalize a 2×2 symmetric matrix. Givens [145, 1954] was the first to use them to introduce a zero at a critical point in a matrix; hence they are often called *Givens rotations*.

For error analyses of plane rotations see [346, pp. 131–143], [142], and especially [177, §18.5].

The superior performance of plane rotations on graded matrices is part of the folklore. As Example 1.8 shows, there are no rigorous general results. In special cases, however, it may be possible to show something. For example, Demmel and Veselić [93] have shown that Jacobi's method applied to a positive define matrices is superior to Householder tridiagonalization followed by the QR algorithm. Mention should also be made of the analysis of Anda and Park [8].

SEC. 1. THE QR DECOMPOSITION

Storing rotations

In most applications plane rotations are used to refine or update an existing decomposition. In this case the rotations are accumulated in the orthogonal part of the decomposition. However, rotations can also be used as an alternative to Householder transformations to triangularize an $n \times p$ matrix. If $n \gg p$, then the reducing matrix must be stored in factored form — i.e., the rotations must be stored. If we store both the sine and cosine, the storage requirement is twice that of Householder transformations. We could store, say, the cosine c, and recover the sine from the formula $s = \sqrt{1 - c^2}$. However, this formula is unstable when c is near one. Stewart [289] shows how to compute a single number from which both s and c can be stably retrieved.

Fast rotations

The operation counts for the application of a plane rotation to a matrix X can be reduced by scaling the rotation. For example, if $c \geq s$ we can write the rotation

$$P = \begin{pmatrix} c & s \\ -s & c \end{pmatrix}$$

in the form

$$\sigma Q \equiv c \begin{pmatrix} 1 & \frac{s}{c} \\ -\frac{s}{c} & 1 \end{pmatrix}.$$

On the other had if $c < s$ we can write the rotation in the form

$$\sigma Q \equiv s \begin{pmatrix} \frac{c}{s} & 1 \\ -1 & \frac{c}{s} \end{pmatrix}.$$

Consequently, the product $P_k \cdots P_1$ of plane rotations can be written in the form

$$P_k \cdots P_1 = (\sigma_k \cdots \sigma_1)(Q_k \cdots Q_1).$$

Thus we can apply the scaled rotations Q_k to X — at reduced cost because two of the elements of Q are now one. The product of the scaling factors — one product for each row of the matrix — can be accumulated separately.

This is the basic idea behind the *fast rotations* of Gentleman [141] and Hammarling [170]. By a careful arrangement of the calculations it is possible to avoid the square roots in the formation of fast rotations. The principal difficulty with the scaling strategy is that the product of the scaling factors decreases monotonically and may underflow. It is therefore necessary to monitor the product and rescale when it becomes too small. Anda and Park [7] give a more flexible scheme that avoids this difficulty. See also [214], [41, §2.3.3].

The Gram–Schmidt algorithm

As we have mentioned (§4.6, Chapter 1), the Gram–Schmidt algorithm originated as a method for orthogonalizing sequences of functions. The method seems to have been first applied to finite-dimensional vectors by Kowalewski [203, 1909]. The origins of the modified Gram–Schmidt algorithm are obscure, but Wilkinson [348, 1971] said he had used the modified method for years. (Wilkinson's natural computational parsimony would cause him to gravitate to the modified Gram–Schmidt algorithm, which performs better when the matrix is contained on a backing store—it has better locality of reference.)

Although we have motivated the Gram–Schmidt algorithm as a device for getting an explicit Q-factor, it plays an important role in Krylov sequence methods, in which it is necessary to orthogonalize a sequence of the form $x, Ax, A^2, \ldots$. It turns out that this is equivalent to orthogonalizing Aq_k against $q_1, \ldots, q_k$, which is most conveniently done by the Gram–Schmidt algorithm. When A is symmetric, the terms $q_1, \ldots, q_{k-2}$ drop out, giving a very economical three-term recurrence. For more see [253] and [163].

The observation that the modified Gram–Schmidt algorithm is superior to the classical algorithm is due to Rice [268]. Björck [37] first showed that the deterioration in orthogonality for the modified algorithm is proportional to the condition number of the matrix. According to Björck [40], the relation of modified Gram–Schmidt and Householder triangularization was apparently mentioned to Gene Golub by Charles Sheffield (a person of many talents, who has gone on to write excellent science fiction). Björck and Paige [44] used the fact to simplify Björck's original error analysis. Also see [177, §18.7].

Reorthogonalization

Rice [268] experimented with reorthogonalization to keep the Gram–Schmidt algorithms on track. Error analyses have been given by Abdelmalek [2]; Daniel, Gragg, Kaufman, and Stewart [84] and by Hoffman [179]. In particular, Hoffman investigates the effect of varying the value of α and concludes that for $\alpha = \frac{1}{2}$ both the classical and modified Gram–Schmidt algorithms give orthogonality to working accuracy.

The twice-is-enough algorithm is due to Kahan and Parlett and is described in [253, §6-9].

2. LINEAR LEAST SQUARES

In this section we will be concerned with the following problem.

Given an $n \times p$ matrix X of rank p and an n-vector y, find a vector b such that

$$\|y - Xb\|_2^2 = \min. \tag{2.1}$$

SEC. 2. LINEAR LEAST SQUARES

This problem, which goes under the name of the *linear least squares problem*, occurs in virtually every branch of science and engineering and is one of the mainstays of statistics.

Historically, least squares problems have been solved by forming and solving the normal equations—a simple and natural procedure with much to recommend it (see §2.2). However, the problem can also be solved using the QR decomposition—a process with superior numerical properties. We will begin with the QR approach in §2.1 and then go on to the normal equations in §2.2. In §2.3 we will use error analysis and perturbation theory to assess the methods. We then consider least squares problems with linear equality constraints. We conclude with a brief treatment of iterative refinement of least squares solutions.

2.1. THE QR APPROACH

In some sense the tools for solving the least squares problem have already been assembled. As the vector b in (2.1) varies over $\mathbb{R}^p$, the vector Xb varies over $\mathcal{R}(X)$. Consequently, the least squares problem (2.1) amounts to finding the vector $\hat{y}$ in $\mathcal{R}(X)$ that is nearest y. By Theorem 4.26, Chapter 1, $\hat{y}$ is the orthogonal projection of y onto $\mathcal{R}(X)$. Since $\hat{y}$ is in $\mathcal{R}(X)$ and the columns of X are linearly independent we can express $\hat{y}$ uniquely in the form Xb.

These observations solve the problem. They even bring us part of the way to a computational solution, because we already know how to compute projections (Algorithm 1.3). However, a direct derivation from the QR decomposition highlights the computational aspects of this approach.

Least squares via the QR decomposition

Let

$$Q^{\mathrm{T}} X = \begin{pmatrix} R \\ 0 \end{pmatrix}$$

be a QR decomposition of X, and partition

$$Q = (Q_X \; Q_\perp),$$

where Q_X has p columns. Also partition

$$z \equiv Q^{\mathrm{T}} y = \begin{pmatrix} Q_X^{\mathrm{T}} y \\ Q_\perp^{\mathrm{T}} y \end{pmatrix} \equiv \begin{pmatrix} z_X \\ z_\perp \end{pmatrix}.$$

Since the 2-norm is unitarily invariant,

$$\begin{aligned}\|y - Xb\|_2^2 &= \|Q^{\mathrm{T}}(y - Xb)\|_2^2 \\ &= \left\|\begin{pmatrix} z_X \\ z_\perp \end{pmatrix} - \begin{pmatrix} R \\ 0 \end{pmatrix}\right\|_2^2 \\ &= \left\|\begin{pmatrix} z_X - Rb \\ z_\perp \end{pmatrix}\right\|_2^2 \\ &= \|z_X - Rb\|_2^2 + \|z_\perp\|_2^2.\end{aligned} \qquad (2.2)$$

Now the second term in the sum $\|z_X - Rb\|_2^2 + \|z_\perp\|_2^2$ is constant. Hence the sum will be minimized when $\|z_X - Rb\|_2^2$ is minimized. Since R is nonsingular, the minimizer is the unique solution of the equation $Rb = z_X$ and the norm at the minimum is $\|y - Xb\|_2 = \|z_\perp\|_2$.

Since $P_X = Q_X Q_X^{\mathrm{T}}$, we may calculate $\hat{y} = Xb$ in the $\hat{y} = Q_X z_X$. Similarly we may calculate the residual vector $r = y - Xb$ in the form $r = Q_\perp z_\perp$. We summarize the results in the following theorem.

Theorem 2.1. *Let X be of full column rank and have a QR decomposition of the form*

$$X = (Q_X \ Q_\perp)\begin{pmatrix} R \\ 0 \end{pmatrix}.$$

Then the solution of the least squares problem of minimizing $\|y - Xb\|_2^2$ is uniquely determined by the QR EQUATION

$$Rb = Q_X^{\mathrm{T}} y \equiv z_X.$$

The LEAST SQUARES APPROXIMATION TO y *is given by*

$$\hat{y} = Xb = P_X y = Q_X Q_X^{\mathrm{T}} y = Q_X z_X.$$

The RESIDUAL VECTOR *is given by*

$$r = y - Xb = P_\perp y = Q_\perp Q_\perp^{\mathrm{T}} y \equiv Q_\perp z_\perp,$$

and the RESIDUAL SUM OF SQUARES *is*

$$\|r\|_2^2 = \|z_\perp\|_2^2.$$

Moreover, the residual at the minimum is orthogonal to the column space of X.

The way we have established this theorem is worth some comments. In Theorem 4.26, Chapter 1, where we proved that the projection of a vector y onto a space $\mathcal{R}(X)$ minimizes the distance between y and $\mathcal{R}(X)$, we wrote an equation of the form

$$\begin{aligned}\|y - x\|_2^2 &= \|P_X(y - x)\|_2^2 + \|P_\perp(z - x)\|_2^2 \\ &= \|P_X y - x\|_2^2 + \|P_\perp y\|_2^2,\end{aligned} \qquad (2.3)$$

Sec. 2. Linear Least Squares

This algorithm takes a QR decomposition

$$X = (Q_X \; Q_\perp) \begin{pmatrix} R \\ 0 \end{pmatrix}$$

of X and a vector y and computes the solution b of the problem of minimizing $\|y - Xb\|_2$. It also returns the least squares approximation $\hat{y} = Xb$ and the residual $r = b - Ax$.

1. $z_X = Q_X^T y$
2. $z_\perp = Q_\perp^T y$
3. Solve the system $Rb = z_X$
4. $\hat{y} = Q_X z_X$
5. $r = Q_\perp z_\perp$

Algorithm 2.1: Least squares from a QR decomposition

---⋄---

where x is an arbitrary vector in $\mathcal{R}(X)$. This equations corresponds to equation (2.2) with Xb playing the part of x. But whereas (2.3) works with projections in the original coordinate system, (2.2) transforms the coordinate axes to lie along the columns of Q. The application of the Pythagorean equality in (2.3) corresponds to splitting the sum of squares that defines $\|Q^T(y - Xb)\|_2^2$. Thus a geometric argument is reduced to an obvious computation.

This simplification is typical of the decompositional approach to matrix analysis. By changing the coordinate system, complex geometric arguments often become a matter of inspection. And the approach often suggests computational methods. But this is not to disparage the use of projections. In many instances they are the natural language for a problem — sometimes the only language — and they often summarize the mathematical essence of a problem more succinctly.

—

Algorithm 2.1 is a computational summary of Theorem 2.1. Here are some comments.

- The algorithm is generic in that it does not specify how the matrix Q is represented. One possibility is that the matrix Q is known explicitly, in which case the formulas in the algorithm can be implemented as shown. However, for n at all large this alternative is inefficient in terms of both storage and operations.

- The solution of the triangular system in statement 3 can be accomplished by the BLAS *xebuib* (see Figure 2.2, Chapter 2).

- If Householder triangularization is used, the products in the algorithm can be calculated as in Algorithm 1.2. In this case, if $n \gg p$, the algorithm requires $6np$ flam.

This algorithm takes the output of Algorithm 1.9 and computes the solution b of the problem of minimizing $\|y - Hb\|_2$. It also returns the least squares approximation $\hat{y} = Hb$ and the residual $r = b - Hx$.

1. $\hat{y} = y$
2.   **for** $j = 1$ **to** n
3. $rotapp(c[j], s[j], \hat{y}[j], \hat{y}[j+1])$
4.   **end for** j
5. $xeuib(b, H[1{:}n, 1{:}n], \hat{y}[1{:}n])$
6. $r = 0;\ r[n+1] = \hat{y}[n+1];\ \hat{y}[n+1] = 0$
7.   **for** $j = n$ **to** 1 **by** -1
8. $rotapp(c[j], -s[j], \hat{y}[j], \hat{y}[j+1])$
9. $rotapp(c[j], -s[j], r[j], r[j+1])$
10. **end for** j

Algorithm 2.2: Hessenberg least squares

Unless p is very small, this is insignificant compared to the original reduction. This algorithm is often called the *Golub–Householder* algorithm. For more see the notes and references.

- If plane rotations have been used to reduce the matrix, the details of the computation will depend on the details of the original reduction. For example, Algorithm 2.2 illustrates how the computations might proceed when plane rotations have been used to triangularize an augmented Hessenberg matrix as in Algorithm 1.9. Since Q^{T} is represented in the form

$$Q^{\mathrm{T}} = P_{n-1,n} \cdots P_{23} P_{12},$$

where P_{ij} is a rotation in the (i,j)-plane, the vector $\hat{y}$ must be computed in the form

$$\hat{y} = P_{12}^{\mathrm{T}} P_{23}^{\mathrm{T}} \cdots P_{n-1,n}^{\mathrm{T}} \begin{pmatrix} z_X \\ 0 \end{pmatrix}.$$

Unlike Householder transformations, plane rotations are not symmetric, and this fact accounts for the argument of $-s[i]$ in statements 8 and 9.

- However the computations are done, $\hat{y}$ and r will be nearly orthogonal. But if X is ill conditioned, they may be far from $\mathcal{R}(X)$ and $\mathcal{R}(X)^{\perp}$.

Least squares via the QR factorization

We have seen that with reorthogonalization the classical and modified Gram–Schmidt algorithms can be coaxed to compute a stable QR factorization $X = Q_X R$, in which

SEC. 2. LINEAR LEAST SQUARES

the columns of Q_X are orthogonal to working accuracy. This factorization can be used to solve least squares problems. The QR equations remain the same, i.e.,

$$Rb = Q_X^T y \equiv z_X,$$

and the approximation can be computed in the form

$$\hat{y} = Q_X z_X.$$

Since we have no basis for the orthogonal complement of $\mathcal{R}(X)$, we must compute the residual in the form

$$r = y - \hat{y}.$$

Note that this is equivalent to using the classical Gram–Schmidt algorithm to orthogonalize y against the columns of Q_X. By the analysis of reorthogonalization [see (1.32) and (1.33)], we cannot guarantee the orthogonality of r to the columns of Q_X when r is small. It is important to keep in mind that here "small" means with respect to the rounding unit. In real-life problems, where y is contaminated with error, r is seldom small enough to deviate much from orthogonality. However, if orthogonality is important — say r is used in an algorithm that presupposes it — reorthogonalization will cure the problem.

Least squares via the modified Gram–Schmidt algorithm

The modified Gram–Schmidt algorithm produces a matrix Q_X and a triangular matrix R such that $Q_X R$ reproduces X to working accuracy. In principal we could proceed as we did above with the QR factorization and solve the equation $Rb = Q_X^T y$. Unfortunately, if X is ill conditioned, this procedure is unstable.

An alternate procedure is suggested by the following theorem. Its proof is left as an exercise.

Theorem 2.2. *Let the QR factorization of the matrix $(X \ x)$ be partitioned in the form*

$$(X \ x) = (Q \ q) \cdot \begin{pmatrix} R & r \\ 0 & \rho \end{pmatrix}.$$

Then

$$\rho q = P_X^\perp x \quad \text{and} \quad r = Q^T x.$$

In terms of our least squares problem, this theorem says that if we compute the QR factorization of the *augmented least squares matrix* $(X \ y)$, we get

$$(X \ y) = (Q \ r/\|r\|_2) \begin{pmatrix} R & z_X \\ 0 & \|r\|_2 \end{pmatrix}. \tag{2.4}$$

> Let
> $$(X\ y) = (Q\ q) \begin{pmatrix} R & z_X \\ 0 & \rho \end{pmatrix}$$
> be a QR factorization of the augmented least squares matrix $(X\ y)$ computed by the modified Gram–Schmidt algorithm. This algorithm produces a solution of the least squares problem of minimizing $\|y - Xb\|_2^2$.
>
> 1. Solve the system $Rb = z_X$
> 2. $\hat{y} = Qz_X$
> 3. $r = y - \hat{y}$

Algorithm 2.3: Least squares via modified Gram–Schmidt

⎯⎯⎯⎯⎯⎯⎯⎯◇⎯⎯⎯⎯⎯⎯⎯⎯

Hence if we apply the Gram–Schmidt algorithm to the augmented least squares matrix, we get right-hand side of the QR equation. For stability we must use the modified form of the algorithm. These considerations yield Algorithm 2.3.

This algorithm is stable in the sense that the computed solution b comes from a small perturbation of X that satisfies the bounds of the usual form. The reason is the connection between modified Gram–Schmidt and Householder triangularization (Theorem 1.11). In fact, least squares solutions produced by Algorithm 2.3 are generally more accurate than solutions produced by orthogonal triangularization.

The stability result also implies that $\hat{y} = Qz_X$ will be a good approximation to Xb, where b is the computed solution. Thus we do not have to save X and compute $\hat{y}$ in the form Xb. Unfortunately, the vector ρq can consist entirely of rounding error, and the residual is best computed in the form $y - \hat{y}$.

2.2. THE NORMAL AND SEMINORMAL EQUATIONS

The natural way to approach least squares is to differentiate the residual sum of squares with respect to components of b and set the results to zero — the textbook way to minimize a function. The result is a linear system called the *normal equations*. It is safe to say that a majority — a great majority — of least squares problems are solved by forming and solving the normal equations. In this subsubsection we will treat the system of normal equations, discussing some of its advantages and drawbacks. We will also consider a hybrid method that combines the QR equation with the normal equation.

The normal equations

The normal equations are the mathematical embodiment of the statement that the least squares residual is orthogonal to the column space of the least squares matrix. We can

SEC. 2. LINEAR LEAST SQUARES

express this statement in the form

$$X^T r = 0.$$

Since $r = y - Xb$, we must have

$$X^T y - X^T X b = 0.$$

Thus we have established the following theorem.

Theorem 2.3. *The least squares solution of the problem of minimizing $\|y - Xb\|_2$ satisfies the* NORMAL EQUATIONS

$$(X^T X)b = X^T y.$$

In what follows we will write

$$A = X^T X$$

for the *cross-product matrix* and set

$$C = X^T y.$$

Thus the normal equations assume the form

$$Ab = c,$$

which is as simple as abc.

Because the normal equations are equivalent to saying that the residual is orthogonal to $\mathcal{R}(X)$, by Theorem 4.26, Chapter 1, the normal equations always have a solution, even when X is not of full column rank. Moreover, any solution of the normal equations solves the least squares problem. This should be contrasted with the QR equation derived from Householder triangularization, which may fail to have a solution if X is not of full rank.

In this chapter, however, we assume that X is of full rank. In that case the cross-product matrix A is positive definite (Theorem 2.2, Chapter 3). Consequently, we can use the Cholesky algorithm (Algorithm 2.1, Chapter 3) to solve the normal equations at a cost of $\frac{1}{6}p^3$ flam.

Forming cross-product matrices

In general, it will cost more to form the normal equations than to solve them. The bulk of the work will be in forming the cross-product matrix A. There are two customary to do this — by inner products and by outer products.

For the inner-product method, partition X by columns:

$$X = (x_1 \ x_2 \ \cdots \ x_p).$$

Then

$$A = X^T X = \begin{pmatrix} x_1^T x_1 & x_1^T x_2 & \cdots & x_1^T x_p \\ x_2^T x_1 & x_2^T x_2 & \cdots & x_2^T x_p \\ \vdots & \vdots & & \vdots \\ x_p^T x_1 & x_p^T x_2 & \cdots & x_p^T x_p \end{pmatrix}.$$

Thus the element a_{ij} of A can be formed by computing the inner products $x_i^T x_j$. By symmetry we need only form the upper (or lower) part of A. It follows that:

The cross-product matrix can be formed in $\frac{1}{2}np^2$ flam.

This should be compared with the count of np^2 flam for Householder triangularization or the modified Gram–Schmidt algorithm.

For the outer-product method, partition X by rows:

$$X = \begin{pmatrix} x_1^T \\ x_2^T \\ \vdots \\ x_n^T \end{pmatrix}.$$

Then

$$A = x_1 x_1^T + x_2 x_2^T + \cdots + x_n x_n^T.$$

Thus the cross-product matrix can be formed by accumulating the sum of outer products in A.

An implementation of this method is given in Algorithm 2.4. If X is full, the algorithm requires $\frac{1}{2}np^2$ flam, just like the inner-product algorithm. But if X contains zero elements, we can skip some of the computations, which is what the test in statement 4 accomplishes. It is not hard to show that

If the jth column of X contains m_j zero elements, Algorithm 2.4 requires

$$(\tfrac{1}{2}np^2 - m_1 p - m_2(p-1) - \cdots - m_1) \text{ flam.}$$

This count shows that one should order the columns so that those having the most zeros appear first. It also suggests that the potential savings in forming the cross-product of a sparse least squares matrix can be substantial. This is one reason why the normal equations are often preferred to their more stable orthogonal counterparts, which gain less from sparseness.

There are variations on this algorithm. If, for example, on specialized computers X is too large to hold in main memory, it can be brought in from a backing store several rows at a time. Care should be taken that X is organized with proper locality on the backing store. (See the discussion of hierarchical memories in §3.3, Chapter 2.)

SEC. 2. LINEAR LEAST SQUARES

Given an $n \times p$ matrix X this algorithm computes the cross-product matrix $A = X^T X$.

1. $A = 0$
2.   **for** $k = 1$ **to** n
3.   **for** $i = 1$ **to** p
4.   **if** $(x_{ki} \neq 0)$
5.   **for** $j = i$ **to** p
6. $a_{ij} = a_{ij} + x_{ki} * x_{kj}$
7.   **end for** j
8.   **end if**
9.   **end for** i
10.   **end for** k

Algorithm 2.4: Normal equations by outer products

―――――◇―――――

The augmented cross-product matrix

A common approach to least squares is to work with the *augmented cross-product matrix*

$$(X \ y)^T (X \ y) = \begin{pmatrix} A & c \\ c^T & \|y\|_2^2 \end{pmatrix}.$$

Since the Cholesky factor of this matrix is the R-factor of the augmented least squares matrix, we have from (2.4) that

> The Cholesky factor of the augmented-cross product matrix is the matrix
> $$\begin{pmatrix} R & z_X \\ 0 & \|r\|_2 \end{pmatrix}.$$
(2.5)

Thus decomposing the augmented cross-product matrix gives the matrix and right-hand side of the QR equation as well as the square root of the residual sum of squares.

The instability of cross-product matrices

A simple error analysis of the inner product gives the following result for the computed cross-product matrix.

> If $\tilde{A}$ is the computed cross-product matrix, then
> $$|\tilde{a}_{ij} - a_{ij}| \leq n \|x_i\|_2 \|x_j\|_2 \epsilon'_M,$$
> where ϵ'_M is the adjusted rounding unit [see (4.12), Chapter 2].

Thus the computed cross-product is a small perturbation of the true matrix. If we go on to solve the normal equations, we find from the error analysis of Gaussian elimination (Theorem 4.9, Chapter 3) that the computed solution satisfies

$$(A + G)b = c,$$

where G is small compared with A. (For simplicity we have omitted the error introduced by the computation of $c = X^T y$.) Thus the method of normal equations has a backward error analysis in terms of the cross-product matrix A.

However, the cross-product matrix is an intermediate, computed quantity, and it is reasonable to ask if the errors in the cross-product matrix can be propagated back to the original matrix X. The answer is: Not in general. It is instructive see why.

The problem is to determine a matrix E such that

$$(X + E)^T (X + E) = A + G. \tag{2.6}$$

To give a sense of scale, we will assume that $\|X\|_2$, and hence $\|A\|_2$, is equal to one.

The first thing to note is that the very structure of the left-hand side of (2.6) puts limits on the perturbation G. For any cross-product matrix $(X+E)^T(X+E)$ must be at least positive semidefinite. Thus the perturbation G must be restricted to not make any eigenvalue of A negative. (See §2.1, Chapter 3.)

Let σ_p be the smallest singular value of X, so that σ_p^2 is the smallest eigenvalue of A. Let $\tilde{\sigma}_p$ and $\tilde{\sigma}_p^2$ be the perturbed quantities. Then by the fourth item in Theorem 4.34, Chapter 1, we must have

$$|\tilde{\sigma}_p^2 - \sigma_p^2| \leq \|G\|_2,$$

and this bound is attainable. Consequently, to guarantee that $\tilde{A}$ is positive semidefinite, the matrix G must satisfy

$$\|G\|_2 \leq \sigma_p^2. \tag{2.7}$$

Even if G satisfies (2.7), the backward error E in (2.6) may be much larger than G. To see this, we use the singular value decomposition to calculate an approximation to E.

Let

$$U^T X V = \begin{pmatrix} \Sigma \\ 0 \end{pmatrix}$$

be the singular value decomposition of X. If we set

$$\begin{pmatrix} F_1 \\ F_2 \end{pmatrix} = U^T E V \quad \text{and} \quad H = V^T G V,$$

SEC. 2. LINEAR LEAST SQUARES

then (2.6) can be written in the form

$$\Sigma^2 + H = \begin{pmatrix} \sigma + F_1 \\ F_2 \end{pmatrix}^{\mathrm{T}} \begin{pmatrix} \sigma + F_1 \\ F_2 \end{pmatrix} = \Sigma^2 + \Sigma^{\mathrm{T}} F_1 + F_1^{\mathrm{T}} \Sigma + F_1^{\mathrm{T}} F_1 + F_2^{\mathrm{T}} F_2.$$

If we set $F_2 = 0$ and assume that F_1 is small enough so that we can ignore the term $F_1^{\mathrm{T}} F_1$, then we get

$$\Sigma^{\mathrm{T}} F_1 + F_1^{\mathrm{T}} \Sigma \cong H.$$

Writing out the (i,j)-element of this relation, we get

$$\sigma_i \varphi_{ij} + \sigma_j \varphi_{ji} \sigma_j \cong \eta_{ij}.$$

For $i = j$ this gives immediately

$$\varphi_{ii} \cong \frac{\eta_{ii}}{2\sigma_i}.$$

On the other hand, if $i \neq j$ by the symmetry of H we have

$$\sigma_i \varphi_{ij} + \sigma_j \varphi_{ji} \sigma_j \cong \sigma_j \varphi_{ij} + \sigma_i \varphi_{ji} \sigma_j,$$

from which if follows that if $\sigma_i \neq \sigma_j$ then $\varphi_{ij} \cong \varphi_{ji}$. Hence

$$\varphi_{ij} \cong \frac{\eta_{ij}}{\sigma_i + \sigma_j}.$$

Note that this relation gives the proper result for $i = j$. It also works when $\sigma_i = \sigma_j$, in which case it gives the solution for which $\varphi_{ij}^2 ! \varphi_{ji}^2$ is minimal.

It follows that $\|F_1\|_{\mathrm{F}} \lesssim \|H\|_{\mathrm{F}}/2\sigma_p$. In terms of the original matrices (remember that the 2-norm is unitarily invariant),

$$\|E\|_{\mathrm{F}} \lesssim \frac{\|G\|_{\mathrm{F}}}{2\sigma_p}.$$

Dividing by $\|X\|_{\mathrm{F}}$ and remembering that $\|X\|_{\mathrm{F}}^2 \geq \|A\|_{\mathrm{F}}$ and $\|X^\dagger\|_{\mathrm{F}} \geq \sigma_p^{-1}$, we get

$$\frac{\|E\|_{\mathrm{F}}}{\|X\|_{\mathrm{F}}} \lesssim \frac{\kappa_{\mathrm{F}}(X)}{2} \frac{\|G\|_{\mathrm{F}}}{\|A\|_{\mathrm{F}}},$$

where $X^\dagger$ is the pseudoinverse of X [see (1.4)] and

$$\kappa_F(X) = \|X\|_F \|X^\dagger\|_F$$

is the condition number of X [see (1.28)]. Thus when we throw the error G in A back on X, the error can grow by as much as $\kappa_F(X)/2$.

To summarize:

If the cross-product matrix $A = X^T X$ is perturbed by an error G and $\rho_q = \|G\|_p/\|A\|_q$ ($q = 2, \mathrm{F}$), then to guarantee that $A + G$ is positive definite we must have

$$\rho_2 \lesssim \frac{1}{\kappa_2(X)}.$$

Moreover, if we attempt to project the error G back onto the least squares matrix X, the resulting error E satisfies

$$\frac{\|E\|_\mathrm{F}}{\|X\|_\mathrm{F}} \lesssim \frac{\rho_\mathrm{F} \kappa_\mathrm{F}(X)}{2}.$$

In particular, as ρ_F approaches $\kappa_\mathrm{F}(X)$, the bound on the normwise relative error approaches one.

The seminormal equations

In computing a QR decomposition or factorization it is customary to allow the Q-factor (or a representation of it) to overwrite the original matrix X. However, if X is sparse, the Q factor will, in general, be less sparse than X. In fact, it may be impossible to store the Q-factor. If the concern is with the one-time solution of a specific least squares problem, one can decompose the augmented least squares matrix and throw away the transformations. (This can be accomplished by building up the decomposition row by row. See Algorithm 3.6.) In some instances, however, one must solve a sequence of least squares problems involving the same matrix and whose y vectors depend on the previous solutions. It this case it would be convenient to have some of the advantages of an orthogonal decomposition without having to store the Q-factor.

The method of *seminormal equations* starts from the observation that although the cross-product matrix A may not have a satisfactory backward error analysis, the R-factor of X does. Since $A^T A = R^T R$, we may hope for a higher quality solution from solving the system

$$R^T(Rb) = X^T y,$$

where R has been computed by orthogonal triangularization. (The parentheses are placed to indicate that the system is to be solved by two triangular solves.) If $n \gg p$, comparatively little memory is required to store R, which may be used again and again.

Unfortunately, solutions computed from the seminormal equations are not better than solutions computed from the normal equations. However, if we add one step of iterative refinement in fixed precision we get a higher quality solution. This is the method of *corrected seminormal equations*, which is implemented in Algorithm 2.5.

Under appropriate conditions—roughly when $\kappa(X) \leq \sqrt{\epsilon_\mathrm{M}}$—the method turns out to be weakly stable in the sense of §4.3, Chapter 2. Namely, the error in the computed solution is no larger than would be obtained by a small perturbation of X. For larger condition numbers, a repetition of the refinement step will reduce the error further.

Sec. 2. Linear Least Squares

Let R be a stably computed R-factor of X. The following algorithm returns a solution to the least squares problem of minimizing $\|y - Xb\|_2^2$.

1. Solve the system $R^\mathrm{T}(Rb) = X^\mathrm{T} y$
2. $s = y - Xb$
3. Solve the system $R^\mathrm{T}(Rd) = X^\mathrm{T} s$
4. $b = b + d$

Algorithm 2.5: Least squares by corrected seminormal equations

———◊———

2.3. Perturbation theory and its consequences

The purpose of this subsection is to compare the two principal methods of solving least squares problems: the normal equations and the QR equation. After summarizing the results of rounding-error analyses, we will develop the perturbation theory of least squares solutions. The combination of the two will give us the wherewithal to compare the methods.

The effects of rounding error

We have seen that the least squares solution b_{NE} calculated from the normal equations satisfies the perturbed equation

$$(A + G)b_{\mathrm{NE}} = c.$$

The error matrix G, which combines the error in forming the normal equations and the error in solving the system, satisfies

$$\frac{\|G\|_2}{\|A\|_2} \leq \gamma_{\mathrm{NE}} \left(1 + \frac{\|y\|_2}{\|X\|_2 \|b\|_2}\right) \epsilon_{\mathrm{M}}. \tag{2.8}$$

(This bound includes the error in calculating $c = X^\mathrm{T} y$, which has been thrown onto A via Theorem 3.14, Chapter 3.) The constant γ_{NE} depends on n and p. As we have seen, the error G cannot be projected back on X without magnification.

A more satisfactory result holds for a least squares solution b_{QR} computed from a QR factorization obtained from either Householder triangularization or the modified Gram–Schmidt algorithm. Specifically it is the solution of a perturbed problem

$$\|(y + f) - (X + E)b_{\mathrm{QR}}\|_2^2 = \min,$$

where E and f satisfy

$$\frac{\|f\|_2}{\|y\|_2}, \frac{\|E\|_2}{\|X\|_2} \leq \gamma_{\mathrm{QR}} \epsilon_{\mathrm{M}}. \tag{2.9}$$

Again the constant γ_{QR} depends on the dimensions of the problem. The error E includes the errors made in computing the QR factorization and the errors made in solving the QR equation.

Perturbation of the normal equations

The perturbation theory of the normal equations is simply the perturbation theory of linear systems treated in §3, Chapter 3. However, we are going to use first-order perturbation theory to analyze the least squares problem, and it will be convenient to have a parallel treatment of the normal equations. Since the errors in the perturbation analyses can come from any source, we will drop the notation b_{NE} and use the customary $\tilde{b}$ for the perturbed quantity.

We begin with a useful result.

Theorem 2.4. *Let A be nonsingular. Then for all sufficiently small G and any consistent norm $\|\cdot\|$*

$$(A+G)^{-1} = A^{-1} - A^{-1}GA^{-1} + O(\|G\|^2). \tag{2.10}$$

Proof. By Corollary 4.19, Chapter 1, we know that for all sufficiently small G the matrix $A+G$ is nonsingular and $\|(A+G)^{-1}\|$ is uniformly bounded. Hence the result follows on taking norms in the equation

$$(A+G)^{-1} = (A^{-1} - A^{-1}GA^{-1}) + (A^{-1}G)^2(A+G)^{-1}. \quad \blacksquare$$

As a convenient shorthand for (2.10) we will write

$$(A+G)^{-1} \cong A^{-1} - A^{-1}GA^{-1}. \tag{2.11}$$

A disadvantage of such first-order expansions is that we lose information about the size of the second-order terms. In this case, we can see from the proof that the relative error in the approximation goes to zero like $\|A^{-1}G\|^2$, so that if $\|A^{-1}G\|$ is reasonably less than one, the first-order expansion is a reasonable approximation to reality.

Observe that (2.11) could be obtained from the expansion

$$(A+G)^{-1} = (I+A^{-1}G)^{-1}A^{-1} = I - (A^{-1}G)A^{-1} + (A^{-1}G)^2A^{-1} - \cdots$$

by dropping terms of second order or higher in G. This approach to obtaining approximations to perturbations matrix functions is called *first-order perturbation theory*. It is widely used in matrix perturbation theory. However, they break down if the perturbation is too large, and the consumer of the results of first-order perturbation theory should have some idea of when they are applicable. In our applications the approximations will be based on the Neumann expansion

$$(I+P)^{-1} = I - P + P^2 - \cdots,$$

which is accurate if $\|P\|$ is reasonably less than one [see (4.17), Chapter 1].

Now let $Ab = c$ and $(A+G)\tilde{b} = c$. Then

$$\tilde{b} = (A+G)^{-1}c \cong A^{-1}c - A^{-1}GA^{-1}c = b - A^{-1}Gc.$$

SEC. 2. LINEAR LEAST SQUARES

Hence

$$\frac{\|\tilde{b} - b\|_2}{\|b\|_2} \lesssim \|A^{-1}G\|_2 \leq \kappa_2(A)\frac{\|G\|_2}{\|A\|_2},$$

where, as usual, $\kappa_2(A) = \|A\|\|A^{-1}\|$. Here we have used the notation "$\lesssim$" to stress that we have left out higher-order terms in E in the inequality.

To make comparisons easier, it will be convenient to write this bound in terms of $\kappa_2(X) = \|X\|_2\|X^\dagger\|_2$. From (1.6), we have $\kappa_2(X) = \sigma_1/\sigma_p$, where σ_1 and σ_p are the largest and smallest singular values of X. Similarly, the condition number of A is the ratio of its largest to smallest singular value. But the singular values of A are the squares of the singular values of X. Hence

$$\kappa_2(A) = \kappa_2^2(X).$$

Thus we have the following result.

Let b be the solution of the equation $Ab = c$, where $A = X^{\mathrm{T}}X$ and let $\tilde{b}$ be the solution of the perturbed equation $(A + G)\tilde{b} = c$. Then

$$\frac{\|\tilde{b} - b\|_2}{\|b\|_2} \lesssim \kappa_2^2(X)\frac{\|G\|_2}{\|A\|_2}. \tag{2.12}$$

Since $\kappa_2(X) \geq 1$, squaring $\kappa_2(X)$ can only increase it. This suggests that the perturbation theory of the normal equations may be less satisfactory than the perturbation theory of the original least squares problem, to which we now turn.

The perturbation of pseudoinverses

The solution of the least squares problem can be written in the form

$$b = A^{-1}X^{\mathrm{T}}y = X^\dagger y,$$

where $X^\dagger$ is the pseudoinverse of X [See (1.4)]. Thus we can base the perturbation theory for least squares solutions on a perturbation expansion for pseudoinverses—just as we based our analysis of the normal equations on an expansion for the inverses.

We are looking for a first-order expansion of the matrix

$$(X + E)^\dagger = [(X + E)^{\mathrm{T}}(X + E)]^{-1}(X + E)^{\mathrm{T}}.$$

By (2.10)

$$\begin{aligned}
[(X + E)^{\mathrm{T}}&(X + E)]^{-1} \\
&= (A + X^{\mathrm{T}}E + E^{\mathrm{T}}X + E^{\mathrm{T}}E)^{-1} \\
&\cong (A + X^{\mathrm{T}}E + E^{\mathrm{T}}X)^{-1} && \text{ignoring } E^{\mathrm{T}}E \\
&\cong A^{-1} - A^{-1}(E^{\mathrm{T}}X + X^{\mathrm{T}}E)A^{-1} && \text{by (2.11)} \\
&= A^{-1} - A^{-1}E^{\mathrm{T}}(X^\dagger)^{\mathrm{T}} - X^\dagger E A^{-1}.
\end{aligned}$$

It follows that

$$(X+E)^\dagger \cong (A^{-1} - A^{-1}E^T(X^\dagger)^T - X^\dagger E A^{-1})(X^T + E^T)$$
$$\cong A^{-1}X^T + A^{-1}E^T - A^{-1}E^T(X^\dagger)^T X^T - X^\dagger E A^{-1} X^T$$
$$= X^\dagger + A^{-1}E^T - A^{-1}E^T P_X - X^\dagger E X^\dagger, \text{ since } P_X = XX^\dagger,$$
$$= X^\dagger - X^\dagger E X^\dagger + A^{-1}E^T P_\perp, \quad P_\perp = I - P_X.$$

Observing that $X^\dagger P_X = X^\dagger$, we obtain the following result.

Let

$$E_X = P_X E \quad \text{and} \quad E_\perp = P_\perp E. \tag{2.13}$$

Then

$$(X+E)^\dagger \cong X^\dagger - X^\dagger E_X X^\dagger + A^{-1} E_\perp^T.$$

Note that when X is square and nonsingular, $X^\dagger = X^{-1}$ and $X_\perp = 0$. In this case the expansion (2.13) reduces to the expansion of an ordinary matrix inverse.

The perturbation of least squares solutions

Turning now to the perturbation of least squares solutions, we consider the problem

$$\|(y+f) - (X+E)\tilde{b}\|_2^2 = \min.$$

Since $E_\perp^T X = 0$,

$$E_\perp y = E_\perp^T(y - Xb) = E_\perp^T r.$$

Hence if $\tilde{b} = (X+E)^\dagger y$ is the solution of the perturbed least squares problem, we have

$$\tilde{b} = (X+E)^\dagger(y+f)$$
$$\cong (X^\dagger - X^\dagger E_X X^\dagger + A^{-1} E_\perp^T)(y+f)$$
$$= b - X^\dagger E_X b + X^\dagger f_X + A^{-1} E_\perp^T r,$$

where $f_X = P_X f$. On taking norms and dividing by $\|b\|_2$ we get

$$\frac{\|\tilde{b} - b\|_2}{\|b\|_2} \lesssim \|X^\dagger\|_2 \|E_X\|_2 + \|X^\dagger\|_2 \frac{\|f_X\|_2}{\|b\|_2} + \frac{\|A^{-1}\|_2 \|E_\perp\|_2 \|r\|_2}{\|b\|_2}$$
$$\leq \|X^\dagger\|_2 \|E_X\|_2 + \|X^\dagger\|_2 \frac{\|f_X\|_2}{\|b\|_2} + \frac{\|X^\dagger\|_2^2 \|E_\perp\|_2 \|r\|_2}{\|b\|_2}.$$

After a little rearranging, we get the following theorem.

SEC. 2. LINEAR LEAST SQUARES

Theorem 2.5. *Let $b = X^\dagger y$ and $\tilde{b} = (X+E)^\dagger(y+f)$. Then for sufficiently small E*

$$\frac{\|\tilde{b}-b\|_2}{\|b\|_2} \lesssim \kappa_2(X)\left(\frac{\|E_X\|_2}{\|X\|_2} + \frac{\|f_X\|_2}{\|X\|_2\|b\|_2}\right) \tag{2.14}$$
$$+ \kappa_2^2(X)\frac{\|r\|_2}{\|X\|_2\|b\|_2}\frac{\|E_\perp\|_2}{\|X\|_2}.$$

Here the subscript E_X and f_X are the projections of E and f onto the column space of X.

The first term on the right-hand side of (2.14) is analogous to the bound for linear systems. It says that the part of the error lying in $\mathcal{R}(X)$ is magnified by $\kappa_2(X)$ in the solution.

The second term depends on $\kappa_2^2(X)$ and is potentially much larger than the first term. However, it is multiplied by $\|r\|_2/\|X\|_2\|b\|_2$, which can be small if the least squares residual is small. But if the residual is not small, this term will dominate the sum. The following example shows how strong this effect can be.

Example 2.6. *Consider the matrix*

$$X = \begin{pmatrix} 5.8628\mathrm{e}{-01} & 5.8633\mathrm{e}{-01} \\ 1.5256\mathrm{e}{+00} & 1.5257\mathrm{e}{+00} \\ 2.1432\mathrm{e}{+00} & 2.1432\mathrm{e}{+00} \\ -7.4604\mathrm{e}{-01} & -7.4600\mathrm{e}{-01} \\ -1.5315\mathrm{e}{+00} & -1.5316\mathrm{e}{+00} \\ -2.1322\mathrm{e}{-01} & -2.1332\mathrm{e}{-01} \end{pmatrix}$$

whose condition number is about $4.2 \cdot 10^4$. *Of the two vectors*

$$y_1 = \begin{pmatrix} 1.17261\mathrm{e}{+00} \\ 3.05130\mathrm{e}{+00} \\ 4.28640\mathrm{e}{+00} \\ -1.49204\mathrm{e}{+00} \\ -3.06310\mathrm{e}{+00} \\ -4.26540\mathrm{e}{-01} \end{pmatrix} \quad \text{and} \quad y_2 = \begin{pmatrix} -3.899239927661189\mathrm{e}{+00} \\ 4.588498435160376\mathrm{e}{+00} \\ 6.197824590162996\mathrm{e}{+00} \\ 8.122083519301400\mathrm{e}{-01} \\ -1.723814540708695\mathrm{e}{+00} \\ -1.842852647264738\mathrm{e}{+00} \end{pmatrix}$$

the first is exactly $X\mathrm{e}$ and has norm about 6.4. The second is $y_1 + r$, where r is in $\mathcal{R}(X)^\perp$ and has the same norm as y_1. Figure 2.1 shows the effects of perturbations in the least squares solutions. The first column gives the norm of the error. Following the error are the relative errors in the perturbed solutions for y_1 and y_2 and below them are the error bounds (computed without projecting the errors). Since y_1 and r are of a size and $\kappa_2(X)$ is around 10^4, we would expect a deterioration of about 10^8 in the solution with the large residual — exactly what we observe.

$\|E\|_2$	y_1	y_2
2.2e−01	1.3e+00	1.7e+01
	2.0e+03	8.5e+07
3.2e−02	4.9e−02	3.6e+01
	3.0e+02	1.2e+07
2.5e−03	3.1e−01	1.4e+03
	2.3e+01	9.7e+05
4.2e−04	6.8e−01	2.5e+03
	3.9e+00	1.6e+05
4.0e−05	1.2e−02	1.9e+03
	3.7e−01	1.5e+04
2.6e−06	7.2e−03	4.4e+02
	2.4e−02	9.8e+02
2.2e−07	3.9e−04	5.9e+01
	2.0e−03	8.5e+01
2.6e−08	1.2e−04	2.3e−01
	2.4e−04	9.8e+00
2.9e−09	6.2e−06	4.9e−02
	2.6e−05	1.1e+00
2.9e−10	3.5e−07	3.6e−02
	2.7e−06	1.1e−01
1.8e−11	9.3e−08	3.2e−03
	1.7e−07	7.1e−03
3.3e−12	5.8e−09	8.0e−05
	3.0e−08	1.3e−03
2.6e−13	8.0e−10	8.3e−06
	2.4e−09	1.0e−04
3.0e−14	6.6e−11	7.1e−06
	2.8e−10	1.2e−05
2.1e−15	8.1e−12	9.7e−07
	1.9e−11	7.9e−07

Figure 2.1: The κ^2 effect

———————◇———————

Accuracy of computed solutions

The summing up of our results consists of combining (2.8) and (2.12) for the normal equations and (2.9) and (2.14) for the QR approach.

Theorem 2.7. *Let b be the solution of the least squares problem of minimizing $\|y - Xb\|_2^2$. Let b_{NE} be the solution obtained by forming and solving the normal equations*

SEC. 2. LINEAR LEAST SQUARES

in floating-point arithmetic with rounding unit ϵ_M. Then b_{NE} satisfies

$$\frac{\|b_{NE} - b\|_2}{\|b\|_2} \leq \gamma_{NE}\kappa_2^2(X)\left(1 + \frac{\|y\|_2}{\|X\|_2\|b\|_2}\right)\epsilon_M.$$

Let b_{QR} be the solution obtained from a QR factorization in the same arithmetic. Then

$$\frac{\|b_{QR} - b\|_2}{\|b\|_2} \leq 2\gamma_{QR}\kappa_2(X)\epsilon_M + \gamma_{QR}\kappa_2^2(X)\frac{\|r\|_2}{\|X\|_2\|b\|_2}\epsilon_M,$$

where $r = y - Xb$ is the residual vector. The constants γ are slowly growing functions of the dimensions of the problem.

Comparisons

We are now in a position to make an assessment of the two chief methods of solving least squares problems: the normal equations and backward stable variants of the QR equation. We will consider three aspects of these methods: speed, stability, and accuracy.

- **Speed.** Here the normal equations are the undisputed winner. They can be formed from a dense $n \times p$ matrix X at a cost of $\frac{1}{2}np^2$ flam. On the other hand, Householder triangularization or the modified Gram–Schmidt algorithms require np^2 flam. It is easier to take advantage of sparsity in forming the normal equations (see Algorithm 2.4).

- **Stability.** Here unitary triangularization is the winner. The computed solution is the exact solution of a slightly perturbed problem. The same cannot be said of the normal equations. As the condition number of X increases, ever larger errors must be placed in X to account for the effects of rounding error on the normal equations. When $\kappa_2(X) \cong \sqrt{\epsilon_M}$, we cannot even guarantee that the computed normal equations are positive definite.

- **Accuracy.** Here the QR approach has the edge — but not a large one. The perturbation theory for the normal equations shows that $\kappa_2^2(X)$ controls the size of the errors we can expect. The bound for the solution computed from the QR equation also has a term multiplied by $\kappa_2^2(X)$, but this term is also multiplied by the scaled residual, which can diminish its effect. However, in many applications the vector y is contaminated with error, and the residual can, in general, be no smaller than the size of that error.

—

To summarize, if one is building a black-box least squares solver that is to run with all sorts of problems at different levels of precision, orthogonal triangularization is the way to go. Otherwise, one should look hard at the class of problems to be solved to see if the more economical normal equations will do.

2.4. LEAST SQUARES WITH LINEAR CONSTRAINTS

In this subsection we will consider the following problem:

$$\begin{aligned} \text{minimize} \quad & \|y - Xb\|_2^2 \\ \text{subject to} \quad & Cb = d. \end{aligned} \quad (2.15)$$

As usual X is an $n \times p$ matrix of rank p. We will assume that C is an $m \times p$ matrix of rank m. This implies that the number of rows of C is not greater than the number of columns. We will consider three ways of solving this problem: the null-space method, the elimination method, and the weighting method.

The null-space method

The null-space method begins by computing an orthogonal matrix V such that

$$CV \equiv C(V_1 \; V_2) = (L \; 0),$$

where L is lower triangular. This reduction is just VR decomposition of C^T and can be computed by orthogonal triangularization.

If we set

$$\hat{b} = V^T b = \begin{pmatrix} V_1^T b \\ V_2^T b \end{pmatrix} \equiv \begin{pmatrix} \hat{b}_1 \\ \hat{b}_2 \end{pmatrix}, \quad (2.16)$$

then the constraint $Cb = d$ can be written in the form

$$(L \; 0) \begin{pmatrix} \hat{b}_1 \\ \hat{b}_2 \end{pmatrix} = d. \quad (2.17)$$

Thus $\hat{b}_1$ is the solution of the system

$$L\hat{b}_1 = d.$$

Moreover, we can vary $\hat{b}_2$ in any way we like and the result still satisfies the transformed constraint (2.17).

Now set

$$\hat{X} = XV = (XV_1 \; XV_2) = (\hat{X}_1 \; \hat{X}_2).$$

Then

$$\|y - Xb\|_2 = \|y - XVV^T b\|_2 = \|(y - \hat{X}_1 \hat{b}_1) - \hat{X}_2 \hat{b}_2\|_2.$$

Since $\hat{b}_1$ is fixed and $\hat{b}_2$ is free to vary, we may minimize $\|y - Xb\|_2$ by solving the least squares problem

$$\|(y - \hat{X}_1 \hat{b}_1) - \hat{X}_2 \hat{b}_2\|_2^2 = \min.$$

SEC. 2. LINEAR LEAST SQUARES

Given an $n \times p$ matrix X of rank p and an $m \times p$ matrix C of rank m, this algorithm solves the constrained least squares problem

$$\text{minimize} \quad \|y - Xb\|_2^2$$
$$\text{subject to} \quad Cb = d.$$

1. Determine an orthogonal matrix V such that
   $$CV = (L \ \ 0),$$
   where L is lower triangular
2. Solve the system $L\hat{b}_1 = d$
3. Partition $V = (V_1 \ \ V_2)$, where $V_1 \in \mathbb{R}^{p \times m}$, and set
   $$\hat{X}_1 = XV_1 \text{ and } \hat{X}_2 = XV_2$$
4. Solve the least squares problem
   $$\|(y - \hat{X}_1\hat{b}_1) - \hat{X}_2\hat{b}_2\|_2^2 = \min$$
5. $b = V_1\hat{b}_1 + V_2\hat{b}_2$

Algorithm 2.6: The null space method for linearly constrained least squares

◇

Once this problem has been solved we may undo the transformation (2.16) to get b in the form

$$b = V_1\hat{b}_1 + V_2\hat{b}_2.$$

The algorithm just sketched is called the *null-space method* because the matrix V_2 from which the least squares matrix $\hat{X}_2 = XV_2$ is formed is a basis for the null space of the constraint matrix C. Algorithm 2.6 summarizes the method. Here are some comments.

- We have left open how to determine the matrix V. A natural choice is by a variant of orthogonal triangularization in which C is reduced to lower triangular form by postmultiplication by Householder transformations. If this method is adopted, the matrix $(\hat{X}_1 \ \hat{X}_2) = XV$ can be formed by postmultiplying X by Householder transformations.

- We have also left open how to solve the least squares problem in statement 4. Any convenient method could be used.

- If Householder triangularization is used throughout the algorithm, the operation count for the algorithm becomes

1. $(pm^2 - \frac{1}{3}m^3)$ flam to reduce C,
2. $n(2mp - m^2) - m^2(p - \frac{2}{3}m)$ flam for the formation of XV, \hfill (2.18)
3. $[n(p-m)^2 - \frac{1}{3}(p-m)^3]$ flam for the solution of the least squares problem in statement 4.

Thus if $n \gg p \gg m$ the method requires about np^2 flam, essentially the same number of operations required to solve the unconstrained problem.

- If the problem is to be solved repeatedly for different constraints but with X remaining constant, it may pay to transform the problem into QR form:

$$\left\| \begin{pmatrix} z_X \\ z_\perp \end{pmatrix} - \begin{pmatrix} R \\ 0 \end{pmatrix} b \right\|_2^2$$

[see (2.2)]. Then one can solve the smaller problem

$$\text{minimize} \quad \|z_X - Rb\|_2^2$$
$$\text{subject to} \quad Cb = d.$$

- Two factors control the accuracy of the result. The first is the condition of the constraint matrix C. If C is ill conditioned the vector $\hat{b}_1$ will be inaccurately determined. The second factor is the condition of $\hat{X}_2$, which limits our ability to solve the least squares problem in statement 4. For more, see the notes and references.

- The condition of X itself does not affect the solution. In fact, X can be of rank less than p—just as long as $\hat{X}_2$ is well conditioned. For this reason constraints on the solution are often used to improve degenerate problems. This technique and its relatives go under the generic name of *regularization methods*.

It is worth noting that on our way to computing the solution of the constrained least squares problem, we have also computed the solution to another problem. For the vector

$$b_{\min} = V_1 \hat{b}_1$$

satisfies the constraint $Cb = d$. Now any solution satisfying the constraint can be written in the form

$$b = V_1 \hat{b}_1 + V_2 \hat{b}_2 \equiv b_{\min} + b_\perp.$$

But $b_{\min}$ and $b_\perp$ are orthogonal. Hence by the Pythagorean theorem,

$$\|b_{\min}\|_2^2 \le \|b_{\min}\|_2^2 + \|b_\perp\|_2^2 = \|b\|_2^2,$$

with equality if and only if $b_\perp = 0$. This shows that:

The vector $b_{\min} = V_1 \hat{b}_1$ is the unique minimal norm solution of the underdetermined system $C^T b = d$. \hfill (2.19)

The method of elimination

The *method of elimination* is the most natural method for solving constrained least squares problems. Specifically, the constraint $Cb = d$ determines m of the components of b in terms of the remaining $p-m$ components. Solve for those components, substitute the expression in the formula $y - Xb$, and solve for the remaining components in the least squares sense.

To give this sketch mathematical substance, partition

$$C = (C_1 \; C_2), \quad X = (X_1 \; X_2), \quad \text{and} \quad b = \begin{pmatrix} b_1 \\ b_2 \end{pmatrix}.$$

Assume for the moment that C_1 is nonsingular. Then we can write

$$b_1 = C_1^{-1}d - C_1^{-1}C_2 b_2.$$

If we substitute this in the expression $y - Xb$, we obtain the reduced least square problem

$$\|(y - X_1 C_1^{-1}d) - (X_2 - X_1 C_1^{-1} C_2)b_2\|_2^2 = \min. \tag{2.20}$$

This problem can be solved in any way we like.

A slightly different way of looking at this process will suggest an elegant way of arranging the calculations and protect us from the chief danger of the method. Consider the augmented matrix

$$W = \begin{pmatrix} C_1 & C_2 & d \\ X_1 & X_2 & y \end{pmatrix}.$$

If we perform m steps of Gaussian elimination on this matrix, the result is

$$\begin{pmatrix} U_1 & U_2 & u \\ 0 & X_2 - X_1 C_1^{-1} X_2 & y - X_1 C_1^{-1} d \end{pmatrix} \equiv \begin{pmatrix} U_1 & U_2 & u \\ 0 & \bar{X}_2 & \bar{y} \end{pmatrix}, \tag{2.21}$$

in which U_1 is upper triangular. Comparing the second row of this matrix with (2.20), we see that pieces of the reduced least squares problem come from the Schur complement of C_1 in W. Thus we can solve the constrained problem by performing m steps of Gaussian elimination on W and solving the least squares problem

$$\|\bar{y} - \bar{X}_2 b_2\|_2^2 = \min$$

for b_2. We then compute b_1 as the solution of the triangular system

$$U_1 b_1 = u - U_2 b_2.$$

As we have said the procedure is elegant—perform m steps of Gaussian elimination on the augmented matrix to get a reduced least squares problem which may be

Given an $n \times p$ matrix X of rank p and an $m \times p$ matrix of rank m, this algorithm solves the constrained least squares problem

$$\text{minimize} \quad \|y - Xb\|_2^2$$
$$\text{subject to} \quad Cb = d.$$

1. Form the augmented matrix
$$W = \begin{pmatrix} C_1 & C_2 & d \\ X_1 & X_2 & y \end{pmatrix}$$
2. Perform m steps of Gaussian elimination with complete pivoting for size on W. The pivots are to be chosen from the elements of C.
Call the result
$$\begin{pmatrix} U_1 & U_2 & u \\ 0 & \bar{X}_2 & \bar{y} \end{pmatrix}$$
3. Solve the least squares problem
$$\|\bar{y} - \bar{X}_2 b_2\|_2^2 = \min$$
4. Solve the system $U_1 b_1 = u - U_2 b_1$
5. Undo the effects of the column permutations on b

Algorithm 2.7: Constrained least squares by elimination

⎯⎯⎯⎯⎯⎯⎯⎯⎯⎯ ◇ ⎯⎯⎯⎯⎯⎯⎯⎯⎯⎯

solved for b_2. But the process as it stands is dangerous. If C_1 is ill conditioned, the matrix $X_1 C_1^{-1} X_2$ in the formula

$$\bar{X}_2 = X_2 - X_1 C_1^{-1} X_2$$

will be large, and its addition to X_2 will cause information to be lost. (For more on this point see Example 2.8, Chapter 3, and §4, Chapter 3.)

The ill-conditioning of C_1 does not necessarily reflect an ill-conditioning of the constrained least squares problem, which depends on the condition of the entire matrix C. Instead it reflects the bad luck of having nearly dependent columns at the beginning of the constraint matrix. The cure is to pivot during the Gaussian elimination. Since we must move columns around, the natural pivoting strategy is complete pivoting for size. However, we may choose pivots only from C, since we must not mix the constraints with the matrix X.

These considerations lead to Algorithm 2.7. Some comments.

- The success of the algorithm will depend on how successful the pivoting strategy is in getting a well-conditioned matrix in the first m columns of the array (assuming that such a matrix exists). The folklore says that complete pivoting is reasonably good at this.

SEC. 2. LINEAR LEAST SQUARES

- If orthogonal triangularization is to be used to solve the reduced least squares problem, it can be done in situ. The final result will be an augmented triangular system which can be back-solved for b_1 and b_2.

- The operation count for the Gaussian elimination part of the algorithm is

$$[(n+m)pm - \tfrac{1}{2}(m+n+p)m^2 + \tfrac{1}{3}m^3] \text{ flam.}$$

When $n \gg p$ this amounts to $n(mp - \tfrac{1}{2}m^2)$ flam, which should be compared to the count $n(2mp + m^2)$ flam from item two in the count (2.18) for the null-space method.

When $m \cong p$ — i.e., when the number of constraints is large — the count becomes approximately $(\tfrac{1}{2}np^2 + \tfrac{1}{3}p^3)$ flam. This should be compared with a count of $(np^2 + \tfrac{1}{3}p^3)$ flam from items one and two in the count for the null-space method. This suggests that elimination is to be especially preferred when the number of constraints is large.

The weighting method

The weighting method replaces the constrained problem with an unconstrained problem of the form

$$\left\| \begin{pmatrix} \tau d \\ y \end{pmatrix} - \begin{pmatrix} \tau C \\ X \end{pmatrix} \right\|_2^2 = \min, \tag{2.22}$$

where τ is a suitably large number. The rationale is that as τ increases, the size of the residual $d - Cb$ must decrease so that the weighted residual $\tau(d - Cb)$ remains of a size with the residual $y - Xb$. If τ is large enough, the residual $d - Cb$ will be so small that the constraint is effectively satisfied.

The method has the appeal of simplicity — weight the constraints and invoke a least squares solver. In principle, the weighted problem (2.22) can be solved by any method. In practice, however, we cannot use the normal equations, which take the form

$$(X^T X + \tau^2 C^T C)b = X^T y + \tau^2 C^T y.$$

The reason is that as τ increases, the terms $\tau^2 C^T C$ and $\tau^2 C^T y$ will dominate $X^T X$ and $X^T y$ and, in finite precision, will eventually obliterate them. For this reason orthogonal triangularization is the preferred method for solving (2.22).

To determine how large to take τ, we will give an analysis of the method based on orthogonal triangularization. Suppose that we have triangularized the first m columns of the matrix

$$\begin{pmatrix} \tau C_1 & \tau C_2 & \tau d \\ X_1 & X_2 & y \end{pmatrix} \tag{2.23}$$

to get the matrix

$$\begin{pmatrix} Q_{11} & Q_{12} \\ Q_{21} & Q_{22} \end{pmatrix} \begin{pmatrix} \tau C_1 & \tau C_2 & \tau d \\ X_1 & X_2 & y \end{pmatrix} = \begin{pmatrix} S_1 & S_2 & s \\ 0 & \hat{X}_2 & \hat{y} \end{pmatrix}, \tag{2.24}$$

where

$$Q = \begin{pmatrix} Q_{11} & Q_{12} \\ Q_{21} & Q_{22} \end{pmatrix}$$

is orthogonal. If we were to stop here, we could solve the reduced least squares problem

$$\|\hat{y} - \hat{X}_2 \hat{b}_2\|_2^2 = \min \qquad (2.25)$$

by any method and obtain an approximation $\hat{b}_2$ to b_2.

It turns out that the reduced least squares problem (2.25) is closely related to the reduced least squares problem for the elimination method. To see this, let us compute the $(2,1)$-element of (2.24) to get

$$\tau Q_{21} C_1 + Q_{22} X_1 = 0,$$

or

$$Q_{21} = -\tau^{-1} Q_{22} X_1 C_1^{-1}. \qquad (2.26)$$

Computing the $(2,2)$-element, we find that

$$\tau Q_{21} C_2 + Q_{22} X_2 = \hat{X}_2,$$

or from (2.26)

$$\hat{X}_2 = Q_{22}(X_2 - X_1 C_1^{-1} C_2) = Q_{22} \bar{X}_1, \qquad (2.27)$$

where $\bar{X}_1$ is from the elimination method [see (2.21)]. Likewise,

$$\hat{y} = Q_{22}(y - X_1 C_1^{-1} d) = Q_{22} \bar{y}, \qquad (2.28)$$

where again $\bar{y}$ is from the elimination method.

Thus the reduced least squares problem (2.25) has the form

$$\|Q_{22}(\bar{y} - \bar{X} \hat{b}_2)\|_2^2 = \min.$$

If Q_{22} were orthogonal, the solution of this problem would be the correct answer b_2 to the constrained problem. The following assertion shows that the deviation of Q_{22} from orthogonality decreases as τ increases.

There is an orthogonal matrix $\hat{Q}_{22} = Q_{22}(I - Q_{21}^T Q_{21})^{-\frac{1}{2}}$ such that

$$\|\hat{Q}_{22} - Q_{22}\|_2 \leq \|Q_{21}\|_2^2 \leq (\gamma/\tau)^2,$$

where

$$\gamma = \|X_1\|_2 \|C_1^{-1}\|_2. \qquad (2.29)$$

SEC. 2. LINEAR LEAST SQUARES

If this result is applied to (2.27) and (2.28), we find that there is an orthogonal matrix $\hat{Q}_{22}$ such that

$$\hat{X}_2 = \hat{Q}_{22}\bar{X}_2 + G \quad \text{and} \quad \hat{y} = \hat{Q}_{22}\bar{y} + g,$$

where

$$\|G\|_2 \leq (\gamma/\tau)^2 \|\bar{X}_2\|_2 \quad \text{and} \quad \|g\|_2 \leq (\gamma/\tau)^2 \|\bar{y}\|_2. \tag{2.30}$$

These results can be used to determine τ. The bounds (2.30) are relative to the quantities that g and G perturb. Thus the perturbations will be negligible if $(\gamma/\tau)^2$ is less than the rounding unit ϵ_M. In view of the definition (2.29) of γ, we should take

$$\tau^2 \geq \frac{\|X_1\|_2^2 \|C_1^{-1}\|_2^2}{\epsilon_M}.$$

Now we cannot know C_1^{-1} without calculating it. But if the columns of C are all of a size, it is unlikely that C_1 would have a singular value smaller than $\|C\|_2 \epsilon_M$, in which case $\|C_1^{-1}\|_2 \leq 1/\|C\|_2 \epsilon_M$. Replacing $\|C_1^{-1}\|_2$ by this bound and replacing $\|X_1\|_2$ by $\|X\|_2$, we get the criterion

$$\tau \geq \frac{\|X\|_2}{\|C\|_2} \epsilon_M^{-1}. \tag{2.31}$$

Stated in words, we must choose τ so that X is smaller that C by a factor that is smaller than the rounding unit.

The above analysis shows that increasing the weight τ increases the orthogonality of Q_{22}. The criterion (2.31) insures that even when C_1 is very ill conditioned Q_{22} will be orthogonal to working accuracy. However, as we have observed in connection with the method of elimination, when C_1 is ill conditioned the matrix X_2 will be inadequately represented in the Schur complement $\bar{X}_2$ and hence in the matrix $\hat{X}_2 = Q_{22}\bar{X}_2$.

We must therefore take precautions to insure that C_1 is well conditioned. In the method of elimination we used complete pivoting. The appropriate strategy for Householder triangularization is called column pivoting for size. We will treat this method in more detail in the next chapter (Algorithm 2.1, Chapter 5), but in outline it goes as follows. At the kth stage of the reduction in Algorithm 1.2, exchange the column of largest norm in the working matrix (that is, $X[k:n, k:p]$) with the first column. This procedure is a very reliable, though not completely foolproof, method of insuring that C_1 is as well conditioned as possible. Most programs for Householder triangularization have this pivoting option.

Algorithm 2.8 summarizes the weighting method. Three comments.

- In statement 1, we have used the easily computable Frobenius norm in place of the 2-norm. This substitution makes no essential difference.

Given an $n \times p$ matrix X of rank p and an $m \times p$ matrix of rank m, this algorithm solves the constrained least squares problem

$$\text{minimize} \quad \|y - Xb\|_2^2$$
$$\text{subject to} \quad Cb = d.$$

1. Let τ be chosen so that
$$\tau\|C\|_F \geq \|X\|_F/\epsilon_M$$
2. Apply Householder triangularization with column pivoting to the matrix
$$\begin{pmatrix} \tau C & \tau d \\ X & y \end{pmatrix}$$
to get the matrix
$$\begin{pmatrix} R & z \\ 0 & \rho \end{pmatrix}$$
(do not pivot on the last column)
3. Solve the system $Rb = z$

Algorithm 2.8: Constrained least squares by weights

───────◇───────

- The placement of τC and τd at the top of the augmented matrix causes the grading of the matrix to be downward and is essential to the success of the Householder triangularization. For more see (1.14).

- The operation count for the algorithm is $[(n+m)p^2 - \frac{1}{3}p^3]$ flam.

- Our analysis shows that we do not have to worry about unusually small elements creating a balanced transformation that obliterates X_2, as in Example 1.8. It is not that such a situation cannot occur — it occurs when C_1 is ill conditioned. But our pivoting strategy makes the condition of C_1 representative of the condition of C as a whole. Consequently, we can lose information in X_2 only when the problem itself is ill conditioned.

2.5. ITERATIVE REFINEMENT

In this subsection we will give a method for refining least squares solutions. The natural algorithm is to mimic the method for linear systems. Specifically, given an approximate solution d, perform the following computations.

1. $s = y - Xb$
2. Solve the least squares problem $\|r - Xb\|_2^2 = \min$ (2.32)
3. $b = b + d$

SEC. 2. LINEAR LEAST SQUARES

Let b and r be approximations to the solution and residual of the least squares problem $\|y - Xb\|_2 = \min$. This algorithm performs one step of iterative refinement via the residual system.

1. $g = y - r - Xb$
2. $h = -X^T r$
3. Solve the system
$$\begin{pmatrix} I_n & X \\ X^T & 0 \end{pmatrix} \begin{pmatrix} s \\ t \end{pmatrix} = \begin{pmatrix} g \\ h \end{pmatrix}$$
4. $r = r + s;\ b = b + t$

Algorithm 2.9: Iterative refinement for least squares (residual system)

⸺⸺⸺⸺⟡⸺⸺⸺⸺

This is the procedure we used to correct the seminormal equations (Algorithm 2.5).

Unfortunately, this algorithm does not perform well unless r is very small. To see why, let $b + h$ be the true solution. Then $s = y - Xb = r + Xh$, where r is the residual at the solution. Since $X^T r = 0$, in exact arithmetic we have

$$d = X^\dagger r + X^\dagger Xh = -h,$$

so that $x + d$ is the exact solution. However, in inexact arithmetic, the vector s will assume the form

$$s = r + Xh + e,$$

where $\|e\|_2 / \|r\|_2 = \gamma \epsilon_M$ for some small constant γ. Thus we are committed to solving a problem in which Xh has been perturbed to be $Xh + e$. The relative error in Xh is then

$$\frac{\|e\|_2}{\|Xh\|_2} = \gamma \frac{\|r\|_2}{\|Xh\|_2} \epsilon_M.$$

If r is large compared to Xh, e will overwhelm Xh.

The cure for this problem is to work with the *residual system*

$$\begin{pmatrix} I_n & X \\ X^T & 0 \end{pmatrix} \begin{pmatrix} r \\ b \end{pmatrix} = \begin{pmatrix} y \\ 0 \end{pmatrix}. \tag{2.33}$$

The first equation in this system, $r + Xb = y$, defines the residual. The second equation $X^T r = 0$ says that the residual is orthogonal to the column space of X.

Iterative refinement applied to the residual system takes the form illustrated in Algorithm 2.9. The chief computational problem in implementing this method is to solve the *general residual system*

$$\begin{pmatrix} I_n & X \\ X^T & 0 \end{pmatrix} \begin{pmatrix} s \\ t \end{pmatrix} = \begin{pmatrix} g \\ h \end{pmatrix}, \tag{2.34}$$

in which the zero on the right-hand side of (2.33) is replaced by a nonzero vector. Fortunately we can do this efficiently if we have a QR decomposition of X.

Let

$$X = (Q_X \; Q_\perp) \begin{pmatrix} R \\ 0 \end{pmatrix},$$

and consider the following transformation of the system (2.34):

$$\begin{pmatrix} Q^{\mathrm{T}} & 0 \\ 0 & I \end{pmatrix} \begin{pmatrix} I_n & X \\ X^{\mathrm{T}} & 0 \end{pmatrix} \begin{pmatrix} Q & 0 \\ 0 & I \end{pmatrix} \begin{pmatrix} Q^{\mathrm{T}} & 0 \\ 0 & I \end{pmatrix} \begin{pmatrix} s \\ t \end{pmatrix} = \begin{pmatrix} Q^{\mathrm{T}} & 0 \\ 0 & I \end{pmatrix} \begin{pmatrix} g \\ h \end{pmatrix}.$$

If we set

$$\hat{s}_X = Q_X^{\mathrm{T}} s, \quad \hat{s}_\perp = Q_\perp^{\mathrm{T}} s, \quad \hat{g}_X = Q_X^{\mathrm{T}} g, \quad \text{and} \quad \hat{g}_\perp = Q_\perp^{\mathrm{T}} g,$$

this system becomes

$$\begin{pmatrix} I_p & 0 & R \\ 0 & I_{n-p} & 0 \\ R^{\mathrm{T}} & 0 & 0 \end{pmatrix} \begin{pmatrix} \hat{s}_X \\ \hat{s}_\perp \\ t \end{pmatrix} = \begin{pmatrix} \hat{g}_X \\ \hat{g}_\perp \\ h \end{pmatrix}.$$

From this system we obtain the equations $R^{\mathrm{T}} \hat{s}_X = h$, $\hat{s}_\perp = \hat{g}_\perp$, and $Rt = \hat{g}_X - \hat{s}_X$. Once these quantities have been computed, we can compute $s = Q_X \hat{s}_X + Q_\perp \hat{s}_\perp$.

Algorithm 2.10 implements this scheme. Two comments.

- We have written the algorithm as if a full QR decomposition were available, perhaps in factored form. However, we can get away with a QR factorization. The key observation is that s in statement 4 can be written in the form

$$s = Q_X \hat{s}_X + P_\perp g.$$

Hence to reconstitute s all we need do is orthogonalize g against the columns of Q_X to get $P_\perp g$. This can be done by the Gram–Schmidt method with reorthogonalization (Algorithm 1.13).

- If Householder triangularization is used to implement the algorithm, the computed solution satisfies

$$\begin{pmatrix} I_n & X + E_1 \\ X^{\mathrm{T}} + E_2^{\mathrm{T}} & 0 \end{pmatrix} \begin{pmatrix} s \\ t \end{pmatrix} = \begin{pmatrix} g \\ h \end{pmatrix},$$

where $\|E_i\|/\|X_i\| \leq \gamma \epsilon_{\mathrm{M}}$ ($i = 1, 2$) for some constant γ depending on the norm and the dimensions of the problem. This is not quite backward stability, since the matrix X is perturbed in two different ways.

—

SEC. 2. LINEAR LEAST SQUARES

Let
$$X = (Q_X \; Q_\perp) \begin{pmatrix} R \\ 0 \end{pmatrix}$$

be a QR decomposition of the $n \times p$ matrix X. The following algorithm solves the general residual system

$$\begin{pmatrix} I_n & X \\ X^T & 0 \end{pmatrix} \begin{pmatrix} s \\ t \end{pmatrix} = \begin{pmatrix} g \\ h \end{pmatrix}.$$

1. $\hat{g}_X = Q_X^T g$; $\hat{g}_\perp = Q_\perp^T g$
2. Solve the system $R^T \hat{s}_X = h$
3. Solve the system $Rt = \hat{g}_X - \hat{s}_X$
4. $s = Q_X \hat{s}_X + Q_\perp \hat{g}_\perp$

Algorithm 2.10: Solution of the general residual system

⸺⋄⸺

Iterative refinement via the residual system works quite well. Unfortunately our general analysis does not apply to this variant of iterative refinement because the condition number of the system is approximately $\kappa^2(X)$. But if the special structure of the system is taken into account, the refinement can be shown to converge at a rate governed by $\kappa(X)\epsilon_M$. The method can be used with double precision calculation of the g and h, in which case the iteration will converge to a solution of working accuracy. The behavior of the fixed precision version is more problematic, but it is known to improve the solution.

2.6. NOTES AND REFERENCES

Historical

The principle of least squares arose out of the attempt to solve overdetermined systems in astronomy, where the number of observations can easily exceed the number of parameters. Gauss discovered the principle in 1794 and arrived at a probabilistic justification of the principle in 1798. He subsequently used least squares in astronomical calculations.

In 1805 Legendre [216] published the method in an appendix to a work on the determination of orbits. When Gauss published his astronomical opus in 1809 [130], he gave a detailed treatment of least squares and claimed it as "my principle" (*principium nostrum*). Legendre took exception, and a famous priority dispute followed.

The subject was subsequently taken up by Laplace, who used his central limit theorem to justify least squares when the number of observations is large [209, 1810] [210,

1812]. Gauss returned to the subject in the 1820s with an extended memoir in three parts [132, 133, 135]. In the first of these he proved the optimality of least squares estimates under suitable assumptions. The result is commonly known as the Gauss–Markov theorem, although Markov's name is spurious in this connection.

There is an extensive secondary literature on the history of least squares. Stigler [311] gives an excellent account of the problem of combining observations, although I find his treatment of Gauss deficient. Plackett [263] gives a balanced and entertaining treatment of the priority controversy accompanied by many passages from the correspondence of the principals. For a numerical analyst's view see the afterword in [140].

The QR approach

In his paper on unitary triangularization [188], Householder observed that when the method was applied to a rectangular matrix the result was the Cholesky factor of the matrix of the normal equations. But he did not give an algorithm for solving least squares problems. I first heard of the QR approach from Ross Burris, who had been using plane rotations in the early 1960s to solve least squares problems [47]. But the real birth year of the approach is 1965, when Gene Golub published a ground-breaking paper on the subject [148, 1965]. In it he showed in full detail how to apply Householder triangularization to least squares. But more important, he pioneered the QR approach as a general technique for solving least squares problems.

Gram–Schmidt and least squares

That the output of the Gram–Schmidt algorithm with reorthogonalization can be used to solve least squares problems is a consequence of our general approach via the QR decomposition. What is surprising is that the modified Gram–Schmidt can be used without orthogonalization. This was established by Björck [37, 1967], who proved weak stability. In 1992 Björck and Paige [44] used the connection of modified Gram–Schmidt with Householder triangularization (Theorem 1.11) to establish the backward stability of the method. For the accuracy of solutions computed by the modified Gram–Schmidt method see [261].

The augmented least squares matrix

The idea of augmenting the least squares matrix with the vector of y before computing the QR factorization [see (2.4)] is analogous to computing the Cholesky factor of the augmented cross-product matrix (2.5), a technique widely used by statisticians.

The normal equations

The normal equations are the natural way — until recently, the only way — of solving discrete least squares problems. As we have seen, the technique trades off computational economy against numerical stability. This suggests trying to first transform the least squares problem so that it is better conditioned. The method of Peters and Wilkin-

SEC. 2. LINEAR LEAST SQUARES

son [261] computes an LU factorization of the least squares matrix with pivoting to insure that L is well conditioned and then applies the method of normal equations to the problem $\|y - L(Ub)\|_2^2 = \min$. Once the solution Ub has been computed b may be found by back substitution.

For more on techniques involving Gaussian elimination, see [243].

The seminormal equations

According to Björck [41, §2.4.6], The seminormal equations were introduced by Saunders [271] to find minimum norm solutions of underdetermined systemsp [see (2.19)], a problem for which the technique is weakly stable. Björck [39] showed that it was not satisfactory for the linear least squares problem but that iterative refinement can improve it.

Rounding-error analyses

Except for the additional error made in forming the normal equations, the analysis of the method of normal equations parallels that of Theorem 4.9, Chapter 3. The error analysis of the QR solution depends on what flavor of decomposition is used. For orthogonal triangularization the basic tool is Theorem 1.5 combined with an error analysis of the triangular solve. It is interesting to note that, in contrast to linear systems, the error is thrown back on y as well as X. For a formal analysis and further references, see [177, Ch. 10].

Perturbation analysis

The first-order perturbation analysis of least squares solutions was given by Golub and Wilkinson [155], where the κ^2-effect was first noted. There followed a series of papers in which the results were turned into rigorous bounds [37, 213, 260, 286]. Special mention should be made of the paper by Wedin [336], who treats rank deficient problems. For surveys see [290] and [310]. For the latest on componentwise and backward perturbation theorems, see [177, Ch. 19].

Constrained least squares

Björck [41, Ch. 5] gives a survey of methods for solving constrained least squares problems, including linear and quadratic inequality constraints. For a view from the optimization community, where the null-space method is widely used, see the books of Gill, Murray, and Wright [144] and Nash and Sofer [239].

The optimizers would call the elimination method a null-space method. For the reduced least squares matrix can be obtained from the relation

$$(X_1 \ X_2) \begin{pmatrix} I \\ -C_1^{-1}C_2 \end{pmatrix} = X_2 - X_1 C_1^{-1} C_2,$$

and the matrix

$$\begin{pmatrix} I \\ -C_1^{-1}C_2 \end{pmatrix}$$

clearly spans the null space of C.

Björck and Golub [42] give a variant of the elimination method in which C is reduced to upper trapezoidal form by orthogonal triangularization as a preliminary to forming the reduced problem (see also [41, §5.1.2]). The fact that the elimination method can be implemented as Gaussian elimination followed by orthogonal triangularization has been used by Shepherd and McWhirter to design a pipeline for constrained recursive least squares [277].

The origins of the weighting method are unknown. Björck [35, 1967] commented on the convergence of the weighted problem to the constrained solution. The method has been analyzed by Lawson and Hanson [213], Van Loan [328], and Barlow [15]. The analysis given here is new — devised to highlight the relation between the elimination method and the weighting method. This relation was first pointed out in [296].

Iterative refinement

The natural algorithm (2.32) was proposed by Golub [148] and analyzed by Golub and Wilkinson [154]. Iterative refinement using the residual system is due to Björck [36, 35, 38]. For an error analysis of iterative refinement see [177, §19.5].

The term "residual system" is new. The matrix of the system is often called the *augmented matrix,* an unfortunate name because it takes a useful general phrase out of circulation. But the terminology is firmly embedded in the literature of optimization.

3. UPDATING

> *The moving Finger writes; and, having writ,*
> *Moves on: nor all thy Piety nor Wit*
> *Shall lure it back to cancel half a Line:*
> *Nor all thy Tears wash out a Word of it.*
>
> Omar Kayyam

We all make mistakes. And when we do, we wish we could go back and change things. Unfortunately, Omar Kayyam said the last word on that.

By contrast, we can sometimes go back and undo mistakes in decompositions. From the decomposition itself we can determine how it is changed by an alteration in the original matrix — a process called *updating*. In general, an update costs far less than recomputing the decomposition from scratch. For this reason updating methods are the computational mainstays of disciplines, such as optimization, that must operate with sequences of related matrices.

SEC. 3. UPDATING

This section is devoted to describing updating algorithms for the QR decomposition, which of all decompositions most lends itself to this kind of treatment. However, we will begin with the problem of updating matrix inverses, which is important in many applications and illustrates in a particularly simple manner some of the numerical problems associated with updating algorithms.

The QR updating problem has several aspects depending on how much of the decomposition we have and how we alter the original matrix. Here is a table of the algorithms we are going to treat.

	Permute Columns	Column Update	Row Update	Rank-One Update
QR Decomposition	*	±	±	*
QR Factorization	*	±	±	*
R-Factor	*	−	±	

(3.1)

A nonblank entry means that there is an algorithm for the corresponding problem. The symbol "±" means that we will treat both adding or deleting a row or column. The lonely "−" means that there is no algorithm for adding a column to an R-factor — the R-factor alone does not contain enough information to allow us do it. Likewise, we cannot perform a general rank-one update on an R-factor. We will go through this table by columns.

Two general observations. First, in addition to its generic sense, the term "updating" is used in contrast with *downdating* — the process of updating an R-factor after a row has been removed from the original matrix. As we shall see, this is a hard problem. The term downdating is sometimes used to refer to removing a row or a column from a QR decomposition or factorization. However, these problems have stable solutions. For this reason we will confine the term downdating to R-factor downdating.

Second, the R-factor of a QR decomposition X is the Cholesky factor of the cross-product matrix $A = X^T X$, and updating an R-factor can be formulated in terms of A alone. For this reason the problem of updating R-factors is also called *Cholesky updating* or *downdating* according to the task.

The algorithms in this section make heavy use of plane rotations, and the reader may want to review the material in §1.3.

3.1. UPDATING INVERSES

In this subsection we will consider two algorithms for updating matrix inverses. The first is a formula for updating the inverse of a matrix when it is modified by the addition of a rank k matrix. The technique has many applications — notably in optimization, where the inverse in question is of an approximate Hessian or of a basis of active constraints. The second algorithm is a method for generating in situ the inverses of submatrices of a given positive definite matrix. It finds application in least squares problems in which variables must be moved in and out of the problem.

Woodbury's formula

Our goal is to compute the inverse of a matrix of the form $A - UV^T$. It will be convenient to begin with the case $A = I$.

Theorem 3.1 (Woodbury). Let $U, V \in \mathbb{R}^{p \times k}$. If $I - UV^T$ is nonsingular, then so is $I - V^T U$ and

$$(I - UV^T)^{-1} = I + U(I - V^T U)^{-1} V^T. \tag{3.2}$$

Proof. Suppose $I - UV^T$ is nonsingular, but $I - V^T U$ is singular. Then there is a nonzero vector x such that

$$(I - V^T U)x = x - V^T U x = 0. \tag{3.3}$$

Let $y = Ux$. Then $y \neq 0$, for otherwise (3.3) would imply that $x = 0$. Multiplying the relation $x - V^T y = 0$ by U, we find that $y - UV^T y = (I - UV^T)y = 0$; i.e., $I - UV^T$ is singular, contrary to hypothesis.

The formula (3.2) can now be verified by multiplying right-hand side by $I - UV^T$ and simplifying to get I. ∎

In most applications of this theorem we will have $p > k$ and U and V will be of full column rank. But as the proof of the theorem shows, neither condition is necessary.

Turning now to the general case, suppose that A is nonsingular. Then

$$(A - UV^T)^{-1} = (I - A^{-1} UV^T)^{-1} A^{-1}.$$

By Theorem 3.1 (with $A^{-1} U$ replacing U)

$$(A - UV^T)^{-1} = [I + A^{-1} U(I - V^T A^{-1} U)^{-1} V^T] A^{-1}.$$

Hence we have the following corollary.

Corollary 3.2. If A and $A - UV^T$ are nonsingular, then

$$(A - UV^T)^{-1} = A^{-1} + A^{-1} U(I - V^T A^{-1} U)^{-1} V^T A^{-1}. \tag{3.4}$$

If $U = u$ and $V = v$ are vectors, then

$$(A - uv^T)^{-1} = A^{-1} + \frac{A^{-1} uv^T A^{-1}}{1 - v^T A^{-1} u}. \tag{3.5}$$

As an example of the application of (3.5), suppose that we have computed a factorization of A, so that we can solve any system of the form

$$Ax = b.$$

In addition suppose that we need to compute the solution of the system

$$(A - uv^T)y = b.$$

SEC. 3. UPDATING

Given x satisfying $Ax = b$, this algorithm computes the solution of the modified equation $(A - uv^T)y = b$.

1. Solve $Aw = u$
2. $\tau = v^T x/(1 - v^T w)$
3. $y = x + \tau w$

Algorithm 3.1: Updating $Ax = b$ to $(A - uv^T)y = b$

———◇———

To refactor $A - uv^T$ from scratch would require $O(n^3)$ operations. Instead we can compute the solution in $O(n^2)$ operations as follows.

From (3.5) we have

$$y = (A - uv^T)^{-1}b = A^{-1}b + \frac{A^{-1}uv^T A^{-1}b}{1 - v^T A^{-1}u} = x + \frac{A^{-1}uv^T x}{1 - v^T A^{-1}u}.$$

If we replace multiplication by an inverse with the solution of linear systems, we get Algorithm 3.1. The most expensive part of this algorithm is the solution of $Aw = u$, which in our example requires $O(n^2)$ operations. The remaining steps of the algorithm require only $O(n)$ operations.

In some applications—e.g., optimization or signal processing—one is presented with a sequence of matrices, each differing from its predecessor by a low-rank modification. If, as is usually the case, it is required to solve linear systems involving these matrices, one can use (3.4) to keep track of their inverses. However, this procedure can be dangerous.

Example 3.3. *Consider the matrix*

$$A = \begin{pmatrix} 1 & 2 \\ 3 & 4 \end{pmatrix},$$

whose inverse is

$$A^{-1} = \begin{pmatrix} -2.0 & 1.0 \\ 1.5 & -0.5 \end{pmatrix}.$$

Let $u = (0.499 \ 0)^T$ *and* $v = e_1$, *so that*

$$\hat{A} = A + uv^T = \begin{pmatrix} 1.499 & 2.000 \\ 3.000 & 4.000 \end{pmatrix}.$$

Note that changing the element 1.499 to 1.500 makes $\hat{A}$ exactly singular.

If we apply the formula (3.5) to compute an approximation to $\hat{A}^{-1}$ and round the result to four digits, we get

$$B = \begin{pmatrix} 1000.0 & 500.0 \\ 750.0 & -374.7 \end{pmatrix}.$$

If we now apply the formula again to compute $(B - uv^T)^{-1}$, which should be A^{-1}, we get

$$C = \begin{pmatrix} -2.000 & 1.000 \\ 1.500 & -0.450 \end{pmatrix}.$$

The $(2,2)$-elements of A^{-1} and C agree to no significant figures.

The size of B in this example is a tip-off that something has gone wrong. It says that B was obtained by adding a large rank-one matrix to A^{-1}. When this matrix is rounded information about A is lost, a loss which is revealed by cancellation in the passage from B to C. It should be stressed the loss is permanent and will propagate through subsequent updates. The cure for this problem is to update not the inverse but a decomposition of the matrix that can be used to solve linear systems.

The sweep operator

The sweep operator is an elegant way of generating inverses and solutions of linear systems inside a single array. To derive it consider the partitioned system

$$\begin{pmatrix} A_{11} & A_{12} \\ A_{21} & A_{22} \end{pmatrix} \begin{pmatrix} x_1 \\ x_2 \end{pmatrix} = \begin{pmatrix} y_1 \\ y_2 \end{pmatrix}, \tag{3.6}$$

in which A_{11} is nonsingular. If the first row of this system is solved for x_1, the result is

$$x_1 = A_{11}^{-1} y_1 - A_{11}^{-1} A_{12} x_2. \tag{3.7}$$

If this value of x_1 is substituted into the second row of (3.6), the result is

$$A_{21} A_{11}^{-1} y_1 + (A_{22} - A_{21} A_{11}^{-1} A_{12}) x_2 = y_2. \tag{3.8}$$

Combining (3.7) and (3.8) we obtain

$$\begin{pmatrix} A_{11}^{-1} & -A_{11}^{-1} A_{12} \\ A_{21} A_{11}^{-1} & A_{22} - A_{21} A_{11}^{-1} A_{12} \end{pmatrix} \begin{pmatrix} y_1 \\ x_2 \end{pmatrix} = \begin{pmatrix} x_1 \\ y_2 \end{pmatrix}. \tag{3.9}$$

Thus the matrix in (3.9) reflects the interchange of the vectors x_1 and y_1.

The sweep operator results from exchanging two corresponding components of x and y. If, for example, the first components are interchanged, it follows from (3.9) that the matrix of the system transforms as follows:

$$\begin{pmatrix} a_{11} & a_{12}^T \\ a_{21} & A_{22} \end{pmatrix} \longrightarrow \begin{pmatrix} a_{11}^{-1} & -a_{11}^{-1} a_{12}^T \\ a_{11}^{-1} a_{21} & A_{22} - \dfrac{a_{21} a_{12}^T}{a_{11}} \end{pmatrix} \stackrel{\text{def}}{=} \text{sweep}(A, 1).$$

Sec. 3. Updating

Given a matrix A of order p and a *pivot* k, consider the system (in northwest indexing)

$$\begin{pmatrix} A_{11} & a_{1k} & A_{1,k+1} \\ a_{k1}^T & \alpha_{kk} & a_{k,k+1}^T \\ A_{k+1,1} & a_{k+1,k} & A_{k+1,k+1} \end{pmatrix} \begin{pmatrix} x_1 \\ \xi_k \\ x_{k+1} \end{pmatrix} = \begin{pmatrix} y_1 \\ \eta_k \\ y_{k+1} \end{pmatrix}.$$

The following function overwrites A with the matrix of the system

$$\begin{pmatrix} B_{11} & b_{1k} & B_{1,k+1} \\ b_{k1}^T & \beta_{kk} & b_{k,k+1}^T \\ B_{k+1,1} & b_{k+1,k} & B_{k+1,k+1} \end{pmatrix} \begin{pmatrix} x_1 \\ \eta_k \\ x_{k+1} \end{pmatrix} = \begin{pmatrix} y_1 \\ \xi_k \\ y_{k+1} \end{pmatrix}.$$

1. sweep(A, k)
2. $A[k, k] = 1/A[k, k]$
3. $A[k, 1{:}k{-}1] = -A(k, k){*}A[k, 1{:}k{-}1]$
4. $A[k, k{+}1{:}p] = -A(k, k){*}A[k, k{+}1{:}p]$
5. $A[1{:}k{-}1, 1{:}k{-}1] = A[1{:}k{-}1, 1{:}k{-}1] + A[1{:}k{-}1, k]{*}A[k, 1{:}k{-}1]$
6. $A[1{:}k{-}1, k{+}1{:}p] = A[1{:}k{-}1, k{+}1{:}p] + A[1{:}k{-}1, k]{*}A[k, k{+}1{:}p]$
7. $A[k{+}1{:}p, 1{:}k{-}1] = A[k{+}1{:}p, 1{:}k{-}1] + A[k{+}1{:}p, k]{*}A[k, 1{:}k{-}1]$
8. $A[k{+}1{:}p, k{+}1{:}p] = A[k{+}1{:}p, k{+}1{:}p] + A[k{+}1{:}p, k]{*}A[k, k{+}1{:}p]$
9. $A[1{:}k{-}1, k] = A(k, k){*}A[1{:}k{-}1, k]$
10. $A[k{+}1{:}p, k] = A(k, k){*}A[k{+}1{:}p, k]$
11. **end** sweep

Algorithm 3.2: The sweep operator

Algorithm 3.2 defines the sweep operator for an arbitrary *pivot* k. There are several appealing facts about the sweep operator. In what follows we assume that the indicated sweeps can actually be performed.

- Since two exchanges of the same two components of x and y leave the system unchanged, *sweep*(A, k) is its own inverse.

- The sequence *sweep*$(A, 1)$, *sweep*$(A, 2)$, ..., *sweep*(A, k) yields a matrix of the form (3.9). In fact, these sweeps can be performed in any order. A set of sweeps on an arbitrary sequence of pivots yields a matrix of the form (3.9) but with its parts distributed throughout the matrix according to the sequence of pivots. In particular, after the sweeps the submatrix corresponding to the sweeps will contain its inverse, and the complementary submatrix will contain its Schur complement.

- If we sweep through the entire matrix, the result is the inverse matrix at a cost of n^3 flam.

- One sweep requires n^2 flam.

- If we generalize Algorithm 3.2 to sweep the augmented matrix $(A\ y)$, the solution of the subsystem $A_{11}x_1 = y_1$ corresponding to the pivots will be found in the components of the last column corresponding to the pivots.

- If A is positive definite then any sweep can be performed at any time. Specifically, the principal submatrix B corresponding to the pivots swept in is the inverse of a positive definite matrix and hence is positive definite (Corollary 2.4, Chapter 3). The complementary principal submatrix is the Schur complement of B and is also positive definite (Theorem 2.6, Chapter 3). Hence the diagonals of a swept matrix are positive — i.e., the pivot elements are nonzero. Of course, we can take advantage of the symmetry of A to save storage and operations.

The sweep operator is widely used in least squares calculations when it is necessary to move variables in and out of the problem. To see why, consider the partitioned augmented cross-product matrix

$$(X_1\ X_2\ y)^{\mathrm{T}}(X_1\ X_2\ y) = \begin{pmatrix} A_{11} & A_{12} & c_1 \\ A_{21} & A_{22} & c_2 \\ c_1^{\mathrm{T}} & c_2^{\mathrm{T}} & \eta^2 \end{pmatrix},$$

where A_{11} is of order k. If we sweep on the first k diagonals of this matrix we get

$$\begin{pmatrix} A_{11}^{-1} & -A_{11}^{-1}A_{12} & -A_{11}^{-1}c_1 \\ A_{21}A_{11}^{-1} & S & d_2 \\ c_1^{\mathrm{T}}A_{11}^{-1} & d_2^{\mathrm{T}} & \rho^2 \end{pmatrix}.$$

Now $A_{11}^{-1}c_1$ is the solution of the least squares problem $\|y - X_1\hat{b}_1\|_2^2 = \min$. Moreover, ρ^2 is the Schur complement of A_{11} in

$$\begin{pmatrix} A_{11} & c_1 \\ c_1^{\mathrm{T}} & \eta^2 \end{pmatrix}$$

and hence is the square of the (k,k)-element of the Cholesky factor of the same matrix [see (2.2), Chapter 3]. Hence by (2.5), ρ^2 is the residual sum of squares. Under a classical statistical model, $\rho^2 A_{11}^{-1}$ is an estimate of the covariance matrix of $\hat{b}_1$. Given these facts, it is no wonder that the sweep operator is a statistician's delight.

The stability properties of the sweep operator are incompletely known. It is closely related to a method called Gauss–Jordan elimination [see (1.32), Chapter 3], and for positive definite matrices it is probably at least as good as invert-and-multiply. But such a statement misses an important point about updating algorithms. We not only require that our algorithm be stable — at least in some sense — but we demand that the stability be preserved over a sequence of updates.

Example 3.4. *Suppose we use the sweep operator to compute the inverse of a matrix with a condition number of 10^t. Then we might reasonably expect a loss of t digits*

SEC. 3. UPDATING

in the solution (see Example 3.4, Chapter 3). Since the condition of the inverse is the same as the condition of the original matrix, if we use the sweep operator to recompute the original matrix, we would expect the errors in the inverse to be magnified by 10^t giving a total error of 10^{-2t}. By repeating this process several times, we should be able to obliterate even a well-conditioned matrix.

I tried this experiment with a positive definite matrix A whose condition is 10^{10} using arithmetic with a rounding unit of about $2 \cdot 10^{-16}$. Below are the relative (normwise) errors in the succession of computed A's.

$$5.9\text{e}{-}09 \quad 7.4\text{e}{-}09 \quad 5.4\text{e}{-}09 \quad 5.0\text{e}{-}09 \quad 8.0\text{e}{-}09 \quad 1.8\text{e}{-}08$$

As can be seen there is no rapid growth in the error.

The results of this example are confirmed by the fact that the sweep operator has been used successfully in problems (e.g., subset selection) for which any significant magnification of the error would quickly show itself. The sweep operator does not have the stability of methods based on orthogonal transformations, but its defects remain bounded.

3.2. MOVING COLUMNS

We now begin deriving the QR updating algorithms promised in the table (3.1). Since some of the updates require that we be able to move columns around in a QR decomposition, we will consider the following problem.

> Given a QR decomposition (QR factorization, R-factor) of a matrix X, determine the QR decomposition (QR factorization, R-factor) of XP, where P is a permutation.

A general approach

The key idea is very simple. Let

$$Q^\mathrm{T} X = \begin{pmatrix} R \\ 0 \end{pmatrix}$$

be a QR decomposition of X. Then for any permutation matrix P

$$Q^\mathrm{T} X P = \begin{pmatrix} RP \\ 0 \end{pmatrix}.$$

Let $U^\mathrm{T}(RP) = \hat{R}$ be a QR decomposition of RP. Then

$$\mathrm{diag}(U, I)^\mathrm{T} Q^\mathrm{T} X P = \begin{pmatrix} \hat{R} \\ 0 \end{pmatrix}$$

is a QR decomposition of XP. To summarize:

1. Perform the permutations on the columns of R
2. Reduce R by orthogonal transformations to triangular form, accumulating the transformations in Q. (3.10)

It is worth noting that the permutation matrix P can be an arbitrary matrix. Thus our procedure is a general updating procedure for any changes that can be cast as a postmultiplication by a matrix. As a general algorithm, this process can be quite expensive. But it can often be simplified when we know the structure of the problem. This turns out to be the case if P represents an interchange of two columns. Since any permutation can be reduced to a sequence of interchanges, we will focus on updating after a column interchange.

Interchanging columns

Suppose we want to apply the strategy in (3.10) to interchange columns two and five of a QR decomposition of an $n \times 6$ matrix X. The first step is to interchange columns two and five of R to get the matrix

$$\begin{pmatrix} X & X & X & X & X & X \\ 0 & X & X & X & X & X \\ 0 & X & X & X & X & X \\ 0 & X & 0 & X & X & X \\ 0 & X & 0 & 0 & X & X \\ 0 & 0 & 0 & 0 & 0 & X \end{pmatrix}.$$

(The replacement of the fifth column by the second introduces some inconsequential zeros not shown in the above diagram.)

We must now reduce this matrix to triangular form. It is done in two steps. First use plane rotations to eliminate the elements in the spike below the first subdiagonal in column two, as shown in the following diagram.

$$\begin{pmatrix} X & X & X & X & X & X \\ 0 & X & X & X & X & X \\ 0 & X & X & X & X & X \\ 0 & X & 0 & X & X & X \\ 0 & \widehat{X} & 0 & 0 & X & X \\ 0 & 0 & 0 & 0 & 0 & X \end{pmatrix} \xRightarrow{P_{45}} \begin{pmatrix} X & X & X & X & X & X \\ 0 & X & X & X & X & X \\ 0 & X & X & X & X & X \\ 0 & \widehat{X} & 0 & X & X & X \\ 0 & 0 & 0 & X & X & X \\ 0 & 0 & 0 & 0 & 0 & X \end{pmatrix} \xRightarrow{P_{34}}$$

$$\begin{pmatrix} X & X & X & X & X & X \\ 0 & X & X & X & X & X \\ 0 & X & X & X & X & X \\ 0 & 0 & X & X & X & X \\ 0 & 0 & 0 & X & X & X \\ 0 & 0 & 0 & 0 & 0 & X \end{pmatrix}$$

As we eliminate the spike, we introduce nonzero elements on the subdiagonal of R. We now proceed to eliminate these elements as follows.

Sec. 3. Updating

Let
$$Q^T X = \begin{pmatrix} R \\ 0 \end{pmatrix}$$
be a QR decomposition of the $n \times p$ matrix X. Given integers $1 \leq \ell < m \leq p$ this algorithm calculates the QR decomposition of the matrix obtained from X by interchanging columns ℓ and m.

1. $qrdxch(R, Q, \ell, m)$
2. $R[1{:}m, \ell] \leftrightarrow R[1{:}m, m]$
3.   **for** $k = m-1$ **to** $\ell+1$ **by** -1
4. $rotgen(R[k,\ell], R[k+1,\ell], c, s)$
5. $rotapp(c, s, R[k, k{:}p], R[k+1, k{:}p])$
6. $rotapp(c, s, Q[:, k], Q[:, k+1])$
7.   **end for** k
8.   **for** $k = \ell$ **to** $m-1$
9. $rotgen(R[k,k], R[k+1,k], c, s)$
10. $rotapp(c, s, R[k, k+1{:}p], R[k+1, k+1{:}p])$
11. $rotapp(c, s, Q[:, k], Q[:, k+1])$
12.   **end for** k
13. **end** $qrdxch$

Algorithm 3.3: QR update: exchanging columns

$$\begin{pmatrix} X & X & X & X & X & X \\ 0 & X & X & X & X & X \\ 0 & \hat{X} & X & X & X & X \\ 0 & 0 & X & X & X & X \\ 0 & 0 & 0 & X & X & X \\ 0 & 0 & 0 & 0 & 0 & X \end{pmatrix} \xRightarrow{P_{23}} \begin{pmatrix} X & X & X & X & X & X \\ 0 & X & X & X & X & X \\ 0 & 0 & X & X & X & X \\ 0 & 0 & \hat{X} & X & X & X \\ 0 & 0 & 0 & X & X & X \\ 0 & 0 & 0 & 0 & 0 & X \end{pmatrix} \xRightarrow{P_{34}}$$

$$\begin{pmatrix} X & X & X & X & X & X \\ 0 & X & X & X & X & X \\ 0 & 0 & X & X & X & X \\ 0 & 0 & 0 & X & X & X \\ 0 & 0 & 0 & \hat{X} & X & X \\ 0 & 0 & 0 & 0 & 0 & X \end{pmatrix} \xRightarrow{P_{45}} \begin{pmatrix} X & X & X & X & X & X \\ 0 & X & X & X & X & X \\ 0 & 0 & X & X & X & X \\ 0 & 0 & 0 & X & X & X \\ 0 & 0 & 0 & 0 & X & X \\ 0 & 0 & 0 & 0 & 0 & X \end{pmatrix}$$

If the rotations are accumulated in Q, the result is the updated decomposition.

Algorithm 3.3 is an implementation of this procedure. Some comments.

- An operation count can be derived as usual by integration. However, because the loops are only two deep, it is easier to use the *method of areas*. Specifically, consider the following diagram.

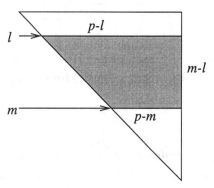

The shaded portion represents the part of the matrix to which the plane rotations are applied during the reduction of the spike and the return to triangular form. Since each application represents a flrot (2 fladd+4 flmlt), the number of flrots is equal to twice the area of the shaded portion—i.e., $(2p-\ell-m)(m-\ell)$ flrot. Finally, the algorithm generates a total of $2(m-\ell)$ plane rotations, each of which must be applied to the columns of Q for a count of $2(m-\ell)n$ flrot. To summarize:

Algorithm 3.3 requires

1. $(2p-\ell-m)(m-\ell)$ flrot *to update R,* (3.11)
2. $2(m-\ell)n$ flrot *to update Q.*

- According to the table (3.1) in the introduction to this section, we need to also show how to update the QR factorization and the R-factor under column interchanges. However, Algorithm 3.3 works for the QR factorization, since the rotations are applied only to the first p columns of Q. The operation count remains the same. For the R-factor all we have to do is suppress the updating of Q (statements 6 and 11). The operation count is now equal to item 1 in (3.11).

- When the algorithm is applied to two contiguous columns—say columns ℓ and $\ell+1$—the operation count is $(p-\ell+n)$ flrot, about one half the value given in (3.11). This is because the reduction of the spike is bypassed. In particular, if we want to move a column to a position in X and do not care what becomes of the other columns, it will be faster to implement it as a sequence of contiguous interchanges. For example, in moving column ℓ to the east side of the matrix, the code

   1. **for** $k = l$ **to** $p-1$
   2.   *qrdexch*(R, Q, k, $k+1$)
   3. **end for** k

(3.12)

is preferable to

   1. *qrdexch*(R, Q, k, p)

SEC. 3. UPDATING

Of course, if we have to do this task frequently, it may pay to code it explicitly to avoid the overhead in invoking *qrdexch*—especially if p is small.

The following is an application of column exchanges.

Example 3.5. We have already observed [see (2.4)] that the R-factor

$$\begin{pmatrix} R & z \\ 0 & \rho \end{pmatrix} \qquad (3.13)$$

of the augmented least squares matrix $(X \; y)$ contains the solution of the least squares problem $\|y - Xb\|_2^2 = \min$ in the form $b = R^{-1}z$. Moreover, the residual sum of squares is ρ^2.

Actually, we can use the same R-factor to solve several related problems. For if we partition the augmented least squares matrix in the form $(X_1 \; X_2 \; y)$ and the R-factor correspondingly in the form

$$\begin{pmatrix} R_{11} & R_{12} & z_1 \\ 0 & R_{22} & z_2 \\ 0 & 0 & \rho \end{pmatrix},$$

then it is easy to see that $R_{11}^{-1} z_1$ is the solution of the least squares problem $\|y - X_1 \hat{b}\|_2^2 = \min$. Moreover, the residual sum of squares is $\rho^2 + \|z_2\|_2^2$. Thus from the single R-factor (3.13) we can compute any least squares solution corresponding to an initial set of columns. By our updating procedures, we can make that initial set anything we like. In particular, column exchanges in the augmented R-factor represent a backward stable alternative to the sweep operator (Algorithm 3.2).

3.3. REMOVING A COLUMN

In this section we will consider the problem of updating a QR decomposition after a column X has been removed. It is easy to remove the last column. Let the QR decomposition of X be partitioned in the form

$$Q^T(X_1 \; x_p) = \begin{pmatrix} R_{11} & r_{1p} \\ 0 & \rho_{pp} \\ 0 & 0 \end{pmatrix}.$$

Then

$$Q^T X_1 = \begin{pmatrix} R_{11} \\ 0 \end{pmatrix}$$

is a QR decomposition of X. Computationally, we just throw away the last column of R.

Given a QR decomposition this algorithm returns the QR decomposition after the ℓth column has been removed.

1.   *qrdrmcol*(R, Q, ℓ)
2.   **for** $j = l$ **to** $p-1$
3.   *qrdxch*(R, Q, j, $j+1$)
4.   **end for** j
5. $R = R[1{:}p{-}1, 1{:}p{-}1]$
6.   **end** *qrdrc*

Algorithm 3.4: QR update: removing columns

———————⋄———————

The general procedure for removing the ℓth column is to move it to the extreme east and drop the last column of R. Algorithm 3.4 uses *qrdexch* to interchange the ℓth row into the last position [see (3.12)] and then adjust R. Two comments.

- The method of areas gives an operation count of

  1. $\frac{1}{2}(p-\ell)^2$ flrot to reduce R,
  2. $(p-\ell)n$ flrot to update Q.

- The same algorithm will update a QR factorization $X = Q_X R$; however, the last column of Q_X must also be dropped from the factorization. In Cholesky updating, where only R is available, forget about Q.

3.4. APPENDING COLUMNS

Up to now our algorithms have been essentially the same for a full decomposition, a factorization, and an R-factor. The two algorithms for appending a column differ in detail, and we must treat them separately.

Appending a column to a QR decomposition

Let us suppose we want to append a column x to a QR decomposition of X. We have

$$Q^{\mathrm{T}}(X \;\; x) = \begin{pmatrix} R & Q_X^{\mathrm{T}} x \\ 0 & Q_\perp^{\mathrm{T}} x \end{pmatrix}.$$

If H is a Householder transformation such that $H(Q_\perp^{\mathrm{T}} x) = \rho \mathbf{e}_1$, then

$$\mathrm{diag}(I, H) Q^{\mathrm{T}}(X \;\; x) = \begin{pmatrix} R & Q_X^{\mathrm{T}} x \\ 0 & \rho \\ 0 & 0 \end{pmatrix}$$

is a QR decomposition of $(X \;\; x)$.

SEC. 3. UPDATING

Given a QR decomposition of X and a vector X, this algorithm computes a QR decomposition of $(X\ x)$.

1. **qrdappcol**(R, Q, x)
2. $x = Q^\mathrm{T} * x$
3. $R[1{:}p, p{+}1] = x[1{:}p]$
4.   **housegen**($x[p{+}1{:}n]$, u, $R[p{+}1, p{+}1]$)
5. $v = Q[:, p{+}1{:}n] * u$
6. $Q[:, p{+}1{:}n] = Q[:, p{+}1{:}n] - v * u^\mathrm{T}$
7. **end** *qrddappcol*

Algorithm 3.5: Append a column to a QR decomposition

———◇———

Algorithm 3.5 is an implementation of this scheme. It requires

$n(3n-2p)$ flam.

Once a column has been appended it can be moved to anywhere in the matrix using Algorithm 3.3.

Appending a column to a QR factorization

A column can be appended to a column by applying the Gram–Schmidt algorithm with reorthogonalization. Specifically, if q, r, and ρ are the output of Algorithm 1.13, then

$$(X\ x) = (Q\ q)\begin{pmatrix} R & r \\ 0 & \rho \end{pmatrix}$$

is the updated QR factorization.

3.5. APPENDING A ROW

The problem considered here is given a QR decomposition of X, compute the QR decomposition of X when a row x^T has been appended to X. To solve it, augment the QR decomposition of X as follows:

$$\begin{pmatrix} Q_X^\mathrm{T} & 0 \\ Q_\perp^\mathrm{T} & 0 \\ 0 & 1 \end{pmatrix}\begin{pmatrix} X \\ x^\mathrm{T} \end{pmatrix} = \begin{pmatrix} R \\ 0 \\ x^\mathrm{T} \end{pmatrix}.$$

If we generate an orthogonal transformation P such that

$$P\begin{pmatrix} R \\ 0 \\ x^\mathrm{T} \end{pmatrix} = \begin{pmatrix} \hat{R} \\ 0 \\ 0 \end{pmatrix},$$

where $\hat{R}$ is upper triangular, then $\hat{R}$ is the R-factor of the updated decomposition and

$$\begin{pmatrix} Q_X & Q_\perp & 0 \\ 0 & 0 & 1 \end{pmatrix} P^{\mathrm{T}}$$

is the updated Q factor.

We will use plane rotations to update R. The procedure is described in the following sequence of diagrams. (In them we ignore the zero rows between R and x^{T}.)

$$\begin{pmatrix} X & X & X & X \\ 0 & X & X & X \\ 0 & 0 & X & X \\ 0 & 0 & 0 & X \\ \hat{x} & X & X & X \end{pmatrix} \xrightarrow{P_{15}} \begin{pmatrix} X & X & X & X \\ 0 & X & X & X \\ 0 & 0 & X & X \\ 0 & 0 & 0 & X \\ 0 & \hat{x} & X & X \end{pmatrix} \xrightarrow{P_{25}} \begin{pmatrix} X & X & X & X \\ 0 & X & X & X \\ 0 & 0 & X & X \\ 0 & 0 & 0 & X \\ 0 & 0 & \hat{x} & X \end{pmatrix} \xrightarrow{P_{35}}$$

$$\begin{pmatrix} X & X & X & X \\ 0 & X & X & X \\ 0 & 0 & X & X \\ 0 & 0 & 0 & X \\ 0 & 0 & 0 & \hat{x} \end{pmatrix} \xrightarrow{P_{45}} \begin{pmatrix} X & X & X & X \\ 0 & X & X & X \\ 0 & 0 & X & X \\ 0 & 0 & 0 & X \\ 0 & 0 & 0 & 0 \end{pmatrix}$$

The transformations P_{ij} are accumulated in the matrix

$$\begin{pmatrix} Q_X & Q_\perp & 0 \\ 0 & 0 & 1 \end{pmatrix}.$$

Algorithm 3.6 implements this scheme. The last column of the updated Q is accumulated in a temporary scratch vector to make the algorithm easy to modify. Three comments.

- The algorithm requires

    1. $\frac{1}{2}p^2$ flrot to update R,
    2. np flrot to update Q.

- The algorithm can be adapted to update the QR factorization — simply delete statements 3 and 10. The operation count remains the same.

- If we remove statements 2, 3, and 10, the algorithm updates the R-factor at a cost of $\frac{1}{2}p^2$ flrot. This algorithm, also called Cholesky updating, is probably the most widely used QR updating algorithm and is worth setting down for later reference.

    1. **for** $k = 1$ **to** p
    2. $\quad$ rotgen($R[k,k]$, $x[k]$, c, s)
    3. $\quad$ rotapp($R[k, k{+}1{:}p]$, $x[k{+}1{:}p]$, c, s)
    4. **end for** k

(3.14)

Given a QR decomposition of X, this algorithm computes the QR decomposition of
$$\begin{pmatrix} X \\ x^{\mathrm{T}} \end{pmatrix}.$$

1. *qrdapprow*(R, Q, x)
2. $\quad Q[n+1, 1{:}p] = 0$
3. $\quad Q[n+1, p+1{:}n] = 0$
4. $\quad t = \mathbf{e}_{n+1}$
5. $\quad$ **for** $k = 1$ **to** p
6. $\quad\quad$ *rotgen*($R[k,k]$, $x[k]$, c, s)
7. $\quad\quad$ *rotapp*($R[k, k+1{:}p]$, $x[k+1{:}p]$, c, s)
8. $\quad\quad$ *rotapp*($Q[{:}; k]$, t, c, s)
9. $\quad$ **end for** k
10. $\quad Q[{:}, n+1] = t$
11. **end** *qrdapprow*

Algorithm 3.6: Append a row to a QR decomposition

⎯⎯⎯⎯⎯⎯⎯⎯⎯⎯◇⎯⎯⎯⎯⎯⎯⎯⎯⎯⎯

An important variant of the problem of appending a row is to append a block of rows — a problem called *block updating*. The general idea is the same as for Algorithm 3.6: append the block to R and reduce to triangular form, accumulating the transformations in Q (if required). The transformations of choice will generally be Householder transformations.

3.6. REMOVING A ROW

In this subsection we will consider the problem of QR updating when a row is removed from the original matrix. Because a row permutation in X corresponds to the same row interchange in Q, we may assume without loss of generality that the row in question is the last. Here it will be convenient to change notation slightly and let

$$\begin{pmatrix} X \\ x^{\mathrm{T}} \end{pmatrix}$$

be the matrix whose last row is to be removed. The algorithms differ depending on what is being updated — the full decomposition, the factorization, or the R-factor — and we will treat each case separately.

Removing a row from a QR decomposition

To derive an algorithm, let

$$\begin{pmatrix} X \\ x^T \end{pmatrix} = Q \begin{pmatrix} R \\ 0 \\ 0 \end{pmatrix},$$

in which the bottom zero in the matrix on the right is a row vector. Suppose we can determine an orthogonal matrix P such that

$$\mathbf{e}_n^T Q P = \mathbf{e}_n^T$$

(i.e., the last row of QP is $\mathbf{e}_n^T$), and

$$P^T \begin{pmatrix} R \\ 0 \\ 0 \end{pmatrix} = \begin{pmatrix} \hat{R} \\ 0 \\ w^T \end{pmatrix},$$

where $\hat{R}$ is upper triangular. The first of these conditions implies that QP^T has the form

$$\begin{pmatrix} \hat{Q} & v \\ 0 & 1 \end{pmatrix},$$

and since Q is orthogonal, v must be zero. It follows that

$$\begin{pmatrix} X \\ x^T \end{pmatrix} = (QP)P^T \begin{pmatrix} R \\ 0 \\ 0 \end{pmatrix} = \begin{pmatrix} \hat{Q} & 0 \\ 0 & 1 \end{pmatrix} \begin{pmatrix} \hat{R} \\ 0 \\ w^T \end{pmatrix}.$$

From this it follows that

$$X = \hat{Q} \begin{pmatrix} \hat{R} \\ 0 \end{pmatrix}$$

is the required QR decomposition of X.

The problem then is how to reduce the last row of Q in such a way that the transformations do not destroy the triangularity of R. It turns out that if we introduce zeros from left to right — that is, with a sequence of plane rotations $P_{n,n-1}, \ldots, P_{n,1}$, where $P_{n,j}$ annihilates the jth element — the matrix $\hat{R}$ will be upper triangular. (The verification of this fact is left as an exercise.)

Algorithm 3.7 implements this scheme. The reduction of the last row of Q is divided into two parts: the reduction $Q[n, p{+}1{:}n]$, in which R does not enter, and the reduction of $Q[n, 1{:}p]$, in which it does. Two comments.

- The algorithm requires

  $(n^2 + \tfrac{1}{2}p^2)$ flrot,

Sec. 3. Updating 343

Given a QR decomposition of

$$\begin{pmatrix} X \\ x^{\mathrm{T}} \end{pmatrix}$$

this algorithm computes a QR decomposition of X.

1. $qrdrmrow(R, Q)$
2.   for $j = n-1$ to $p+1$ by -1
3. $rotgen(Q[n,n], Q[n,j], c, s)$
4. $rotapp(c, s, Q[1{:}n{-}1, n], Q[1{:}n{-}1, j])$
5.   end for j
6. $w[1{:}p] = 0$
7.   for $j = p$ to 1 by -1
8. $rotgen(Q[n,n], Q[n,j], c, s)$
9. $rotapp(c, s, Q[n, 1{:}n{-}1], Q[j, 1{:}n{-}1])$
10. $rotapp(c, s, w[j{:}p], R[j, j{:}p])$
11.   end for j
12. $Q = Q[1{:}n{-}1, 1{:}n{-}1]$
13. end $qrdrmrow$

Algorithm 3.7: Remove the last row from a QR decomposition

———————⋄———————

with $n(n-p)$ flrot for the first loop.

- In an industrial strength implementation, the reduction in the first loop would be done by a Householder transformation. Specifically, we would replace the loop by

1. $housegen(Q[n, p{+}1{:}n]*\mathcal{f}, u, trash)$
2. $u = \mathcal{f}*u$
3. $v = Q[:, p{+}1{:}n]*u$
4. $Q[:, p{+}1{:}n] = Q[:, p{+}1{:}n] - v*u^{\mathrm{T}}$

Here $\mathcal{f}$ is the identity with its columns in reverse order [see (2.10), Chapter 1]. Its effect is to reverse the order of the components of any vector it multiplies. It is necessary because we are performing a backward reduction. The substitution saves $2n(n-p)$ flmlt.

Removing a row from a QR factorization

The algorithm derived here is a variant of the algorithm for removing a row from a QR decomposition. Let

$$\begin{pmatrix} X \\ x^{\mathrm{T}} \end{pmatrix} = Q_X R$$

be the factorization in question. Suppose that we can find a vector q and an orthonormal matrix P such that

1. $(Q_X \ q)$ is orthonormal,

2. $\mathbf{e}_{p+1}^T (Q_X \ q) P = \mathbf{e}_{p+1}^T$,

3. $P \begin{pmatrix} R \\ 0 \end{pmatrix} = \begin{pmatrix} \hat{R} \\ w^T \end{pmatrix}$, where $\hat{R}$ is upper triangular.

Then arguing as above

$$(Q \ q) P = \begin{pmatrix} \hat{Q}_X & 0 \\ 0 & 1 \end{pmatrix}. \tag{3.15}$$

Hence

$$\begin{pmatrix} X \\ x^T \end{pmatrix} = (Q \ q) P P^T \begin{pmatrix} R \\ 0 \end{pmatrix} = \begin{pmatrix} \hat{Q}_X & 0 \\ 0 & 1 \end{pmatrix} \begin{pmatrix} \hat{R} \\ w^T \end{pmatrix},$$

and $X = \hat{Q}_X \hat{R}$ is the QR factorization of X.

To find q observe that by (3.15) the column space of the augmented matrix $(Q \ q)$ contains the vector $\mathbf{e}_n$. Now Q_X already contains $P_X \mathbf{e}_n$ (since $P_X \mathbf{e}_n = Q_X(Q_X^T \mathbf{e}_n)$ is a linear combination of the columns of Q_X). Thus if we take $q = P_\perp \mathbf{e}_n / \|P_\perp \mathbf{e}_n\|_2$, then q will be orthogonal to the column space of Q_X, and the column space of $(Q \ q)$ will contain the vector $\mathbf{e}_n$. We can use the Gram–Schmidt algorithm with reorthogonalization to compute q (Algorithm 1.13).

Algorithm 3.8 implements this method. Two comments.

- The algorithm has the following operation count.

    1. $2knp$ flam for the generation of q. Here k is the number of orthogonalizations performed in *gsreorthog*.
    2. $(np + \frac{1}{2}p^2)$ flrot for the updating of Q and R.

If two orthogonalizations are performed in *gsreorthog*, the total operation count is

$$(4np + p^2) \text{ fladd} + (6np + 2p^2) \text{ flmlt}.$$

- If $P_\perp \mathbf{e}_n = 0$, the procedure sketched above breaks down, since $P_\perp \mathbf{e}_n$ cannot be normalized. However, in that case $\mathbf{e}_n$ is already in the column space of Q_X, and any normalized vector that is orthogonal to $\mathcal{R}(Q_X)$ will do. That is precisely what is returned by *gsreorthog*.

SEC. 3. UPDATING

Given a QR factorization of

$$\begin{pmatrix} X \\ x^{\mathrm{T}} \end{pmatrix}$$

this algorithm computes a QR factorization of X. It uses Algorithm 1.13.

1.   *qrfrmrow*(R, Q)
2.   *gsreorthog*(Q, $\mathbf{e}_n$, q, r, ρ)
3. $w[1{:}p] = 0$
4.   **for** $j = p$ **to** 1 **by** -1
5.   *rotgen*($q[n]$, $Q[n,j]$, c, s)
6.   *rotapp*(c, s, $q[1{:}n{-}1]$, $Q[1{:}n{-}1,j]$)
7.   *rotapp*(c, s, $w[j{:}p]$, $R[j,j{:}p]$)
8.   **end for** j
9. $Q = Q[1{:}n{-}1,:]$
10.   **end** *qrfrmrow*

Algorithm 3.8: Remove the last row from a QR factorization

---◇---

Removing a row from an R-factor (Cholesky downdating)

We are now going to consider the problem of updating the R-factor of a QR decomposition of X when a row x^{T} is removed from X — a problem generally called *downdating*. This is a difficult problem, and what makes it hard is the absence of any information other than the matrix R and the vector x^{T}. In fact, we can make X disappear altogether. For if $A = X^{\mathrm{T}}X = R^{\mathrm{T}}R$, then R is the Cholesky factor of A. Since the removal of x^{T} from X results in the cross-product matrix $A - xx^{\mathrm{T}}$, the problem is equivalent to updating a Cholesky factor when the original matrix is altered by subtracting a rank-one semidefinite matrix. In this form of the problem, which is known as *Cholesky downdating*, there is no matrix X, and the vector x^{T} can be arbitrary, as long as $A - xx^{\mathrm{T}}$ is positive definite.

Let us suppose that $\hat{A} = A - xx^{\mathrm{T}}$ is positive definite, so that it has a Cholesky decomposition $\hat{A} = \hat{R}^{\mathrm{T}}\hat{R}$. The key to deriving an algorithm for calculating $\hat{R}$ from R is to recognize that we can obtain R from $\hat{R}$ by updating. Specifically, if we use Algorithm 3.6 to append x^{T} to $\hat{R}$, we get an orthogonal matrix P such that

$$P^{\mathrm{T}} \begin{pmatrix} \hat{R} \\ x^{\mathrm{T}} \end{pmatrix} = \begin{pmatrix} R \\ 0 \end{pmatrix}.$$

Let us consider the first step of this algorithm, in which we generate a rotation from $\hat{\rho}_{11}$ and ξ_1 and apply it to the first row of X and $\hat{R}$. We can represent this step in

partitioned form as follows:
$$\begin{pmatrix} c & 0 & s \\ 0 & I & 0 \\ -s & 0 & c \end{pmatrix} \begin{pmatrix} \hat{\rho}_{11} & \hat{r}_{12}^T \\ 0 & R_{22} \\ \xi_1 & x_2^T \end{pmatrix} = \begin{pmatrix} \rho_{11} & r_{12}^T \\ 0 & \hat{R}_{22} \\ 0 & \tilde{x}_2^T \end{pmatrix}.$$

This relation, it turns out, is sufficient to allow us to derive $\hat{\rho}_{11}$, c, s, $\hat{r}_{12}^T$, and $\tilde{x}_2$ from $(\rho_{11} \; r_{12}^T)$ and x^T.

We begin by observing that because we are working with an orthogonal transformation, we have $\hat{\rho}_{11}^2 + \xi_1^2 = \rho_{11}^2$ or

$$\hat{\rho}_{11} = \sqrt{\rho_{11}^2 - \xi_1^2}.$$

Knowing $\hat{\rho}_{11}$, we may determine c and s in the form

$$c = \frac{\hat{\rho}_{11}}{\rho_{11}} \quad \text{and} \quad s = \frac{\xi_1}{\rho_{11}}.$$

From the relation

$$r_{12}^T = c\hat{r}_{12}^T + sx_2^T,$$

we get

$$\hat{r}_{12}^T = \frac{r_{12}^T - sx_2^T}{c}.$$

Finally,

$$\tilde{x}_2^T = cx_2^T - s\hat{r}_{12}^T.$$

Thus we have computed the first row $\hat{R}$. Since we know $\tilde{x}_2$, we may repeat the process on the matrix

$$\begin{pmatrix} R_{22} \\ \tilde{x}_2^T \end{pmatrix}$$

to get the second row of $\hat{R}$. And so on.

Algorithm 3.9 is an implementation of this scheme. Three comments.

- This algorithm is sometimes called *the method of mixed rotations*. For more on nomenclature, see the notes and references.

- The algorithm takes p^2 fladd and $2p^2$ flmlt to compute $\hat{R}$. This is exactly the count for the updating method in Algorithm 3.6.

- The algorithm breaks down if the quantity $R[k,k]^2 - x[k]^2$ in statement 3 is negative, in which case the matrix $R^T R - xx^T$ is indefinite. The algorithm also fails if $R[k,k]^2 - x[k]^2$ is zero, since in that case the divisor c is also zero. It might be thought that the appearance of a small c is associated with instability; however, it can only hurt if the problem itself is ill conditioned. We will return to the stability of the algorithm at the end of this section.

Sec. 3. Updating

Given a triangular matrix R with positive diagonal elements and a vector x such that $\hat{A} = R^T R - xx^T$ is positive definite, this algorithm overwrites R with the Cholesky factor of $\hat{A}$.

```
1.    chdd(R, x)
2.        for k = 1 to p
3.            hrkk = √(R[k,k]² - x[k]²)
4.            c = hrkk/R[k,k];  s = x[k]/R[k,k]
5.            R[k,k] = hrkk
6.            R[k, k+1:p] = c⁻¹*(R[k, k+1:p] - s*x[k+1:p])
7.            x[k+1:p] = c*x[k+1:p] - s*R[k, k+1:p]
8.        end for k
9.    end chdd
```

Algorithm 3.9: Cholesky downdating

Downdating a vector

A special case of Cholesky downdating occurs when $R = \rho$ is a scalar and $X = x$ is a vector. In this case $\rho = \|x\|_2$, and downdating this quantity amounts to recomputing the norm after a component ξ has been removed from x. An important difference between vector downdating and downdating a general Cholesky factor is that it is sometimes feasible to retain x, which can be used to recover from a failure of the algorithm. A second difference is that in applications we may not need a great deal of accuracy in the norm. (Pivoted orthogonal triangularization is an example. See §2.1, Chapter 5.)

In principle the norm can be downdated by the formula

$$\hat{\rho} = \sqrt{\rho^2 - \xi^2}.$$

However, in a sequence of downdates this procedure may break down. To see why, let $\bar{\rho}$ be the norm of the original vector and let ρ be the norm of the current vector x. If y is the vector of components we have already removed from x then

$$\rho^2 = \bar{\rho}^2 - \|y\|_2^2 = \bar{\rho}^2 \left(1 - \frac{\|y\|_2^2}{\bar{\rho}^2}\right).$$

From this equation we see that any attempt to reduce $\rho^2/\bar{\rho}^2$ to a number near the rounding unit must produce inaccurate results. For in that case, the quantity $\|y\|_2^2/\bar{\rho}^2$ must be very near one, and the slightest change in either $\|y\|_2$ or $\bar{\rho}$ will completely change the result.

The cure for this problem is to make a tentative computation of $\hat{\rho}$. If the ratio of $\hat{\rho}/\bar{\rho}$ is satisfactory, use the computed value. Otherwise recompute the norm from $\hat{x}$. The details are contained in Algorithm 3.10.

This algorithm takes the norm ρ of a vector x and overwrites it with the norm of the vector $\hat{x}$ which is obtained by a deleting the component ξ of x. The algorithm uses and updates a quantity $\bar{\rho}$, which should be initialized to $\|x\|_2$ on first calling.

1. $vecdd(\rho, \bar{\rho}, \xi, \hat{x})$
2.   **if** $(\rho = 0)$ **return** ; **fi**
3. $\mu = \max\{0, 1-(\xi/\rho)^2\}$
4.   **if** $(\mu*(\rho/\bar{\rho})^2 \geq 100*\epsilon_M)$
5. $\rho = \rho*\sqrt{\mu}$
6.   **else**
7. $\bar{\rho} = \rho = \|\hat{x}\|_2$
8.   **end if**
9. **end** *vecdd*

Algorithm 3.10: Downdating the norm of a vector

The quantity μ in the algorithm is a reduction factor that tells how much $\|x\|_2^2$ is reduced by the deletion of ξ. The total reduction of the square of the norm of the original vector is then $\mu*(\rho/\bar{\rho})^2$. If this quantity is sufficiently greater than the rounding unit, the value of ρ is downdated. Otherwise, ρ is computed directly from $\hat{x}$, and $\bar{\rho}$ is reinitialized to ρ. The number 100 in statement 4 defines what is meant by "sufficiently great." It enforces about two decimal digits of accuracy in ρ.

3.7. GENERAL RANK-ONE UPDATES

In this subsection we will consider the problem of computing the QR decomposition and QR factorization of the matrix $X + uv^T$. As might be expected, the problem of updating the factorization is more complicated than updating the full decomposition. Accordingly we will present the former in detail and sketch the latter.

Updating a factorization

Let $X = Q_X R$ be the QR factorization of X. Let

$$(X \ u) = (Q_X \ q) \begin{pmatrix} R & t \\ 0 & \tau \end{pmatrix}$$

be the QR factorization of $(X \ u)$. This factorization can be obtained by applying Gram–Schmidt orthogonalization to the vector u (Algorithm 1.13). It follows that

$$X + uv^T = (Q_X \ q) \left[\begin{pmatrix} R \\ 0 \end{pmatrix} + \begin{pmatrix} t \\ \tau \end{pmatrix} v^T \right]. \tag{3.16}$$

SEC. 3. UPDATING

Suppose we determine plane rotations $P_{k,k+1}$ in the $(k, k+1)$-plane such that

$$P_{12} \cdots P_{p-1,p} P_{p,p+1} \begin{pmatrix} t \\ \tau \end{pmatrix} = \nu e_1.$$

Then

$$X + uv^T = (Q_X \; q) P P^T \left[\begin{pmatrix} R \\ 0 \end{pmatrix} + \begin{pmatrix} t \\ \tau \end{pmatrix} v^T \right] = (Q_X \; q) P (H + \nu e_1 v^T).$$

Since $P^T = P_{12} \cdots P_{p-1,p} P_{p,p+1}$, the matrix $H + \nu e_1 v^T$ has the form illustrated below for $p = 5$:

$$H + \nu e_1 v^T = \begin{pmatrix} X & X & X & X & X \\ X & X & X & X & X \\ 0 & X & X & X & X \\ 0 & 0 & X & X & X \\ 0 & 0 & 0 & X & X \\ 0 & 0 & 0 & 0 & X \end{pmatrix}.$$

If we now determine an orthogonal matrix U such that

$$U^T(H + \tau e_1 v^T) = \begin{pmatrix} \hat{R} \\ 0 \end{pmatrix},$$

where $\hat{R}$ is upper triangular and set $\hat{Q}_X$ equal to the first p columns of $(Q_X \; q) P U$, then

$$X + uv^T = \hat{Q}_X \hat{R}$$

is the required factorization. The matrix U can be calculated as in Algorithm 1.8 as a product of plane rotations.

Algorithm 3.11 is a straightforward implementation of this scheme. It requires $(2np + p^2)$ flrot.

Updating a decomposition

The method for updating a decomposition is similar to the method for updating a factorization. In analogy with (3.16), write

$$X + uv^T = Q \left[\begin{pmatrix} R \\ 0 \end{pmatrix} + tv^T \right],$$

where $t = Q^T u$. We then proceed to reduce t by plane rotations from the bottom up to a multiple of e_1, accumulating the transformations in Q and R. The result is a decomposition of the form

$$X + uv^T = QP(H + \nu e_1 v^T),$$

where $H + \nu e_1 v^T$ is zero below its first subdiagonal. If we reduce this matrix to upper triangular form, accumulating the transformations in QP, we get the updated decomposition.

Let $X = QR$ be the QR factorization of X. This algorithm overwrites Q and R with the QR factorization of $X + uv^T$.

1.   *qrfrnk1*(R, Q, u, v)
2.   *gsreorthog*(Q, u, $Q[:,p+1]$, t, τ)
3. $R[p+1,:] = 0$; $t[p+1] = \tau$
4.   **for** $k = p$ **to** 1 **by** -1
5.   *rotgen*($t[k]$, $t[k+1]$, c, s)
6.   *rotapp*(c, s, $R[k,k{:}p]$, $R[k+1,k{:}p]$)
7.   *rotapp*(c, s, $Q[:,k]$, $Q[:,k+1]$)
8.   **end for** k
9. $R[1,:] = R[1,:] + t[1] * v^T$
10.   **for** $k = 1$ **to** p
11.   *rotgen*($R[k,k]$, $R[k+1,k]$, c, s)
12.   *rotapp*(c, s, $R[k,k+1{:}p]$), $R[k+1,k+1{:}p]$)
13.   *rotapp*(c, s, $Q[:,k]$), $Q[:,k+1]$)
14.   **end for** k
15. $Q = Q[:,1{:}p]$; $R = R[1{:}p, 1{:}p]$
16.   **end** *qrfrnk1*

Algorithm 3.11: Rank-one update of a QR factorization

3.8. NUMERICAL PROPERTIES

In discussing the numerical properties of our updating algorithms, we must distinguish between the algorithm for downdating an R-factor and all the rest. The reason can be seen from a simple example.

Consider the matrix

$$X = \begin{pmatrix} 1 \\ \eta \end{pmatrix},$$

where $\eta < \sqrt{\epsilon_M}$ so that

$$\text{fl}(1 + \eta^2) = 1.$$

Then the correctly rounded QR factorization of X is

$$X = \begin{pmatrix} 1 \\ \eta \end{pmatrix} (1).$$

The R-factor (1) contains no information about the number γ. Hence any attempt to downdate by removing the first row of X must fail. On the other hand γ is fully present in the Q-factor, and we can safely use Algorithm 3.8 to remove the first row of X.

Updating

The error results for updating QR decomposition, QR factorizations, and R-factors all have the same flavor. For definiteness, we will consider the QR factorization in detail and then describe briefly how the results apply to the full decomposition and the R-factor.

It is important to keep in mind that updating is not a one-time thing. The decomposition that presents itself to the algorithm will have been computed with error, presumably from a sequence of previous updates. Thus our problem is not to bound the error from a single update but to show how much the update adds to the error already present.

To start things off, we will assume that we have a computed factorization $Q_X R$ that satisfies the following conditions (in some suitable norm $\|\cdot\|$).

1. $\|I - Q_X^T Q_X\| \leq \alpha \epsilon_M$
2. $\|X - Q_X R\| \leq \beta \rho \epsilon_M$

The first inequality bounds the deviation of the columns of Q_X from orthogonality. The second bounds the backward error in the computed decomposition. The number ρ measures the size of the problem in a sense that will become clear a little later.

Now assume that this QR factorization is updated by any of the algorithms of this section to give the new QR factorization $\hat{Q}_X \hat{R}$. Then this factorization satisfies

1. $\|I - \hat{Q}_X^T \hat{Q}_X\| \leq (\alpha + \gamma)\epsilon_M$ \hfill (3.17)
2. $\|\hat{X} - \hat{Q}_X \hat{R}\| \leq (\beta + \delta)\max\{\rho, \|\hat{R}\|\}\epsilon_M$

Here γ and δ are constants that depend on the dimensions of the problem.

To interpret these bounds, let us suppose we perform a sequence of updates on the matrices $X_0, X_1, \ldots$. Then if Q_{X_k} and R_k denote the kth computed factorization, we have

$$\|I - Q_{X_k}^T Q_{X_k}\| \leq (\alpha + \gamma_0 + \cdots + \gamma_k)\epsilon_M.$$

This says that the Q-factors suffer a slow loss in orthogonality. Specifically, if γ is an upper bound on the γ_i from the individual updates, then the loss of orthogonality is bounded by $(\alpha + k\gamma)\epsilon_M$. Thus the deterioration in orthogonality grows at most linearly in the number of updates k.

The bound for the backward error shows that the process has a memory. Specifically, if δ bounds the δ_k for the individual updates, then

$$\|\hat{X} - Q_{X_k} R_k\| \leq (\beta + k\delta)\max\{\|\hat{R}_0\|, \ldots, \|R_k\|\}\epsilon_M.$$

Thus the backward error is small compared not to $\|R_k\|$ but to the norm of the largest R_i encountered in the sequence of updates. If one encounters a very large R-factor, it will introduce a large backward error that stays around to harm subsequent smaller updates.

This situation is analogous to the situation in Example 3.3, where Woodbury's formula was used to update inverses. There a large inverse introduced large errors that propagated to a subsequent inverse. However, there is an important difference. The large inverse resulted from the presence of an ill-conditioned matrix — the Woodbury update cannot pass through an ill-conditioned problem without loosing accuracy. On the other hand, an ill-conditioned R-factor does not have to be large. Consequently, our QR updating algorithms can pass through ill-conditioning with no bad effects.

The same results hold for updating the QR decomposition. The orthogonality deteriorates linearly with k as does the backward error. Large intermediate matrices magnify the error, which propagates to subsequent updates.

At first glance it would seem that these results cannot apply to updating the R-factor, since no matrix Q_X is computed. However, it can be shown that there is an *exactly orthonormal* Q_X such that the backward error bound holds. Thus the above comments also apply to updating R-factors.

For convenience we have presented normwise bounds involving entire matrices. However, if we exclude the general rank-one updating algorithms, the backward error has columnwise bounds analogous to those in Theorem 1.5. For these algorithms the statement about remembering large matrices may be modified to say that large columns are remembered. But one large column does not affect the backward error in the other columns.

Downdating

Algorithm 3.9 for downdating a Cholesky or R-factor is not backward stable. Nonetheless, it has a useful error analysis. Specifically, let the vector x^T be downdated from R to give the matrix $\hat{R}$. Then there is an orthogonal matrix Q such that

$$Q^\mathrm{T} \begin{pmatrix} \hat{R} \\ x^\mathrm{T} \end{pmatrix} = \begin{pmatrix} R \\ 0 \end{pmatrix} + E,$$

where

$$\|E\|_2 \leq \alpha \epsilon_\mathrm{M} \|R\|_2$$

for some constant α depending on the dimensions of the problem.

This is not a backward error analysis, since the last row of E is not associated with the original data. If we define $F = -QE$, then we can write

$$Q^\mathrm{T} \left[\begin{pmatrix} \hat{R} \\ x^\mathrm{T} \end{pmatrix} + F \right] = \begin{pmatrix} R \\ 0 \end{pmatrix}$$

and associate the last row of F with x^T. But in that case the rest of F must be associated with $\hat{R}$, which is not a backward error analysis. We call this kind of a result *relational stability* because a mathematical relation that must hold in exact arithmetic holds approximately in the presence of rounding error.

SEC. 3. UPDATING

Two facts make relational stability important. First, it continues to hold through a sequence of downdates and updates. As with the updating algorithms, the error grows slowly and is proportional to the largest R-factor in the sequence.

Second, if we pass to cross-product matrices, we have that

$$\hat{R}^T\hat{R} = R^T R - xx^T + R^T E_1 + E_1^T R + E^T E,$$

where E_1 consists of the first p rows of E. It follows that the computed R-factor is the R-factor of a perturbation G of the exact downdated matrix $R^T R - xx^T$. The norm of G is bounded by

$$\|G\|_2 \leq (2\alpha + \alpha^2 \epsilon_M)\epsilon_M \|R\|_2^2. \tag{3.18}$$

It follows that if the R-factor is not sensitive to perturbations in the cross-product matrix, the result will be accurate.

The result also suggests a fundamental limitation on downdating. We have seen [see (2.7)] that for $R^T R - xx^T$ to remain positive definite under a perturbation G, the norm of G must be less than the square of the smallest singular value of $R^T R - xx^T$ — call it σ_p^2. From the bound (3.18) this will be true if

$$(2\alpha + \alpha^2 \epsilon_M)\epsilon_M \|R\|_2^2 \leq \sigma_p^2.$$

If $\|R\|_2 \cong \|\hat{R}\|_2$, this relation is essentially the same as

$$\kappa_2^2(\hat{R})(2\alpha + \alpha^2 \epsilon_M)\epsilon_M \leq 1.$$

If $\kappa_2(\hat{R}) \geq 1/\sqrt{\epsilon_M}$, then this inequality fails. In other words, one should not expect to successfully downdate matrices whose condition number is greater that the reciprocal of the square root of the rounding error. In IEEE double-precision arithmetic, this means that one should beware of matrices whose condition number is greater than about 10^8. In fact, with such matrices the downdating Algorithm 3.9 may fail in statement 3 attempting to take the square root of a negative number.

3.9. NOTES AND REFERENCES

Historical

When Legendre introduced the method of least squares [216, 1805], he recommended discarding observations with unacceptably large residuals. He pointed out that it was not necessary to recompute the normal equations from scratch; all one had to do is subtract out the cross products corresponding to the offending residuals. This correction of the normal equations is the first example of an updating algorithm in numerical linear algebra.

In 1823, Gauss [133] showed how to update a least squares solution when an observation is appended or deleted from the problem. However, Gauss did not update the triangular decomposition from the normal equations, and consequently his updating technique could only be applied once.

Modern updating seems to have begun with the simplex method for linear programing, in which the inverse of the matrix of active constraints is updated as the constraints are swapped in and out (e.g., see [85]). Inverse updating is also used in quasi-Newton methods for nonlinear optimization [87, 1959]. The first example of QR updating was given by Golub in the same paper in which he showed how to use Householder transformations to solve least squares problems [148, 1965].

Updating inverses

According to Zielke [355], Woodbury's formula can be found as incidental formulas in papers by Duncan [109, 1944] and Guttman [164, 1946]. Woodbury's formula appeared explicitly in a technical report in 1950 [351]. Earlier Sherman and Morrison gave formulas for special cases [278, 279, 280], and the general method is sometimes called the Sherman–Morrison–Woodbury formula. Although the formula has its numerical drawbacks (Example 3.3), it is an indispensable theoretical tool.

The sweep operator was introduced by Beaton [23, 1964]. For a tutorial see [156]. The method was used by Furnival and Wilson [128] to select optimal subsets of regression variables. The observation that the errors do not grow exponentially—as would be suggested by a naive analysis—is due to the author; but a formal analysis is lacking. The operator is closely related to Gauss–Jordan elimination, which is discussed in §1.6, Chapter 3.

Updating

The collection of updating algorithms has been assembled from various sources. Algorithms for moving around columns may be found in LINPACK [99]. The algorithms for updating a QR factorization are due to Daniel, Gragg, Kaufman, and Stewart [84]. The algorithm for appending a row to an R-factor is due to Golub [148], although he used 2×2 Householder transformations rather than plane rotations.

Yoo and Park [352] give an alternative method, based on the relation of Gram–Schmidt orthogonalization and Householder triangularization (Theorem 1.11), for removing a row from a QR factorization.

It is not surprising that stable algorithms should exist for updating QR decompositions and factorizations. If we begin with a stable QR factorization—say $Q_X R = X + E$—we can compute an update stably by reconstituting $X + E$ from Q_X and R, making the modification, and recomputing the factorization. Thus the problem is not one of the existence of stable updating algorithms but of finding algorithms that are both stable and efficient.

There is no formal error analysis of all the updating algorithms presented here, and the results in (3.17) are largely my own concoction. For appending a row, the result follows from the standard error analyses of plane rotations; e.g., [346, pp. 131–143], [142], and [177, §18.5].

Exponential windowing

In signal processing, the rows of X represent a time series. Since only the most recent rows are pertinent to the problem, it is important to discard old rows. This can be done by interleaving updates and downdates, a process called windowing. However, a widely used alternative is to update the configuration

$$\begin{pmatrix} \beta R \\ x^T \end{pmatrix},$$

where $\beta < 1$ is a positive "forgetting factor." This process is known as *exponential windowing*, since the influence of a row decays as β^k.

An error analysis of exponential windowing has been given by Stewart [300]. It is shown that the effects of rounding error remain bounded no matter how many updates have been performed. In a related analysis of a method for updating an approximate singular value decomposition, Moonen [233] considers the tricky problem of maintaining the orthogonality of the Q-factor. See also [234].

Cholesky downdating

There are three algorithms for downdating an R-factor: Saunders' method, the method of hyperbolic rotations, and the method of mixed rotations presented here.

Saunders' method [271, 1972] is the algorithm used by LINPACK [99]. It was shown to be relationally stable by Stewart [293], who introduced the term "downdating."

The method of hyperbolic rotations originated in an observation of Golub [149, 1969] that downdating could be regarded as updating with the row in question multiplied by the square root of -1. When this result is cast in terms of real arithmetic, it amounts to multiplying by transformations of the form

$$\hat{P} = \begin{pmatrix} \hat{c} & -\hat{s} \\ -\hat{s} & \hat{c} \end{pmatrix},$$

where $\hat{c}^2 - \hat{s}^2 = 1$. This implies that $\hat{c} = \cosh t$ and $\hat{s} = \sinh t$ for some t. For this reason matrices of the form $\hat{P}$ are called *hyperbolic rotations*. The method of hyperbolic rotations is not relationally stable, and in sequential application it can give unnecessarily inaccurate results [307].

The method of mixed rotations is due to Chambers [62, 1971], who in transcribing the method of hyperbolic rotations wrote the formulas in the form used here. It is called the method of mixed rotations because one updated component is computed from a hyperbolic rotation and the other from an ordinary rotation. The proof that the method is relationally stable is due to Bojanczyk, Brent, Van Dooren, and de Hoog [48]. The implications of relational stability for sequences of updates and downdates are due to Stewart [307].

The notion of hyperbolic rotations can be extended to Householder-like transformations. For the use of mixed Householder transformations in block updating, see [49].

Downdating a vector

The algorithm given here first appeared as undocumented code in the LINPACK routine SQRDC [99]. It seems to have lived on in other programs as a black box which no one dared tamper with.

5

RANK-REDUCING DECOMPOSITIONS

In this chapter we will be concerned with approximating a matrix X by a matrix whose rank is less than that of X. This problem arises in many applications. Here are three.

- We have seen that a full-rank factorization (Theorem 3.13, Chapter 1) of a matrix X can be stored and manipulated more economically than the matrix itself. When X is not actually deficient in rank, we may be able to substitute a sufficiently accurate low-rank approximation.

- The matrix X may be a perturbation of a "true" matrix $\ddot{X}$ of rank m. The problem here is to determine m along with an approximation $\hat{x}$ of rank m to X. Such problems arise, for example, in signal processing. An important feature of these problems is that the matrix $\ddot{X}$ and its rank m may vary, so that it is necessary to update the approximation.

- The matrix X may be a discretization of a continuous problem that is inherently singular — *ill-posed problems* they are called. In such instances X will have small singular values that will magnify errors in the original problem. One cure is to solve a problem of smaller rank whose singular values are satisfactorily large.

The singular value decomposition provides an elegant solution to our approximation problem. By Theorem 4.32, Chapter 1, we can obtain an optimal approximation of rank m by setting the singular values beyond the mth to zero. Moreover, the singular values themselves will often guide us in choosing m. For example, a deficiency in rank may be signaled by a gap in the singular values (see Example 1.2).

However, the singular value decomposition is expensive to compute and resists updating. Consequently, other decompositions are often used in their place. The purpose of this chapter is to describe algorithms for computing and manipulating these decompositions.

The success of most rank-reducing decompositions is measured against the optimal singular value decomposition. Therefore, in §1 we will consider the mathematical properties of the singular value decomposition, with particular attention being paid to how certain subspaces corresponding to the largest and smallest singular values behave under perturbations. In §2, we will consider decompositions based on orthogonal triangularization. In §3 we will digress and examine algorithms for determining approximate null vectors of a matrix — a process that in the literature goes under the name of *condition estimation*. In §4 we will consider two kinds of updatable, rank-reducing decompositions — the URV and the ULV decompositions. Computational algorithms for the singular value decomposition itself will be treated in the second volume of this series.

Throughout this chapter we will assume that:

X is an $n \times p$ $(n \geq p)$ matrix with the singular value decomposition

$$X = U\Sigma V^T,$$

where

$$\Sigma = \begin{pmatrix} \mathrm{diag}(\sigma_1, \ldots, \sigma_p) \\ 0 \end{pmatrix}, \quad \text{with} \quad \sigma_1 \geq \cdots \geq \sigma_p \geq 0.$$

(1)

Note that we have changed notation slightly from the more conventional notation of §4.3, Chapter 1. There Σ was a diagonal matrix of order p. Here Σ is an $n \times p$ matrix, with $\Sigma[1{:}p, 1{:}p]$ a diagonal matrix containing the singular values. The change is convenient because it puts partitions of U and V on an equal footing.

1. Fundamental Subspaces and Rank Estimation

As we have mentioned, the singular value decomposition is the crème de la crème of rank-reducing decompositions — the decomposition that all others try to beat. Since these decompositions can often be regarded as perturbations of a block singular value decomposition, we will begin with the perturbation theory of the singular value decomposition. We will then apply the results to the important problem of rank determination.

1.1. The perturbation of fundamental subspaces

We begin this subsection by introducing some nomenclature to describe the subspaces associated with the singular value decomposition. We then turn to assessing the accuracy of certain approximations to these subspaces.

Superior and inferior singular subspaces

Let X have the singular value decomposition (1). Let U_1 be formed from some subset of columns of U. Then we will say that $\mathcal{R}(U_1)$ is a *left singular subspace* of X. Analogously, if V_1 is formed from some subset of the columns of V, we say that $\mathcal{R}(V_1)$ is a *right singular subspace* of X. (When X has a multiple singular value, any linear combination of the corresponding singular vectors may also be included in the subspace.)

SEC. 1. FUNDAMENTAL SUBSPACES AND RANK ESTIMATION

We will be particularly concerned with the case where U_1 and V_1 come from the partitions of the form

$$U = (U_1 \ U_2) \quad \text{and} \quad V = (V_1 \ V_2),$$

in which U_1 and V_1 have $m < p$ columns. Then

$$(U_1 \ U_2)^\mathrm{T} X (V_1 \ V_2) = \begin{pmatrix} \Sigma_1 & 0 \\ 0 & \Sigma_2 \end{pmatrix},$$

where Σ_1 contains the m largest singular values of X and Σ_2 the $p-m$ smallest. We will call the subspace spanned by U_1 the *left superior subspace* and call the subspace spanned by U_2 the *left inferior subspace*. Together they will be called the *left fundamental subspaces*. Similarly we will call the subspaces spanned by V_1 and V_2 the *right superior and inferior subspaces*—collectively, the *right fundamental subspaces*.

It should be stressed that the notion of fundamental subspaces is relative to the integer m and requires a gap between σ_m and σ_{m+1} to be well defined (see Theorem 4.28, Chapter 1). However, in rank-reduction problems we will generally have such a gap.

In what follows we will use a pair of simple algebraic relations. Specifically, it is easy to verify that

$$U_1^\mathrm{T} X = \Sigma_1 V_1^\mathrm{T} \quad \text{and} \quad X V_1 = U_1 \Sigma_1. \tag{1.1}$$

The matrix Σ_1 is square, and if $\mathrm{rank}(X) \geq m$ it is nonsingular. Thus (1.1) provides a way of passing from a basis for a left (right) superior subspace to a basis for a left (right) superior subspace.

Approximation of fundamental subspaces

The rank-reducing algorithms of this chapter reduce X to the form

$$X = \begin{pmatrix} S & H \\ G & F \end{pmatrix}, \tag{1.2}$$

where S is of order m and F and G or H (possibly both) are small. If G and H were zero, the left and right fundamental subspaces of X would be spanned by

$$\begin{pmatrix} I_k \\ 0 \end{pmatrix}, \ \begin{pmatrix} 0 \\ I_{n-k} \end{pmatrix}, \quad \text{and} \quad \begin{pmatrix} I_k \\ 0 \end{pmatrix}, \ \begin{pmatrix} 0 \\ I_{p-k} \end{pmatrix}. \tag{1.3}$$

If G or H is nonzero but small, the bases at best approximate the fundamental subspaces. Our concern here will be with assessing their accuracy. In addition we will relate the singular values of S and F to those of X.

We will begin by partitioning the singular vectors of X in the form

$$U = \begin{pmatrix} U_{11} & U_{12} \\ U_{21} & U_{22} \end{pmatrix} \quad \text{and} \quad V = \begin{pmatrix} V_{11} & V_{12} \\ V_{21} & V_{22} \end{pmatrix}, \tag{1.4}$$

where U_{11} and V_{11} are of order m. The columns of these partitions span the fundamental subspaces of X. By Theorem 4.37, Chapter 1, the singular values of U_{12} or U_{21}—they are the same—are the sines of the canonical angles between these subspaces and the column spaces of (1.3). Similarly for V_{12} and V_{12}. We will now show how to bound these quantities.

Theorem 1.1. *Let X be partitioned as in (1.2), where S is of order m, and let the singular vectors of X be partitioned as in (1.4), where U_{11} and V_{11} are of order m. Let*

$$\tau = \inf(\Sigma_1) \text{ or } \inf(S),$$
$$\gamma = \|G\|_2,$$
$$\eta = \|H\|_2,$$
$$\varphi = \|F\|_2.$$

Let

$$s_u = \|U_{21}\|_2 = \|U_{12}\|_2, \quad c_u = \sqrt{1 - s_u^2}$$

be the sine and cosine between the left fundamental subspaces of X and their approximations from (1.3). Similarly, let

$$s_v = \|V_{21}\|_2 = \|V_{12}\|_2, \quad c_v = \sqrt{1 - s_v^2}$$

be the sine and corresponding cosine for the right fundamental subspaces. If

$$\rho = \frac{\varphi}{\tau} < 1,$$

then

$$s_u \leq \frac{1}{1-\rho^2}\left(\frac{\gamma}{\tau} + \rho\frac{\eta}{\tau}\right) \quad \text{and} \quad s_v \leq \frac{1}{1-\rho^2}\left(\frac{\eta}{\tau} + \rho\frac{\gamma}{\tau}\right). \tag{1.5}$$

Moreover, if s_u and s_v are less than one, then $U_{11}, U_{22}, V_{11},$ and V_{22} are nonsingular, and

$$\Sigma_1 = U_{11}^{-1}(S + HV_{21}V_{11}^{-1})V_{11} = U_{11}^T(S + U_{11}^{-T}U_{21}^T G)V_{11}^{-T} \tag{1.6}$$

and

$$\Sigma_2 = U_{22}^{-1}(F + GV_{21}V_{22}^{-1})V_{22} = U_{22}^T(F + U_{22}^{-T}U_{12}H)V_{22}^{-T}. \tag{1.7}$$

Proof. By the max-min characterization (4.41), Chapter 1, of singular values, we have

$$\inf(S) \leq \inf(\Sigma_1).$$

Consequently, the choice $\tau = \inf(\Sigma_1)$ gives the smaller bounds, and it is sufficient to prove the theorem for that choice.

SEC. 1. FUNDAMENTAL SUBSPACES AND RANK ESTIMATION

Since [cf. (1.1)]

$$X \begin{pmatrix} V_{11} \\ V_{21} \end{pmatrix} = \begin{pmatrix} U_{11} \\ U_{21} \end{pmatrix} \Sigma_1,$$

it follows that

$$U_{21}\Sigma_1 = (0 \ I) \begin{pmatrix} S & H \\ G & F \end{pmatrix} \begin{pmatrix} V_{11} \\ V_{21} \end{pmatrix} = GV_{11} + FV_{21}. \tag{1.8}$$

On multiplying by Σ_1^{-1} and taking norms we find that

$$s_u \leq \frac{\gamma c_v + \varphi s_v}{\tau}. \tag{1.9}$$

Similarly, from the relation

$$(U_{11}^T \ U_{21}^T)X = \Sigma_1(V_{11}^T \ V_{21}^T)$$

it follows that

$$s_v \leq \frac{\eta c_u + \varphi s_u}{\tau}. \tag{1.10}$$

If we substitute (1.10) into (1.9) and replace c_u and c_v with the upper bound one, we get

$$s_u \leq \frac{\gamma}{\tau} + \rho \frac{\eta}{\tau} + \rho^2 s_u.$$

Solving this inequality for s_u we get the first bound in (1.5). The second inequality follows similarly.

To establish (1.6) and (1.7), first note that if $s_u, s_v < 1$ then the canonical cosines are all greater than zero. Since the canonical cosines are the singular values of the matrices U_{11}, U_{22}, V_{11}, and V_{22}, these matrices are nonsingular.

We will now establish the first expression in (1.6). Multiply out the relation

$$\begin{pmatrix} U_{11}^T & U_{21}^T \\ U_{12}^T & U_{22}^T \end{pmatrix} \begin{pmatrix} S & H \\ G & F \end{pmatrix} \begin{pmatrix} V_{11} & V_{21} \\ V_{12} & V_{22} \end{pmatrix} = \begin{pmatrix} \Sigma_1 & 0 \\ 0 & \Sigma_2 \end{pmatrix}$$

to get

1. $\Sigma_1 = U_{11}^T SV_{11} + U_{11}^T HV_{21} + U_{21}^T GV_{11} + U_{21}^T FV_{21},$
2. $\Sigma_2 = U_{12}^T SV_{12} + U_{12}^T HV_{22} + U_{22}^T GV_{12} + U_{22}^T FV_{22},$
3. $0 = U_{11}^T SV_{12} + U_{11}^T HV_{22} + U_{21}^T GV_{12} + U_{21}^T FV_{22},$
4. $0 = U_{12}^T SV_{11} + U_{12}^T HV_{21} + U_{22}^T GV_{11} + U_{22}^T FV_{21}.$

From the fourth equation we find that

$$GV_{11} + FV_{21} = U_{22}^{-T} U_{12}^T (SV_{11} + HV_{21}).$$

If we substitute this expression into the first equation, we get

$$\Sigma_1 = (U_{11}^T - U_{21}^T U_{22}^{-T} U_{12}^T)(S + HV_{21}V_{11}^{-1})V_{11}.$$

Now $U_{11}^T - U_{21}^T U_{22}^{-T} U_{12}^T$ is the Schur complement of U_{22}^T in U^T, and by Theorem 1.6, Chapter 3, it is the inverse of the (1,1)-block of U^{-T}. But by the orthogonality of U^T that block is simply U_{11}. Hence $U_{11}^T - U_{21}^T U_{22}^{-T} U_{12}^T = U_{11}^{-1}$, and the first expression in (1.6) follows directly.

The other expressions for Σ_1 and Σ_2 follow by similar arguments. ∎

Let us examine what this theorem says in more detail. In what follows, we will let

$$\epsilon = \left\| \begin{pmatrix} 0 & H \\ G & 0 \end{pmatrix} \right\|_2$$

and consider the behavior of our bounds as $\epsilon \to 0$.

- The quantity ρ essentially represents a relative gap in the singular values of X. Since X is a perturbation of $\text{diag}(S, F)$ of norm ϵ, by Corollary 4.31, Chapter 1, the singular values of S and F lie within ϵ of those of X. In particular, if $\sigma_{m+1} + \epsilon < \sigma_m$, then the singular values of F consist of ϵ-perturbations of the $p-m$ smallest singular values of X, and hence $|\sigma_{m+1} - \|F\|_2| \leq \epsilon$. It follows that as $\epsilon \to 0$,

$$\rho \cong \frac{\sigma_{m+1}}{\sigma_m}.$$

Thus if ρ is small, there is a strong gap between the mth and $(m+1)$th singular values of X.

- The bounds (1.5) say that the fundamental subspaces of X are $O(\epsilon)$ perturbations of those of $\text{diag}(S, F)$. Moreover, since

$$1 - c_u = 1 - \sqrt{1 - s_u^2} \cong \frac{1}{2}s_u^2,$$

the cosines of the canonical angles between the subspaces are $O(\epsilon^2)$ approximations to one. In particular, since the canonical cosines are the singular values of the matrices U_{11}, U_{22}, V_{11}, and V_{22}, these matrices are $O(\epsilon^2)$ approximations to the orthogonal matrices obtained by setting their singular values to one.

- An important phenomenon occurs when X is block triangular. Suppose, for example, that $\eta = 0$ so that X is block lower triangular. Then the bounds (1.5) become

$$s_u \leq \frac{1}{1-\rho^2}\frac{\gamma}{\tau} \quad \text{and} \quad s_u \leq \frac{\rho}{1-\rho^2}\frac{\gamma}{\tau}.$$

Thus the approximate left singular subspaces are better than the approximate right singular subspaces by a factor of ρ—the relative gap in the singular values.

SEC. 1. FUNDAMENTAL SUBSPACES AND RANK ESTIMATION

- The expressions (1.6) and (1.7) for Σ_1 and Σ_2 imply that the singular values of S and F are $O(\epsilon^2)$ approximations to those of Σ_1 and Σ_2 respectively. For example, we have already observed that the matrices U_{11} and V_{11} in the expression

$$\Sigma_1 = U_{11}^{-1}(S + HV_{21}V_{11}^{-1})V_{11}$$

are within $O(\epsilon^2)$ of orthogonal matrices. It follows that $S + HV_{21}V_{11}^{-1}$ contain $O(\epsilon^2)$ approximations to the singular values of Σ_1. But $\|HV_{21}V_{11}^{-1}\|_2 = O(\epsilon^2)$, so that S also contains $O(\epsilon^2)$ approximations to those of Σ_1. It is straightforward to evaluate bounds on the error given ϵ.

—

The statement of Theorem 1.1 represents a convenient summary of results we will need later. But one should not lose sight of the basic relation (1.8), which can be massaged in various ways. For example, if we observe that $(V_{11}^T \ V_{12}^T)^T$ has orthonormal columns, we may conclude that

$$\|U_{21}\|_F \leq \|(G \ F)\|_F,$$

which gives an easily computable bound on the square root of the sum of squares of the sines of the canonical angles.

1.2. RANK ESTIMATION

In this subsection we will sketch out what our theory says about the problem of estimating rank. We will assume that we have in the background an $n \times p$ matrix $\ddot{X}$ of rank $m < p$. Instead of $\ddot{X}$, however, we observe $X = \ddot{X} + E$, where E is some unknown error matrix. The problem is to recover m from X.

It should be stressed that this problem is intractable without further information. For example, because $\ddot{X}$ is of rank m it has a gap at the mth singular value $\ddot{\sigma}_m$. This suggests that we try to determine rank by looking for gaps in the singular values of X. But if $\|E\|_2$ is too large, it will perturb the zero singular values of $\ddot{X}$ to be of a size with σ_m, thus filling in the gap at $\ddot{\sigma}_m$.

These considerations show that we must fulfill two conditions to detect rank by looking at gaps.

1. We must have an estimate of the size of, say, $\|E\|_2$ — call it ϵ.
2. We must know that $\ddot{\sigma}_m$ is substantially greater than ϵ. (1.11)

The first of these conditions insures that $\sigma_{m+1}, \ldots, \sigma_p$ will all be less than ϵ. The second condition insures that σ_m, which is never less than $\ddot{\sigma}_m - \epsilon$, will be greater than ϵ. Thus to determine rank we may look for the largest integer m such that $\sigma_m > \epsilon$.

Unfortunately, the knowledge needed to satisfy these conditions can only come from the science of the underlying problem. An important implication of this observation is that rank determination is not an exercise in pure matrix computations; instead it is a collaborative effort between the matrix algorithmist and the originator of the problem.

In some cases a little common sense will help, as in the following example.

Example 1.2. *The matrix*

$$X = \begin{pmatrix} 2.75264923421521 & -1.38596590633453 & 0.43555588928583 \\ -0.62398940923427 & 0.48222097427979 & -0.27820182478314 \\ -0.36467070170849 & 0.16827170447293 & -0.04131833712094 \\ 3.86434789850664 & -2.38968291208709 & 1.08562457860561 \\ 2.14348788652727 & -1.30888575262670 & 0.58441599476806 \\ -0.59728895540373 & 0.17172714649047 & 0.04327170850873 \end{pmatrix}$$

has singular values

$$6.2625\mathrm{e}{+00}, \quad 4.3718\mathrm{e}{-01}, \quad 2.6950\mathrm{e}{-16}.$$

The third singular value is suspiciously near the rounding unit for IEEE double precision, and it is reasonable to conjecture that X started life as a matrix of rank two and acquired double-precision rounding errors. In fact, X was generated as the product of random matrices of dimensions 6×2 and 2×3.

The reason that this problem is easy is that we can reasonably infer from the third singular value that the error in X is due to rounding error, which gives us an estimate of the size of $\|E\|_2$. It is also reasonable to assume that $\ddot{X}$ is unlikely to have singular values smaller than the rounding unit. Thus the two conditions in (1.11) are satisfied.

Having determined m, we may go on to use our perturbation theory to assess the quality of the fundamental subspaces of X as approximations to those of $\ddot{X}$. Specifically, let $\ddot{X}$ have the singular value decomposition

$$\begin{pmatrix} \ddot{U}_1^\mathrm{T} \\ \ddot{U}_2^\mathrm{T} \end{pmatrix} \ddot{X} (\ddot{V}_1 \; \ddot{V}_2) = \begin{pmatrix} \ddot{\Sigma}_1 & 0 \\ 0 & 0 \end{pmatrix},$$

where $\ddot{\Sigma}_1$ is of order m. When the transformations $\ddot{U}$ and $\ddot{V}$ are applied to X, we get

$$\begin{pmatrix} \ddot{U}_1^\mathrm{T} \\ \ddot{U}_2^\mathrm{T} \end{pmatrix} (X+E)(\ddot{V}_1 \; \ddot{V}_2) = \begin{pmatrix} \ddot{\Sigma}_1 + \ddot{U}_1^\mathrm{T} E \ddot{V}_1 & \ddot{U}_1^\mathrm{T} E \ddot{V}_2 \\ \ddot{U}_2^\mathrm{T} E \ddot{V}_1 & \ddot{U}_2^\mathrm{T} E \ddot{V}_2 \end{pmatrix} \equiv \begin{pmatrix} S & H \\ G & F \end{pmatrix}. \quad (1.12)$$

Since both $\ddot{X}$ and X have been transformed by $\ddot{U}$ and $\ddot{V}$, the bounds in Theorem 1.1 bound the canonical angles between the singular subspaces of $\ddot{X}$ and X.

At first glance we seem not to have gained much, since the right-hand side of (1.12) is unknowable. However, because our transformations are orthogonal we may bound the quantities in the theorem by any bound ϵ on the norm of E:

$$\phi, \gamma, \eta \leq \epsilon.$$

Moreover, from the min-max characterization of singular values (Corollary 4.30, Chapter 1) we have

$$\tau = \inf(S) \geq \sigma_m,$$

SEC. 1. FUNDAMENTAL SUBSPACES AND RANK ESTIMATION

a quantity which we have already computed and used. Thus we may apply Theorem 1.1 to obtain completely rigorous bounds on the canonical angles between the singular subspaces of $\tilde{X}$ and X.

The bounds obtained from this procedure may be pessimistic. The reason is that $\|E\|_2$ will generally be an overestimate for the quantities in the theorem. One cure is to make probabilistic assumptions about the error and calculate estimates of the quantities. However, it would take us too far afield to develop this approach here. For more, see the notes and references.

1.3. NOTES AND REFERENCES

Rank reduction and determination

Rank-reduction and rank-determination problems arise in all quantitative fields. For example, in array signal processing the rank of a certain matrix is the number of objects being tracked by the array (e.g., see [270]). A feature of this problem is that the signal varies over time, so that one must track changing singular subspaces [219]. In addition the rank of the matrix in question may be small — one or two — and it is desirable to take computational advantage of this situation.

Another example comes from the analysis of multivariate data. Hotelling's *principal component analysis* [276, §5.2] decomposes the data into factors corresponding to the dominant singular subspace. Another technique for accomplishing the same thing is *factor analysis* [276, §5.4]. Since the factors may represent controversial constructs — general intelligence, for example — and since the two methods do not always give the same answers, there is considerable disagreement over which one is better (see, e.g., [46]).

Approximations of lower rank are also used in data analysis. For example, the least squares problem of minimizing $\|y - Xb\|_2$ can be regarded as finding the smallest change in y such that the matrix $(X\ y)$ is reduced in rank. If we allow both X and y to vary, we obtain a method known as *total least squares* [151, 325]. It is also equivalent to one of the procedures statisticians use to analyze measurement error models [127].

For more on the use of low-rank approximations to regularize ill-posed problems see [171].

Singular subspaces

Singular subspaces are a natural analogue of the invariant subspace associated with eigendecompositions. The term itself is comparatively new but is now well established. The use of "superior" and "inferior" to denote singular subspaces associated with leading and trailing sets of singular values is new. When X is of rank m and the breaking point in the singular values is set at m, these spaces are the row, column, and null spaces of X. Strang [313] calls these particular subspaces fundamental subspaces. Following Per Christian Hansen, I have applied the term "fundamental" to the four superior and inferior subspaces at any fixed break point m. In array signal processing the right superior and inferior subspaces are called the signal and noise subspaces.

The perturbation theory developed here is based on an elegant paper of Wedin [335] (also see [310, §V.4]), with some modifications to make explicit the role of the individual blocks S, F, G, and H of the partition (1.2). The fact that block triangular matrices behave exceptionally—with the subspaces on one side being more accurate than those on the other—was first noted by Mathias and Stewart [226] and independently by Fierro [118]. The expressions (1.6) and (1.7) for Σ_1 and Σ_2 were first given by Stewart [287], though in a rather different form.

Rank determination

The approach taken here is basic common sense, refined somewhat to show its limitations. The idea of looking for gaps in the singular values is natural and often recommended. The assumptions (1.11) are less often emphasized by numerical analysts—perhaps through overacquaintance with easy problems like the one in Example 1.2. It is worth stressing that the gap must be reasonably large compared with the particular error estimate ϵ that one is actually using. Too large an ϵ can cause a gap to be missed.

Error models and scaling

Our rank-determination strategy tacitly assumes that the norm of the error E in some sense represents the size of E. There are two ways this assumption can fail. First, the elements of E may differ greatly in size. Second, we may be concerned with only a part of E. For example, the norm of $S = \ddot{U}_2^T E \ddot{V}_2$ in (1.12) may be smaller than the norm of E itself—especially when m is large so that U_2 and V_2 have few columns.

The first problem can be alleviated to some extent by scaling the problem so that the components of E are roughly equal—if that is possible. The second problem requires that we assume something further about E. A common assumption is that E represents white noise—i.e., its components are uncorrelated with mean zero and common standard deviation. In this case, it is possible to develop formulas for estimating the size of matrices like $S = \ddot{U}_2^T E \ddot{V}_2$ (e.g., see [299]).

Another frequently occurring model is to assume that the rows of E are uncorrelated random vectors with mean zero and common covariance matrix D. (Statisticians would use Σ, but we have preempted that letter for the singular value decomposition.) If D is positive definite, then the error $ED^{-\frac{1}{2}}$ in $XD^{-\frac{1}{2}}$ has uncorrelated elements with common variance one. The process of postmultiplying by $D^{-\frac{1}{2}}$ is called *whitening the noise*.

A shortcoming of whitening is that it breaks down entirely when D is singular and can lead to numerical inaccuracies when D is ill conditioned. There are various fixes, and it is worthwhile to examine each in turn.

- **Enforce grading.** Let $D = W\Lambda W^T$ be the spectral decomposition of D (see §4.4, Chapter 1) with the eigenvalues appearing in Λ in *ascending* order. In practice, there will be no zero eigenvalues; but if there are, they may be replaced by the rounding unit times the largest eigenvalue—or an even smaller number. Then the noise in the matrix

$XW\Lambda^{-\frac{1}{2}}$ is effectively whitened, and the columns of this matrix are graded downward. Although we cannot guarantee that our computational procedures will work well with such matrices, by and large they do.

- **Project out errorless columns.** Suppose that the first k columns of X are errorless. It has been shown in [150] that the appropriate way to handle this situation is to project the last $p-k$ columns onto the orthogonal complement of the space spanned by the first k and to work with that matrix (see also [92]). In the general case, if we compute the spectral decomposition of D then the initial columns of XW — the ones corresponding to zero eigenvalues — are error free, and we can use the same procedure.

- **Solve a generalized eigenvalue problem.** It can be shown that the squares of the singular values are the eigenvalues of the *generalized eigenvalue problem* $X^{\mathrm{T}}Xv = \mu Dv$. Consequently, we can form the cross-product matrix $X^{\mathrm{T}}X$ and solve the generalized eigenvalue problems. This is the way statisticians do it in treating measurement error models [127]. The procedure is open to the same objections that apply to forming the normal equations (§2.3, Chapter 4).

- **Compute a generalized singular value decomposition.** It can be shown that there are orthogonal matrices Q_X and Q_D and a nonsingular matrix B such that the matrices $(Q_X^{\mathrm{T}}XB)[1:p, 1:p]$ and $Q_D^{\mathrm{T}}DB$ are diagonal. The ratios of their diagonal elements are the singular values of the whitened X. The computation of the generalized singular value decomposition avoids the need to compute cross-product matrices. (The generalized singular value decomposition was introduced by Van Loan [326, 1975] and was reformulated in a more convenient form by Paige and Saunders [249].)

—

All these methods have their advantages and drawbacks. In practice, zero eigenvalues of X are usually part of the structure of the problem and can be projected out of it. The remaining eigenvalues are not usually small, at least compared to the double-precision rounding unit, and one can use whatever method one finds convenient.

Another approach is to use first-order perturbation theory to compute test statistics directly from the unwhitened data in which D but not its inverse appears. I know of no systematic exposition of this approach, although I gave one in an earlier version of this section. For more see [299, 302].

2. PIVOTED ORTHOGONAL TRIANGULARIZATION

As a rank-reducing method the singular value decomposition has two drawbacks. In the first place, it is expensive to compute. The second drawback is more subtle. In many applications it is sufficient to have orthonormal bases for the fundamental subspaces, something which the singular value decomposition provides. In other applications, however, it is desirable to have *natural bases* that consist of the rows or columns of a matrix. For example, in statistical regression problems the columns of X represent distinct variables in a model, and a basis that consists of a subset of the columns of X represents not just a computational economization but a simplification of the model itself.

In this section we will consider two decompositions based on pivoted orthogonal triangularization. The first — orthogonal triangularization with column pivoting — is cheap to compute and tends to isolate independent columns of X. We will treat this algorithm in §2.1. The R-factor computed by this algorithm can also be computed from the cross-product matrix $A = X^T X$ by a pivoted version of the Cholesky algorithm, which we treat in §2.2. These decompositions tend to reveal gaps in the singular values, especially when the gap is large. However, they can be improved by a subsequent reduction to lower triangular form, which we treat in §2.3.

2.1. THE PIVOTED QR DECOMPOSITION

We have already met the pivoted QR decomposition in the weighting method for constrained least squares (Algorithm 2.8, Chapter 4). The decomposition is computed by a variation of orthogonal triangularization by Householder transformations. At the kth stage of the algorithm, column k is swapped with some other column of index greater than k before the reduction proceeds. Our concern is to choose the pivot columns so as to isolate a set of independent columns of the matrix X.

There are many pivoting strategies for accomplishing this goal. In this subsection we are going to describe the method of column pivoting for size. It is the oldest and simplest of pivoting strategies, yet in many respects it is still the best. We will begin by describing the algorithm itself. We will then move on to discuss its properties.

Pivoted orthogonal triangularization

The algorithm is based on Algorithm 1.2, Chapter 4 — triangularization by Householder transformations. At the first stage of the algorithm, before computing the first Householder transformation, we determine the column having largest norm and interchange it with the first. If Π_1 denotes the exchange matrix that accomplishes the interchange and H_1 denotes the first Householder transformation, then the result of the first step of the algorithm (in northwest indexing) is

$$H_1 X \Pi_1 = \begin{pmatrix} r_{11} & r_{12}^T \\ 0 & X_{22} \end{pmatrix}. \tag{2.1}$$

At the kth stage of the algorithm we will have computed $k-1$ Householder transformations H_i and $k-1$ interchanges Π_i such that

$$H_{k-1} \cdots H_1 X \Pi_1 \cdots \Pi_{k-1} = \begin{pmatrix} R_{11} & R_{1k} \\ 0 & X_{kk} \end{pmatrix}, \tag{2.2}$$

where R is upper triangular. We now repeat the pivoting strategy of the first step: Find the column of X_{kk} of largest 2-norm, interchange it with the initial column, and proceed with the reduction.

SEC. 2. PIVOTED ORTHOGONAL TRIANGULARIZATION

Given an $n \times p$ matrix X, let $\ell = \min\{n, p\}$. This algorithm computes Householder transformations $H_1, \ldots, H_\ell$ and exchange matrices $\Pi_1, \ldots, \Pi_\ell$ such that

$$H_\ell \cdots H_1 X \Pi_1 \cdots \Pi_\ell = \begin{pmatrix} R \\ 0 \end{pmatrix},$$

where R is upper triangular. The generators of the Householder transformation are stored in the array U. At the kth step the column of largest 2-norm is exchanged with the kth column and its index stored in $pvt[k]$. The algorithm uses the routine *vecdd* (Algorithm 3.10, Chapter 4) to keep track of the column norms of the submatrices $X[k:n, k:p]$.

1. $hpqrd(X, U, R, pvt)$
2.   **for** $j = 1{:}p$
3. $nrm[j] = oldnrm[j] = \|X[:,j]\|_2$
4.   **end for** j
5.   **for** $k = 1$ **to** $\min\{n, p\}$
6.   Find $pvt[k] \geq k$ so that $nrm[pvt[k]]$ is maximal
7. $X[k:n, k] \leftrightarrow X[k:n, pvt[k]]$
8. $R[1{:}k{-}1, k] \leftrightarrow R[1{:}k{-}1, pvt[k]]$
9.   **if** $(k \neq n)$
10. $housegen(X[k:n, k], U[k:n, k], R[k, k])$
11. $v^{\text{T}} = U[k:n, k]^{\text{T}} * X[k:n, k{+}1, p]$
12. $X[k:n, k{+}1{:}p] = X[k:n, k{+}1{:}p] - U[k:n, k] * v^{\text{T}}$
13. $R[k, k{+}1{:}p] = X[k, k{+}1{:}p]$
14.   **for** $j = k{+}1{:}p$
15. $vecdd(nrm[j], oldnrm[j], R[k, j], X[k{+}1{:}p, j])$
16.   **end for** j
17.   **end if**
18.   **end for** k
19.   **end** $hqrd$

Algorithm 2.1: Pivoted Householder triangularization

The principal computational difficulty with this strategy is the expense of computing the norms of the columns of X_{kk}. We can solve this problem by using Algorithm 3.10, Chapter 4, to downdate the vectors. The result is Algorithm 2.1. Note that the columns of R are interchanged along with the columns of X to preserve the integrity of the final decomposition [cf. (2.2)]. Here are some comments on the algorithm.

- The current pivot column is chosen at statement 6. The strategy given here is column pivoting for size, but we could substitute any other strategy at this point.

- The downdating function *vecdd* requires $O(1)$ time, except in the rare cases when the norm must be recomputed from scratch. Thus the pivoted algorithm takes essentially the same amount of time as the unpivoted algorithm.

- For $j = k, \ldots, n$ let $x_j^{(k)}$ denote the vector contained in $X[k{:}n, j]$ at the kth step of the algorithm. These vectors are transformed by the subsequent Householder transformations into the vectors $R[k{:}j, j]$. Since Householder transformations are orthogonal, after some rearrangement to account for the subsequent pivoting, we have

$$\|x_j^{(k)}\|_2 = \|R[k{:}j, j]\|_2.$$

But since we pivot the largest column into $X[k{:}n, k]$, it follows that

$$|R[k, k]| \geq \|R[k{:}j, j]\|_2, \qquad j = k, \ldots, n. \tag{2.3}$$

Thus the kth diagonal element of R dominates the trailing principal submatrix of R. In this light the pivoting strategy can be regarded as a greedy algorithm keeping R well conditioned by keeping its diagonal elements large.

- If, after the interchange (statement 8), the vector $X[k{:}n, k]$ is zero, the entire matrix $X[k{:}n, k{:}p]$ must also be zero, and the algorithm can be terminated. With inexact data, we are more likely to encounter a column of small norm. By (2.3), the trailing elements of R will be dominated by $|R[k, k]| = \|X[k{:}n, k]\|_2$. Thus if we have a criterion for determining when elements of R can be regarded as zero, we can terminate the algorithm simply by inspecting the norms $\|X[k{:}n, k]\|_2$.

- The algorithm is frequently used to extract a well-conditioned, square matrix from a collection of p n-vectors. In this case, $n > p$, and we must take special action in processing the last column — hence the **if** in statement 9.

Bases for the fundamental subspaces

In determining whether a decomposition is suitable for rank determination, it is useful to consider its behavior on a matrix that is exactly rank degenerate. For the pivoted QR decomposition, we have the following theorem.

Theorem 2.1. *Let the pivoted QR decomposition that is computed by Algorithm 2.1 be partitioned in the form*

$$\hat{X} \equiv X\Pi = (Q_1 \; Q_2 \; Q_\perp) \begin{pmatrix} R_{11} & R_{12} \\ 0 & R_{22} \\ 0 & 0 \end{pmatrix},$$

where R_{11} is of order m and $\Pi = \Pi_1 \cdots \Pi_\ell$. If $\mathrm{rank}(X) \geq m$, then R_{11} is nonsingular; if $\mathrm{rank}(X) = m$, then $R_{22} = 0$.

SEC. 2. PIVOTED ORTHOGONAL TRIANGULARIZATION

Proof. The proof is by induction on m.

For rank$(X) \geq m = 1$, note that r_{11} must be nonzero, since it is the norm of the largest column of X, which is nonzero. But if $r_{11} \neq 0$ and rank$(X) = 1$, the elements of R_{22} must all be zero. For if some element of R_{22} is zero, the column containing it is independent of the first column.

Now let $m > 1$ be given and consider the partitioned factorization computed by the algorithm after $m-1$ steps:

$$H_{m-1} \cdots H_1 X \Pi_1 \cdots \Pi_{m-1} = \begin{pmatrix} \bar{R}_{11} & \bar{R}_{12} \\ 0 & X_{mm} \end{pmatrix},$$

where $\bar{R}$ is of order $m-1$. By the induction hypothesis, if rank$(X) \geq m > k$, then $\bar{R}_{11}$ is nonsingular. Moreover, X_{mm} is nonzero, for otherwise the rank of X would be $m-1$. Consequently, r_{mm}, which is the norm of the largest columns of X_{mm} is nonzero, and R_{11} is nonsingular.

If R_{11} is nonsingular, then for rank(X) to be equal to m we must, as above, have $R_{22} = 0$. ∎

Thus if X is of rank m, the pivoted QR decomposition (computed exactly) will reveal the rank of X by the presence of a zero trailing principal submatrix of order $p-m$. In this case, Q_1 provides an orthonormal basis for the column space of X (the left superior subspace), and $(Q_2 \; Q_\perp)$ provides a basis for the orthogonal complement (the left inferior subspace). Moreover, if we partition the pivoted matrix $\hat{X} = X\Pi$ in the form

$$\hat{X} = (\hat{X}_1 \; \hat{X}_2),$$

where X_1 has m columns, then

$$\hat{X}_1 = Q_1 R_{11}, \tag{2.4}$$

so that $\hat{X}_1$ is a natural basis for $\mathcal{R}(X)$.

An advantage of the particular basis $\hat{X}_1$ is that its column space tends to be insensitive to perturbations in its elements — or rather, as insensitive as any basis consisting of columns of X can be. To see this, suppose that we perturb $\hat{X}_1$ to get the matrix $\tilde{X}_1 \equiv \hat{X}_1 + E_1$, and suppose that $\tilde{X}_1$ has the QR factorization

$$\tilde{X}_1 = \tilde{Q}_1 \tilde{R}_{11}.$$

Let $\tilde{Q}_{2\perp}$ be an orthonormal basis for the orthogonal complement of the column space of $\tilde{Q}_1$. Then

$$s(\hat{X}_1, \tilde{X}_1) = \|\tilde{Q}_{2\perp}^T Q_1\|_2$$

is the sine of the largest canonical angle between the spaces spanned by $\hat{X}_1$ and $\tilde{X}_1$.

We may evaluate $s(\hat{X}_1, \tilde{X}_1)$ as follows. Write (2.4) in the form

$$\tilde{X}_1 - E = Q_1 R_{11}.$$

Since $\tilde{Q}_{2\perp}^T \tilde{X} = 0$, we have

$$-\tilde{Q}_{2\perp}^T E = \tilde{Q}_{2\perp}^T Q_1 R_{11},$$

from which it follows that

$$s(\hat{X}_1, \tilde{X}_1) \leq \|R_{11}^{-1}\|_2 \|E\|_2 = \frac{\|E\|_2}{\inf(R_{11})}.$$

Thus if the singular values of R_{11} are large compared with E, $\mathcal{R}(\hat{X}_1)$ will be insensitive to perturbations in $\hat{X}_1$. The fact that column pivoting for size tends to keep R_{11} as well conditioned as possible is the reason why $\hat{X}_1$ tends to be a stable basis for $\mathcal{R}(X)$.

The decomposition does not directly provide orthonormal bases for the right fundamental subspaces of X. However, it is easy to see that

$$\begin{pmatrix} R_{11}^T \\ R_{12}^T \end{pmatrix} \quad \text{is a basis for the row space of } \hat{X} \qquad (2.5)$$

and

$$\begin{pmatrix} I \\ -R_{11}^{-1} R_{12} \end{pmatrix} \quad \text{is a basis for the null space of } \hat{X}. \qquad (2.6)$$

Bases for the original matrix X may be computed by premultiplying (2.5) and (2.6) by Π — i.e., by undoing the interchanges. Thus the algorithm provides (nonorthogonal) bases for the right fundamental subspaces of X, at the cost of some additional calculation for the right inferior subspace.

When X is near a matrix of rank m we may hope that Algorithm 2.1 will return an R-factor with R_{22} small. In this case the decomposition

$$\hat{X} \cong Q_1 (R_{11} \; R_{12})$$

is a low-rank approximation to X with $\|\hat{X} - X\| = \|R_{22}\|$ in the spectral or Frobenius norms. Moreover, the spaces

$$\mathcal{U}_1 = \mathcal{R}(Q_1) \quad \text{and} \quad \mathcal{U}_2 = \mathcal{R}[(Q_2 \; Q_\perp)] \qquad (2.7)$$

only approximate the left fundamental subspaces of X. Similarly, the column spaces $\mathcal{V}_1$ and $\mathcal{V}_2$ of the bases (2.5) and (2.6) only approximate the right fundamental subspaces. Unfortunately, because R_{12} need not be small we cannot apply Theorem 1.1 directly to bound the accuracy of these approximations. However, we can show the following.

SEC. 2. PIVOTED ORTHOGONAL TRIANGULARIZATION

Assume that

$$\rho \equiv \frac{\|R_{22}\|_2}{\inf(R_{11})} < 1.$$

The sine s_u of the largest canonical angle between $\mathcal{U}_1$ and the right superior subspace of X is bounded by

$$s_\mathrm{u} \leq \frac{\rho}{1-\rho}. \tag{2.8}$$

The sine s_v of the largest canonical angle between $\mathcal{V}_1$ and the left superior subspace of X is bounded by

$$s_\mathrm{v} \leq \frac{\rho^2}{(1-\rho)^2}.$$

The bounds for the inferior subspaces are the same.

We will defer the proof of this result to §2.3, where it follows as an easy corollary to the analysis of the related QLP decomposition.

When the break point m is small compared with p we can obtain these approximations at very little cost. For, as we have noted, we can detect when R_{22} is small after the mth step of Algorithm 2.1 and terminate the computations, leaving us with R_{11}, R_{12}, and a set of Householder transformations that generate Q_1 and an orthonormal basis for the orthogonal complement of $\mathcal{R}(Q_1)$.

Pivoted QR as a gap-revealing decomposition

The pivoted QR algorithm has a good reputation as a gap-revealing algorithm. If the singular values of X have a substantial gap at σ_m, the diagonals of R will generally exhibit a gap at r_{mm} — although the gap may not be as substantial. To simplify the exposition, we will refer to the diagonal elements of R as *R-values*.

To illustrate the gap-revealing properties of the pivoted QR decomposition, a matrix X of order 100 was generated in the form

$$X = U\Sigma V^\mathrm{T} + 0.1\sigma_{50}E, \tag{2.9}$$

where
1. Σ is a diagonal matrix with diagonals decreasing geometrically from one to 10^{-3} with the last 50 values replaced by zero,
2. U and V are random orthogonal matrices,
3. E is a matrix of standard normal deviates.

(2.10)

Thus X represents a matrix of rank 50 perturbed by an error whose elements are one-tenth the size of the last nonzero singular value.

Figure 2.1 plots the common logarithms of the singular values of X (solid line)

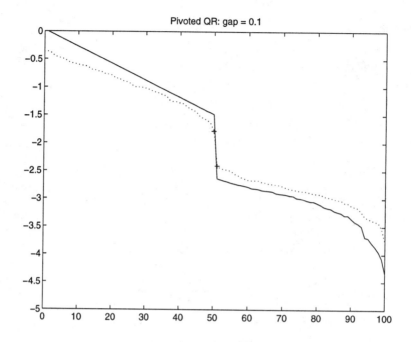

Figure 2.1: Gap revelation in pivoted QR

and R-values of X (dotted line) against their indices. The +'s indicate the values of $r_{50,50}$ and $r_{51,51}$. It is seen that there is a well-marked gap in the R-values, though not as marked as the gap in the singular values.

Unfortunately, the pivoted QR decomposition is not foolproof, as the following example shows.

Example 2.2. Let K_n be the upper triangular matrix illustrated below for $n = 6$:

$$K_6 = \begin{pmatrix} 1 & 0 & 0 & 0 & 0 & 0 \\ 0 & s & 0 & 0 & 0 & 0 \\ 0 & 0 & s^2 & 0 & 0 & 0 \\ 0 & 0 & 0 & s^3 & 0 & 0 \\ 0 & 0 & 0 & 0 & s^4 & 0 \\ 0 & 0 & 0 & 0 & 0 & s^5 \end{pmatrix} \begin{pmatrix} 1 & -c & -c & -c & -c & -c \\ 0 & 1 & -c & -c & -c & -c \\ 0 & 0 & 1 & -c & -c & -c \\ 0 & 0 & 0 & 1 & -c & -c \\ 0 & 0 & 0 & 0 & 1 & -c \\ 0 & 0 & 0 & 0 & 0 & 1 \end{pmatrix},$$

where $c^2 + s^2 = 1$. All the columns of the matrix have the same 2-norm—namely, one—so that if ties in the pivoting process are broken by choosing the first candidate, the first step of Algorithm 2.1 leaves the matrix unchanged. Similarly for the the remaining steps. Thus Algorithm 2.1 leaves K_n unchanged, and the smallest R-value is s^{n-1}.

However, the matrix can have singular values far smaller than s^{n-1}. The follow-

SEC. 2. PIVOTED ORTHOGONAL TRIANGULARIZATION

ing table

c	σ_{99}	σ_{100}	$r_{99,99}$	$r_{100,100}$
0.0	1.0e+00	1.0e+00	1.0e+00	1.0e+00
0.1	6.4e−01	9.5e−05	6.1e−01	6.1e−01
0.2	1.5e−01	3.7e−09	1.4e−01	1.3e−01
0.3	1.1e−02	9.3e−14	9.8e−03	9.4e−03
0.4	2.3e−04	1.1e−18	1.9e−04	1.8e−04

(2.11)

presents the 99th and 100th singular and R-values of K_{100} for various values of c. When $c = 0$, $K_n = I$, and the R-values and singular values coincide. As c departs from zero, however, there is an increasingly great gap between the next-to-last and last singular values, while the ratio of the corresponding R-values remains near one.

This example, which is closely allied to Example 4.2, Chapter 3, shows that the R-values from the pivoted QR decomposition can fail by orders of magnitude to reveal gaps in the singular values. Although such dramatic failures seem not to occur in practice, the possibility has inspired a great deal of work on alternative pivoting strategies, for which see the notes and references.

Assessment of pivoted QR

It is important to appreciate that a pivoted QR decomposition, whatever the pivoting strategy, has fundamental limitations when it comes to revealing the properties of singular values. For example, we would hope that the first R-value r_{11} of X would approximate the first singular value σ_1. But r_{11} is the 2-norm of the first columns of X, while σ_1 is the 2-norm of the entire matrix. When X is of order n, the latter can exceed the former by a factor of $\sqrt{n}$. For example, if $X = \mathbf{e}\mathbf{e}^{\mathrm{T}}$, then $\|X\|_2 = n$, while the norm of the first column of X is $\sqrt{n}$. Moreover, all the columns of X have the same norm, so that no pivoting strategy can make the first R-value a better approximation to the first singular value.

The graph in Figure 2.1 shows that, in a modest way, the problem occurs without our looking for it. For the largest R-value in the graph underestimates the largest singular value by a factor of greater than two. Moreover, the smallest R-value overestimates the smallest singular value by a factor of almost four.

The pivoted QR decomposition finds its most important application in rank-reduction problems where there is a strong gap in the singular values. Excluding artificial examples like K_n, it is cheap and effective, and it isolates a set of independent columns of X. However, the R-values from the decomposition tend to be fuzzy approximations to the singular values. In §2.3 we will show how to sharpen the approximations by a subsequent reduction to lower triangular form.

2.2. THE PIVOTED CHOLESKY DECOMPOSITION

We have seen that the R-factor of a matrix X is the Cholesky factor of the cross-product matrix $A = X^{\mathrm{T}}X$. A corresponding relation holds for the pivoted factorization.

Specifically, if

$$X\Pi = Q_X R$$

is a pivoted QR factorization of X then

$$\Pi^T A \Pi = R^T R,$$

so that the pivoted R-factor is the Cholesky factor of the permuted cross-product matrix $\Pi^T A \Pi$. Thus if we can find some way of adaptively determining pivots as we compute the Cholesky factor of A, we can compute the pivoted R-factor of X directly from A.

The problem has a nice solution when we pivot for size. At the kth step of the pivoted Householder reduction, we have

$$H_{k-1} \cdots H_1 X \Pi_1 \cdots \Pi_{k-1} = \begin{pmatrix} R_{11} & R_{1k} \\ 0 & X_{kk} \end{pmatrix}$$

[see (2.2)]. The pivot column is determined by examining the norms of the columns $x_j^{(k)}$ of X_{kk}. Now if we were to continue the computation without pivoting, we would obtain a cross-product matrix of the form

$$\Pi_{k-1}^T \cdots \Pi_1^T A \Pi_1 \cdots \Pi_{k-1} = \begin{pmatrix} \hat{A}_{11} & \hat{A}_{12} \\ \hat{A}_{12}^T & \hat{A}_{22} \end{pmatrix}$$
$$= \begin{pmatrix} R_{11} & R_{1k} \\ 0 & X_{kk} \end{pmatrix}^T \begin{pmatrix} R_{11} & R_{1k} \\ 0 & X_{kk} \end{pmatrix}$$
$$= \begin{pmatrix} R_{11} & R_{1k} \\ 0 & R_{kk} \end{pmatrix}^T \begin{pmatrix} R_{11} & R_{1k} \\ 0 & R_{kk} \end{pmatrix}.$$

Thus

$$X_{kk}^T X_{kk} = R_{kk}^T R_{kk}, \tag{2.12}$$

and the quantities $\|x_j^{(k)}\|_2^2$ are the diagonals of $R_{kk}^T R_{kk}$. But by Theorem 1.6, Chapter 3, the matrix $R_{kk}^T R_{kk}$ is the Schur complement of $\hat{A}_{11}$. Thus if we compute the Cholesky decomposition of A by the classical variant of Gaussian elimination, which generates the full Schur complement at each stage, we will find the numbers we need to determine the pivots on the diagonals of the Schur complement.

Algorithm 2.2 is an implementation of this procedure. Here are some comments.

- Only the upper half of the matrix A is stored and manipulated, and the lower half of the array A can be used for other purposes.

- In many ways the trickiest part of the algorithm is to perform the interchanges, which is done in statements 4–12. The problem is that only the upper half of the matrix

Sec. 2. Pivoted Orthogonal Triangularization

Given a positive definite matrix stored in the upper half of the array A, this algorithm overwrites it with its pivoted Cholesky decomposition.

1.   **for** $k = 1$ **to** n
2.   Determine $p_k \geq k$ for which $A[p_k, p_k]$ is maximal
3.   **if** $(A[p_k, p_k] = 0)$ quit **fi**
4.   **for** $i = 1$ **to** k
5. $A[i, k] \leftrightarrow A[i, p_k]$
6.   **end for** i
7.   **for** $i = k+1$ **to** $p_k - 1$
8. $A[k, i] \leftrightarrow A[i, p_k]$
9.   **end for** i
10.   **for** $i = p_k$ **to** n
11. $A[k, i] \leftrightarrow A[p_k, i]$
12.   **end for** i
13. $temp[k{:}n] = A[k, k{:}n] = A[k, k{:}n]/\sqrt{A[k,k]}$
14.   **for** $j = k+1$ **to** n
15.   **for** $i = k+1$ **to** j
16. $A[i, j] = A[i, j] - temp[i]*A[k, j]$
17.   **end for** i
18.   **end for** j
19. **end for** k

Algorithm 2.2: Cholesky decomposition with diagonal pivoting

---

is stored, so we cannot simply interchange rows and then interchange columns. There is no really good way to explain this code; the reader should verify by example that it works.

- We have written out the inner loops that update the Schur complement (statements 14–18) because our notation does not provide a compact means of specifying that only the upper part of a matrix is to be modified. In practice, however, this computation would be done by a level-two BLAS.

- The scratch array *temp* in statements 13 and 16 has been introduced to preserve column orientation. Without it statement 16 in the inner loop would become

$$A[i, j] = A[i, j] - A[k, i]*A[k, j]$$

which repeatedly traverses the kth row of A.

- As we showed above, the pivoting strategy implemented in statement 4 is equivalent to column pivoting for size in the matrix X. However, an alternative pivoting strategy can be incorporated at this point.

- An important difference between this algorithm and pivoted orthogonal triangularization is that no norms have to be downdated. Their proxies — the diagonals of A — are calculated automatically in the course of the elimination. However, any lost accuracy cannot be regained by a recomputation of a norm, as in pivoted orthogonal triangularization. Hence, it is necessary to test for the case when all the diagonals in the Schur complement are zero or negative.

- The algorithm does not assume that A is a cross-product matrix; it can be applied to any positive definite matrix A. In fact, it can be used to test whether an arbitrary symmetric matrix A is positive definite.

- The operation count for the algorithm is the same as for the Cholesky algorithm — $\frac{1}{6}p^3$ flam. The effects of rounding error are essentially the same as for Gaussian elimination. See §4, Chapter 3.

Since the pivoted Cholesky algorithm applied to the cross-product matrix $A = X^T X$ produces the same R-factor as pivoted orthogonal triangularization applied to X, we can in principal use the latter whenever only the R-factor is needed. In particular, when X is sparse, we will usually save operations by forming the cross-product matrix and reducing it. The price to be paid is that the singular values of A are the squares of the singular values of X, so that the range of singular values that can be handled at a given precision is reduced. For more on this, see the comparison of the QR and the normal equations in §§2.2–2.3, Chapter 4.

2.3. THE PIVOTED QLP DECOMPOSITION

Although the pivoted QR decomposition is reasonably good at revealing gaps in the singular values of a matrix, we have seen in Figure 2.1 that it could be better. Moreover, the associated R-values tend to underestimate the large singular values and overestimate the small ones. In this subsection we will consider a postprocessing of the pivoted QR decomposition that yields a new decomposition with better gap-revealing properties.

The pivoted QLP decomposition

To motivate the QLP decomposition, consider the partitioned R-factor

$$R = \begin{pmatrix} r_{11} & r_{12}^T \\ 0 & R_{22} \end{pmatrix}$$

of the pivoted QR decomposition

$$Q^T X \Pi_R = \begin{pmatrix} R \\ 0 \end{pmatrix}.$$

We have observed that r_{11} is an underestimate of $\|X\|_2$. A better estimate is the norm $\ell_{11} = \sqrt{r_{11}^2 + r_{12}^T r_{12}}$ of the first row of R. We can calculate that norm by postmul-

SEC. 2. PIVOTED ORTHOGONAL TRIANGULARIZATION

Given an $n \times p$ matrix X this algorithm computes the pivoted decomposition

$$X = Q \Pi_L \begin{pmatrix} L \\ 0 \end{pmatrix} P^T \Pi_R^T,$$

where Π_L and Π_R are permutations, Q and P are orthogonal, and L is lower triangular. The matrices Q and P are each the products of p Householder transformations stored in the arrays Q and P. The matrices Π_L and Π_R are the products of interchanges whose indices are stored in the arrays pl and pr. (See Algorithm 2.1 for more details.)

1. $hpqlp(X, pl, Q, L, pr, P)$
2. $\quad hpqrd(X, Q, R, pr)$
3. $\quad hpqrd(R^T, P, L, pl)$
4. $\quad L = L^T$
5. **end** $hpqlp$

Algorithm 2.3: The pivoted QLP decomposition

―――――――◇―――――――

tiplying R by a Householder transformation H_1 that reduces the first row of R to a multiple of $\mathbf{e}_1$:

$$RH_1 = \begin{pmatrix} \ell_{11} & 0 \\ \ell_{12} & \hat{R}_{22} \end{pmatrix}.$$

We can obtain an even better value if we interchange the largest row of R with the first:

$$\Pi_1 R H_1 = \begin{pmatrix} \ell_{11} & 0 \\ \ell_{12} & \hat{R}_{22} \end{pmatrix}. \tag{2.13}$$

Now if we transpose (2.13), we see that it is the first step of pivoted Householder triangularization applied to R^T [cf. (2.1)]. If we continue this reduction and transpose the result, we obtain a triangular decomposition of the form

$$\Pi_L^T Q^T X \Pi_R P = \begin{pmatrix} L \\ 0 \end{pmatrix}.$$

We will call this the *pivoted QLP decomposition* of X and will call the diagonal elements of L the *L-values* of X.

Computing the pivoted QLP decomposition

It turns out that nothing more that Algorithm 2.1 is required to compute the pivoted QLP decomposition, as is shown in Algorithm 2.3. Here are some comments.

- The algorithm consists essentially of two applications of the routine *hpqrd* to compute pivoted QR decompositions. Since this kind of routine is widely implemented, the pivoted QLP decomposition can be computed using off-the-shelf software.

- The operations count for *hpqlp* applied to an $n \times p$ matrix is approximately $(np^2 - \frac{1}{3}p^3)$ flam. In the above algorithm it is applied once to the $n \times p$ matrix X and once to the $p \times p$. Thus:

    Algorithm 2.3 requires $(np^2 + \frac{1}{3}p^3)$ *flam.*

    If $n = p$, the computation of L doubles the work over the initial computation of R. If $n \gg p$ the additional work is negligible.

- It might be thought that one could take advantage of the triangular form of R in its subsequent reduction to L. But the reduction of the first row of R [see (2.13)] destroys the triangularity.

- The decomposition also requires an additional p^2 words of storage to contain L and the generating vectors of the Householder transformations used to compute L. This should be compared with the np words for the initial reduction.

Although the pivoted QLP decomposition costs more that a pivoted QR decomposition, there are good reasons for bearing the expense. First, the pivoted QLP decomposition tracks the singular values better. Second, it furnishes good approximations to orthonormal bases for all four fundamental subspaces at any reasonable break point m. We will consider each of these points in turn.

Tracking properties of the QLP decomposition

The way we motivated the pivoted QLP decomposition suggests that it might provide better approximations to the singular values of the original matrix X than does the pivoted QR decomposition. The top two graphs in Figure 2.2 compare performance of the two decompositions on the matrix generated as in (2.10). The solid lines, as above, indicate singular values and the dotted lines represent R-values on the left and L-values on the right. It is seen that in comparison with the R-values, the L-values track the singular values with remarkable fidelity.

The lower pair of graphs shows the behavior of the decomposition when the gap is reduced from a ratio of 0.1 to 0.25 [see (2.9)]. Here the L-values perform essentially as well as the singular values. The gap in the R-values is reduced to the point where an automatic gap-detecting algorithm might fail to see it.

Figure 2.3 presents a more taxing example — called the devil's stairs — in which the singular values have multiple gaps. When the gaps are small, as in the top two graphs, neither decomposition does well at exhibiting their presence, although the L-values track the general trend of the singular values far better than the R-values. In the pair of graphs at the bottom, the gaps are fewer and bigger. Here the L-values clearly reveal the gaps, while the R-values do not.

SEC. 2. PIVOTED ORTHOGONAL TRIANGULARIZATION

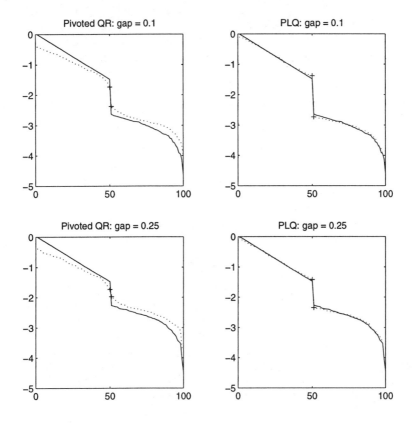

Figure 2.2: Pivoted QR and QLP decompositions compared

The above examples show that the pivoted QLP decomposition is better at tracking singular values and revealing gaps than the pivoted QR decomposition. That the improvement is so striking is an empirical observation, unsupported at this time by adequate theory.

Fundamental subspaces

If we incorporate the pivots in the QLP decomposition into the orthogonal transformations by defining

$$\hat{Q} = Q\Pi_L \quad \text{and} \quad \hat{P} = \Pi_R P,$$

then the decomposition can be written in the partition form

$$X = (\hat{Q}_1 \ \hat{Q}_2 \ Q_\perp) \begin{pmatrix} L_{11} & 0 \\ L_{21} & L_{22} \\ 0 & 0 \end{pmatrix} \begin{pmatrix} \hat{P}_1^T \\ \hat{P}_2^T \end{pmatrix},$$

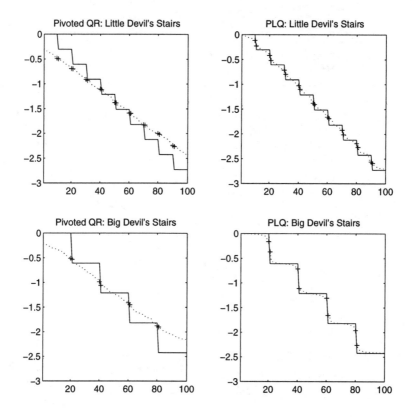

Figure 2.3: The devil's stairs

where L_{11} is of order m. Since the L-values tend to track the singular values, if there is a gap in the latter at m, the partition of $\hat{P}$ and $\hat{Q}$ provides orthonormal bases approximating the four fundamental subspaces of X at m. Specifically,

1. $\mathcal{R}(\hat{Q}_1)$ approximates the left superior subspace of X,
2. $\mathcal{R}[(\hat{Q}_2 \; Q_\perp)]$ approximates the left inferior subspace of X,
3. $\mathcal{R}(\hat{P}_1)$ approximates the right superior subspace of X,
4. $\mathcal{R}(\hat{P}_2)$ approximates the right inferior subspace of X.

Thus the pivoted QLP decomposition, like the pivoted QR decomposition, furnishes orthonormal approximations to the left fundamental subspaces, but unlike the latter, it also furnishes orthonormal approximations to the right fundamental subspaces.

We can apply Theorem 1.1 to bound the accuracy of these approximations. Specifically, we have the following theorem.

Theorem 2.3. *Let s_u be the sine of the largest canonical angle between the left superior subspace of X and $\mathcal{R}(\hat{Q}_1)$, and let s_v be the sine of the largest canonical angle*

SEC. 2. PIVOTED ORTHOGONAL TRIANGULARIZATION

between the right superior subspace of X and $\mathcal{R}(\hat{P}_1)$. If

$$\rho = \frac{\|L_{22}\|_2}{\inf(L_{11})} < 1,$$

then

$$s_\mathrm{u} \leq \frac{1}{1-\rho^2}\frac{\|L_{21}\|}{\inf(L_{11})} \quad \text{and} \quad s_\mathrm{v} \leq \frac{\rho}{1-\rho^2}\frac{\|L_{21}\|}{\inf(L_{11})}. \tag{2.14}$$

The bounds for the inferior subspaces are the same.

Proof. In Theorem 1.1 set

$$\tau = \inf(L_{11}),$$
$$\gamma = \|L_{21}\|_2,$$
$$\eta = 0,$$
$$\varphi = \|L_{22}\|_2. \quad \blacksquare$$

Two comments on this theorem.

- The theorem shows that we can expect the approximations to the right singular subspaces to be more accurate by a factor of ρ than the left singular subspaces.

- Because the L values tend to track the singular values, we may estimate the bound on the error by replacing $\inf(L_{11})$ by ℓ_{mm} and $\|L_{22}\|_2$ by $\ell_{m+1,m+1}$. The quantity $\|L_{21}\|_2$ can be bounded by $\|L_{21}\|_\mathrm{F}$. In the next section we will show how to obtain other estimates for the quantities in the theorem.

In the last section we stated the perturbation bounds [namely, (2.8)] for the QR decomposition but deferred their proof until this section. In fact, the bounds follow directly from Theorem 2.3.

Specifically, suppose that the reduction of R to L is done without pivoting. Then Q is left unaltered, so that $\mathcal{U}_1$ of (2.7) is the column space of Q_1 from the QLP decomposition. Moreover,

$$P^\mathrm{T}\begin{pmatrix}R_{11}^\mathrm{T}\\R_{12}^\mathrm{T}\end{pmatrix} = \begin{pmatrix}L_{11}^\mathrm{T}\\0\end{pmatrix}.$$

Consequently $\mathcal{R}(P_1)$ is the approximation to the right superior subspace

$$\mathcal{V}_1 = \mathcal{R}\left[\begin{pmatrix}R_{11}^\mathrm{T}\\R_{12}^\mathrm{T}\end{pmatrix}\right]$$

from the pivoted QR decomposition. Our theorem now applies to give the bounds (2.14).

To obtain the bounds in (2.8), we simply note that because

$$\begin{pmatrix} L_{11} & 0 \\ L_{21} & L_{22} \end{pmatrix} = \begin{pmatrix} R_{11} & R_{12} \\ 0 & R_{22} \end{pmatrix} P$$

and P is orthogonal, we have

$$\inf(R_{11}) \leq \inf(L_{11}) \quad \text{and} \quad \|L_{12}\|_2, \|L_{22}\|_2 \leq \|R_{22}\|_2.$$

The matrix $\hat{Q}$ and the columns of X

In the introduction to this section we stressed the desirability of having a natural basis for the left superior subspace at a gap — that is, a basis consisting of columns of X. In the pivoted QR decomposition, if we partitioned

$$Q = (Q_1 \ Q_2 \ Q_\perp) \quad \text{and} \quad \hat{X} \equiv X\Pi_R = (\hat{X}_1 \ \hat{X}_2), \tag{2.15}$$

then $\mathcal{R}(Q_1) = \mathcal{R}(X_1)$.

In the pivoted QRP decomposition we must replace Q by

$$\hat{Q} \equiv Q\Pi_L = (\hat{Q}_1 \ \hat{Q}_2 \ Q_\perp).$$

In general, the pivoting mixes up the columns of Q so that $\hat{Q}_1$ cannot be associated with a set of columns of X. However, if the partition (2.15) corresponds to a substantial gap in the R-values, it is unlikely that the pivoting process will interchange columns between Q_1 and Q_2 [see (2.3)]. In this case the column spaces of Q_1 and $\hat{Q}_1$ are the same and are spanned by the columns of $\hat{X}_1$. Thus in the applications we are most interested in, the left superior subspace will be associated with a specific set of columns of X.

Low-rank approximations

In some applications the matrix X is a perturbation of a matrix of rank m, and it is desired to compute a full-rank approximation of X of rank m. One way to do this is to start computing the pivoted QR decomposition via Algorithm 2.1 and stop after the mth stage — or, if the rank is initially unknown, stop after the step m at which a gap appears. The resulting decomposition will have the form

$$Q^T X \Pi_R = \begin{pmatrix} R_{11} & R_{12} \\ 0 & E \end{pmatrix},$$

where E can be regarded as negligible. If we partition Q in the form

$$Q = (Q_1 \ Q_2)$$

SEC. 2. PIVOTED ORTHOGONAL TRIANGULARIZATION

and set $E = 0$ we obtain the full-rank approximation

$$\tilde{X} = Q_1[(R_{11} \; R_{12})\Pi^T].$$

In any unitary invariant norm we will have

$$\|\hat{X} - X\| = \|E\|.$$

This approximation could of course be obtained from the entire decomposition. However, if m is small and p is large, the savings in stopping the reduction are substantial.

There is a QLP variant of the full-rank decomposition. If we go on to compute the pivoted QR decomposition

$$P^T \begin{pmatrix} R_{11}^T \\ R_{12} \end{pmatrix} \Pi_L = \begin{pmatrix} L_{11} \\ 0 \end{pmatrix}$$

and set

$$\Pi_R P = (\hat{P}_1 \; \hat{P}_2) \quad \text{and} \quad Q\Pi_L = (\hat{Q}_1 \; \hat{Q}_2),$$

then

$$\tilde{X} = \hat{Q}_1 L_{11} \hat{P}_1^T.$$

This is not necessarily the same decomposition as we would obtain from the full pivoted QLP decomposition, since the range of pivots is restricted. However, as we observed above, if the gap is substantial, the approximations to the fundamental subspaces will likely be the same for both.

This procedure is particularly attractive in cases where gaps in the singular values are narrow. It this case we use the pivoted QR decomposition as an exploratory tool to locate potential gaps, and the QLP decomposition as a confirmatory tool. If it fails to confirm the gap, the QR decomposition can be advanced until another potential gap is found, and the QLP is advanced to check it. The details of this interleaving are tedious but straightforward.

2.4. NOTES AND REFERENCES

Pivoted orthogonal triangularization

Pivoting for size in the Householder reduction was first proposed by Golub [148, 1965] and implemented by Businger and Golub [55]. The LINPACK version of the algorithm [99, SQRDC] incorporated the recomputation of vector norms when they drop below the level of rounding error (see Algorithm 3.10, Chapter 4). The routine also has provisions, useful to statisticians, to move columns to the beginning or end of the matrix and freeze them there. These procedures have also been incorporated in the LAPACK routine SGEQPF [9].

An unfortunate aspect of column pivoting is that it is incompatible with blocking in the style of Algorithm 1.5, Chapter 4. The problem is that after one column is chosen another candidate cannot be examined until the Householder transformation associated with the former has been applied to it.

The pivoted Cholesky decomposition

That diagonal pivoting for size in the cross-product matrix $A = X^\mathrm{T}X$ corresponds to column pivoting for size in X is obvious to anyone acquainted with both algorithms. Lawson and Hanson [213, 1974] seem to be the first to have suggested diagonal pivoting to bring the Cholesky decomposition into rank-revealing form. For implementation details see the LINPACK routine SCHDC [99, Ch. 8].

Column pivoting, rank, and singular values

The connection between column pivoting for size, rank, and singular values emerged rather slowly. In introducing his pivoting strategy Golub simply observed that it gave a modest increase in the accuracy of least squares solutions. (The reason seems to be that it tends to make any ill-conditioning in the R-factor artificial. See [309].) By 1974, Lawson and Hanson [213, Ch. 14] speak of the "pseudorank" of a matrix, by which they mean the result of replacing X with an approximation $\tilde{X}$ of defective rank. In the case of the pivoted QR decomposition this means setting a trailing principal submatrix of R to zero. Stewart [294, 1980] gave an empirical investigation of how well the smallest R value approximates the smallest singular value. Much of this and other work is colored by the tacit assumption that the matrix in question is a perturbation by rounding error of a matrix that is exactly rank deficient.

Example 2.2 is due to Kahan [195, 1966]. The key feature of the matrix K_n is that its columns all have 2-norm one — a property which many consider essential in such an example. However, the same property makes the example sensitive to small perturbations. For example, if one actually generates and reduces K_{100} with rounding error, one obtains the table

c	σ_{99}	σ_{100}	$r_{99,99}$	$r_{100,100}$
0.0	1.0e+00	1.0e+00	1.0e+00	1.0e+00
0.1	6.4e−01	9.5e−05	6.7e−01	2.3e−04
0.2	1.5e−01	3.7e−09	1.6e−01	7.1e−08
0.3	1.1e−02	9.3e−14	1.3e−02	1.4e−13
0.4	2.3e−04	1.1e−18	2.7e−04	2.4e−17

in which the gap in the singular values is quite evident [cf. (2.11)].

Rank-revealing QR decompositions

The impact of Kahan's example has been to fuel a search for better pivoting strategies. An early strategy, first suggested in [297], is equivalent to computing the Cholesky factor with pivoting of the inverse cross-product matrix. In 1987 Chan [63] proposed a method in which the R-factor is postprocessed to produce a small block in the southeast corner. Although the theoretical bounds for the method were disappointing, its obvious worth and its elegant name — Chan coined the term "rank-revealing QR decomposition" — set off a search for alternatives [31, 32, 64, 161, 180, 298]. For some of these algorithms to work the putative rank m must be known ahead of time. Others will find a gap at an unknown point but do not provably reveal multiple gaps.

The QLP decomposition

The pivoted QLP decomposition arose in the writing of this book. Unhappy with the poor approximation to σ_1 by the first R-value in Figure 2.1, I noted that the norm of the first row of R is a much improved estimate. Since that norm is just the $(1,1)$-element of the lower triangular matrix that you get by reducing R from the right, I decided to go the whole way — with results that surprised me, among others.

Less the pivoting, the reduction from R to L represents half an iteration in an algorithm for computing the singular values of R. The asymptotic behavior of this algorithm has been analyzed [226]. However, the asymptotic rates, which depend on the ratios of neighboring singular values, cannot account for the dramatic improvement of the pivoted L-values over the R-values.

The pivoted QLP decomposition is a special case of the ULV decompositions to be treated in §4. It is distinguished from these by the way in which it is computed and by the fact that it is not updatable.

The decomposition has been introduced independently by Hosoda [184], who uses it to regularize ill-posed problems. However, he seems to have missed the interleaving property that allows the early termination of the algorithm — something especially desirable in regularization.

3. NORM AND CONDITION ESTIMATION

The condition number

$$\kappa(A) = \|A\|\|A^{-1}\|$$

of a matrix A of order n involves two quantities — the norm of a matrix A and the norm of its inverse. The norm of A may or may not be difficult to calculate. Of the commonly used norms, the 1-norm, the ∞-norm, and the Frobenius norm are easy to calculate. The 2-norm, which is the largest singular value of A, is expensive to calculate. Computing the norm of A^{-1} introduces the additional problem of calculating the inverse, which we have seen is an expensive undertaking.

In this section we will consider techniques by which norms of matrices and their inverses can be estimated at a reasonable cost. We will begin with the LAPACK algorithm for estimating the 1-norm of a matrix. This algorithm requires only the ability to form the products of A and A^T with a vector x. To estimate the 1-norm of A^{-1}, the necessary products can be formed from a suitable decomposition of A. In §3.2 we will consider some LINPACK-type estimators for $\|T^{-1}\|$, where T is triangular. Unlike the LAPACK estimator these estimators require a knowledge of the elements of the matrix in question. Finally, in §3.3 we will consider a method for estimating the 2-norm of a general matrix based on the QLP decomposition.

3.1. A 1-NORM ESTIMATOR

Since the matrix 1-norm of A is the maximum of the 1-norms of its columns, there is an index j for which

$$\|A\|_1 = \|A\mathbf{e}_j\|_1.$$

The LAPACK 1-norm estimator is based on a technique for finding indices $j_1, j_2, \ldots$ such that the quantities $\|A\mathbf{e}_{j_i}\|_1$ are strictly increasing. What makes the technique especially suitable for condition estimation is that it does not require that we know the elements of A or a decomposition of A — only that we be able to multiply arbitrary vectors by A and A^T.

To derive the algorithm, suppose we have a vector v of 1-norm one for which we hope $\|Av\|_1$ approximates $\|A\|_1$. We would like to determine if there is vector $\mathbf{e}_j$ that gives a better approximation. One way is to compute $A\mathbf{e}_j$ and compare its 1-norm with the 1-norm of Av. This is fine for a single vector, but if we wish to investigate all possible vectors $\mathbf{e}_j$, the overhead becomes unacceptable.

To circumvent this problem, we use a weaker test that can fail to recognize when $\|A\mathbf{e}_j\|_1 > \|Av\|_1$. Set

$$u = Av \quad \text{and} \quad w = \text{sign}(u)$$

[i.e., $w_i = \text{sign}(u_i)$], so that

$$\|u\|_1 = w^T u.$$

Since the components of w are 0, or ± 1, we have

$$\|A\mathbf{e}_j\|_1 \geq |w^T A\mathbf{e}_j| = |x^T \mathbf{e}_j|, \qquad (3.1)$$

where

$$x = A^T w.$$

It follows that

$$\max_j \|A\mathbf{e}_j\|_1 \geq \max_j |x^T \mathbf{e}_j| = \|x\|_\infty.$$

Thus if $\|x\|_\infty > \|u\|_1$ and $\|x\|_\infty = |x_j|$, then $\|A\mathbf{e}_j\|_1 > \|u\|_1$. Hence we can restart our search by replacing v with $\mathbf{e}_j$.

These considerations lead to the following algorithm.

1. $v =$ an initial vector with $\|v\|_1 = 1$
2. **for** $k = 1, 2, \ldots$
3. $u = Av$
4. $w = \text{sign}(u)$
5. $x = A^T w$
6.   **if** $(\|x\|_\infty \leq \|u\|_1)$ **leave** k **fi**
7.   Choose j so that $|x_j| = \|x\|_\infty$
8. $v = \mathbf{e}_j$
9. **end for** k

SEC. 3. NORM AND CONDITION ESTIMATION

The program must terminate, since the norms $\|u\|_1$ are strictly increasing. On termination $\|u\|_1 = \|Av\|_1$ is the estimate of the 1-norm.

Because we have replaced $\|Ae_j\|_1$ by the lower bound $|w^T Ae_j|$ from (3.1), the test in statement 6 can bypass a vector e_j that gives a better estimate. In fact, examples can be constructed for which the algorithm underestimates the 1-norm of A by an arbitrary amount (see the notes and references). Nonetheless, the algorithm is very good and can be made even better by the following modifications.

1. Rather than starting with a unit vector, the initial vector is taken to be $n^{-1}\mathbf{e}$. This mixes the columns and avoids a chance choice of an uncharacteristically small column.
2. The number of iterations is restricted to be at least two and no more than five.
3. The sequence of norm estimates is required to be strictly increasing (to avoid cycling in finite precision arithmetic).
4. If the vector w is the same as the previous w convergence is declared.
5. On convergence the estimate is compared with the estimate obtained with the vector

$$\begin{pmatrix} 1 & -1 - \frac{1}{n-1} & 1 + \frac{2}{n-1} & \cdots & (-1)^{n-1} \cdot 2 \end{pmatrix}$$

and the larger of the two taken as the estimate. This provides an additional safeguard against an unfortunate starting vector.

Algorithm 3.1 implements this scheme. Here are some comments.

- The major source of work in the algorithm is the formation of matrix-vector products. The algorithm requires a minimum of four such products and a maximum of eleven. The average is between four and five.

- The algorithm can be fooled, but experience shows that the estimate is unlikely to be less than the actual norm by more than a factor of three. In fact, rounding errors cause the algorithm to perform rather well on examples specifically designed to cause it to fail.

- If A is replaced by A^T, the algorithm estimates the ∞-norm of A.

Turning to applications of Algorithm 3.1, we begin with condition estimation. To estimate $\kappa_1(A)$, we must estimate $\|A\|_1$ and $\|A^{-1}\|_1$. If we have A, we can calculate $\|A\|_1$ directly. However, if we have overwritten A with a factorization, say for definiteness

$$A = LU,$$

we can use Algorithm 3.1 to estimate $\|A\|_1$ by computing products in the forms

$$Ax = L(Ux) \quad \text{and} \quad A^T x = U^T(L^T x).$$

Given a matrix A this algorithm returns an estimate *nrm* of the 1-norm of A and a vector v such that $\|Av\|_1 = nrm$.

1. $v = n^{-1}\mathbf{e}$; $w_{\text{old}} = 0$; $nrm_{\text{old}} = 0$
2. **for** $k = 1, \ldots, 5$
3. $u = Av$
4. $nrm = \|u\|_1$
5. $w = \text{sign}(u)$
6.   **if** ($nrm \leq nrm_{\text{old}}$ or $w = w_{\text{old}}$) **leave** k **fi**
7. $w_{\text{old}} = w$; $nrm_{\text{old}} = nrm$
8. $x = A^{\text{T}} w$
9.   **if** ($\|x\|_\infty \leq nrm$ and $k \neq 1$) **leave** k **fi**
10.   Choose j so that $|x_j| = \|x\|_\infty$
11. $v = \mathbf{e}_j$
12. **end for** k
13. $y_i = (-1)^{i+1}\left(1 + \frac{i-1}{n-1}\right)$ $(i = 1, \ldots, n)$
14. $nrm_{\text{alt}} = \frac{2}{3n}\|Ay\|_1$
15. **if** ($nrm_{\text{alt}} > nrm$)
16. $nrm = nrm_{\text{alt}}$; $v = y$
17. **end if**

Algorithm 3.1: A 1-norm estimator

───────── ◇ ─────────

Similarly, to estimate $\|A^{-1}\|_1$ we can compute products with the inverse in the forms

$$A^{-1}x = U^{-1}(L^{-1}x) \quad \text{and} \quad A^{-\text{T}}x = L^{-\text{T}}(U^{-\text{T}}x),$$

where the multiplications by L^{-1} etc. are accomplished by solving triangular systems.

If A is ill conditioned and we apply Algorithm 3.1 to estimate $\|A^{-1}\|_1$, we also get an approximate null vector of A. To see this assume that A has been scaled so that $\|A\|_1 = 1$. Then the algorithm returns a vector v and $u = A^{-1}v$ such that

$$\|u\|_1 \cong \|A^{-1}\|_1 = \kappa_1(A).$$

It follows that

$$\frac{\|Au\|_1}{\|u\|_1} \cong \frac{\|v\|_1}{\kappa_1(A)} = \kappa_1^{-1}(A),$$

which is small because A is ill conditioned. Thus $u/\|u\|_1$ is an approximate null vector of A.

SEC. 3. NORM AND CONDITION ESTIMATION

The algorithm is especially useful in computing mixed perturbation bounds of the form

$$\frac{\|\tilde{x} - x\|_\infty}{\|\tilde{x}\|_\infty} \leq \frac{\||A^{-1}||E||\tilde{x}|\|_\infty}{\|\tilde{x}\|_\infty}$$

(see Corollary 3.13, Chapter 3). Specifically, if we let D be the diagonal matrix whose diagonal entries are the components of $|E||\tilde{x}|$, then

$$\||A^{-1}||E||\tilde{x}|\|_\infty = \||A^{-1}D|\mathbf{e}|\|_\infty = \|A^{-1}D\|_\infty. \tag{3.2}$$

The last norm can be estimated by applying the algorithm to the matrix DA^{-T}.

3.2. LINPACK-STYLE NORM AND CONDITION ESTIMATORS

The heart of a LINPACK-style condition estimator is a method for estimating the norm of the inverse of a triangular matrix. Suppose, for example, that L is a lower triangular matrix and $\|\cdot\|$ is an operator norm. Then

$$\|L^{-1}\| = \max_{\|u\|=1} \|L^{-1}u\|.$$

To put it in another way that does not involve the explicit inverse of L,

$$\|L^{-1}\| = \max\{\|v\| : Lv = u, \|u\| = 1\}.$$

This suggests that one way to approximate $\|L\|$ is to choose a suitable u with $\|u\| = 1$ and solve the system

$$Lv = u. \tag{3.3}$$

By "suitable" we mean a vector u for which $\|v\|$ is large. In a LINPACK-style estimator we choose the components of u on the fly as we solve the triangular system $Lv = u$.

To illustrate the ideas we will start with a simple, cheap, but useful norm estimator. The estimator in LINPACK is an enhancement of this one, designed to overcome certain counterexamples. Because this enhanced estimator has been widely discussed in the literature, we will not present it here. Instead we will go on to treat a lesser known 2-norm estimator that uses an analogous enhancement.

A simple estimator

In the forward substitution algorithm for solving (3.3), the ith component of v is given by

$$v[i] = \frac{u[i] - L[1{:}i{-}1, i] * v[1{:}i{-}1]}{L[i, i]}. \tag{3.4}$$

Given a triangular matrix T of order n, this routine returns an estimate *inf* of $\inf(T)$ and a vector v of 2-norm one such that $\|Lv\|_2 = inf$

1.   *rightinf*(T, v, *nrm*)
2.   **if** (T is lower triangular)
3.   **for** $i = 1$ **to** n
4. $d = T[1{:}i{-}1, i] * v[1{:}i{-}1]$
5.   **if** ($d \geq 0$)
6. $v[i] = -(1{+}d)/T[i, i]$
7.   **else**
8. $v[i] = (1{-}d)/T[i, i]$
9.   **end if**
10.   **end for** i
11.   **else**! T is upper triangular
12.   **for** $i = n$ **to** 1 **by** -1
13. $d = T[i{+}1{:}n] * v[i{+}1{:}nv]$
14.   **if** ($d \geq 0$)
15. $v[i] = -(1{+}d)/T[i, i]$
16.   **else**
17. $v[i] = (1{-}d)/T[i, i]$
18.   **end if**
19.   **end for** i
20.   **end if**
21. $v = v/\sqrt{n}$
22. $inf = 1/\|v\|_2$
23. $v = inf * v$
24.   **end** *rightinf*

Algorithm 3.2: A simple LINPACK estimator

———————⋄———————

The simple strategy is to chose

$$u[i] = \begin{cases} -1, & \text{if } L[1{:}i{-}1, i] * v[1{:}i{-}1] \geq 0, \\ 1, & \text{if } L[1{:}i{-}1, i] * v[1{:}i{-}1] < 0. \end{cases}$$

This choice encourages growth in v by assuring that there is no cancellation in the denominator of the right-hand side of (3.4).

This scheme is implemented in Algorithm 3.2. There are several comments.

- Although we have chosen to estimate the spectral norm, we could have worked in any operator norm. Provided the norm is reasonably balanced with respect to the coordinate axes, the algorithm will generally give a ballpark estimate.

Sec. 3. Norm and Condition Estimation

- The implementation handles both upper and lower triangular systems.

- The algorithm requires $\frac{1}{2}n^2$ flam — the same as for the solution of a triangular system.

- In a quality implementation, the vector v would be scaled adaptively to prevent overflow. This can easily happen with innocuous-looking matrices. For example, in the matrix illustrated below for $n = 5$

$$\begin{pmatrix} 0.1 & 0 & 0 & 0 \\ -1 & 0.1 & 0 & 0 \\ 0 & -1 & 0.1 & 0 \\ 0 & 0 & -1 & 0.1 \end{pmatrix}$$

the components of v will grow at a rate of about 10^i. However, when T is a triangular factor from a decomposition of a balanced matrix A, such growth is unlikely. The reason is that the initial rounding of the matrix will increase the small singular values to approximately $\|A\|_2 \epsilon_M$.

- One could write a corresponding program to approximate left inferior vectors. But if transposing is cheap, then *rightinf*(T^T) will do the same job.

The following example reflects the ability of *rightinf* to reveal a rank deficiency.

Example 3.1. *The routine rightinf was used to estimate the smallest singular value of the matrix K_{100} of Example 2.2 for various values of the cosine of c. The following table shows the results.*

c	σ_{100}	inf
0.0	1.0e+00	1.0e+00
0.1	9.5e−05	2.1e−04
0.2	3.7e−09	1.2e−08
0.3	9.3e−14	3.6e−13
0.4	1.1e−18	5.2e−18

It is seen that the output of rightinf gives a good estimate of the smallest singular value of K_{100}.

An enhanced estimator

Algorithm 3.2 is a greedy algorithm. It attempts to increase each v_i as much as possible as it is generated. Many greedy algorithms can be made to fail because in their greed they eat up resources that are needed later. Algorithm 3.2 is no exception. In particular, a greedy choice at one point can make subsequent values of d in statements 4 and 13 too small. An important enhancement in the LINPACK estimator is a device to look ahead at the effects of the choice on these quantities. For variety, however, we will consider a different estimator, in which the components of u are allowed to vary, subject to the constraint that $\|u\|_2 = 1$.

To derive the algorithm, partition the lower triangular matrix L in the form

$$L = \begin{pmatrix} L_{11} & 0 & 0 \\ \ell_{k1}^T & \ell_{kk} & 0 \\ L_{k+1,1} & \ell_{k+1,k} & L_{k+1,k+1} \end{pmatrix},$$

and suppose that we have determined u_1 with $\|u_1\|_2$ such that the solution of the equation

$$L_{11} v_1 = u_1$$

is suitably large. For any c and s with $c^2 + s^2 = 1$ the right-hand side of the system

$$\begin{pmatrix} L_{11} & 0 \\ \ell_{k1}^T & \ell_{kk} \end{pmatrix} \begin{pmatrix} sv_1 \\ \frac{c - s\ell_{k1}^T v_1}{\ell_{kk}} \end{pmatrix} = \begin{pmatrix} su_1 \\ c \end{pmatrix} \tag{3.5}$$

has 2-norm one. Hence the solution $\hat{v}_1$ of (3.5) is a candidate for the next vector.

A greedy algorithm would choose c and s to maximize $\|\hat{v}_1\|_2$. However, such a strategy overlooks the fact that we will also want the components of

$$sL_{k+1,1} v_1 + \frac{c - s\ell_{k1}^T v_1}{\ell_{kk}} \ell_{k+1,k}$$

to be large when it comes time to determine subsequent components of v. Thus we might select a diagonal matrix D of weights and demand that we choose c and s so that

$$\left\| \begin{pmatrix} sv_1 \\ \frac{c - s\ell_{k1}^T v_1}{\ell_{kk}} \end{pmatrix} \right\|_2^2 + \left\| D \left(sL_{k+1,1} v_1 + \frac{c - s\ell_{k1}^T v_1}{\ell_{kk}} \ell_{k+1,k} \right) \right\|_2^2 \tag{3.6}$$

is maximized subject to $c^2 + s^2 = 1$.

At first sight this appears to be a formidable problem. But it simplifies remarkably. To keep the notation clean, set

$$\sigma = \ell_{kk}, \quad \tau = \ell_{k1}^T v_1, \quad p = L_{k+1,1} v_1, \quad \text{and} \quad q = \ell_{k+1,k}. \tag{3.7}$$

Then a tedious but straightforward calculation shows that the problem of maximizing (3.6) is equivalent to the following problem:

$$\text{maximize} \quad (c \; s) \begin{pmatrix} \alpha & \gamma \\ \gamma & \beta \end{pmatrix} \begin{pmatrix} c \\ s \end{pmatrix}$$

$$\text{subject to} \quad c^2 + s^2 = 1,$$

Sec. 3. Norm and Condition Estimation

Given a lower triangular matrix L of order n, this algorithm returns an estimator *invnorm* of $\|L^{-1}\|_2$ and a vector v of 2-norm one such that $\|Lv\|_2 = \text{invnorm}^{-1}$.

1. normlinv(L, v, *invnorm*)
2. $v[1] = 1$
3. $p = L[3{:}n, 1]$
4.   **for** $k = 2$ **to** n
5.   Choose D
6. $\sigma = L[k, k]$
7. $\tau = L[k, 1{:}k{-}1] * v[1{:}k{-}1]$
8. $q = L[k{+}1{:}n, k]$
9. $\alpha = 1 + q^T D^2 q$
10. $\beta = \sigma^2 * \|v[1{:}k{-}1]\|^2 + \tau^2 + \sigma^2 * p^T D^2 p$
                $- 2\sigma * \tau * p^T D^2 q + \tau^2 * q^T D^2 q$
11. $\gamma = -(\tau + \sigma * p^T D^2 q + \tau * q^T D^2 q)$
12.   Let $(c\ s)$ be the normalized eigenvector corresponding to the largest eigenvalue of
    $$\begin{pmatrix} \alpha & \gamma \\ \gamma & \beta \end{pmatrix}$$
13. $v[1{:}k{-}1] = s * v[1{:}k{-}1]$
14. $v[k] = (c - s*\tau)/\sigma$
15. $p = s*p[2{:}n{-}k] + v[k]*q[2{:}n{-}k]$
16.   **end for** k
17.   *invnorm* $= \|v\|_2$
18. $v = v/\textit{invnorm}$
19. **end** *normlinv*

Algorithm 3.3: An estimator for $\|L^{-1}\|_2$

———————◇———————

where

$$\alpha = 1 + q^T D^2 q,$$
$$\beta = \sigma^2 \|v_1\|_2^2 + \tau^2 + \sigma^2 p^T D^2 p - 2\sigma\tau p^T D^2 q + \tau^2 q^T D^2 q,$$
$$\gamma = -(\tau + \sigma p^T D^2 q + \tau q^T D^2 q).$$

Referring to (4.43), Chapter 1, we see that the solution is the normalized eigenvector corresponding to the largest eigenvalue of $\begin{pmatrix} \alpha & \gamma \\ \gamma & \beta \end{pmatrix}$.

With the exception of the formation of p, the above operations require only $O(n)$ work. The formation of p by the formula in (3.7) requires a higher order of work. Fortunately, p can be updated after c and s have been determined as follows:

$$p[k{+}1{:}n] = s*p[k{+}1{:}n] + c*L[k{+}1{:}n, k]$$

Algorithm 3.3 is an implementation of this scheme. Here are some comments.

- The behavior of the algorithm depends on the choice of D. Two alternatives have appeared in the literature.

  1. $D = I$. This choice has been found to give very good estimates of $\|L^{-1}\|_2$, sometimes accurate to a significant figure or more. However, in this case the algorithm costs about $3\frac{1}{2}n^2$ flam.
  2. $D = 0$. This gives an algorithm known as *incremental condition estimation*. It is cheap (n^2 flam) and has the advantage that L can be brought in a row at a time (whence its name).

For most matrices there is not much to choose between these alternatives. For the matrix K_n^T (Example 2.2), the case $D = I$ does a little worse. Perhaps a good compromise is to take

$$D = \operatorname{diag}(I_j, 0)$$

for j equal to one or two. This is effectively as fast as incremental condition estimation and provides a degree of additional protection.

- It is not recommended that the novice try to code the solution of the eigenproblem in statement 12. Conceptually, the solution of 2×2 eigenvalue problems is trivial. Practically, the details are difficult to get right. LAPACK has a routine (SLAEV2) to do the job.

- The comments made about scaling in Algorithm 3.2 apply here.

Condition estimation

In LINPACK the routines to estimate the norm of the inverse of a triangular matrix are used as part of a more extensive algorithm to estimate condition. To motivate the algorithm, consider the system

$$(A^T A)z = w. \tag{3.8}$$

Since $A^T A$ is symmetric and positive definite, it has a spectral decomposition

$$V^T (A^T A) V = \Sigma^2,$$

where $V = (v_1 \cdots v_n)$ is orthogonal and Σ is a diagonal matrix consisting of the singular values of A [see (4.45), Chapter 1]. If we set

$$\nu_i = v_i^T w,$$

then it is easily verified that

$$z = \frac{\nu_1}{\sigma_1^2} v_1 + \cdots + \frac{\nu_n}{\sigma_n^2} v_n.$$

Sec. 3. Norm and Condition Estimation

Thus if ν_n is not unusually small, the last term in the above sum will dominate, and z will grow in proportion to σ_n^{-2}.

In LINPACK the triangular estimators are used to insure that the number ν_n is not too small. Specifically, if we have decomposed $A = LU$, the first step in solving the system (3.8) is to solve the triangular system

$$U^T x = w.$$

If we use a triangular estimator to determine the right-hand side w, then x will reflect the ill-conditioning of U. Now if the LU factorization of A has been computed using pivoting, the matrix L will tend to be well conditioned and hence x will also reflect the ill-conditioning of A — that is, it will have significant components along the inferior singular vectors. These considerations lead to the following condition estimator.

1. Solve the system $U^T x = w$ using a triangular estimator to encourage growth in x
2. Solve $L^T y = x$
3. Solve $Az = y$
4. Estimate $\|A^{-1}\|$ by $\|z\|/\|y\|$

3.3. A 2-norm estimator

We have see that the first L-value in the pivoted QLP decomposition of a matrix X is generally a good estimate of $\|X\|_2$. We have also indicated that it is possible to interleave the computation of the R-factor and the L-factor. This suggests that we estimate $\|X\|_2$ by computing a few rows of the pivoted R-factor. Usually (but not always) the norm of the largest row will be the $(1,1)$-element of the pivoted L-factor — the first L value. In this subsection we will describe how to implement this estimation scheme.

The pivoted R-factor of X is the pivoted Cholesky factor of the cross-product matrix $A = X^T X$. We will now show how to compute the factor row by row. We begin with the unpivoted version.

Suppose we have computed the first $k-1$ rows of R:

$$(R_{11} \ R_{12}),$$

where R_{11} is triangular of order $k-1$. Then (cf. Theorem 1.4, Chapter 3)

$$A - \begin{pmatrix} R_{11}^T \\ R_{12}^T \end{pmatrix} (R_{11} \ R_{12}) = \begin{pmatrix} 0 & 0 \\ 0 & S \end{pmatrix},$$

where S is the Schur complement of $A_{11} = R_{11}^T R_{11}$ in A. Since the kth row of R is the first row of S divided by the square root of the $(1,1)$-element of S, we can compute the elements of the kth row as follows.

1. **for** $j = k$ **to** p
2. $\quad r_{kj} = x_k^T x_j$
3. $\quad r_{kj} = r_{kj} - r_{1k} r_{1j} - \cdots - r_{k-1,k} r_{k-1,j}$
4. $\quad r_{kj} = r_{kj}/\sqrt{r_{kk}}$
5. **end for** k

The first statement in the loop generates an element in the kth row of A. The second statement computes the Schur complement of that element. The third statement scales the element so that it becomes an element of the R-factor.

Turning now to the pivoted algorithm, we must answer two questions.

1. How do we keep track of the norms to determine the pivots?
2. How do we organize the interchanges?

The answer to the first question is that the squares of the norms that determine the pivots are the diagonals of the Schur complement [see the discussion surrounding (2.12)]. These quantities can be formed initially and downdated as we add rows. The answer to the second question is that we perform no interchanges. Instead we keep track of indices of the columns we have selected as pivots and skip operations involving them as we add the kth row. For example, with a pivot sequence of 3, 2, 4, 1 in a 4×4 matrix we would obtain an "R-factor" of the form

$$\begin{pmatrix} r_{14} & r_{12} & r_{11} & r_{13} \\ r_{24} & r_{22} & 0 & r_{23} \\ r_{34} & 0 & 0 & r_{33} \\ r_{44} & 0 & 0 & 0 \end{pmatrix}.$$

Algorithm 3.4 is an implementation of this scheme. The heart of it is the loop on j, in which $R[k, j]$ is initialized, transformed into the corresponding element of the Schur complement, and normalized. The norms of the reduced columns of X, which are contained in the array *normx2*, are also updated at this point. The square of the norm of the row is accumulated in *nr2* and then compared with *norm2est*. Here are some comments.

- The variable *kmax* is an upper bound on the number of rows of R to compute. Two or perhaps three is a reasonable number. Another possibility is to compute rows of R until a decrease in norm occurs.

- The bulk of the work is in computing the norms of the columns of X and initializing R. Specifically if X is $n \times p$ and *kmax* is not too large, the algorithm requires about $(kmax+1)np$ flam.

- The algorithm can produce an underestimate. For example, consider the $n \times n$ matrix

$$X = \begin{pmatrix} I_{kmax} & 0 \\ 0 & \frac{1}{\sqrt{n}} ee^T \end{pmatrix}.$$

SEC. 3. NORM AND CONDITION ESTIMATION

Given an $n \times p$ matrix X, this algorithm returns an estimate *norm2est* of $\|X\|_2$.

1. $normx2[j] = \|X[:,j]\|_2^2 \ (j = 1, \ldots, p)$
2. $\mathcal{P} = \emptyset$
3. $norm2est = 0$
4.   **for** $k = 1$ **to** *kmax*
5.   Choose $pvt \notin \mathcal{P}$ so that $nrmx2[pvt] \geq nrmx2[j] \ (j \notin \mathcal{P})$
6. $nr2 = nrmx2[pvt]$
7. $rkk = \sqrt{nr2}$
8.   **for** $j = 1$ **to** $p, j \notin \mathcal{P}$
9. $R[k,j] = X[:,pvt]^\text{T} * X[:,j]$
10.   **for** $i = 1$ **to** $k-1$
11. $R[k,j] = R[k,j] - R[i,pvt] * R[i,j]$
12.   **end for** i
13. $R[k,j] = R[k,j]/rkk$
14. $nr2 = nr2 + R[k,j]^2$
15. $normx2[j] = normx2[j] - R[k,j]^2$
16.   **end for** j
17. $norm2est = \max\{norm2est, nr2\}$
18.   **end for** k
19. $norm2est = \sqrt{norm2est}$

Algorithm 3.4: A 2-norm estimator

⸺⬦⸺

Because the first *kmax* columns of X dominate the rest, the estimator will return a value of *norm2est* of one. But the norm of the matrix is $(n - kmax)/\sqrt{n}$.

3.4. NOTES AND REFERENCES

General

The subject of this section goes under the rubric of "condition estimation." The terminology is unfortunate because it obscures the fact that the heart of a condition estimator is an algorithm that computes a norm and provides an approximate null vector when the norm is small. In many applications these quantities are more useful than the condition number itself.

    Higham [174], [177, Ch. 14] surveys condition estimation, giving many references and historical comments. The former reference contains detailed numerical experiments. The latter reference discusses several estimators not treated here: a general p-norm estimator due to Boyd [50], a probabilistic estimator due to Dixon [97], and condition numbers for tridiagonal matrices.

    The technique for estimating componentwise bounds [see (3.2)] is due to Arioli,

Demmel, and Duff [12].

LINPACK-style condition estimators

The heart of a LINPACK-style condition estimator is an algorithm for approximating an inferior singular vector of a triangular matrix. The first such algorithm — essentially Algorithm 3.2 — was proposed by Gragg and Stewart [157, 1976] to detect and remedy ill-conditioning in the secant method for nonlinear equations. The authors of LINPACK adopted the method with modifications to get around certain counterexamples [69]. The actual LINPACK condition estimator returns the reciprocal of the condition number, which is allowed to underflow when it becomes too large. In fact, scaling to avoid overflow while solving the triangular system can be quite expensive. Grimes and Lewis [160] show how to reduce the scaling overhead.

Experimental results [174, 294] indicate that the LINPACK estimator is quite reliable. However, counterexamples exist [70] and improvements and variations have been published. Cline and Rew [70] question whether variations to handle contrived, unlikely counterexamples are really necessary — surely a defensible position.

One of the variations is the enhanced estimator (Algorithm 3.3) of Cline, Conn, and Van Loan [68] (see also [329]). As we have presented it here, it includes Bischof's incremental condition estimation [30] as a special case ($D = 0$ in statement 5 of the algorithm). Bischof gives an analysis of how the algorithm can fail.

The 1-norm estimator

Although we have presented the 1-norm estimator first, it was proposed by Hager [167, 1984] as an alternative to the LINPACK estimator. Since it depends only on matrix-vector multiplications it is more widely applicable than LINPACK-style estimators. Higham [175], recognizing this fact, added the improvements listed on page 389 and gave a quality implementation. The estimator is used in LAPACK.

Hager derived his estimator using optimization theory for nondifferentiable convex functions. The simpler derivation given here is new.

The 2-norm estimator

The 2-norm estimator based on the pivoted QLP decomposition is new. Unfortunately, there seems to be no way to adapt it to estimate the 2-norm of an inverse.

4. UTV DECOMPOSITIONS

In some applications we must track the rank and the fundamental subspaces of a matrix which changes over time. For example, in signal processing one must deal with a sequence of matrices defined by the recursion

$$X_{k+1} = \begin{pmatrix} \beta X_k \\ x_{k+1}^{\mathrm{T}} \end{pmatrix}, \tag{4.1}$$

SEC. 4. UTV DECOMPOSITIONS

where $\beta < 1$ is a positive *forgetting factor* that damps out the effects of old information contained in X_k (this way of damping is called *exponential windowing*). In general, the matrix X_k will be a perturbation of a matrix $\ddot{X}_k$ whose rank is exactly m. Thus the problem of tracking rank amounts to updating a decomposition that reveals the small singular values due to the perturbation.

Unfortunately, the singular value decomposition itself is not updatable — or rather not cheaply updatable. An alternative is to update a pivoted QR or QLP decomposition; however, no satisfactory gap-revealing algorithm to do this is known. The approach taken in this section is to update a decomposition of the form

$$U^{\mathrm{T}} X V = \begin{pmatrix} T \\ 0 \end{pmatrix},$$

where T is triangular. Such a decomposition, which we will call a *UTV decomposition*, lies on a continuum with the singular value decomposition at one end and the QR decomposition at the other. By relaxing the condition that the final matrix be diagonal, we obtain a decomposition that can be updated cheaply. By replacing the permutation matrix in a pivoted QR decomposition with the orthogonal matrix V we insure that the updated matrix continues to reveal gaps.

In this section, we will consider *URV decompositions*, in which T is upper triangular, and *ULV decompositions*, in which T is lower triangular. The former is especially useful in low-rank problems. The latter, although more expensive, produces a higher quality subspace.

We shall loosely speak of these decomposition as gap revealing. However, it is more precise to say that the decomposition reveals the presence of small singular values. In particular, as the rank changes, there may be no gap in the singular values.

Since it is important to preserve the error structure in updating our decompositions, the first subsection discusses plane rotations and errors. We then go on to describe the updating algorithms for the two decompositions.

4.1. ROTATIONS AND ERRORS

In §1.3, Chapter 4, we treated plane rotations as a combinatorial game in which X's and 0's were preserved or annihilated depending on how they were combined. In this section we will be concerned with a third category of small elements — the errors in X. Here we will sketch the rules of the extended rotation game, recapitulating the rules of the original game.

Figure 4.1 shows two rows of a matrix before and after the application of a plane rotation. As usual in a Wilkinson diagram, the X's represent nonzero elements and the 0's represent zero elements. The E's represent small elements. The plane rotation has been chosen to introduce a zero into the position occupied by the hatted X in column 2. When the rotation is applied the following rules hold.

```
                Before
    1   2   3   4   5   6   7   8
    X   X   X   0   0   X   E   E
    X   X̂   X   X   0   E   E   0

                After
    1   2   3   4   5   6   7   8
    X   X   X   X   0   X   E   E
    X   0   X   X   0   X   E   E
```

Figure 4.1: Application of a plane rotation

───────────────◇───────────────

1. A pair of X's remains a pair of X's (columns 1 and 3).
2. An X and an 0 are replaced by a pair of X's (column 4).
3. A pair of 0's remains a pair of 0's (column 5).
4. An X and an E are replaced by a pair of X's (column 6).
5. A pair of E's remains a pair of E's (column 7).
6. An E and an 0 are replaced by a pair of E's (column 8).

(4.2)

The fact that a pair of small elements remains small (column 7) follows from the fact that a plane rotation is orthogonal and cannot change the norm of any vector to which it is applied. As usual a zero paired with a nonzero element is annihilated, but if the other element is an E, the zero is replaced by an E (column 8).

The importance of these observations for our algorithms is that it is possible to organize calculations with plane rotations in such a way as to preserve patterns of small elements. In particular, we can preserve the gap-revealing structure of URV and ULV decompositions.

A little additional nomenclature will prove useful. Premultiplication by a plane rotation operates on the rows of the matrix. We will call such rotations *left rotations*. Postmultiplication by *right rotations* operates on the columns. Rules analogous to those in (4.2) hold when a right rotation combines two columns of a matrix. We will denote left and right rotations in the (i, j)-plane by Q_{ij} and P_{ij} respectively.

4.2. UPDATING URV DECOMPOSITIONS

In this subsection we will show how to update URV decompositions. We will begin with the basic algorithm and then add extensions and modifications.

URV decompositions

Let X be an $n \times p$ matrix. A URV decomposition of X is a decomposition of the form

$$U^T X V = \begin{pmatrix} R \\ 0 \end{pmatrix}, \qquad (4.3)$$

where U and V are orthogonal and R is upper triangular. As we mentioned in the introduction to this section, there are many URV decompositions — including the singular value decomposition and the QR decomposition.

Suppose X has a gap in its singular values at m. We will say that a URV decomposition of X is gap revealing if it can be partitioned in the form

$$(U_1 \; U_2)^T X (V_1 \; V_2) = \begin{pmatrix} S & H \\ 0 & F \\ 0 & 0 \end{pmatrix},$$

where

1. S is of order m,
2. $\inf(S) \cong \sigma_m(X)$,
3. $\|F\|_2 \cong \sigma_{m+1}(X)$,
4. $\|H\|_2$ is suitably small.

The last condition insures that the blocks of the partitioned matrices U and V will approximate the fundamental subspaces of X. We are going to show how to update a gap-revealing URV decomposition when a row is added to X. Although it is possible to update both the matrices U and V, the order of U can grow beyond reasonable bounds as more and more rows are added. Fortunately, in most applications we are interested in only the right fundamental subspaces. Consequently, we will ignore U in what follows.

The basic algorithm consists of two steps: the updating proper and the adjustment of the gap. We will consider each in turn.

Incorporation

We will suppose that we have a rank revealing URV decomposition of the form (4.3) and that we wish to updated it by adding a row x^T to X. The first step is to transform it into the coordinate system corresponding to V. This is done as follows

1. $y^T = x^T V$

We are then left with the problem of incorporating y^T into the decomposition — i.e., of reducing the matrix

$$\begin{pmatrix} S & H \\ 0 & F \\ y_1^T & y_2^T \end{pmatrix}$$

to triangular form.

To see the chief difficulty in effecting this reduction, imagine that H and F are exactly zero, so that the matrix is of rank m. If y_2 is nonzero, the updated matrix will be of rank $m+1$. Now if we use plane rotations to fold y_1 into S [see (3.14), Chapter 4], quantities from y_2 will fill H and F, and it will be impossible to tell that the matrix is of rank $m+1$. In the general case, where H and F are small but y_2 is large, the gap that should be at position $m+1$ will be obliterated.

Thus we must distinguish two cases. In the first, where y_2 is small enough, we can simply perform Cholesky updating as in (3.14), Chapter 4. This will cost $\frac{1}{2}p^2$ flrot for the reduction. Since only left rotations are performed, no rotations have to be accumulated in V. We will call this form of incorporation *simple incorporation*.

If y_2 is too large, we zero out the last $p-m$ components of y^T, so that the effect of y_2 on the updating is restricted to the $(m+1)$th column. We will call this process *constrained incorporation*. The process, which affects only H and F, is illustrated below.

$$
\begin{pmatrix} h & h & h & h \\ f & f & f & f \\ 0 & f & f & f \\ 0 & 0 & f & f \\ 0 & 0 & 0 & f \\ y & y & y & \widehat{y} \end{pmatrix}
\overset{P_{34}}{\Longrightarrow}
\begin{pmatrix} h & h & h & h \\ f & f & f & f \\ 0 & f & f & f \\ 0 & 0 & f & f \\ 0 & 0 & \widehat{f} & f \\ y & y & y & 0 \end{pmatrix}
\overset{Q_{45}}{\Longrightarrow}
\begin{pmatrix} h & h & h & h \\ f & f & f & f \\ 0 & f & f & f \\ 0 & 0 & f & f \\ 0 & 0 & 0 & f \\ y & y & \widehat{y} & 0 \end{pmatrix}
$$

$$
\overset{P_{23}}{\Longrightarrow}
\begin{pmatrix} h & h & h & h \\ f & f & f & f \\ 0 & f & f & f \\ 0 & \widehat{f} & f & f \\ 0 & 0 & 0 & f \\ y & y & 0 & 0 \end{pmatrix}
\overset{Q_{34}}{\Longrightarrow}
\begin{pmatrix} h & h & h & h \\ f & f & f & f \\ 0 & f & f & f \\ 0 & 0 & f & f \\ 0 & 0 & 0 & f \\ y & \widehat{y} & 0 & 0 \end{pmatrix}
$$

$$
\overset{P_{12}}{\Longrightarrow}
\begin{pmatrix} h & h & h & h \\ f & f & f & f \\ \widehat{f} & f & f & f \\ 0 & 0 & f & f \\ 0 & 0 & 0 & f \\ y & 0 & 0 & 0 \end{pmatrix}
\overset{Q_{23}}{\Longrightarrow}
\begin{pmatrix} h & h & h & h \\ f & f & f & f \\ 0 & f & f & f \\ 0 & 0 & f & f \\ 0 & 0 & 0 & f \\ y & 0 & 0 & 0 \end{pmatrix}
$$

Note that in these diagrams the h's represent the last row of H, and the planes of the rotations are relative to the northwest corner of the diagram. Code for this reduction is given below.

SEC. 4. UTV DECOMPOSITIONS 405

1. **for** $k = p-1$ **to** $m+1$ **by** -1
2. $\quad rotgen(y[k], y[k+1], c, s)$
3. $\quad rotapp(R[1{:}k+1, k], R[1{:}k+1, k+1], c, s)$
4. $\quad rotapp(V[:, k], V[:, k+1], c, s)$ (4.4)
5. $\quad rotgen(R[k, k]\ R[k+1, k], c, s)$
6. $\quad rotapp(R[k, k+1{:}p], R[k+1, k+1{:}p], c, s)$
7. **end for** k

Note that we must accumulate the right rotations in V. If we were updating U we would also have to accumulate the left rotations.

We now have a simple incorporation problem of the form

$$\begin{pmatrix} s & s & s & h & h & h \\ 0 & s & s & h & h & h \\ 0 & 0 & s & h & h & h \\ 0 & 0 & 0 & f & f & f \\ 0 & 0 & 0 & 0 & f & f \\ 0 & 0 & 0 & 0 & 0 & f \\ y & y & y & y & 0 & 0 \end{pmatrix}.$$

The reduction of this matrix to lower triangular form proceeds as follows.

$$\begin{pmatrix} s & s & s & h & h & h \\ 0 & s & s & h & h & h \\ 0 & 0 & s & h & h & h \\ 0 & 0 & 0 & f & f & f \\ 0 & 0 & 0 & 0 & f & f \\ 0 & 0 & 0 & 0 & 0 & f \\ \widehat{y} & y & y & y & 0 & 0 \end{pmatrix} \stackrel{Q_{17}}{\Longrightarrow} \begin{pmatrix} s & s & s & y & h & h \\ 0 & s & s & h & h & h \\ 0 & 0 & s & h & h & h \\ 0 & 0 & 0 & f & f & f \\ 0 & 0 & 0 & 0 & f & f \\ 0 & 0 & 0 & 0 & 0 & f \\ 0 & \widehat{y} & y & y & e & e \end{pmatrix} \stackrel{Q_{27}}{\Longrightarrow}$$

$$\begin{pmatrix} s & s & s & y & h & h \\ 0 & s & s & y & h & h \\ 0 & 0 & s & h & h & h \\ 0 & 0 & 0 & f & f & f \\ 0 & 0 & 0 & 0 & f & f \\ 0 & 0 & 0 & 0 & 0 & f \\ 0 & 0 & \widehat{y} & y & e & e \end{pmatrix} \stackrel{Q_{37}}{\Longrightarrow} \begin{pmatrix} s & s & s & y & h & h \\ 0 & s & s & y & h & h \\ 0 & 0 & s & y & h & h \\ 0 & 0 & 0 & f & f & f \\ 0 & 0 & 0 & 0 & f & f \\ 0 & 0 & 0 & 0 & 0 & f \\ 0 & 0 & 0 & \widehat{y} & e & e \end{pmatrix} \stackrel{Q_{47}}{\Longrightarrow}$$

$$\begin{pmatrix} s & s & s & y & h & h \\ 0 & s & s & y & h & h \\ 0 & 0 & s & y & h & h \\ 0 & 0 & 0 & y & f & f \\ 0 & 0 & 0 & 0 & f & f \\ 0 & 0 & 0 & 0 & 0 & f \\ 0 & 0 & 0 & 0 & \widehat{e} & e \end{pmatrix} \stackrel{Q_{57}}{\Longrightarrow} \begin{pmatrix} s & s & s & y & h & h \\ 0 & s & s & y & h & h \\ 0 & 0 & s & y & h & h \\ 0 & 0 & 0 & y & f & f \\ 0 & 0 & 0 & 0 & f & f \\ 0 & 0 & 0 & 0 & 0 & f \\ 0 & 0 & 0 & 0 & 0 & \widehat{e} \end{pmatrix} \stackrel{Q_{67}}{\Longrightarrow}$$

$$\begin{pmatrix} s & s & s & y & h & h \\ 0 & s & s & y & h & h \\ 0 & 0 & s & y & h & h \\ 0 & 0 & 0 & y & f & f \\ 0 & 0 & 0 & 0 & f & f \\ 0 & 0 & 0 & 0 & 0 & f \\ 0 & 0 & 0 & 0 & 0 & 0 \end{pmatrix}$$

By items 5 and 6 in the list (4.2) of rules for plane rotations, the elements in the last $m-p$ columns of H and F must remain small. However, if y_{m+1} is large, its effect will spread through the first columns of H and F. In other words, the addition of x^T to the decomposition has the potential to increase the estimated rank by one. Code for this reduction is given below.

1. **for** $k = 1$ **to** p
2. $rotgen(R[k,k], y[k], c, s)$
3. $rotapp(R[k, k+1{:}p], y[k+1{:}p], c, s)$ (4.5)
4. **end for** k

This completes the incorporation process.

Adjusting the gap

Mathematically, the addition of a row to a matrix can only increase the rank. However, if we are using a forgetting factor to damp out past information [see (4.1)], the point m at which the decomposition is to be split can increase, decrease, or remain unchanged. Thus after updating we must adjust the gap. In what follows, we will assume the existence of a user-supplied function *splitat*(m) that returns a value of true if the decomposition can be split at m. This may happen if there is a gap in the singular values at m or if the matrix F is below the error level.

The first step in our adjustment process is to tentatively increase m until a gap is found.

1. **while** (not *splitat*(m))
2. $m = m+1$ (4.6)
3. **end while**

Deflation

We must now consider the possibility that m is too large, i.e., that S has a small singular value that is not revealed by the current decomposition. To do this we estimate the smallest singular value of S by a LINPACK-style estimator. Call the estimate σ. A byproduct of this process is a vector w of 2-norm one such that $\sigma = \|Sw\|_2$. If σ is suitably small, we conclude that m is too large.

In the event that m is too large, we must alter the decomposition to reveal the small singular value — a process we will call *deflation*. The basic idea is simple. Suppose

Sec. 4. UTV decompositions

we have an orthogonal transformation P such that $Pw = \mathbf{e}_m$. If we determine another orthogonal matrix Q such that $\hat{S} = QSP^\mathrm{T}$ is upper triangular, then

$$\sigma = \|Sw\|_2 = \|(QSP^\mathrm{T})Pw\|_2 = \|\hat{S}\mathbf{e}_m\|.$$

Thus the last column of S has 2-norm σ and can be moved into H and F by reducing m.

In practice we will determine P and Q as the product of plane rotations. The reduction of w goes as follows.

$$\begin{pmatrix}\hat{w}\\w\\w\\w\end{pmatrix} \stackrel{P_{21}}{\Longrightarrow} \begin{pmatrix}0\\\hat{w}\\w\\w\end{pmatrix} \stackrel{P_{32}}{\Longrightarrow} \begin{pmatrix}0\\0\\\hat{w}\\w\end{pmatrix} \stackrel{P_{43}}{\Longrightarrow} \begin{pmatrix}0\\0\\0\\1\end{pmatrix}$$

As the rotations P_{ij} are generated they are applied to S, and the resulting deviation from triangularity is undone by a left rotation.

$$\begin{pmatrix}s&s&s&s\\0&s&s&s\\0&0&s&s\\0&0&0&s\end{pmatrix} \stackrel{P_{21}^\mathrm{T}}{\Longrightarrow} \begin{pmatrix}s&s&s&s\\\hat{s}&s&s&s\\0&0&s&s\\0&0&0&s\end{pmatrix} \stackrel{Q_{12}}{\Longrightarrow} \begin{pmatrix}s&s&s&s\\0&s&s&s\\0&0&s&s\\0&0&0&s\end{pmatrix} \stackrel{P_{32}^\mathrm{T}}{\Longrightarrow}$$

$$\begin{pmatrix}s&s&s&s\\0&s&s&s\\0&\hat{s}&s&s\\0&0&0&s\end{pmatrix} \stackrel{Q_{23}}{\Longrightarrow}$$

$$\begin{pmatrix}s&s&s&s\\0&s&s&s\\0&0&s&s\\0&0&0&s\end{pmatrix} \stackrel{P_{43}^\mathrm{T}}{\Longrightarrow} \begin{pmatrix}s&s&s&s\\0&s&s&s\\0&o&s&s\\0&0&\hat{s}&s\end{pmatrix} \stackrel{Q_{34}}{\Longrightarrow} \begin{pmatrix}s&s&s&g\\0&s&s&g\\0&0&s&g\\0&0&0&f\end{pmatrix}$$

The labeling of the last column of the result of this reduction reflects where the elements will end up when m is reduced.

The following is code for the reduction. We place it in a loop in which the process is repeated until m cannot be reduced.

1. **while** (1=1)
2.   Determine σ and w such that σ is an estimate of the smallest singular value of $R[1{:}m, 1{:}m]$, $\|w\|_2 = 1$, and $\|R[1{:}m, 1{:}m]{*}w\|_2 = \sigma$
3.   **if** (σ is large) **leave** the while loop **fi**
4.   **for** $k = 1$ **to** $m-1$
5. $rotgen(w[k{+}1], w[k], c, s)$
6. $rotapp(R[1{:}k{+}1, k{+}1], R[1{:}k{+}1, k], c, s)$ (4.7)
7. $rotapp(V[:, k{+}1], V[:, k], c, s)$
8. $rotgen(R[k, k], R[k{+}1, k], c, s)$
9. $rotapp(R[k, k{+}1{:}p], R[k{+}1, k{+}1{:}p], c, s)$
10.   **end for** k
11.   Refine the decomposition (optional)
12. $m = m-1$;
13. **end while**

We have added an additional refinement step, which will be treated later.

The URV updating algorithm

We are now in a position to assemble the basic URV updating algorithm, which is displayed in Algorithm 4.1. The display numbers refer to the program fragments in the derivation of the algorithm. Here are some comments.

- This is not an algorithm about which we can prove a great deal. In practice it is found to track gaps with considerable fidelity.

- The quality of the decomposition depends on the algorithm for estimating an inferior vector. In practice the simple estimator in Algorithm 3.2 seems to work well.

- The algorithm is quite stable. If it is used with a forgetting factor as in (4.1), old errors in R damp out. However, over a long period errors can accumulate in V.

- A particularly nice feature of the algorithm is that it is not necessary to start with a precomputed URV decomposition. Instead take $U = I$, $R = 0$, and $V = I$, and use the algorithm to add rows of X.

- There is no one operation count for this complicated algorithm. Here are counts for the various pieces.

  1. p^2 flam to compute $y^T = x^T V$.
  2. $\frac{1}{2}p^2$ flrot for simple updating.
  3. $[\frac{5}{2}(p-m)^2 + p(p-m)]$ flrot for constrained updating.
  4. $\frac{1}{2}m^2$ flam to estimate an inferior vector (more if something more elaborate than Algorithm 3.2 is used).
  5. $(\frac{1}{2}mp + m^2)$ flrot for each step of the deflation process.

Sec. 4. UTV decompositions

Given a gap-revealing URV decomposition of the form

$$(U_1 \ U_2)^T X (V_1 \ V_2) = \begin{pmatrix} S & H \\ 0 & F \\ 0 & 0 \end{pmatrix}$$

and a vector x, this algorithm computes a gap-revealing decomposition of

$$\begin{pmatrix} X \\ x^T \end{pmatrix}.$$

1. $y^T = x^T V$
2.   **if** (the last $p-m$ components of y are too large)
3.   Annihilate the last $p-m$ components of y: (4.4)
4.   **end if**
5.   Update the vector y into R: (4.5)
6.   Increase m until a gap is found or until $m = p$: (4.6)
7.   Decrease m by deflation: (4.7)

Algorithm 4.1: URV updating

The most expensive task in the algorithm is constrained updating for small m, in which case the count is approximately $\frac{7}{2}p^2$.

In understanding how URV updating works, it is important to keep in mind that the appearance of a large y_2 need not represent an increase in rank. Instead it can represent a drifting of the right fundamental subspaces. Specifically, the matrix V does not change as long as we do only simple updating. As x drifts out of the space spanned by the first m columns of V, y_2 becomes larger until it triggers a step of constrained updating. The right rotations in this step then cause V to change. The putative rank is then reduced in a subsequent deflation step.

Refinement

We can apply Theorem 1.1 to assess the accuracy of the column spaces of V_1 and V_2 as approximations to the right fundamental subspaces of X. Specifically, the sines of the canonical angles between the spaces are bounded by

$$s_V \leq (1 - \rho^2)^{-1} \frac{\|H\|_2}{\inf(S)}, \tag{4.8}$$

where
$$\rho = \frac{\|F\|_2}{\inf(S)}$$

is the gap ratio. In a gap-revealing decomposition, the quantity $\|F\|_2 \cong \sigma_{m+1}$ is effectively fixed. Consequently, if we are unhappy with the accuracy of the approximations we must reduce the size of $\|H\|_2$. At the cost of some additional work we can do just that.

To motivate our refinement step, suppose that R has a gap at m and partition the leading $(m+1)\times(m+1)$ principal submatrix of R in the form

$$\begin{pmatrix} S & h \\ 0 & \varphi \end{pmatrix}.$$

Now suppose that we generate an orthogonal matrix that reduce this submatrix to block lower triangular form:

$$\begin{pmatrix} S & h \\ 0 & \varphi \end{pmatrix} \begin{pmatrix} P_{11} & p_{12} \\ p_{21}^T & \pi_{22} \end{pmatrix} = \begin{pmatrix} \hat{S} & 0 \\ g^T & \hat{\varphi} \end{pmatrix}. \tag{4.9}$$

We are going to show that if there is a good gap ratio, the norm of g will be less than that of h.

From (4.9) we have

$$g = \varphi p_{21} \tag{4.10}$$

and

$$Sp_{12} + h\pi_{22} = 0.$$

It follows that if σ is the smallest singular value of S then

$$\|p_{12}\|_2 \leq \frac{\|h\|_2}{\sigma}$$

(remember $|\pi_{22}| \leq 1$). Since $\|p_{12}\|_2 = \|p_{21}^T\|_2$, we have from (4.10)

$$\|g\|_2 \leq \frac{\varphi}{\sigma}\|h\|_2.$$

In other words, the norm of g is smaller than the norm of h by a factor no larger than the gap ratio. If we now reduce the left-hand side of (4.9) back to upper triangular form, we will obtain a URV decomposition in which h is reduced by about the square of the gap ratio.

Algorithmically, the two reductions are easily implemented. Specifically, we can generate the lower triangular matrix by the sequence of transformations illustrated below.

Sec. 4. UTV decompositions

Given a URV decomposition with a gap at m, this algorithm reduces the size of $R[1{:}m, m{+}1]$.

1. **for** $k = m$ **to** 1 **by** -1
2. $rotgen(R[k,k], R[k,m{+}1], c, s)$
3. $rotapp(R[1{:}k{-}1, k], R[1{:}k{-}1, m{+}1], c, s)$
4. $rotapp(R[m{+}1, k], R[m{+}1, m{+}1], c, s)$
5. $rotapp(V[:, k], V[:, m{+}1], c, s)$
6. **end for** k
7. **for** $k = 1$ **to** m
8. $rotgen(R[k,k], R[m{+}1, k], c, s)$
9. $rotapp(R[k, k{+}1{:}p], R[m{+}1, k{+}1{:}p], c, s)$
10. **end for** k

Algorithm 4.2: URV refinement

---◇---

$$\begin{pmatrix} s & s & s & s & h \\ 0 & s & s & s & h \\ 0 & 0 & s & s & \widehat{h} \\ 0 & 0 & 0 & s & \widehat{h} \\ 0 & 0 & 0 & 0 & \varphi \end{pmatrix} \stackrel{P_{45}}{\Longrightarrow} \begin{pmatrix} s & s & s & s & h \\ 0 & s & s & s & h \\ 0 & 0 & s & s & \widehat{h} \\ 0 & 0 & 0 & s & 0 \\ 0 & 0 & 0 & g & \varphi \end{pmatrix} \stackrel{P_{35}}{\Longrightarrow} \begin{pmatrix} s & s & s & s & h \\ 0 & s & s & s & \widehat{h} \\ 0 & 0 & s & s & 0 \\ 0 & 0 & 0 & s & 0 \\ 0 & 0 & g & g & \varphi \end{pmatrix} \stackrel{P_{25}}{\Longrightarrow}$$

$$\begin{pmatrix} s & s & s & s & \widehat{h} \\ 0 & s & s & s & 0 \\ 0 & 0 & s & s & 0 \\ 0 & 0 & 0 & s & 0 \\ 0 & g & g & g & \varphi \end{pmatrix} \stackrel{P_{15}}{\Longrightarrow} \begin{pmatrix} s & s & s & s & 0 \\ 0 & s & s & s & 0 \\ 0 & 0 & s & s & 0 \\ 0 & 0 & 0 & s & 0 \\ g & g & g & g & \varphi \end{pmatrix}$$

The return to upper triangular form is analogous to the basic Cholesky update. Algorithm 4.2 implements this refinement. It requires approximately

$2mp$ flrot.

For small m the additional work is insignificant. For $m \cong p$, the algorithm requires about $2p^2$ flrot. This should be compared with the count of $\frac{1}{2}p^2$ flrot for the basic updating.

The decision to refine must be based on the application. If one expects m to be small, there is no reason not to take advantage of the benefits of refinement. On the other hand, if m is near p refinement quadruples the work over the basic update step.

Low-rank splitting

We have seen that when m is small the constrained URV updating scheme requires about $\frac{7}{2}p^2$ flrot — seven times the amount of work required for a simple update. It turns out that if we are willing to update only the bases for the right superior subspace, we can reduce the work considerably.

To see how this comes about, let us examine the decomposition computed by Algorithm 4.1 right after statement 3, where the last $t-m$ components of the vector $y^T = V^T x$ are annihilated. At this point the decomposition has the form

$$\begin{pmatrix} U^T X \\ x^T \end{pmatrix} = \begin{pmatrix} S & h & H \\ 0 & \varphi & \hat{f}^T \\ 0 & 0 & F \\ y_1^T & \eta & 0 \end{pmatrix} \begin{pmatrix} V_1^T \\ v_2^T \\ V_3^T \end{pmatrix}.$$

Note that in the course of this reduction the vector $y_1^T = x^T V_1$ remains unaltered — only the last $p-m$ components of y^T are changed. Consequently, the first m steps of the subsequent reduction to triangular form amount to a simple QR update on the matrix

$$\begin{pmatrix} S \\ y_1^T \end{pmatrix}.$$

Since only left rotations are used in this part of the reduction, the matrix V_1 does not change. The total count for this algorithm is $\frac{1}{2}m^2$ flrot.

Although this algorithm allows us to test S and if necessary decrease m, it does not allow us to increase m, which we would have to do if η were large. Fortunately, we can compute η and v_2 directly. For we have

$$x = V_1 y_1 + \eta v_2.$$

By the orthogonality of V it follows that v_2 is just the projection of x onto the orthogonal complement of $\mathcal{R}(V_1)$ and $\eta = v_2^T x$. Thus we can generate v_2 — say by the Gram–Schmidt algorithm with reorthogonalization (Algorithm 1.13, Chapter 4) — then compute η and test its size.

At this point we cannot proceed with a direct update, since we do not know h and φ. However, if we are using exponential windowing, we can update

$$\begin{pmatrix} \beta S & 0 \\ 0 & 0 \\ y_1^T & \eta \end{pmatrix},$$

increase m by one, and take for our new V_1 the matrix $(V_1 \ v_2)$. Although this decomposition will not be exact, the exponential windowing will damp out the inaccuracies, so that in a few iterations it will be good enough for practical purposes.

4.3. UPDATING ULV DECOMPOSITIONS

In this subsection we will treat the problem of updating ULV decompositions — that is, decompositions in which the updated matrix is lower triangular. These decompositions give better approximations to the right fundamental subspaces, but they are more

ULV decompositions

A ULV decomposition of the $n \times p$ matrix X has the form

$$U^T X V = \begin{pmatrix} L \\ 0 \end{pmatrix},$$

where U and V are orthogonal and L is lower triangular. A ULV decomposition is gap revealing at m if it can be partitioned in the form

$$(U_1 \ U_2)^T X (V_1 \ V_2) = \begin{pmatrix} S & 0 \\ G & F \\ 0 & 0 \end{pmatrix},$$

where

1. S is of order m,
2. $\inf(S) \cong \sigma_m(X)$,
3. $\|F\|_2 \cong \sigma_{m+1}(X)$,
4. $\|G\|_2$ is suitably small.

Theorem 1.1 shows that the sines of the canonical angles between the column spaces spanned by V_1 and V_2 and the corresponding right fundamental subspaces of X will be bounded by

$$s_V \leq (1-\rho^2)^{-1} \frac{\rho \|G\|_2}{\inf(S)},$$

where

$$\rho = \frac{\|F\|_2}{\inf(S)}$$

is the gap ratio. This is better by a factor of ρ than the corresponding bound (4.8) for the URV decomposition. This suggests that we try to update ULV decompositions as an alternative to refining URV decompositions.

Updating a ULV decomposition

Given a gap-revealing ULV decomposition of X, we wish to compute the corresponding decomposition of

$$\begin{pmatrix} X \\ x^T \end{pmatrix}.$$

We first transform x into the coordinate system of V by computing

$$y^{\mathrm{T}} = x^{\mathrm{T}} V.$$

This leaves us with an updating problem of the form

$$\begin{pmatrix} s & 0 & 0 & 0 & 0 & 0 \\ s & s & 0 & 0 & 0 & 0 \\ s & s & s & 0 & 0 & 0 \\ g & g & g & f & 0 & 0 \\ g & g & g & f & f & 0 \\ g & g & g & f & f & f \\ y & y & y & y & y & y \end{pmatrix}.$$

As with URV updating, any direct attempt to fold y into the decomposition can make the elements of G and F large. To limit the damage, we reduce the last $m-p$ components to zero as follows.

$$\begin{pmatrix} s & 0 & 0 & 0 & 0 & 0 \\ s & s & 0 & 0 & 0 & 0 \\ s & s & s & 0 & 0 & 0 \\ g & g & g & f & 0 & 0 \\ g & g & g & f & f & 0 \\ g & g & g & f & f & f \\ y & y & y & y & y & y \end{pmatrix} \overset{P_{56}}{\Longrightarrow} \begin{pmatrix} s & 0 & 0 & 0 & 0 & 0 \\ s & s & 0 & 0 & 0 & 0 \\ s & s & s & 0 & 0 & 0 \\ g & g & g & f & 0 & 0 \\ g & g & g & f & f & \widehat{f} \\ g & g & g & f & f & f \\ y & y & y & y & y & 0 \end{pmatrix} \overset{Q_{65}}{\Longrightarrow}$$

$$\begin{pmatrix} s & 0 & 0 & 0 & 0 & 0 \\ s & s & 0 & 0 & 0 & 0 \\ s & s & s & 0 & 0 & 0 \\ g & g & g & f & 0 & 0 \\ g & g & g & f & f & 0 \\ g & g & g & f & f & f \\ y & y & y & y & \widehat{y} & 0 \end{pmatrix} \overset{P_{45}}{\Longrightarrow} \begin{pmatrix} s & 0 & 0 & 0 & 0 & 0 \\ s & s & 0 & 0 & 0 & 0 \\ s & s & s & 0 & 0 & 0 \\ g & g & g & f & \widehat{f} & 0 \\ g & g & g & f & f & 0 \\ g & g & g & f & f & f \\ y & y & y & y & 0 & 0 \end{pmatrix} \overset{Q_{64}}{\Longrightarrow}$$

$$\begin{pmatrix} s & 0 & 0 & 0 & 0 & 0 \\ s & s & 0 & 0 & 0 & 0 \\ s & s & s & 0 & 0 & 0 \\ g & g & g & f & 0 & 0 \\ g & g & g & f & f & 0 \\ g & g & g & f & f & f \\ y & y & y & y & 0 & 0 \end{pmatrix}$$

We next fold in the rest of the vector y using right rotations.

Sec. 4. UTV decompositions

$$\begin{pmatrix} s & 0 & 0 & 0 & 0 & 0 \\ s & s & 0 & 0 & 0 & 0 \\ s & s & s & 0 & 0 & 0 \\ g & g & g & f & 0 & 0 \\ g & g & g & f & f & 0 \\ g & g & g & f & f & f \\ y & y & y & \widehat{y} & 0 & 0 \end{pmatrix} \overset{Q_{47}}{\Longrightarrow} \begin{pmatrix} s & 0 & 0 & 0 & 0 & 0 \\ s & s & 0 & 0 & 0 & 0 \\ s & s & s & 0 & 0 & 0 \\ y & y & y & y & 0 & 0 \\ g & g & g & f & f & 0 \\ g & g & g & f & f & f \\ y & y & \widehat{y} & 0 & 0 & 0 \end{pmatrix} \overset{Q_{37}}{\Longrightarrow}$$

$$\begin{pmatrix} s & 0 & 0 & 0 & 0 & 0 \\ s & s & 0 & 0 & 0 & 0 \\ s & s & s & 0 & 0 & 0 \\ y & y & y & y & 0 & 0 \\ g & g & g & f & f & 0 \\ g & g & g & f & f & f \\ y & \widehat{y} & 0 & 0 & 0 & 0 \end{pmatrix} \overset{Q_{27}}{\Longrightarrow} \begin{pmatrix} s & 0 & 0 & 0 & 0 & 0 \\ s & s & 0 & 0 & 0 & 0 \\ s & s & s & 0 & 0 & 0 \\ y & y & y & y & 0 & 0 \\ g & g & g & f & f & 0 \\ g & g & g & f & f & f \\ \widehat{y} & 0 & 0 & 0 & 0 & 0 \end{pmatrix} \overset{Q_{17}}{\Longrightarrow}$$

$$\begin{pmatrix} s & 0 & 0 & 0 & 0 & 0 \\ s & s & 0 & 0 & 0 & 0 \\ s & s & s & 0 & 0 & 0 \\ y & y & y & y & 0 & 0 \\ g & g & g & f & f & 0 \\ g & g & g & f & f & f \\ 0 & 0 & 0 & 0 & 0 & 0 \end{pmatrix}$$

Note how the reduction introduces a row of y's in the $(m+1)$th row (the fourth row in the diagram). To handle them we increase m by one and attempt to deflate S. The deflation process is analogous to the one for the URV decomposition. We use a condition estimator to determine a vector w of 2-norm one such that $\|w^T X\|_2$ is small. We then reduce w to a multiple of $\mathbf{e}_m$ as follows.

$$\begin{pmatrix} \widehat{w} \\ w \\ w \\ w \end{pmatrix} \overset{Q_{21}}{\Longrightarrow} \begin{pmatrix} 0 \\ \widehat{w} \\ w \\ w \end{pmatrix} \overset{Q_{32}}{\Longrightarrow} \begin{pmatrix} 0 \\ 0 \\ \widehat{w} \\ w \end{pmatrix} \overset{Q_{43}}{\Longrightarrow} \begin{pmatrix} 0 \\ 0 \\ 0 \\ 1 \end{pmatrix}$$

We now apply the transformations to S, undoing their effects as we go.

$$\begin{pmatrix} s & 0 & 0 & 0 \\ s & s & 0 & 0 \\ s & s & s & 0 \\ s & s & s & s \end{pmatrix} \overset{Q_{21}^T}{\Longrightarrow} \begin{pmatrix} s & \widehat{s} & 0 & 0 \\ s & s & 0 & 0 \\ s & s & s & 0 \\ s & s & s & s \end{pmatrix} \overset{P_{12}}{\Longrightarrow} \begin{pmatrix} s & 0 & 0 & 0 \\ s & s & 0 & 0 \\ s & s & s & 0 \\ s & s & s & s \end{pmatrix} \overset{Q_{32}^T}{\Longrightarrow}$$

$$\begin{pmatrix} s & 0 & 0 & 0 \\ s & s & \widehat{s} & 0 \\ s & s & s & 0 \\ s & s & s & s \end{pmatrix} \overset{P_{23}}{\Longrightarrow}$$

$$\begin{pmatrix} s & 0 & 0 & 0 \\ s & s & 0 & 0 \\ s & s & s & 0 \\ s & s & s & s \end{pmatrix} \overset{Q_{43}^{\mathrm{T}}}{\Longrightarrow} \begin{pmatrix} s & 0 & 0 & 0 \\ s & s & 0 & 0 \\ s & s & s & \widehat{s} \\ s & s & s & s \end{pmatrix} \overset{P_{34}}{\Longrightarrow} \begin{pmatrix} s & 0 & 0 & 0 \\ s & s & 0 & 0 \\ s & s & s & 0 \\ s & s & s & s \end{pmatrix}$$

After this step has been performed we may proceed to adjust the gap, as in URV updating.

The ULV update is less flexible than the URV update — less able to take advantage of a vector x that essentially lies in the column space of V_1. For example, we must always reduce the trailing components of y. Moreover, after we have folded y into S, we are left with a large row in G. This makes a deflation step mandatory.

The following is a list of the operation counts for the pieces of the algorithm.

1. p^2 flam to compute $y^{\mathrm{T}} = x^{\mathrm{T}} V$.
2. $[2p(p-m) + \frac{1}{2}m^2]$ flrot to reduce and incorporate y.
3. $\frac{1}{2}m^2$ flam to compute w.
4. $\frac{3}{2}mp$ flam for the deflation step.

Obviously counts like these and the ones for URV updating provide only crude hints concerning which method to use. Experience and experiment will be a better guide.

4.4. Notes and references

UTV decompositions

The URV and ULV decompositions were introduced by Stewart [301, 1992], [305, 1993]. They are also called two-sided orthogonal decompositions, and they are related to complete orthogonal decompositions of the form

$$U^{\mathrm{T}} X V \cong \begin{pmatrix} T & 0 \\ 0 & 0 \end{pmatrix}$$

in which small elements are set to zero (see, e.g., [153, §5.4.2]). The difference is that here the small elements are retained and preserved during the updating. A condition estimator is necessary to keep the operation count for the update to $O(n^2)$. For applications in signal processing see [3, 219].

The UTV decompositions were introduced to overcome the difficulties in updating the singular value decomposition. However, in some circumstances one or two steps of an iterative method will suffice to maintain a sufficiently accurate approximation to the singular value decomposition [235].

The low-rank version of URV updating is due to Rabideau [266], who seems not to have noticed that his numerical algorithms are minor variations on those of [301].

Although we have developed our algorithms in the context of exponential windowing, the algorithms can also be downdated. For more see [16, 219, 251].

REFERENCES

[1] J. O. Aasen. On the reduction of a symmetric matrix to tridiagonal form. *BIT*, 11:233–242, 1971.

[2] N. N. Abdelmalek. Roundoff error analysis for Gram–Schmidt method and solution of linear least squares problems. *BIT*, 11:345–368, 1971.

[3] G. Adams, M. F. Griffin, and G. W. Stewart. Direction-of-arrival estimation using the rank-revealing URV decomposition. In *Proceedings of the IEEE International Conference on Acoustics, Speech and Signal Processing*, pages 1385–1388. IEEE, Washington, DC, 1991.

[4] A. V. Aho, J. E. Hopcroft, and J. D. Ullman. *The Design and Analysis of Computer Algorithms*. Addison–Wesley, Reading, MA, 1974.

[5] A. V. Aho, R. Sethi, and J. D. Ullman. *Compilers: Principles, Techniques, and Tools*. Addison–Wesley, Reading, MA, 1986.

[6] A. C. Aitkin. Studies in practical mathematics I: The evaluation, with applications, of a certain triple product matrix. *Proceedings of the Royal Society, Edinburgh*, 57:172–181, 1937.

[7] A. A. Anda and H. Park. Fast plane rotations with dynamic scaling. *SIAM Journal on Matrix Analysis and Applications*, 15:162–174, 1994.

[8] A. A. Anda and H. Park. Self-scaling fast rotations for stiff least squares problems. *Linear Algebra and Its Applications*, 234:137–162, 1996.

[9] E. Anderson, Z. Bai, C. Bischof, J. Demmel, J. Dongarra, J. Du Croz, A. Greenbaum, S. Hammarling, A. McKenney, S. Ostrouchov, and D. Sorensen. *LAPACK Users' Guide*. SIAM, Philadelphia, second edition, 1995.

[10] ANSI. *American National Standard Programming Language FORTRAN*. American National Standards Institute, New York, 1978.

[11] H. Anton and C. Rorres. *Elementary Linear Algebra with Applications*. John Wiley, New York, 1987.

[12] M. Arioli, J. W. Demmel, and I. S. Duff. Solving sparse linear systems with sparse backward error. *SIAM Journal on Matrix Analysis and Applications*, 10:165–190, 1989.

[13] O. Axelsson. *Iterative Solution Methods*. Cambridge University Press, Cambridge, 1994.

[14] S. Banach. Sur les opérations dans les ensembles abstraits et leur application aux équations integrales. *Fundamenta Mathematicae*, 3:133–181, 1922.

[15] J. L. Barlow. Error analysis and implementation aspects of deferred correction for equality constrained least squares problems. *SIAM Journal on Numerical Analysis*, 25:1340–1358, 1988.

[16] J. L. Barlow, P. A. Yoon, and H. Zha. An algorithm and a stability theory for downdating the ULV decomposition. *BIT*, 36:14–40, 1996.

[17] R. Barrett, M. Berry, T. F. Chan, J. Demmel, J. Donato, J. Dongarra, V. Eijkhout, R. Pozo, C. Romine, and H. van der Vorst. *Templates for the Solution of Linear Systems: Building Blocks for Iterative Methods*. SIAM, Philadelphia, 1994.

[18] A. Barrlund. Perturbation bounds for the LDL^H and the LU factorizations. *BIT*, 31:358–363, 1991.

[19] F. L. Bauer. Optimal scaling of matrices and the importance of the minimal condition number. In C. M. Popplewell, editor, *Proceedings of the IFIP Congress 1962*, pages 198–201. North-Holland, Amsterdam, 1963. Cited in [177].

[20] F. L. Bauer. Optimally scaled matrices. *Numerische Mathematik*, 5:73–87, 1963.

[21] F. L. Bauer. Genauigkeitsfragen bei der Lösung linear Gleichungssysteme. *Zeitschrift für angewandte Mathematik und Mechanik*, 46:409–421, 1966.

[22] F. L. Bauer and C. Reinsch. Inversion of positive definite matrices by the Gauss-Jordan methods. In J. H. Wilkinson and C. Reinsch, editors, *Handbook for Automatic Computation Vol. 2: Linear Algebra*, pages 45–49. Springer-Verlag, New York, 1970.

[23] A. E. Beaton. The use of special matrix operators in statistical calculus. Research Bulletin 64-51, Educational Testing Service, Princeton, NJ, 1964.

[24] R. Bellman. *Introduction to Matrix Analysis*. McGraw–Hill, New York, second edition, 1970.

[25] E. Beltrami. Sulle funzioni bilineari. *Giornale di Matematiche ad Uso degli Studenti Delle Universita*, 11:98–106, 1873. An English translation by D. Boley is available in Technical Report 90–37, Department of Computer Science, University of Minnesota, Minneapolis, 1990.

[26] A. Ben-Israel and T. N. E. Greville. *Generalized Inverses: Theory and Applications*. John Wiley and Sons, New York, 1973.

[27] Commandant Benoît. Note sur une méthode de résolution des équations normales provenant de l'application de la méthode des moindres cárres a un systèm d'équations linéares en nombre inférieur à celui des inconnues. — application de la méthode à la résolution d'un système déini d'équations linéares (Procédé du Commandant Cholesky). *Bullétin Géodésique (Toulouse)*, 2:5–77, 1924.

[28] A. Berman and R. J. Plemmons. *Nonnegative Matrices in the Mathematical Sciences*. Academic Press, New York, 1979. Reprinted by SIAM, Philadelphia, 1994.

[29] R. Bhatia. *Perturbation Bounds for Matrix Eigenvalues*. Pitman Research Notes in Mathematics. Longmann Scientific & Technical, Harlow, Essex, 1987. Published in the USA by John Wiley.

[30] C. H. Bischof. Incremental condition estimation. *SIAM Journal on Matrix Analysis and Applications*, 11:312–322, 1990.

[31] C. H. Bischof and P. C. Hansen. Structure preserving and rank-revealing QR factorizations. *SIAM Journal on Scientific and Statistical Computing*, 12:1332–1350, 1991. Citation communicated by Per Christian Hansen.

[32] C. H. Bischof and P. C. Hansen. A block algorithm for computing rank-revealing QR-factorizations. *Numerical Algorithms*, 2:371–392, 1992. Citation communicated by Per Christian Hansen.

[33] C. H. Bischof and C. F. Van Loan. The WY representation for products of Householder transformations. *SIAM Journal on Scientific and Statistical Computing*, 8:s2–s13, 1987.

[34] A. Bjerhammer. Rectangular reciprocal matrices with special reference to geodetic calculations. *Bulletin Géodésique*, 52:118–220, 1951.

[35] Å. Björck. Contribution no. 22. Iterative refinement of linear least squares solutions by Householder transformations. *BIT*, 7:322–337, 1967.

[36] Å. Björck. Iterative refinement of linear least squares solutions I. *BIT*, 7:257–278, 1967.

[37] Å. Björck. Solving linear least squares problems by Gram–Schmidt orthogonalization. *BIT*, 7:1–21, 1967.

[38] Å. Björck. Iterative refinement of linear least squares solutions II. *BIT*, 8:8–30, 1968.

[39] Å. Björck. Stability analysis of the method of seminormal equations. *Linear Algebra and Its Applications*, 88/89:31–48, 1987.

[40] Å. Björck. Numerics of Gram–Schmidt orthogonalization. *Linear Algebra and Its Applications*, 197–198:297–316, 1994.

[41] Å. Björck. *Numerical Methods for Least Squares Problems*. SIAM, Philadelphia, 1996.

[42] Å. Björck and G. H. Golub. Iterative refinement of linear least squares solution by Householder transformation. *BIT*, 7:322–337, 1967.

[43] Å. Björck and G. H. Golub. Numerical methods for computing angles between linear subspaces. *Mathematics of Computation*, 27:579–594, 1973.

[44] Å. Björck and C. C. Paige. Loss and recapture of orthogonality in the modified Gram–Schmidt algorithm. *SIAM Journal on Matrix Analysis and Applications*, 13:176–190, 1992.

[45] M. M. Blevins and G. W. Stewart. Calculating eigenvectors of diagonally dominant matrices. *Journal of the ACM*, 21:261–271, 1974.

[46] S. F. Blinkhorn. Burt and the early history of factor analysis. In N. J. Mackintosh, editor, *Cyril Burt: Fraud or Framed?*, pages 13–44. Oxford University Press, Oxford, 1994.

[47] D. Bogert and W. R. Burris. Comparison of least squares algorithms. Report ORNL-3499, V.1, §5.5, Neutron Physics Division, Oak Ridge National Laboratory Oak Ridge, TN, 1963.

[48] A. Bojanczyk, R. P. Brent, P. Van Dooren, and F. de Hoog. A note on downdating the Cholesky factorization. *SIAM Journal on Scientific and Statistical Computing*, 8:210–221, 1987.

[49] A. W. Bojanczyk and A. O. Steinhardt. Stability analysis of a Householder-based algorithm for downdating the Cholesky factorization. *SIAM Journal on Scientific and Statistical Computing*, 12:1255–1265, 1991.

[50] D. W. Boyd. The power method for ℓ^p norms. *Linear Algebra and Its Applications*, 9:95–101, 1974.

[51] J. R. Bunch. Analysis of the diagonal pivoting method. *SIAM Journal on Numerical Analysis*, 8:656–680, 1971.

[52] J. R. Bunch. The weak and strong stability of algorithms in numerical linear algebra. *Linear Algebra and Its Applications*, 88/89:49–66, 1987.

[53] J. R. Bunch and L. Kaufman. Some stable methods for calculating inertia and solving symmetric linear systems. *Mathematics of Computation*, 31:163–179, 1977.

[54] J. R. Bunch and B. N. Parlett. Direct methods for solving symmetric indefinite systems of linear equations. *SIAM Journal on Numerical Analysis*, 8:639–655, 1971.

[55] P. Businger and G. H. Golub. Linear least squares solutions by Householder transformations. *Numerische Mathematik*, 7:269–276, 1965. Also in [349, pp. 111–118].

[56] H. G. Campbell. *Linear Algebra*. Addison–Wesley, Reading, MA, second edition, 1980.

[57] A. L. Cauchy. Cours d'analyse de l'école royale polytechnique. In *Oeuvres Complétes (IIe Série)*, volume 3. 1821.

[58] A. L. Cauchy. Sur l'équation á l'aide de laquelle on détermine les inégalités séculaires des mouvements des planètes. In *Oeuvres Complétes (IIe Série)*, volume 9. 1829.

[59] A. Cayley. Remarques sur la notation des fonctions algébriques. *Journal für die reine und angewandte Mathematik*, 50:282–285, 1855. Cited and reprinted in [61, v. 2, pp. 185–188].

[60] A. Cayley. A memoir on the theory of matrices. *Philosophical Transactions of the Royal Socient of London*, 148:17–37, 1858. Cited and reprinted in [61, v. 2, pp. 475–496].

[61] A. Cayley. *The Collected Mathematical Papers of Arthur Cayley*. Cambridge University Press, Cambridge, 1889–1898. Arthur Cayley and A. R. Forsythe, editors. Thirteen volumes plus index. Reprinted 1963 by Johnson Reprint Corporation, New York.

[62] J. M. Chambers. Regression updating. *Journal of the American Statistical Association*, 66:744–748, 1971.

[63] T. F. Chan. Rank revealing QR factorizations. *Linear Algebra and Its Applications*, 88/89:67–82, 1987.

[64] S. Chandrasekaran and I. Ipsen. On rank-revealing QR factorizations. *SIAM Journal on Matrix Analysis and Applications*, 15:592–622, 1991. Citation communicated by Per Christian Hansen.

[65] B. A. Chartres and J. C. Geuder. Computable error bounds for direct solution of linear equations. *Journal of the ACM*, 14:63–71, 1967.

[66] F. Chió. Méoire sur les functions connues sus le nom des résultantes ou de déterminants. *Trurin*. Cited in [339], 1853.

[67] C. W. Clenshaw and F. W. J. Olver. Beyond floating point. *Journal of the ACM*, 31:319–328, 1984.

[68] A. K. Cline, A. R. Conn, and C. F. Van Loan. Generalizing the LINPACK condition estimator. In J.P. Hennart, editor, *Numerical Analysis*, Lecture Notes in Mathematics 909, pages 73–83. Springer-Verlag, Berlin, 1982. Cited in [177].

[69] A. K. Cline, C. B. Moler, G. W. Stewart, and J. H. Wilkinson. An estimate for the condition number of a matrix. *SIAM Journal on Numerical Analysis*, 16:368–375, 1979.

[70] A. K. Cline and R. K. Rew. A set of counter examples to three condition number estimators. *SIAM Journal on Scientific and Statistical Computing*, 4:602–611, 1983.

[71] T. Coleman and C. F. Van Loan. *Handbook for Matrix Computations*. SIAM, Philadelphia, 1988.

[72] John B. Conway. *A Course in Functional Analysis*. Springer-Verlag, New York, 1985.

[73] T. H. Cormen, C. E. Leiserson, and R. L. Rivest. *Introduction to Algorithms*. McGraw–Hill, New York, 1990.

[74] R. B. Costello, editor. *The American Heritage College Dictionary*. Houghton Mifflin, Boston, third edition, 1993.

[75] R. W. Cottle. Manifestations of the Schur complement. *Linear Algebra and Its Applications*, 8:189–211, 1974.

[76] R. Courant. Ueber die Eigenwert bei den Differentialgleichungen der Mathematischen Physik. *Mathematische Zeitschrift*, 7:1–57, 1920.

[77] A. J. Cox and N. J. Higham. Stability of Householder QR factorization for weighted least squares. Numerical Analysis Report 301, Department of Mathematics, University of Manchester, 1997.

[78] H. G. Cragon. *Memory Systems and Pipelined Processors*. Jones and Bartlett, Sudbury, MA, 1996.

[79] P. D. Crout. A short method for evaluating determinants and solving systems of linear equations with real or complex coefficients. *Transactions of the American Institute of Electrical Engineers*, 60:1235–1240, 1941.

[80] M. J. Crowe. *A History of Vector Analysis*. University of Notre Dame Press, Notre Dame, IN, 1967.

[81] C. W. Cryer. The LU factorization of totally positive matrices. *Linear Algebra and Its Applications*, 7:83–92, 1973.

[82] J. K. Cullum and R. A. Willoughby. *Lanczos Algorithms for Large Symmetric Eigenvalue Computations*, volume 1: Theory. volume 2: Programs. Birkhäuser, Stuttgart, 1985. Cited in [41].

[83] A. R. Curtis and J. K. Reid. On the automatic scaling of matrices for Gaussian elimination. *Journal of the Institute for Mathematics and Applications*, 10:118–124, 1972.

[84] J. Daniel, W. B. Gragg, L. Kaufman, and G. W. Stewart. Reorthogonalization and stable algorithms for updating the Gram–Schmidt QR factorization. *Mathematics of Computation*, 30:772–795, 1976.

[85] G. B. Dantzig. *Linear Programming and Extensions*. Princeton University Press, Princeton, NJ, 1963.

[86] B. N. Datta. *Numerical Linear Algebra and Applications*. Brooks/Cole, Pacific Grove, CA, 1995.

[87] W. C. Davidon. Variable metric method for minimization. AEC Research and Development Report ANL-5990, Argonne National Laboratory, Argonne, IL, 1959.

[88] C. Davis and W. Kahan. The rotation of eigenvectors by a perturbation. III. *SIAM Journal on Numerical Analysis*, 7:1–46, 1970.

[89] P. J. Davis. *Interpolation and Approximation*. Blaisdell, New York, 1961. Reprinted by Dover, New York, 1975.

[90] C. de Boor and A. Pinkus. A backward error analysis for totally positive linear systems. *Numerische Mathematik*, 27:485–490, 1977.

[91] J. Demmel. On error analysis in arithmetic with varying relative precision. In *Proceeding of the Eighth Symposium on Computer Arithmetic*, pages 148–152. IEEE Computer Society, Washington, DC, 1987.

[92] J. Demmel. The smallest perturbation of a submatrix which lowers the rank and constrained total least squares problems. *SIAM Journal on Numerical Analysis*, 24:199–206, 1987.

[93] J. Demmel and K. Veselić. Jacobi's method is more accurate than QR. *SIAM Journal on Matrix Analysis and Applications*, 13:1204–1245, 1992.

[94] J. W. Demmel. Three methods for refining estimates of invariant subspaces. *Computing*, 38:43–57, 1987.

[95] Jean Dieudonné. *History of Functional Analysis*. North–Holland, Amsterdam, 1981.

[96] E. W. Dijkstra. Go to statement considered harmful. *Communications of the ACM*, 11:147–148, 1968.

[97] J. D. Dixon. Estimating extremal eigenvalues and condition numbers of matrices. *SIAM Journal on Numerical Analysis*, 20:812–814, 1983.

[98] J. J. Dongarra. Improving the accuracy of computed singular values. *SIAM Journal on Scientific and Statistical Computing*, 4:712–719, 1983.

[99] J. J. Dongarra, J. R. Bunch, C. B. Moler, and G. W. Stewart. *LINPACK User's Guide*. SIAM, Philadelphia, 1979.

[100] J. J. Dongarra, J. Du Croz, I. S. Duff, and S. Hammarling. Algorithm 679: A set of level 3 basic linear algebra subprograms. *ACM Transactions on Mathematical Software*, 16:1–17, 1990.

[101] J. J. Dongarra, J. Du Croz, I. S. Duff, and S. Hammarling. A set of level 3 basic linear algebra subprograms. *ACM Transactions on Mathematical Software*, 16:1–17, 1990.

[102] J. J. Dongarra, J. Du Croz, S. Hammarling, and R. J. Hanson. An extended set of fortran basic linear algebra subprograms. *ACM Transactions on Mathematical Software*, 14:1–17, 1988.

[103] J. J. Dongarra, J. Du Croz, S. Hammarling, and R. J. Hanson. An extended set of fortran basic linear algebra subprograms: Model implementation and test programs. *ACM Transactions on Mathematical Software*, 14:18–32, 1988.

[104] J. J. Dongarra, I. S. Duff, D. C. Sorensen, and H. A. van der Vorst. *Solving Linear Systems on Vector and Shared Memory Computers*. SIAM, Philadelphia, 1991.

[105] J. J. Dongarra, F. G. Gustavson, and A. Karp. Implementing linear algebra algorithms for dense matrices on a vector pipeline machine. *SIAM Review*, 26:91–112, 1984.

[106] J. J. Dongarra, C. B. Moler, and J. H. Wilkinson. Improving the accuracy of computed eigenvalues and eigenvectors. *SIAM Journal on Numerical Analysis*, 20:23–45, 1983.

[107] M. H. Doolittle. Method employed in the solution of normal equations and the adjustment of a triangulation. In *U. S. Coast and Geodetic Survey Report*, pages 115–120. 1878. Cited in [112].

[108] I. S. Duff, A. M. Erisman, and J. K. Reid. *Direct Methods for Sparse Matrices*. Clarendon Press, Oxford, 1986.

[109] W. J. Duncan. Some devices for the solution of large sets of simultaneous linear equations (with an appendix on the reciprocation of partitioned matrices). *The London, Edinburgh, and Dublin Philosophical Magazine and Journal of Science*, 35:660–670, 1944.

[110] P. S. Dwyer. The Doolittle technique. *Annals of Mathematical Statistics*, 12:449–458, 1941.

[111] P. S. Dwyer. A matrix presentation of least squares and correllation theory with matrix justification of improved methods of solution. *Annals of Mathematical Statistics*, 15:82–89, 1944.

[112] P. S. Dwyer. *Linear Computations*. John Wiley, New York, 1951.

[113] C. Eckart and G. Young. The approximation of one matrix by another of lower rank. *Psychometrika*, 1:211–218, 1936.

[114] M. A. Ellis and B. Stroustrup. *The Annotated C++ Rerenence Manual*. Addison–Wesley, Reading, MA, 1990.

[115] D. K. Faddeev and V. N. Faddeeva. *Computational Methods of Linear Alegbra*. W. H. Freeman and Co., San Francisco, 1963.

[116] V. N. Faddeeva. *Computational Methods of Linear Algebra*. Dover, New York, 1959. Translated from the Russian by C. D. Benster.

[117] W. Feller and G. E. Forsythe. New matrix transformations for obtaining characteristic vectors. *Quarterly of Applied Mathematics*, 8:325–331, 1951.

[118] R. D. Fierro. Perturbation theory for two-sided (or complete) orthogonal decompositions. *SIAM Journal on Matrix Analysis and Applications*, 17:383–400, 1996.

[119] E. Fischer. Über quadratische Formen mit reelen Koffizienten. *Monatshefte für Mathematik und Physik*, 16:234–249, 1905.

[120] G. Forsythe and C. B. Moler. *Computer Solution of Linear Algebraic Systems*. Prentice–Hall, Englewood Cliffs , NJ, 1967.

[121] L. V. Foster. Gaussian elimination with partial pivoting can fail in practice. *SIAM Journal on Matrix Analysis and Applications*, 15:1354–1362, 1994.

[122] L. Fox. *An Introduction to Numerical Linear Algebra*. Oxford University Press, New York, 1965.

[123] J. G. F. Francis. The QR transformation, parts I and II. *Computer Journal*, 4:265–271, 332–345, 1961, 1962.

[124] F. G. Frobenius. Über den von L. Bieberbach gefundenen Beweis eines Satzes von C. Jordan. *Sitzungsberichte der Königlich Preusischen Akademie der Wisenschaften zu Berlin*, 3:492–501, 1911. Cited and reprinted in [126, v. 3, pp. 492–501].

[125] F. G. Frobenius. Über die unzerlegbaren diskreten Beweguugsgruppen. *Sitzungsberichte der Königlich Preusischen Akademie der Wissenschaften zu Berlin*, 3:507–518, 1911. Cited and reprinted in [126, v. 3, pp. 507–518].

[126] F. G. Frobenius. *Ferdinand Georg Frobenius. Gesammelte Abhandlungen*, J.-P. Serre, editor. Springer-Verlag, Berlin, 1968.

[127] W. A. Fuller. *Measurement Error Models.* John Wiley, New York, 1987.

[128] G. Furnival and R. Wilson. Regression by leaps and bounds. *Technometrics,* 16:499–511, 1974.

[129] F. R. Gantmacher. *The Theory of Matrices, Vols. I, II.* Chelsea Publishing Company, New York, 1959.

[130] Carl Friedrich Gauss. *Theoria Motus Corporum Coelestium in Sectionibus Conicis Solem Ambientium.* Perthes and Besser, Hamburg, 1809. Cited and reprinted in [138, v. 7, pp. 1–261]. English translation by C. H. Davis [137]. French and German translations of Book II, Part 3 in [136, 139].

[131] Carl Friedrich Gauss. Disquisitio de elementis ellipticis Palladis. *Commentatines societatis regiae scientarium Gottingensis recentiores,* 1, 1810. Cited and reprinted in [138, v. 6, pp. 1–64]. French translation of §§13–14 in [136]. German translation of §§10–15 in [139].

[132] Carl Friedrich Gauss. Anzeige: Theoria combinationis observationum erroribus minimis obnoxiae: Pars prior. *Göttingische gelehrte Anzeigen,* 33:321–327, 1821. Cited and reprinted in [138, v. 4, pp. 95–100]. English translation in [140].

[133] Carl Friedrich Gauss. Anzeige: Theoria combinationis observationum erroribus minimis obnoxiae: Pars posterior. *Göttingische gelehrte Anzeigen,* 32:313–318, 1823. Cited and reprinted in [138, v. 4, pp. 100–104]. English translation in [140].

[134] Carl Friedrich Gauss. Theoria combinationis observationum erroribus minimis obnoxiae: Pars posterior. *Commentatines societatis regiae scientarium Gottingensis recentiores,* 5, 1823. Cited and reprinted in [138, v. 4, pp. 27–53]. French, German, and English translations in [136, 139, 140].

[135] Carl Friedrich Gauss. Supplementum theoriae combinationis observationum erroribus minimis obnoxiae. *Commentatines societatis regiae scientarium Gottingensis recentiores,* 6, 1828. Cited and reprinted in [138, v. 4, pp. 55–93]. French, German, and English translations in [136, 139, 140].

[136] Carl Friedrich Gauss. *Méthode des Moindres Carres.* Ballet–Bachelier, Paris, 1855. Translation by J. Bertrand of various works of Gauss on least squares.

[137] Carl Friedrich Gauss. *Theory of the Motion of the Heavenly Bodies Moving about the Sun in Conic Sections.* Little, Brown, and Company, 1857. Translation by Charles Henry Davis of *Theoria Motus* [130]. Reprinted by Dover, New York, 1963.

[138] Carl Friedrich Gauss. *Werke.* Königlichen Gesellschaft der Wissenschaften zu Göttingen, 1870–1928.

REFERENCES 427

[139] Carl Friedrich Gauss. *Abhandlungen zur Methode der kleinsten Quadrate*. P. Stankeiwica', Berlin, 1887. Translation by A. Borsch and P. Simon of various works of Gauss on least squares.

[140] Carl Friedrich Gauss. *Theory of the Combination of Observations Least Subject to Errors*. SIAM, Philadelphia, 1995. Translation by G. W. Stewart.

[141] W. M. Gentleman. Least squares computations by Givens transformations without square roots. *Journal of the Institute of Mathematics and Its Applications*, 12:329–336, 1973.

[142] W. M. Gentleman. Error analysis of QR decompositions by Givens transformations. *Linear Algebra and Its Applications*, 10:189–197, 1975.

[143] J. A. George and J. W. H. Liu. *Computer Solution of Large Sparse Positive Definite Systems*. Prentice–Hall, Englewood Cliffs, NJ, 1981.

[144] P. E. Gill, W. Murray, and M. H. Wright. *Practical Optimization*. Academic Press, New York, 1981.

[145] W. Givens. Numerical computation of the characteristic values of a real matrix. Technical Report 1574, Oak Ridge National Laboratory, Oak Ridge, TN, 1954.

[146] D. Goldberg. Computer arithmetic. Appendix A in [173], 1990.

[147] D. Goldberg. What every computer scientist should know about floating-point arithmetic. *ACM Computing Surveys*, 23:5–48, 1991.

[148] G. H. Golub. Numerical methods for solving least squares problems. *Numerische Mathematik*, 7:206–216, 1965.

[149] G. H. Golub. Matrix decompositions and statistical computation. In R. C. Milton and J. A. Nelder, editors, *Statistical Computation*, pages 365–397. Academic Press, New York, 1969.

[150] G. H. Golub, A. Hoffman, and G. W. Stewart. A generalization of the Eckart–Young matrix approximation theorem. *Linear Algebra and Its Applications*, 88/89:317–327, 1987.

[151] G. H. Golub and C. F. Van Loan. An analysis of the total least squares problem. *SIAM Journal on Numerical Analysis*, 17:883–893, 1980.

[152] G. H. Golub and C. F. Van Loan. *Matrix Computations*. Johns Hopkins University Press, Baltimore, MD, second edition, 1989.

[153] G. H. Golub and C. F. Van Loan. *Matrix Computations*. Johns Hopkins University Press, Baltimore, MD, third edition, 1996.

[154] G. H. Golub and J. H. Wilkinson. Iterative refinement of least squares solutions. In W. A. Kalenich, editor, *Proceedings of the IFIP Congress 65, New York, 1965*, pages 606–607. Spartan Books, Washington, 1965. Cited in [41].

[155] G. H. Golub and J. H. Wilkinson. Note on the iterative refinement of least squares solution. *Numerische Mathematik*, 9:139–148, 1966.

[156] J. H. Goodnight. A tutorial on the SWEEP operator. *American Statistician*, 33:149–158, 1979.

[157] W. B. Gragg and G. W. Stewart. A stable variant of the secant method for solving nonlinear equations. *SIAM Journal on Numerical Analysis*, 13:880–903, 1976.

[158] J. P. Gram. Über die Entwicklung reeler Functionen in Reihen mittelst der Methode der kleinsten Quadrate. *Journal für die reine und angewandte Mathematik*, 94:41–73, 1883.

[159] W. H. Greub. *Linear Algebra*. Springer-Verlag, New York, third edition, 1967.

[160] R. G. Grimes and J. G. Lewis. Condition number estimation for sparse matrices. *SIAM Journal on Scientific and Statistical Computing*, 2:384–388, 1981.

[161] M. Gu and S. C. Eisenstat. An efficient algorithm for computing a strong rank-revealing QR-factorization. *SIAM Journal on Scientific Computing*, 17:848–869, 1996.

[162] M. Gulliksson. Iterative refinement for constrained and weighted linear least squares. *BIT*, 34:239–253, 1994.

[163] M. H. Gutknecht. *Lanczos-Type Solvers for Nonsymmetric Linear Systems of Equations*. Technical Report TR-97-04, Swiss Center for Scientific Computing, ETH, Zürich, 1997.

[164] L. Guttman. Enlargement methods for computing the inverse matrix. *Annals of Mathematical Statistics*, 17:336–343, 1946.

[165] W. Hackbusch. *Iterative Lösung großer schwachbesetzer Gleichungssysteme*. Teubner, Stuttgart, 1991.

[166] L. A. Hageman and D. M. Young. *Applied Iterative Methods*. Academic Press, New York and London, 1981.

[167] W. W. Hager. Condition estimators. *SIAM Journal on Scientific and Statistical Computing*, 5:311–316, 1984.

[168] P. R. Halmos. *Finite-Dimensional Vector Spaces*. Van Nostrand, Princeton, NJ, second edition, 1958.

[169] V. C. Hamacher, Z. G. Vranesic, and S. G. Zaky. *Computer Organization*. McGraw–Hill, New York, second edition, 1984.

[170] S. Hammarling. A note on modifications to the Givens plane rotation. *Journal of the Institute of Mathematics and Its Applications*, 13:215–218, 1974.

[171] P. C. Hansen. Truncated singular value decomposition solutions to discrete ill-posed problems with ill-determined numerical rank. *SIAM Journal on Scientific and Statistical Computing*, 11:503–518, 1990.

[172] E. V. Haynsworth. Determination of the itertia of a partitioned Hermitian matrix. *Linear Algebra and Its Applications*, 1:73–81, 1968.

[173] J. L. Hennessy and D. A. Patterson. *Computer Architecture: A Quantitative Approach*. Morgan Kaufmann, San Mateo, CA, second edition, 1996.

[174] N. J. Higham. A survey of condition number estimation for triangular matrices. *SIAM Review*, 29:575–596, 1987.

[175] N. J. Higham. Fortran codes for estimating the one-norm of a real or complex matrix, with applications to condition estimation. *ACM Transactions on Mathematical Software*, 14:381–396, 1988.

[176] N. J. Higham. A survey of componentwise perturbation theory in numerical linear algebra. In W. Gautschi, editor, *Mathematics of Computation 1943–1993: A Half-Century of Computational Mathematics,* Volume 48 of *Proceedings of Symposia in Applied Mathematics*, pages 49–77. American Mathematical Society, Providence, RI, 1993.

[177] N. J. Higham. *Accuracy and Stability of Numerical Algorithms*. SIAM, Philadelphia, 1996.

[178] N. J. Higham and D. J. Higham. Large growth factors in Gaussian elimination with pivoting. *SIAM Journal on Matrix Analysis and Applications*, 10:155–164, 1989.

[179] W. Hoffman. Iterative algorithms for Gram–Schmidt orthogonalization. *Computing*, 41:335–348, 1989. Cited in [41].

[180] Y. P. Hong and C.-T. Pan. Rank-revealing QR factorizations and the singular value decomposition. *Mathematics of Computation*, 58:213–232, 1992.

[181] J. Hopcroft and J. D. Ullman. *Introduction to Automata Theory, Language, and Computation*. Addison–Wesley, Reading, MA, 1979.

[182] R. A. Horn and C. R. Johnson. *Matrix Analysis*. Cambridge University Press, Cambridge, 1985.

[183] R. A. Horn and C. R. Johnson. *Topics in Matrix Analysis*. Cambridge University Press, Cambridge, 1991.

[184] Y. Hosoda. A new method for linear ill-posed problems with double use of the QR-decomposition. Paper presented at the Kalamazoo Symposium on Matrix Analysis and Applications, Kalamazoo, MI, October, 1997.

[185] H. Hotelling. Relation between two sets of variates. *Biometrika*, 28:322–377, 1936.

[186] H. Hotelling. Some new methods in matrix calculations. *Annals of Mathematical Statistics*, 14:1–34, 1943.

[187] A. S. Householder. *Principles of Numerical Analysis*. McGraw–Hill, New York, 1953.

[188] A. S. Householder. Unitary triangularization of a nonsymmetric matrix. *Journal of the ACM*, 5:339–342, 1958.

[189] A. S. Householder. *The Theory of Matrices in Numerical Analysis*. Dover, New York, 1964. Originally published by Blaisdell, New York.

[190] C. G. J. Jacobi. Über ein leichtes Verfahren die in der Theorie der Säculärstörungen vorkommenden Gleichungen numerisch aufzulösen. *Journal für die reine und angewandte Mathematik*, 30:51–s94, 1846.

[191] C. G. J. Jacobi. Über eine elementare Transformation eines in Buzug jedes von zwei Variablen-Systemen linearen und homogenen Ausdrucks. *Journal für die reine und angewandte Mathematik*, 53:265–270, 1857, posthumous.

[192] C. Jordan. *Traité des Substitutions et des Équations Algébriques*. Paris, 1870. Cited in [220].

[193] C. Jordan. Mémoire sur les formes bilinéaires. *Journal de Mathématiques Pures et Appliquées, Deuxième Série*, 19:35–54, 1874.

[194] C. Jordan. Essai sur la géométrie à n dimensions. *Bulletin de la Société Mathématique*, 3:103–174, 1875.

[195] W. Kahan. Numerical linear algebra. *Canadian Mathematical Bulletin*, 9:757–801, 1966.

[196] T. Kato. *Perturbation Theory for Linear Operators*. Springer-Verlag, New York, 1966.

[197] B. W. Kernighan and D. M. Ritchie. *The C Programming Language*. Prentice–Hall, Englewood Cliffs, NJ, second edition, 1988.

[198] T. Kilburn, D. B. G. Edwards, M. J. Lanigan, and F. H. Sumner. One-level storage system. *IRE Transactions on Electronic Computers*, EC11:223–235, 1962.

[199] M. Kline. *Mathematical Thought from Ancient to Modern Times*. Oxford University Press, New York, 1972.

[200] D. E. Knuth. Structured programming with go to statements. *Computing Surveys*, 6:261–301, 1974.

[201] D. E. Knuth. *The Art of Computer Programming II: Seminumerical Algorithms*. Addison–Wesley, Reading, MA, second edition, 1981.

[202] Y. Kôsaku. *Functional Analysis*. Springer-Verlag, New York, 1978.

[203] G. Kowalewski. *Einführung in die Determinantentheorie*. Verlag von Veit & Comp., Leipzig, 1909.

[204] L. Kronecker. Algebraische reduction der schaaren bilinearer formen. *Sitzungberichte der Königlich Preußischen Akademie der Wissenschaften zu Berlin*, pages 1225–1237, 1890.

[205] J.-L. Lagrange. Researches sur la métode de maximis et minimis. *Miscellanea Taurinensia*, 1, 1759. Cited and reprinted in [206, v. 1, pp. 1–16].

[206] J.-L. Lagrange. *Œvres de Langrange*. Gauthier–Villars, Paris, 1867–1892.

[207] P. Lancaster. *Lambda-Matrices and Vibrating Systems*. Pergamon, Oxford, 1966.

[208] P. Lancaster and M. Tismenetski. *The Theory of Matrices*. Academic Press, New York, 1985.

[209] P. S. Laplace. Mémoire sur les intégrales définies et sur application aux probabilités. *Memoires de l'Academie des Sciences de Paris*, 11, 1810. Cited and reprinted in [212, v. 12, pp. 355–412]. According to Todhunter [318, §914] the publication date was 1811.

[210] P. S. Laplace. *Théorie Analytique des Probabilités*. Courcier, Paris, 1812. Facsimile edition by Impression Anastaltique Culture et Civilisation, Brussels, 1967.

[211] P. S. Laplace. *Théorie Analytique des Probabilités*. Courcier, Paris, third edition, 1820. Reprinted in [212, v. 7].

[212] P. S. Laplace. *Œvres Compeétes*. Gauthier–Villars, Paris, 1878–1912.

[213] C. L. Lawson and R. J. Hanson. *Solving Least Squares Problems*. Prentice–Hall, Englewood Cliffs, NJ, 1974. Reissued with a survey on recent developments by SIAM, Philadelphia, 1995.

[214] C. L. Lawson, R. J. Hanson, D. R. Kincaid, and F. T. Krogh. Basic linear algebra subprograms for fortran usage. *ACM Transactions on Mathematical Software*, 5:308–323, 1979.

[215] D. C. Lay. *Linear Algebra and Its Applications*. Addison–Wesley, Reading, MA, 1994.

[216] A. M. Legendre. *Nouvelles méthodes pour la détermination des orbites des comètes*. Courcier, Paris, 1805. Cited in [311], where the appendix on least squares is reproduced.

[217] S. J. Leon. *Linear Algebra with Applications*. Macmillan, New York, fourth edition, 1994.

[218] S. B. Lippman. *C++ Primer*. Addison–Wesley, Reading, MA, second edition, 1991.

[219] K. J. R. Liu, D. P. O'Leary, G. W. Stewart, and Y.-J. J. Wu. URV ESPRIT for tracking time-varying signals. *IEEE Transactions on Signal Processing*, 42:3441–3449, 1994.

[220] C. C. Mac Duffee. *The Theory of Matrices*. Chelsea, New York, 1946.

[221] J. P. Mallory. *In Search of the Indo-Europeans: Language, Archaeology, and Myth*. Thames and Hudson, New York, 1989.

[222] M. Marcus. *Basic Theorems in Matrix Theory*. Applied Mathematics Series 57. National Bureau of Standards, Washington, DC, 1960.

[223] M. Marcus and H. Minc. *A Survey of Matrix Theory and Matrix Inequalities*. Allyn and Bacon, Boston, 1964.

[224] M. Marcus and H. Minc. *Introduction to Linear Algebra*. Macmillan, New York, 1965.

[225] R. S. Martin, C. Reinsch, and J. H. Wilkinson. Householder tridiagonalization of a real symmetric matrix. *Numerische Mathematik*, 11:181–195, 1968. Also in [349, pp. 212–226].

[226] R. Mathias and G. W. Stewart. A block QR algorithm and the singular value decomposition. *Linear Algebra and Its Applications*, 182:91–100, 1993.

[227] M. Metcalf and J. Reid. *Fortran 90 Explained*. Oxford Science Publications, Oxford, 1990.

[228] H. Minkowski. Theorie der Konvexen Körper, insbesondere Begründung ihres Oberflächenbegriffs. In David Hilbert, editor, *Minkowski Abhandlung*. Teubner Verlag, 1911, posthumous.

[229] L. Mirsky. Symmetric gauge functions and unitarily invariant norms. *Quarterly Journal of Mathematics*, 11:50–59, 1960.

[230] C. B. Moler. Iterative refinement in floating point. *Journal of the ACM*, 14:316–321, 1967.

[231] C. B. Moler. Matrix computations with Fortran and paging. *Communications of the ACM*, 15:268–270, 1972.

[232] C. B. Moler, J. Little, and S. Bangert. *Pro-Matlab User's Guide*. The Math Works, Natick, MA, 1987.

[233] M. Moonen. *Jacobi-Type Updating Algorithms for Signal Processing, Systems Identification and Control*. PhD thesis, Katholieke Universiteit Leuven, 1990.

[234] M. Moonen, P. Van Dooren, and J. Vandewalle. Combined Jacobi-type algorithms in signal processing. In R. J. Vaccaro, editor, *SVD and Signal Processing, II*, pages 177–188. Elsevier/North-Holland, Amsterdam, 1991.

[235] M. Moonen, P. Van Dooren, and J. Vandewalle. A singular value decomposition updating algorithm for subspace tracking. *SIAM Journal on Matrix Analysis and Applications*, 13:1015–1038, 1992.

[236] E. H. Moore. On the reciprocal of the general algebraic matrix. *Bulletin of the AMS*, 26:394–395, 1920. Abstract.

[237] T. Muir. *A Treatise on the Theory of Determinants*. Dover, New York, 1928. Revised and enlarged by William H. Metzler.

[238] T. Muir. *The Theory of Determinants in the Historical Order of Development*. Macmillan, London, v. 1 1906, v. 2 1911, v. 3 1920, v. 4 1923. Reprinted in two volumes by Dover, New York, 1969.

[239] S. G. Nash and A. Sofer. *Linear and Nonlinear Programming*. McGraw–Hill, New York, 1996.

[240] M. Z. Nashed, editor. *Generalized Inverses and Applications*. Academic Press, New York, 1976.

[241] M. Z. Nashed and L. B. Rall. Annotated bibliography on generalized inverses and applications. In M. Z. Nashed, editor, *Generalized Inverses and Applications. Proceedings of an Advanced Seminar, The University of Wisconsin-Madison, Oct. 1973*, pages 771–1041. Academic Press, New York, 1976.

[242] C. Neumann. *Untersuchungen über das logarithmische und Newtonische Potential*. Leipzig, 1877.

[243] B. Noble. Methods for computing the Moore–Penrose generalized inverse and related matters. In M. Z. Nashed, editor, *Generalized Inverses and Applications. Proceedings of an Advanced Seminar, The University of Wisconsin-Madison, Oct. 1973*, pages 245–301. Academic Press, New York, 1976.

[244] B. Noble and J. W. Daniel. *Applied Linear Algebra*. Prentice–Hall, Englewood Cliffs, NJ, 1977.

[245] W. Oettli and W. Prager. Compatibility of approximate solution of linear equations with given error bounds for coefficients and right-hand sides. *Numerische Mathematik*, 6:405–409, 1964.

[246] J. M. Ortega. The ijk forms of factorization methods I: Vector computers. *Parallel Computing*, 7:135–147, 1988.

[247] O. Østerby and Z. Zlatev. *Direct Methods for Sparse Matrices*. Lecture Notes in Computer Science 157. Springer-Verlag, New York, 1983.

[248] D. V. Ouellette. Schur complement and statistics. *Linear Algebra and Its Applications*, 36:187–295, 1981.

[249] C. C. Paige and M. A. Saunders. Toward a generalized singular value decomposition. *SIAM Journal on Numerical Analysis*, 18:398–405, 1981.

[250] C. C. Paige and M. Wei. History and generality of the CS-decomposition. *Linear Algebra and Its Applicationsn*, 208/209:303–326, 1994.

[251] H. Park and L. Eldén. Downdating the rank-revealing URV decomposition. *SIAM Journal on Matrix Analysis and Applications*, 16:138–155, 1995.

[252] B. N. Parlett. Analysis of algorithms for reflections in bisectors. *SIAM Review*, 13:197–208, 1971.

[253] B. N. Parlett. *The Symmetric Eigenvalue Problem*. Prentice–Hall, Englewood Cliffs, NJ, 1980. Reissued with revisions by SIAM, Philadelphia, 1998.

[254] B. N. Parlett and J. K. Reid. On the solution of a system of linear equations whose matrix is symmetric but not definite. *BIT*, 10:386–397, 1970.

[255] E. Pärt-Enander, A. Sjöberg, B. Melin, and P. Isaksson. *The MATLAB Handbook*. Addison–Wesley, Reading, MA, 1996.

[256] R. V. Patel, A. J. Laub, and P. M. Van Dooren, editors. *Numerical Linear Algebra Techniques for Systems and Control*. IEEE Press, Piscataway, NJ, 1994.

[257] D. A. Patterson and J. L. Hennessy. *Computer Organization and Design: The Hardware/Software Interface*. Morgan Kaufmann, San Mateo, CA, 1994.

[258] G. Peano. Intégration par séries des équations différentielles linéaires. *Mathematische Annallen*, 32:450–456, 1888.

[259] R. Penrose. A generalized inverse for matrices. *Proceedings of the Cambridge Philosophical Society*, 51:406–413, 1955.

[260] V. Pereyra. Stability of general systems of linear equations. *Aequationes Mathematicae*, 2:194–206, 1969.

[261] G. Peters and J. H. Wilkinson. The least squares problem and pseudo-inverses. *The Computer Journal*, 13:309–316, 1970.

[262] G. Peters and J. H. Wilkinson. On the stability of Gauss–Jordan elimination. *Communications of the ACM*, 18:20–24, 1975.

[263] R. L. Plackett. The discovery of the method of least squares. *Biometrika*, 59:239–251, 1972.

[264] M. J. D. Powell and J. K. Reid. On applying Householder's method to linear least squares problems. In *Proceedings IFIP Congress*, pages 122–126, 1968. Cited in [152].

[265] T. W. Pratt and M. V. Zelkowitz. *Programming Languages: Design and Implementation*. Prentice–Hall, Englewood Cliffs, NJ, third edition, 1995.

[266] D. J. Rabideau. Fast, rank adaptive subspace tracking and applications. *IEEE Transactions on Signal Processing*, 44:2229–2244, 1996.

[267] C. R. Rao and S. K. Mitra. *Generalized Inverse of Matrices and Its Applications*. John Wiley, New York, 1971.

[268] J. R. Rice. Experiments on Gram–Schmidt orthogonalization. *Mathematics of Computation*, 20:325–328, 1966.

[269] J. L. Rigal and J. Gaches. On the compatibility of a given solution with the data of a linear system. *Journal of the ACM*, 14:543–548, 1967.

[270] R. Roy and T. Kailath. ESPRIT–estimation of signal parameters via rotational invariance techniques. In F. A. Grünbaum, J. W. Helton, and P. Khargonekar, editors, *Signal Processing Part II: Control Theory and Applications*, pages 369–411. Springer-Verlag, New York, 1990.

[271] M. A. Saunders. Large-scale linear programming using the Cholesky factorization. Technical Report CS252, Computer Science Department, Stanford University, Stanford, CA, 1972. Cited in [41].

[272] E. Schmidt. Zur Theorie der linearen und nichtlinearen Integralgleichungen. I Teil. Entwicklung willkürlichen Funktionen nach System vorgeschriebener. *Mathematische Annalen*, 63:433–476, 1907.

[273] R. Schreiber and C. F. Van Loan. A storage efficient WY representation for products of Householder transformations. *SIAM Journal on Scientific and Statistical Computing*, 10:53–57, 1989.

[274] J. Schur. Über die charakteristischen Würzeln einer linearen Substitution mit einer Anwendung auf die Theorie der Integralgleichungen. *Mathematische Annalen*, 66:448–510, 1909.

[275] J. Schur. Über Potenzreihen, die im Innern des Einkeitskreise beschänkt sind. *Journal für die reine und angewandte Mathematik*, 147:205–232, 1917.

[276] G. A. F. Seber. *Multivariate Observations*. John Wiley, New York, 1983.

[277] T. J. Shepherd and J. G. McWhirter. A pipelined array for linearly constrained least squares optimization. In T. S. Durrani, J. B. Abbiss, J. E. Hudson, R. N. Madan, J. G. McWhirter, and T. A. Moore, editors, *Mathematics in Signal Processing*, pages 457–484. Clarendon Press, Oxford, 1987.

[278] J. Sherman and W. J. Morrison. Adjustment of an inverse matrix corresponding to a change in one element of a given matrix. *Annals of Mathematical Statistics*, 20:317, 1949. Abstract.

[279] J. Sherman and W. J. Morrison. Adjustment of an inverse matrix corresponding to a change in the elements of a given column or a given row of the original matrix. *Annals of Mathematical Statistics*, 20:627, 1949. Abstract.

[280] J. Sherman and W. J. Morrison. Adjustment of an inverse matrix corresponding to a change in one element of a given matrix. *Annals of Mathematical Statistics*, 21:124–127, 1950.

[281] R. D. Skeel. Scaling for numerical stability in Gaussian elimination. *Journal of the ACM*, 26:494–526, 1979.

[282] R. D. Skeel. Iterative refinement implies numerical stability for Gaussian elimination. *Mathematics of Computation*, 35:817–832, 1980.

[283] A. J. Smith. Cache memories. *Computing Surveys*, 14:473–530, 1982.

[284] B. T. Smith, J. M. Boyle, J. J. Dongarra, B. S. Garbow, Y. Ikebe, V. C. Klema, and C. B. Moler. *Matrix Eigensystem Routines–EISPACK Guide. Lecture Notes in Computer Science*. Springer-Verlag, New York, second edition, 1976.

[285] P. H. Sterbenz. *Floating-Point Computation*. Prentice–Hall, Englewood Cliffs, NJ, 1974.

[286] G. W. Stewart. On the continuity of the generalized inverse. *SIAM Journal on Applied Mathematics*, 17:33–45, 1969.

[287] G. W. Stewart. Error and perturbation bounds for subspaces associated with certain eigenvalue problems. *SIAM Review*, 15:727–764, 1973.

[288] G. W. Stewart. *Introduction to Matrix Computations*. Academic Press, New York, 1973.

[289] G. W. Stewart. The economical storage of plane rotations. *Numerische Mathematik*, 25:137–138, 1976.

[290] G. W. Stewart. On the perturbation of pseudo-inverses, projections, and linear least squares problems. *SIAM Review*, 19:634–662, 1977.

[291] G. W. Stewart. Perturbation bounds for the QR factorization of a matrix. *SIAM Journal on Numerical Analysis*, 14:509–518, 1977.

[292] G. W. Stewart. Research development and LINPACK. In J. R. Rice, editor, *Mathematical Software III*, pages 1–14. Academic Press, New York, 1977.

[293] G. W. Stewart. The effects of rounding error on an algorithm for downdating a Cholesky factorization. *Journal of the Institute for Mathematics and Applications*, 23:203–213, 1979.

[294] G. W. Stewart. The efficient generation of random orthogonal matrices with an application to condition estimators. *SIAM Journal on Numerical Analysis*, 17:403–404, 1980.

[295] G. W. Stewart. Computing the CS decomposition of a partitioned orthogonal matrix. *Numerische Mathematik*, 40:297–306, 1982.

[296] G. W. Stewart. On the asymptotic behavior of scaled singular value and QR decompostions. *Mathematics of Computation*, 43:483–489, 1984.

[297] G. W. Stewart. Rank degeneracy. *SIAM Journal on Scientific and Statistical Computing*, 5:403–413, 1984.

[298] G. W. Stewart. Incremental condition calculation and column selection. Technical report 2495, Department of Computer Science, University of Maryland, College Park, MD, 1990.

[299] G. W. Stewart. Stochastic perturbation theory. *SIAM Review*, 32:576–610, 1990.

[300] G. W. Stewart. Error analysis of QR updating with exponential windowing. *Mathematics of Computation*, 59:135–140, 1992.

[301] G. W. Stewart. An updating algorithm for subspace tracking. *IEEE Transactions on Signal Processing*, 40:1535–1541, 1992.

[302] G. W. Stewart. Determining rank in the presence of error. In M. S. Moonen, G. H. Golub, and B. L. R. DeMoor, editors, *Linear Algebra for Large Scale and Real-Time Applications*, pages 275–292. Kluwer Academic Publishers, Dordrecht, 1993.

[303] G. W. Stewart. On the early history of the singular value decomposition. *SIAM Review*, 35:551–566, 1993.

[304] G. W. Stewart. On the perturbation of LU, Cholesky, and QR factorizations. *SIAM Journal on Matrix Analysis and Applications*, 14:1141–1146, 1993.

[305] G. W. Stewart. Updating a rank-revealing ULV decomposition. *SIAM Journal on Matrix Analysis and Applications*, 14:494–499, 1993.

[306] G. W. Stewart. Gauss, statistics, and Gaussian elimination. In J. Sall and A. Lehman, editors, *Computing Science and Statistics: Computationally Intensive Statistical Methods*, pages 1–7. Interface Foundation of North America, Fairfax Station, VA, 1994.

[307] G. W. Stewart. On the stability of sequential updates and downdates. *IEEE Transactions on Signal Processing*, 43:1643–1648, 1995.

[308] G. W. Stewart. On the perturbation of LU and Cholesky factors. *IMA Journal on Numerical Analysis*, 17:1–6, 1997.

[309] G. W. Stewart. The triangular matrices of Gaussian elimination and related decompositions. *IMA Journal on Numerical Analysis*, 17:7–16, 1997.

[310] G. W. Stewart and J.-G. Sun. *Matrix Perturbation Theory*. Academic Press, New York, 1990.

[311] S. M. Stigler. *The History of Statistics*. Belknap Press, Cambridge, MA, 1986.

[312] G. Strang. *Linear Algebra and Its Applications*. Academic Press, New York, third edition, 1988.

[313] G. Strang. Wavelet transforms versus Fourier transforms. *Bulletin of the AMS*, 28:288–305, 1993.

[314] J.-G. Sun. Perturbation bounds for the Cholesky and QR factorizations. *BIT*, 31:341–352, 1991.

[315] J.-G. Sun. Rounding-error and perturbation bounds for the Cholesky and LDL^T factorizations. *Linear Algebra and Its Applications*, 173:77–98, 1992.

[316] A. J. Sutcliffe, editor. *The New York Public Library Writer's Guide to Style and Usage*. Stonesong Press, HarperCollins, New York, 1994.

[317] A. S. Tannenbaum. *Modern Operating Systems*. Prentice–Hall, Englewood Cliffs, NJ, 1992.

[318] I. Todhunter. *A History of the Mathematical Theory of Probability from the Time of Pascal to that of Laplace*. G. E. Stechert, New York, 1865. Reprint 1931.

[319] L. N. Trefethen and D. Bau, III. *Numerical Linear Algebra*. SIAM, Philadelphia, 1997.

[320] L. N. Trefethen and R. S. Schreiber. Average-case stability of Gaussian elimination. *SIAM Journal on Matrix Analysis and Applications*, 11:335–360, 1990.

[321] A. M. Turing. Rounding-off errors in matrix processes. *The Quarterly Journal of Mechanics and Applied Mathematics*, 1:287–308, 1948.

[322] H. W. Turnbull and A. C. Aitken. *An Introduction to the Theory of Canonical Matrices*. Blackie and Son, London, 1932.

[323] A. van der Sluis. Condition, equilibration, and pivoting in linear algebraic systems. *Numerische Mathematik*, 15:74–86, 1970.

[324] A. van der Sluis. Stability of solutions of linear algebraic systems. *Numerische Mathematik*, 14:246–251, 1970.

[325] S. Van Huffel and J. Vandewalle. *The Total Least Squares Problem: Computational Aspects and Analysis*. SIAM, Philadelphia, 1991.

[326] C. F. Van Loan. A general matrix eigenvalue algorithm. *SIAM Journal on Numerical Analysis*, 12:819–834, 1975.

[327] C. F. Van Loan. How near is a stable matrix to an unstable matrix? *Contemporary Mathematics*, 47:465–477, 1985. Reprinted and cited in [256].

[328] C. F. Van Loan. On the method of weighting for equality constrained least squares. *SIAM Journal on Numerical Analysis*, 22:851–864, 1985.

[329] C. F. Van Loan. On estimating the condition of eigenvalues and eigenvectors. *Linear Algebra and Its Applications*, 88/89:715–732, 1987.

[330] R. S. Varga. *Matrix Iterative Analysis*. Prentice–Hall, Englewood Cliffs, NJ, 1962.

[331] J. von Neumann and H. H. Goldstine. Numerical inverting of matrices of high order. *Bulletin of the American Mathematical Society*, 53:1021–1099, 1947.

[332] E. L. Wachspress. *Iterative Solution of Elliptic Systems*. Prentice-Hall, Englewood Cliffs, NJ, 1966.

[333] D. S. Watkins. *Fundamentals of Matrix Computations*. John Wiley & Sons, New York, 1991.

[334] J. H. M. Wedderburn. *Lectures on Matrices*. American Mathematical Society Colloquium Publications, V. XVII. American Mathematical Society, New York, 1934.

[335] P.-Å. Wedin. Perturbation bounds in connection with singular value decomposition. *BIT*, 12:99–111, 1972.

[336] P.-Å. Wedin. Pertubation theory for pseudo-inverses. *BIT*, 13:217–232, 1973.

[337] K. Weierstrass. Zur Theorie der bilinearen und quadratischen Formen. *Monatshefte Akademie Wissenshaften Berlin*, pages 310–338, 1868.

[338] H. Weyl. Das asymptotische Verteilungsgesetz der Eigenwerte linearer partieller Differentialgleichungen (mit einer Anwendung auf die Theorie der Hohlraumstrahlung). *Mathematische Annalen*, 71:441–479, 1912.

[339] E. Whittaker and G. Robinson. *The Calculus of Observations*. Blackie and Son, London, fourth edition, 1944.

[340] N. Wiener. Limit in terms of continuous transformations. *Bulletin de le Société Mathématique de France*, 50:119–134, 1922.

[341] M. Wilkes. Slave memories and dynamic storage allocation. *IEEE Transactions on Electronic Computers*, EC-14:270–271, 1965.

[342] J. H. Wilkinson. Error analysis of floating-point computation. *Numerische Mathematik*, 2:319–340, 1960.

[343] J. H. Wilkinson. Householder's method for the solution of the algebraic eigenvalue problem. *Computer Journal*, 3:23–27, 1960.

[344] J. H. Wilkinson. Error analysis of direct methods of matrix inversion. *Journal of the ACM*, 8:281–330, 1961.

[345] J. H. Wilkinson. *Rounding Errors in Algebraic Processes*. Prentice–Hall, Englewood Cliffs, NJ, 1963.

[346] J. H. Wilkinson. *The Algebraic Eigenvalue Problem*. Clarendon Press, Oxford, 1965.

[347] J. H. Wilkinson. Error analysis of transformations based on the use of matrices of the form $I - 2xx^H$. In L. B. Rall, editor, *Error in Digital Computation*, pages 77–101. John Wiley, New York, 1965.

[348] J. H. Wilkinson. Modern error analysis. *SIAM Review*, 13:548–568, 1971.

[349] J. H. Wilkinson and C. Reinsch. *Handbook for Automatic Computation. Vol. II Linear Algebra*. Springer-Verlag, New York, 1971.

[350] H. Wittmeyer. Einfluß der Änderung einer Matrix auf der Lösung des zugehörigen Gleichungssystems, sowie auf die charakteristischen Zahlen und die Eigenvektoren. *Zeitschrift für angewandte Mathematik und Mechanik*, 16:287–300, 1936.

[351] M. A. Woodbury. Inverting modified matrices. Memorandum Report 42, Statistical Research Group, Princeton, NJ, 1950. Cited in [248].

[352] K. Yoo and H. Park. Accurate downdating of a modified Gram-Schmidt QR decomposition. *BIT*, 36:166–181, 1996.

[353] D. M. Young. *Iterative Solution of Large Linear Systems*. Academic Press, New York, 1971.

[354] H. Zha. A componentwise perturbation analysis of the QR decomposition. *SIAM Journal on Matrix Analysis and Applications*, 14:1124–1131, 1993.

[355] G. Zielke. Inversion of modified symmetric matrices. *Journal of the ACM*, 15:402–408, 1968.

INDEX

Underlined page numbers indicate a defining entry. Slanted page numbers indicate an entry in a notes and references section. Page numbers followed by an "n" refer to a footnote; those followed by an "a," to an algorithm. The abbreviation *me* indicates that there is more information at the main entry for this item. Only authors mentioned explicitly in the text are indexed.

1-norm
    estimator, 388–389a, *400*
    matrix 1-norm, 51
        absoluteness, 53
        vector 1-norm, 44
2-norm, 75
    and Frobenius norm, 65
    and singular values, 64
    estimator, 391–393a, 394–396a, 397–399a, *400*
    matrix 2-norm (spectral norm), 52
        nonabsoluteness, 53
        of orthonormal matrix, 57
        properties of, 52
        unitary invariance, 57
        vector 2-norm, 44

Aasen, J. O., *208*
Abdelmalek, N. N., *292*
absolute error, 122
absolute norm, 52–53
absolute value
    of a matrix, 43
        consistency, 43
    of a scalar, 2, 43
address in memory, 102
adjusted rounding unit, *see* rounding error
Aho, *100*
Aitken, A. C., *80, 289*
Anda, A. A., *290, 291*
angle between vectors, 56, 73
    see also canonical angles

argument of complex number, *see* complex number
Arioli, M., *399*
array contrasted with matrix, 7
    see also storage of arrays
array references
    and matrices, 104
    optimization, 105–107
artificial ill-conditioning, *see* condition

back substitution, 92
backward rounding-error analysis, 81, 130, *143–144*, 225
    backward error, 131
    Gaussian elimination, 229–231, *245*
    general references, *141*
    growth factor, 230, *me*
    Householder triangularization, 261–264, *290*
    least squares, 305, *325*
    linear system, 232–233
    normal equations, 305
    residual vector, 228–229, 233
    sum of scalars, 129–132, 135–136
    triangularization by plane rotations, 275, *290, 354*
    triangular system, 226–227
    see also rounding-error analysis
backward stability, 132, *143*, 225
    componentwise, 244, *247*
    graded matrices, 264–267, *290*
    interpretation, 132–133
    QR updating, 351–352, *354*

sum of scalars, 132
Banach, S., 75
band matrix, <u>12</u>, 150
   linear system, 202–207a, *208*
   storage of, 205, 208
Barlow, J. L., *326*
basic linear algebra subprograms, *see* BLAS
basis, <u>30</u>
   change of, 39–40
   components with respect to, 39
   existence, 31–32
   for $\mathbb{R}^{m \times n}$, 31
   for $\mathbb{R}^n$, 31
   orthonormal, 59, 251
Bau, D. III, 79
Bauer, F. L., 217, *224*, *247*
Bauer–Skeel bound, 217
Beaton, A. E., *354*
Bellman, R., 79
Beltrami, E., 76, 78
Benoît, Commandant, *207*
Berman, A., 79
Bhatia, R., 79
bidiagonal matrix, <u>12</u>
   bidiagonal system, 93a
big O notation, 95, *101*
bilinear form, 78, 78, 180
Bischof, C. H., *290*, *400*
Björck, Å., *77*, *144*, *288*, *292*, *324*, *325*, *326*
Bjerhammer, A., *289*
BLAS, 86, 89, <u>99</u>, <u>107</u>, *120*
   and LINPACK, *120*
   and strides, 107
   axpy, 109
   dot, 106–107
   for triangular systems, 99
   level-three, 118
   level-two, 115
   optimization of array references, 105–107
block algorithm
   choice of block size, 119
   distinct from blocked algorithm, *121*, 161, *182*
blocked algorithm, 117–119

distinct from block algorithm, *121*, 161, *182*
Householder triangularization, 267–270a, *290*
block Gaussian elimination, *see* Gaussian elimination
block triangular matrix, <u>20</u>
   determinant of, 16
   singular value decomposition, 362
Bojanczyk, A., *355*
Brent, R. P., *355*
Bunch, J. R., *208*
Bunyon, J., 208
Burris, W. R., 324
Businger, P, *385*

C, 85, 86
   storage of arrays, 103
cache memory, *see* hierarchical memory
cancellation, <u>129</u>, 136–138, *144*
   exact in standard system, 129
   Householder transformation, 256
   orthogonal projections, 252
   QR decomposition, 251
   revealing loss of information, 137, 330
canonical angles between subspaces, <u>73</u>, 77
   and projections, 74
   computation of, 74
canonical correlations, 77
canonical orthonormal bases for two subspaces, <u>73</u>
Cauchy, A. L., 75, 77
Cauchy inequality, <u>45</u>, 49, 75
   angle between vectors, 55
Cauchy interlacing theorem, 72, 77
Cayley, A., 77, 78
Chambers, J. M., *355*
Chartres, B. A., *246*
Chió, F., *182*
Cholesky algorithm, 189–190a
   pivoted, 190, 375–378a
Cholesky decomposition, *101*, <u>188</u>, *207*
   and QR decomposition, 251
   growth factor, 239, *247*
   tridiagonal matrix, 201a
Cholesky updating, *see* QR updating

classical Gram–Schmidt algorithm, *see* Gram–Schmidt algorithm
Clenshaw, C. W., *145*
Cline, A. K., *400*
Coleman, T., *79*
column-sum norm, *see* 1-norm
column index, *see* matrix
column orientation, *see* orientation of algorithms
column space, 34, 365
    orthonormal basis, 251
complementary subspace, 32, 36
    see also orthogonal complement
complete orthogonal decomposition, *416*
complex n-space ($\mathbb{C}^n$), 4
complexity of matrix algorithms, *100*
complex number, 2
    absolute value, 2, *me*
    argument, 2
    polar representation of complex number, 2
complex symmetric matrix, 15, *26*, 194
component, *see* basis, vector
componentwise perturbation theory, *see* linear system
computer memory, *102*
    address, 102
    storage of arrays, 102–104, *me*
condition, 132
    and scaling, 224, 247
    artificial ill-conditioning, 214–217, 218, 224, 247, 386
    equal error scaling, 216–217, *224*
    ill-conditioning, 133, 237
    linear system, 212
    well-conditioning, 132
    see also norm and condition estimation
condition estimator, *see* norm and condition estimation
condition number, 135, 227
    Bauer–Skeel, *224*
    cross-product matrix, 307
    introduction by Turing, *144*, *224*
    linear system, *211–212*
        and significant digits, 212
        and singular values, 212
        more than one, 144

rectangular matrix, 283
sum of scalars, 135
see also norm and condition estimation
conformability, 13, 14, 20
conjugate transpose, *see* matrix
Conn, A. R., *400*
consistency, *see* absolute value, matrix norm
constrained least squares, 312, *325–326*
    method of elimination, 315–317a
    null-space method, 312–314a
        accuracy, 314
    weighting method, 317–319a
contiguous submatrix, 17
convergence
    componentwise, 46
    equivalence of norms, 46–47
    normwise, 46
    see also Neumann series
Cottle, R. W., *182*
Courant, R., *76*
Cox, A. J., *290*
cross-product matrix, 251, 299
    augmented, 301
    backward instability, 301–304
    condition number, 307
    formation, 299–300a
    positive definiteness, 186
    rounding-error analysis, 301
cross diagonal matrix, 11
cross operator, 23
cross triangular matrix, 11, 23
Crout's method, *see* Gaussian elimination
Cryer, C. W., *247*
CS decomposition, 74–75, *77*
Curtis, A. R., *247*

Daniel, J., *292, 354*
Datta, B. N., *79*
Davis, C., *77*
de Boor, C., *247*
de Hoog, F., *355*
Demmel, J. W., *145, 290, 400*
determinant, 16, *78*
    and nonsingularity, 38
    from LU decomposition, 176–177
    in linear algebra, 16

of positive definite matrix, 187
properties, 16–17
devil's stairs, 380
diagonally dominant matrix, 157, <u>240</u>
    growth factor, 239–241, *247*
diagonal matrix, 10
    row and column scaling, 22
diagonal permutation, 186
Dijkstra, E. W., *86*
dimension of a subspace, <u>32</u>
distance to a subspace, *see* orthogonal projection
Dixon, J. D., *399*
Dongarra, J. J., *101*
dot product, *see* inner product
downdating, <u>327</u>
    2-norm of a vector, 347–348a, *356*, 369
    see also Cholesky downdating *under* QR updating
Duff, I. S., 147, *400*
Duncan, W. J., *354*
Dwyer, P. S., *27*, *80*, *180*

Eckart, C., *76*
Edwards, D. B. G., *120*
eigenpair, <u>71</u>
eigenvalue, <u>71</u>, *101*
    and 2-norm, 52
    and determinant, 16
    and nonsingularity, 38
    and singular values, 71
    Cauchy interlacing theorem, 72
    iterative refinement, 225
    min-max characterization, 72, *76*
    of 2×2 matrix, 396
    of positive definite matrix, 186
    perturbation theory, 72
    similarity transformation, 41
eigenvector, <u>71</u>
    and singular vector, 71
EISPACK, *80*
element, *see* matrix
elementary Hermitian matrix, *see* Householder transformation
elementary matrix, *181*
    lower triangular, <u>152</u>, *181*

elementary reflector, *see* Householder transformation
error
    absolute, <u>122</u>, *me*
    relative, <u>122–124</u>, *me*
    rounding, <u>127–128</u>, *me*
Euclidean norm, 15, 45, 85, 140a
    see also 2-norm, Frobenius norm
exchange matrix, <u>9</u>
exponent exception, 138–140
    coding around, 139–140
    in IEEE standard, 140
    overflow, <u>138</u>
        Clenshaw–Olver proposal, *144*
    underflow, <u>138</u>
        flush to zero, 139
        gradual underflow, 139
exponential windowing, *355*, <u>401</u>, 412
    forgetting factor, <u>401</u>

factor analysis, *365*
Faddeev, V. N., *79*
Faddeeva, D. K., *79*
fast Fourier transform, *101*
Fierro, R. D., *366*
Fischer, E., 72, *76*
floating-point arithmetic, 81, <u>128–129</u>
    cancellation, <u>129</u>, *me*
    elementary functions, 129
    exponent exception, 138–140, *me*
    fl notation, 128, *143*
    general references, *141*
    guard digit, *143*
    IEEE standard, 129, *me*
    nonstandard, *142*
    square root, 129
    standard bound, 128
floating-point number, <u>124–126</u>
    base, 125
    characteristic, 142
    exponent, 125
        biased, 126
    fraction, *141*
    general references, *141*
    IEEE standard, 126, *me*
    mantissa, 125, *142*
    normalization, 125
    precision, 125

INDEX 445

flop, 96, *100*
forgetting factor, *see* exponential windowing
FORTRAN, 85, *86*
    storage of arrays, 103, 115
forward rounding-error analysis, 130
    Gaussian elimination, *245*
forward substitution, 88a
Fourier transform, *101*
Fox, L., 79
Francis, J. G. F., *76*
Frobenius, F. G., *75*
Frobenius norm, *75*
    absoluteness, 53
    and 2-norm, 65
    and singular values, 64
    and trace, 49
    consistency, 49
    consistency with 2-norm, 52
    of identity matrix, 52
    of orthonormal matrix, 57
    unitary invariance, 57
full-rank factorization, 21, <u>34</u>, *42*, 357
    from singular value decomposition, 64
full-rank matrix, <u>37</u>
    and linear systems, 37
fundamental subspaces, *see* singular value decomposition
Furnival, G., *354*

Gaches, J., *225*
GAMS (Guide to Available Mathematical Software), *80*
Gantmacher, F. R., 79
gap in singular values, 359, <u>362</u>, *366*, 401
    revealed by pivoted QLP decomposition, 380–381, *387*
    revealed by pivoted QR decomposition, 373–375, *386*
    revealed by ULV decomposition, 413
    revealed by URV decomposition, 403
Gauss, C. F., *27*, 78, *100*, *181*, *207*, *289*, *353*
    and least squares, *323–324*
Gauss–Jordan elimination, *184*, 332, *354*
Gauss–Markov theorem, *324*

Gaussian elimination, 25, *27–28*, 78, *100*, 113, 147, <u>148–153</u>, *181–184*
    after kth step, 158
    and elementary lower triangular matrices, 152
    and elementary row operations, 150
    and least squares, *324*
    and Schur complements, 150
    as computing LU decomposition, 152–153
    as elimination in linear system, 148–150
    as transformation to triangular form, 151–152, 249
    augmented matrices, *184*
    backward rounding-error analysis, 229–231, *245*
    block elimination, 160–163a, 191
        after k steps, 162
        operation count, 161
        successful completion, 162
    classical, 153–165a, *181*, *183–184*, 376
    compared with Householder triangularization, 259
    comparison of variants, 173
    condition of triangular factors, 231–232, *246*
    conditions for completion, 156
    Crout's method, 117, <u>172–173</u>a, *183–184*
    exotic elimination orders, *183*, 208
    growth factor, <u>230</u>, *me*
    Hessenberg matrix, 195–197a, 237, 247
    in constrained least squares, 315
    partial pivoting, 168a
    patterned matrices, 195
    Pickett's charge, <u>171–172</u>a, *183–184*
    pivoting, 172
    pivoted elimination, 166a
        after k steps, 166
        and LU decomposition, 166
        rank determination, 167
        rectangular matrices, 157
    Sherman's march, <u>169–171</u>a, *183–184*
    tridiagonal matrix, 237, *247*

see also Cholesky decomposition, growth factor, LU decomposition, pivoting, Schur complement
generalized eigenvalue problem, *367*
generalized inverse, *289*
generalized singular value decomposition, *367*
Gentleman, W. M., *291*
Geuder, J. C., *246*
Gill, P. E., *325*
Givens, W., *143*, *245*, *271*, *290*
Givens rotation, *see* plane rotation
Goldberg, D., *141*, *144*
Goldstine, H. H., *143*, *245*
Golub, G. H., *76*, *77*, *79*, *100*, *292*, *296*, *324*, *325*, *326*, *354*, *355*, *385*, *386*
graded matrix, <u>266</u>
    and plane rotation, *290*
    Householder triangularization, 264–267, *290*, *320*
    in rank determination, *366*
    plane rotation, 276
Gragg, W. B., *292*, *354*, *400*
Gram, J. P., *76*, *289*
Gram–Schmidt algorithm, 252, *292*, 339
    and Krylov sequences, *292*
    and least squares, *324*
    classical, 59, <u>277–278</u> a
    loss of orthogonality, 281–284
    modified, <u>278–280</u> a, *292*
        and Householder triangularization, 280, *292*, *354*
        rounding-error analysis, 280–281, *292*, *324*
        weak stability, *144*
    reorthogonalization, 284–288, *292*
    twice-is-enough algorithm, 288, *292*
Greub, W. H., *79*
Grimes, R. G., *400*
growth factor, 230, <u>236</u>, 246–247
    complete pivoting, 238, *247*
    diagonally dominant matrix, 239–241, *247*
    Hessenberg matrix, 237, *247*
    partial pivoting, 236–238, *246*
    positive definite matrix, 239, *247*
    totally positive matrix, 241, *247*
    tridiagonal matrix, 237, *247*
    see also pivoting
Guttman, L., *354*

Hager, W. W., *400*
Halmos, P. R., *79*
Hammarling, S., *291*
Hanson, R. J., *288*, *326*, *386*
Haynsworth, E. V., *182*
Hermitian matrix, <u>15</u>
Hessenberg matrix, <u>11</u>, 150
    growth factor, 237, *247*
    Hessenberg system, 198a
    reduction by Gaussian elimination, 195–197a
    triangularization by plane rotations, 274–275a
Hessian matrix, 207
hierarchical memory, 81, <u>109–110</u>
    cache, <u>112</u>, 120
        blocks, 112
        miss, 112, 115
        write back, 112
        write through, 112, 116
    locality of reference
        in space, 110, 113
        in time, 112
    orientation of algorithms, 113–115
    virtual memory, <u>110</u>, *120*
        page, 110
        page fault, 110, 115
Higham, D. J., *246*, *247*
Higham, N. J., *141*, *182*, *184*, *225*, *246*, *247*, *290*, *399*, *400*
Hoffman, W., *292*
homogeneous system, 25
Hopcroft, J. E., *100*
Horn, R. A., *79*
Hosoda, Y., *387*
Hotelling, H., *77*, *245*, *365*
Householder, A. S., *7*, *27*, *79*, *180*, *181*, *289*, *296*, *324*
Householder transformation, 59, 152, <u>254–258</u> a, 289–290
    elementary Hermitian matrix, *289*
    elementary reflector, 255, *289*

hyperbolic, *355*
introducing zeros, 256
multiplication by, 255
storage of, 255
Householder triangularization, *101*,
 258–260 a, *290*, *324*
 and the modified Gram–Schmidt
  algorithm, 280, *292*, *354*
 backward rounding-error analysis,
  261–264, *290*
 blocked algorithm, 267–270a, *290*
 compared with Gaussian elimination,
  259
 computing projections, 260a
 computing Q-factors, 260
 graded matrices, 264–267, *290*, 320
 in null-space method, 313
 orthogonality of computed Q-factor,
  264
 pivoting, 260

identity matrix, 9, 14
 Frobenius norm, 52
 operator norm, 50
 perturbation theory, 53–54
  componentwise bound, 55
IEEE floating-point standard, 126, *141*
 exponent exceptions, 140
 NaN (not a number), 141
 rounding, 129
 rounding unit, 128
ill-posed problem, 357, 365, 387
inferior subspace, *see* singular value
  decomposition
∞-norm
 matrix ∞-norm, 51
  absoluteness, 53
 vector ∞-norm, 44
inner product, 7, 21
interleaved memory, *119*
inverse matrix, 38, 38
 avoiding computation of, 180, *184*
 determinant of, 16
 from LU decomposition, 177–178a,
  184
 left inverse, 40, *me*
 norm estimator, 389–390, 391–393a,
  394–396a

of orthogonal matrix, 57
perturbation theory, 54, 76
rounding-error analysis, 234–235,
 246
transpose, 38
triangular matrix, 93, 94a, *100*
updating, 328–330
 numerical difficulties, 329–330
 via sweep operator, 331
invert-and-multiply algorithm, 39, 176,
 234–235, *246*, 332
 compared with LU solution of linear
  systems, 178
iterative refinement, 221–223, *225*,
 242–244, *247*
 convergence, 222
 double-precision residual, 244
 eigenvalue problems, *225*
 general algorithm, 222
 least squares, 320–323, *326*
 nonlinear functions, 223, *225*
 seminormal equations, 304
 single-precision residual, 244

Jacobi, C. G. J., *28*, *78*, *181*, *290*
Johnson, C. R., *79*
Jordan, C., *76*, *77*, *78*

Kahan, W., *77*, *292*, *386*
κ^2 effect, *325*
Kato, T., *79*
Kaufman, L., *208*, *292*, *354*
Kayyam, O., *326*
Kilburn, T., *120*
Kline, M., *78*, *80*
Knuth, D. E., *86*, *141*
Kowalewski, G., *292*
Kronecker, L., *78*

L-values, *see* pivoted QLP decomposition
Lagrange, J.-L., *27*, *78*, *207*
Lanigan, M. J., *120*
LAPACK, *80*
 1-norm estimator, 387
 2×2 eigenvalue problem, 396
 choice of block size, *119*
 column orientation, 115
 pivoted orthogonal triangularization,
  385

storage of band matrices, 208
symmetric indefinite system, *208*
Laplace, P. S., *76*, *323*
Lawson, C. L., *288*, *326*, *386*
leading submatrix, <u>17</u>
least squares, *100*, *144*, *181*, <u>*292*</u>
   assessment of computational
      methods, 311
   augmented cross-product matrix, <u>301</u>
   augmented least squares matrix, <u>297</u>,
      324
     QR decomposition, 297
   augmented matrix, *326*
   constrained, *see* constrained least
      squares
   cross-product matrix, <u>299</u>, *me*
   Golub–Householder algorithm, 296
   Hessenberg matrix, 296a
   historical, *323–324*
   iterative refinement, 320–323, *326*
   κ^2 effect, 309
   least squares approximation, <u>294</u>
   normal equations, <u>298–299</u>
     accuracy, 310–311
     backward rounding-error analysis,
       305
     formation, 299–300a
     improving condition, *324*
     in weighting method, 317
     perturbation theory, 306–307
   perturbation theory, 308–309, *325*
   QR equation, 299
   residual sum of squares, <u>294</u>
   residual system, 321–322a, *326*
     rounding-error analysis, 322
   residual vector, <u>294</u>
     effect on perturbation bound, 309
   seminormal equations, <u>304</u>, *325*
   sparse, 300
   statisticians' notation, 249
   subset of columns, 337, *354*
   sweep operator, 332
   updating, 353–354
   via Householder triangularization,
      324
     accuracy, 310–311
     backward stability, 305, *325*
   via modified Gram–Schmidt
      algorithm, 297–298a
     accuracy, 310–311
     backward stability, 305
   via QR decomposition, 293–296a
   via QR factorization, 296–297
left inverse, <u>40</u>
left rotation, *see* plane rotation
Legendre, A. M., *323*, *353*
Lewis, J. G., *400*
linear algebra, 1, <u>28</u>
   general references, *79*
linear combination, <u>6</u>
   matrix representation, 21–22
   of independent vectors, 30
linear independence, <u>29</u>
   and nonsingularity, 38
   matrix characterization, 30
   uniqueness of linear combinations, 30
linear space, 6
linear system
   and nonsingularity, 38
   artificial ill-conditioning, *see*
      condition
   backward componentwise stability,
      244, *247*
   backward error analysis, 232–233
   backward perturbation theory,
      219–221, *225*
     componentwise, 220
     normwise, 219
   band matrix, 202–207a, *208*
   bidiagonal, 93a
   existence of solution, 36
   full rank, 37
   Hessenberg, 198
   invert-and-multiply algorithm, <u>39</u>,
      me
   iterative refinement, 242–244
   matrix representation, 8
   more than one right-hand side, 176
   nonuniqueness of solution, 35
   perturbation theory, *224*
     componentwise bounds, 217–219,
       224, *390–391*
     individual components, 218, *225*
     normwise bounds, 209–210, 224
     right-hand side, 213
   positive definite tridiagonal,

INDEX   449

200–202a
residual system, 321–322a, *326*
residual vector, 233
solution via augmented matrix, *184*
solution via LU decomposition, 147, 174–176a
   transposed system, 175a
symmetric indefinite, *see* symmetric indefinite matrix
transposed system, 147
triangular, 88a, 90, *100*
   accuracy of solutions, 227–228
   backward error analysis, 226–227
   transposed, 92
tridiagonal, 198–200a
underdetermined, 25
   homogeneous, 25
   least norm solution, 314
   uniqueness of solution, 25
updating, 328–329a
see also bidiagonal matrix, positive definite matrix, triangular matrix, etc.
linear transformation
   between subspaces, 41–42
   change of basis, 41
   matrix representation, 41
   representation with respect to a basis, 41
LINPACK, *80, 355*
   BLAS, *120*
   column orientation, 115
   iterative refinement, *247*
   norm and condition estimation, 391–397, *400*
   pivoted Cholesky decomposition, 386
   pivoted orthogonal triangularization, 385
   QR updating, *354*
   storage of band matrices, *208*
   symmetric indefinite system, *183, 208*
low-rank approximation, 357
   from pivoted QLP decomposition, 384–385
   from pivoted QR decomposition, 372
   from the singular value decomposition, 69–70, *76*, 357

in URV updating, 411–412
LU decomposition, 23–25, *27–28*, 34, 63, 90, *181*
   and determinants, 176–177
   and pivoted Gaussian elimination, 166
   and Schur complement, 153, *181*
   block, 156, 161, 163, *181*, 191
      symmetric indefinite matrix, 190–194
   computation of inverse matrix, 177–178a
   computed by Gaussian elimination, 152
   condition of triangular factors, 231–232, *246*
   existence, 157, 159
   normalization, 159
   perturbation theory, *246*
   pivoted, 24, 42
   solution of linear systems, 174–176a
      transposed system, 175a
   uniqueness, 159
   see also Gaussian elimination, Schur complement
LU factorization, *see* LU decomposition

M-matrix, 157
Mac Duffee, C. C., *80*
Marcus, M., *79*
Markov, A. A., *324*
Martin, R. S., *290*
Mathias, R., *366*
MATLAB, *80*
matrix, 8
   absolute value, 43
   and bilinear form, *78*
   basis for $\mathbb{R}^{m \times n}$, 31
   column index, 8
   column space, 34, *me*
   componentwise inequalities, 43
   conjugate transpose, 15
   contrasted with array, 7
   element, 8
   entomology, *77*
   general references, *79*
   history, *77–78*
   indexing conventions, *26*

inverse matrix, 38, *me*
matrix-vector product, 14
nonnegative, 43
nonsingularity, 37, *me*
notational conventions, 8–9
nullity, 35, *me*
null space, 35, *me*
order of a square matrix, 9
positive, 43
product, 14
   determinant of, 16
   noncommutativity, 14
   transpose, 15
rank, 34, *me*
representation of linear systems, 8
row index, 8
row space, 34, *me*
scalar-matrix product, 13
sum, 13
   transpose, 15
transpose, 14
   of sum and product, 15
see also square matrix, identity
   matrix, triangular matrix,
   partitioned matrix, submatrix,
   etc.
matrix decomposition, 78, 147
   and matrix computations, 147, *180*
matrix norm, 48, *75*
   1-norm, 51, *me*
   2-norm, 52, *me*
   absolute norm, 52–53, *me*
   consistency, 48–49
    of Frobenius norm, 49
    of operator norms, 50
   consistent vector norm, 50
   family of norms, 50
   Frobenius norm, 49, *me*
   ∞-norm, 51, *me*
   operator norm, 50
   spectral norm, *see* 2-norm
McWhirter, J. G., *326*
measurement error models, *365*
memory, *see* computer memory,
   hierarchical memory
method of elimination, *see* constrained
   least squares
method of mixed rotation, *see* QR

   updating, Cholesky downdating
min-max characterization, *see* eigenvalue,
   singular value
Minc, H., *79*
Minkowski, H., *75*
Minkowski operations, 3
Mirsky, L., *76*
modulus, *see* absolute value
Moler, C. B., *80, 120*
Moonen, M., *355*
Moore, E. H., *289*
Morrison, W. J., *354*
Muir, T., *78*
Murray, W., *325*

NaN (not a number), 141
Nash, S. G., *325*
natural basis, 367, 371, 384
negative definite matrix, 186
NETLIB, *80*
Neumann series, 54–55, *76*
nonsingularity, 37, 43
   characterizations, 38
   existence of inverse, 38
   of positive definite matrix, 187
   of Schur complements, 155
   of triangular matrix, 87
norm, 1, 44, *75*
   1-norm, 44, 51, *me*
   2-norm, 44, 52, *me*
   absolute norm, 52–53, *me*
   dual norms, 45
   equivalence of norms, 46–47
   family of norms, 50
   Frobenius norm, 49, *me*
   Hölder norms, 45
   ∞-norm, 44, 51, *me*
   Manhattan norm (1-norm), 45
   matrix norm, 48, *me*
   normalization, 44
   on $\mathcal{C}[0, 1]$, 7
   unitary invariance, 57
   vector norm, 44
   see also convergence, norm and
    condition estimation
normal equations, *see* least squares, *100,*
   207

INDEX 451

norm and condition estimation, 212, 219, 399
    1-norm estimator, 388–389a, *400*
    2-norm estimator, 391–393a, 397–399a, *400*
    approximate null vector, 390
    componentwise bounds, 390–391, 399
    condition estimation, 389–390
    in LINPACK, 396–397
    incremental condition estimation, 396, 400
    in UTV updating, 406, 408, 415
    inverse matrix, 389–390
    inverse triangular matrix
        2-norm estimator, 394–396a
    LINPACK-style estimators, 391–397, *400*
normwise relative error, *see* relative error
northwest indexing, *see* partitioned matrix
null-space method, *see* constrained least squares
nullity, 35
    and nonsingularity, 38
null space, 35, 365
null vector, 35

Oettli, W., 220, *225*
Oettli–Prager theorem, 220
Olver, F. W. J., *145*
one-norm, *see* 1-norm
operation count, 81, 94–98, *100–101*
    1-norm estimator, 389
    2-norm estimator, 393, 396, 398
    banded system, 206
    bidiagonal system, 95
    big O notation, 95, 101
    blocked Householder triangularization, 269
    by integration, 97
    Cholesky algorithm, 190
    Cholesky downdating, 346
    comparing algorithms, 98
    complex arithmetic, 96
    complexity of matrix algorithms, *100*
    cross-product matrix, 300
    dominant term, 95
    flop, 96, *100*
    Gaussian elimination, 154
    Gram–Schmidt algorithm, 278
    Hessenberg reduction, 196, 275
    Hessenberg system, 198
    Householder triangularization, 259
    inverse of triangular matrix, 97
    limitations, 98, 105
    linear systems via LU decomposition, 176
    lower bounds on execution time, 98
    matrix inversion, 178
    method of areas, 336
    method of elimination, 317
    method of integrals, *100*
    multiplication by Householder transformation, 255
    nomenclature, 96, *100*
    null-space method, 313
    order, 95
    order constant, 95, *101*
    order n, n^2, n^3 compared, 97
    orthogonal projection, 260
    pivoted QLP decomposition, 380
    positive definite tridiagonal system, 201
    Q-factors, 261
    QR updating
        appending columns, 339
        appending rows, 340
        interchange of columns, 336
        rank-one update, 349
        removing columns, 338
        removing rows from a decomposition, 342
        removing rows from a factorization, 344
    reduction of a symmetric indefinite matrix, 194
    sweep operator, 331
    triangular system, 95
    tridiagonal system, 200
    ULV updating, 416
    URV refinement, 411
    URV updating, 408
    UTU form, 268
    weighting method, 320
operator norm, *see* matrix norm
order constant, *see* operation count

order of a square matrix, 9
orientation of algorithms, *183*
    and hierarchical memories, 113–115
    axpy algorithm, 91
    forward substitution, 91
    Hessenberg reduction, 197, 275
orthogonal complement, 36, 59
orthogonal matrix, <u>56</u>, 238
    inverse, 57
    role in matrix computations, 57
    with specified initial columns, 59
orthogonal projection, <u>60</u>
    and pseudoinverse, 252
    as Hermitian idempotent matrix, 61
    distance to a subspace, 61
    from Householder triangularization, 260a
    from QR decomposition, 251, 253
    from QR factorization, 61
orthogonal triangularization, *see* Householder triangularization, pivoted orthogonal triangularization, plane rotation
orthogonal vectors, <u>56</u>
orthonormal matrix, <u>56</u>
    norm of, 57
orthonormal vectors, <u>56</u>
Ouellette, D. V., *182*
outer product, <u>21</u>
overflow, *see* exponent exception
overwriting
    in Gaussian elimination, 155
    in linear systems, 176
    matrix by its inverse, 94
    of matrix by inverse, 179a
    right-hand side of a linear system, 88

packed storage, 108–109
    and BLAS, 108
    band matrix, 205
    Cholesky algorithm, 189
Paige, C. C., 77, *292*, *324*, *367*
Park, H., *290*, *291*, *354*
Parlett, B. N., 101, *208*, *290*, *292*
partitioned matrix, 9, <u>18</u>, 27
    block, 18
    by columns, 18
    by rows, 18

    in pseudocode, 83
    northwest indexing, 19, 27
Peano, G., *75*
Penrose, R., *289*
permutation matrix, <u>9</u>, 22
permuting rows and columns, 22
perturbation theory, 81, <u>134</u>
    backward perturbation theory, *see* linear system
    componentwise perturbation theory, *see* linear system
    eigenvalue, 72
    first order, <u>306</u>
    fundamental subspaces, 359–363, *365–366*
    identity matrix, 53–54
        componentwise bound, 55
    inverse matrix, 54, *76*
    least squares, 308–309, *325*
    linear system, *see* linear system, condition, condition number
    LU decomposition, *246*
    normal equations, 306–307
    positive definite matrix, 302
    pseudoinverse, 307–308
    QR decomposition, *289*
    singular value decomposition, 69, 359–363, *365–366*
    sum of scalars, 134
Peters, G., *324*
Pickett's charge, *see* Gaussian elimination
Pinkus, A., *247*
pivotal condensation, *182*
pivoted Cholesky decomposition, 190, 375–378a, *386*
    see also pivoted QR decomposition
pivoted orthogonal triangularization, 368–370a, 385
    computation of column norms, 368
    early termination, 370, 373
    incompatibility with blocking, 385
pivoted QLP decomposition, 378–385a, *387*
    2-norm estimation, 397, *400*
    and ULV decomposition, *387*
    bases for fundamental subspaces, 381–384
        accuracy, 382

natural basis, 384
early termination, 384–385, 387
L-values, 379
low-rank approximation, 384–385
revealing gap in singular values, 380–381, 387
pivoted QR decomposition, 42, 319, 368–375
    assessment, 375
    bases for fundamental subspaces, 370–373
        accuracy, 373
        natural basis, 371
    properties of the R-factor, 370
    R-values, 373
    revealing gap in singular values, 373–375, 386
    see also pivoted Cholesky decomposition
pivoting, 160, 165–166, 182
    and scaling, 235
    Cholesky algorithm, 190
    complete, 167, 182, 238, 247, 316
    diagonal, 167
    for size, 167
    for sparsity, 167
    for stability, 167
    Gaussian elimination, 166a
    in Pickett's charge, 172
    nomenclature, 168
    order forced by scaling, 241, 247
    partial, 167, 168a, 182, 230, 236–238, 246
        failure, 247
    pivot element, 155, 165
    positive definite matrix, 239
    symmetric indefinite system, 192–193, 208
    see also growth factor, Householder triangularization
Plackett, R. L., 324
plane rotation, 271, 290
    application, 273a
    backward rounding-error analysis, 275, 290, 354
    fast, *291*
    generation, 272a
    Givens rotation, 271, *290*

graded matrices, 276, *290*
    in (i, j)-plane, 271
    introducing zeros, 271, 273
    left rotation, 402
    preserving and destroying small element, 401–402
    right rotation, 402
    storage, *291*
    triangularization of Hessenberg matrix, 274–275a
Plemmons, R. J., 79
polar representation, *see* complex number
positive definite matrix, 43, 157, *181*, 186, 207
    cross-product matrix, 186
    determinant of, 187
    diagonal permutation, 186
    nonsingularity of, 187
    nonsymmetric, 186
    perturbations of, 302
    pivoting, 239
    positivity of eigenvalues, 186
    principle submatrix, 186
    Schur complements in, 187
    square root of, 187
    sweep operator, 332
    tridiagonal system, 200–202a
    see also Cholesky algorithm, Cholesky decomposition
Powell, M. J. D., *290*
Prager, W., 220, *225*
Pratt, T. W., *86*
principal component analysis, *365*
principal submatrix, 17
    of positive definite matrix, 186
product, *see* vector, matrix
projection, *see* orthogonal projection
    nonorthogonal, 76
pseudocode, 81, 82–85, *86*
    comments, 85
    for statement, 84
    functions and subprograms, 85
    goto statement, 85, *86*
    if statement, 83
    inconsistent dimensions, 83
    inconsistent loops, 84, 88
    leave statement, 85

parameters passed by reference, 85, 106
statement label, 85
while statement, 84
pseudoinverse, 252, *289*
   and orthogonal projections, 252
   and singular value decomposition, 252
   perturbation theory, 307–308
Pythagorean equality, 55, 56, 61, 295

Q-factor, *see* QR decomposition
QLP decomposition, *see* pivoted QLP decomposition
QR algorithm, 76
QR decomposition, 62, 250
   and Cholesky decomposition, 251
   and Schur complement, 253
   and singular value decomposition, 253
   of augmented least squares matrix, 297
   orthogonal projections, 251, 253
   orthonormal bases, 251
   partitioned, 253
   perturbation theory, *289*
   Q-factor, 251, 264
   R-factor, 251
   via cross-product matrix, 251
   see also Gram–Schmidt algorithm, Householder triangularization, least squares, pivoted QR decomposition, QR updating
QR equation, 294
QR factorization, 57, 76
   and Gram–Schmidt algorithm, 59
   and orthonormal bases, 59
   see also QR decomposition
QR updating, 327, *354*
   a general approach, 333–334
   appending columns, 338–339a
   appending rows, 339–340a
   backward stability, 351–352, *354*
   block, 340
   Cholesky downdating, 345–346a, *355*
      hyperbolic rotations, *355*
      method of mixed rotations, 346, *355*
   numerical properties, 352–353
   relational stability, 352
   Saunder's method, *355*
   Cholesky updating, 327, 340
   downdating the vector 2-norm, 347–348a, *356*, 369
   exponential windowing, *355*
   interchanging columns, 334–337a
      contiguous, 336
      least squares, 337
   rank-one update, 348–349
      decomposition, 349
      factorization, 348–349a
   removing a column, 337–338a
   removing rows
      from a decomposition, 342–343a
      from a factorization, 343–344a
quaternion, 77

R-factor, *see* QR decomposition
R-values, *see* pivoted QR decomposition
Rabideau, D. J., 416
rank, 34
   and nonsingularity, 38
   and singular values, 64
   degeneracy, 37
rank determination, 42, 357, *365*, *366*
   by Gaussian elimination, 167
   drawbacks of the singular value decomposition, 367
   error bounds for fundamental subspaces, 364–365
   from the singular value decomposition, 363–364
   scaling, *366–367*
   whitening noise, *366*
   see also gap in singular values
real n-space ($\mathbb{R}^n$), 4
   unit vector basis, 31
recursion, *100*
   in deriving algorithms, 89–90
   LU decomposition, 90
   seldom used in matrix algorithms, 90
register management, 116
Reid, J. K., *208*, *247*, *290*
Reinsch, C, *290*
relational stability, 352
relative error, 122–124

INDEX   455

and rounding error, 127
and significant digits, 124, *141*, 210
normwise, 210
reciprocity, 123, 209
residual system, *see* least squares
Rew, R. K., *400*
Rice, J. R., *292*
Rigal, J. L., *225*
right rotation, *see* plane rotation
rounding-error analysis, 81, 128
    cross-product matrix, 301
    forward error analysis, 130
    general references, *141*
    inverse matrix, 234–235, *246*
    modified Gram–Schmidt algorithm, 280–281, *292, 324*
    residual system, 322
    see also backward rounding-error analysis
rounding-error bounds
    assessment, 136
    pessimism of bounds, 136
    slow growth, 135
rounding error, <u>127–128</u>
    adjusted rounding unit, 131, *143*
    chopping, 128
        effects on a sum, 136
    equivalence of forward substitution and the *axpy* algorithm, 91
    first-order bounds, 131
    fl notation, 127, *143*
    general references, *141*
    inevitability, 121
    in hexadecimal arithmetic, 144
    rigorous bounds, 131
    rounding unit, <u>128</u>, *142*
        approximation of, *142*
    slow accumulation, 135
    truncation, 128
    varieties of rounding, 127
rounding unit, *see* rounding error
row-sum norm, *see* ∞-norm
row index, *see* matrix
row orientation, *see* orientation of algorithms
row space, <u>34</u>, 365
row vector, 9

Saunders, M. A., *77, 325, 355, 367*
scalar, <u>2</u>
    as a 1-vector or a 1×1 matrix, 9
    notational conventions, 2, 7
scalar-matrix product, *see* matrix
scalar-vector product, *see* vector
scalar product, *see* inner product
scaling, *247*
    and condition, 216–217, *224*
    and Gaussian elimination, 241–242
    and pivoting, 235
    approximate balancing, *247*
    equal error, 216–217, 242, *247*
    for minimum condition, *247*
    in rank determination, *366–367*
    whitening noise, 366
scaling a matrix, 22
Schmidt, E., *76, 289*
Schmidt–Mirsky theorem, 69
Schreiber, R. S., *247, 290*
Schur, J., *27, 181, 289*
Schur complement, 150, <u>155</u>
    and block LU decomposition, 163
    and Gaussian elimination, 150
    and LU decomposition, 153, *181*
    and QR decomposition, 253
    generated by k steps of Gaussian elimination, 158
    in positive definite matrix, 187
    nested, 164
    nonsingularity, 155
    via sweep operator, 331
    see also Gaussian elimination, LU decomposition
semidefinite matrix, <u>186</u>
seminormal equations, *see* least squares
    corrected, <u>305</u> a
Sheffield, C., *292*
Shepherd, T. J., *326*
Sherman's march, *see* Gaussian elimination
Sherman, J., *354*
signal processing, 357, 365, 400, *416*
    signal and noise subspaces, *365*
significant digits
    and relative error, 124
similarity transformation, 41
singular matrix, <u>37</u>

singular value, 63
   2-norm, 64
   and eigenvalue, 71
   and Frobenius norm, 64
   and rank, 64
   gap in singular values, 362, *me*
   invariance under unitary equivalence, 66
   min-max characterization, 68
   perturbation theory, 69, 359–363, 365–366
   uniqueness, 65–66
   Weyl's theorem, 67
singular value decomposition, 42, 62, 76
   and 2-norm, 52
   and determinant, 16
   and nonsingularity, 38
   and pseudoinverse, 252
   and QR decomposition, 253
   and the spectral decomposition, 71
   and unitary equivalence, 66
   approximate updating, *416*
   block triangular matrix, 362
   drawbacks, 367, 401
   full-rank factorization, 64
   fundamental subspaces, 358–359, 365
      error bounds, 364–365
      from pivoted QLP decomposition, 381–384
      from pivoted QR decomposition, 370–373
      from URV decomposition, 403
      perturbation theory, 359–363, 365–366
      signal and noise subspaces, 365
   gap in singular values, 362, *me*
   inferior subspace, 358–359, 365
   low-rank approximation, 69–70, *me*
   perturbation theory, 359–363, 365–366
   Schmidt–Mirsky theorem, 69
   singular subspace, 358, 365
   singular value, 63, *me*
   singular value factorization, 64
   singular vector, 63
      and eigenvectors, 71
   superior subspace, 358–359, 365
   uniqueness, 65–66
singular value factorization, *see* singular value decomposition
Skeel, R. D., 217, *224*
skew Hermitian matrix, 15
skew symmetric matrix, 15
Sofer, *325*
software, *see* LAPACK, LINPACK, MATLAB
span of a set of vectors, 29
sparse matrix, 167
   least squares, 300
spectral decomposition, 70–72, 76, 77
   and the singular value decomposition, 71
   in solution of symmetric indefinite systems, 187, *208*
   uniqueness, 71
spectral norm, *see* 2-norm
square matrix, 9
square root of positive definite matrix, 187
stability, *see* backward stability, relational stability, weak stability
Sterbenz, P. H., *141*
Stewart, G. W., 77, 79, *224*, *291*, *292*, *354*, *355*, *366*, *386*, *400*, *416*
Stigler, S. M., *324*
storage of arrays, 102–104
   by blocks, *119*
   by columns, 103
   by rows, 102
   column major order, 103
   in C, 103
   in FORTRAN, 103, 115
   lexicographical order, 103
   packed storage, 108–109
      and BLAS, 108
   row major order, 103
   stride, 103, 107
      interleaved memory, *119*
stride, *see* storage of arrays
submatrix, 17
   extracting and inserting, 23
subspace, 28
   canonical angles between subspaces, 73, *me*
   canonical orthonormal bases for two subspaces, 73

dimension, 32, *me*
direct sum, 29
disjoint subspace, 29
intersection and sum, 28–29
orthonormal basis, 59, 60
sum, *see* vector, matrix
Sumner, F. H., *120*
Sun, J.-G., *79, 224*
superior subspace, *see* singular value
    decomposition
sweep operator, 330–333 a, *354*
    inverse matrix, 331
    least squares, 332
    positive definite matrix, 332
    Schur complement, 331
    stability, 332
Sylvester, J. J., 77
symmetric indefinite matrix, *208*
    block LU decomposition, 190–194
    complete diagonal pivoting, 192, *208*
    partial diagonal pivoting, 192, *208*
    reduction to tridiagonal form, *208*
symmetric matrix, 15
    packed storage, 109

total least squares, *365*
totally positive matrix, 157, 241
    growth factor, 241, *247*
trace, 16
    and Frobenius norm, 49
trailing submatrix, 17
transpose, *see* matrix
trapezoidal matrix, 11
    independence of columns, 30
Trefethen, L. N., *79, 247*
triangle inequality, 43, 44
triangular matrix
    accuracy of solutions, 227–228
    axpy algorithm, 90–91
    back substitution algorithm, 92
    backward rounding-error analysis,
        226–227
    BLAS, 99
    forward substitution algorithm, 88a
    independence of columns, 30
    inverse, 93, 94a, *100*
    lower, 11
    nomenclature, *26*

nonsingularity, 87
orientation of algorithms, 91
packed storage, 108
strict, 11
transposed system, 92
triangular system, *100*
unit, 11
upper, 10
tridiagonal matrix, 12, 150
    growth factor, 237, *247*
    linear system, 198–200a
    packed storage, 109
    positive definite system, 200–202a
    storage of, 200
Turing, A. M., 143, *144, 224, 245*
Turnbull, H. W., *80, 289*
two-norm, *see* 2-norm
two-sided orthogonal decomposition, *416*

Ullman, J. D., *100*
ULV decomposition, *see* UTV
    decomposition
underdetermined system, *see* linear system
    minimum norm solution, *325*
underflow, *see* exponent exception
unitary equivalence, 66
unitary matrix, *see* orthogonal matrix
unit vector, 5
    basis for $\mathbb{R}^n$, 31, 39
updating, 326
    inverse matrix, 328–330
        numerical difficulties, 329–330
    least squares, *353–354*
    linear programming, *354*
    linear system, 328–329a
    optimization, *354*
    sweep operator, 330–333, *me*
URV decomposition, *see* UTV
    decomposition
UTU form, 267–268a
UTV decomposition, 401, *416*
    downdating, 416
    gap in singular values, 401
    ULV decomposition, 401, 413
        and pivoted QLP decomposition,
            387
    gap revealing, 413
    updating, 413–416

URV decomposition, 401, <u>403</u>
 deflation, 406–408
 effects of rounding error, 408
 gap adjustment, 406
 gap revealing, 403
 incorporation, 403–406
 low-rank splitting, 411–412, 416
 refinement, 409–411a
 updating, 402–409
 updating algorithm, 408–409a

Van Dooren, P., *355*
Van Loan, C. F., *76*, *79*, *100*, *290*, *326*, *367*, *400*
Varga, R., *79*
vector, <u>3</u>
 angle between vectors, <u>56</u>, *me*
 as an $n \times 1$ matrix, 9
 component, <u>4</u>
 dimension, <u>3</u>
 inner product, <u>21</u>, *me*
 matrix-vector product, 14
 normalization, 44
 notational conventions, 4, 7
 orthogonal vectors, <u>56</u>, *me*
 orthonormal vectors, <u>56</u>, *me*
 outer product, <u>21</u>, *me*
 scalar-vector product, <u>5</u>, 7
 sum, <u>5</u>
 see also unit vector, zero vector, etc.
vector analysis, 78
vector space, 1, 6, 78
 $C[0, 1]$, 6, 7
 complex n-space ($\mathbb{C}^n$), <u>4</u>, *me*
 function space, 7
 infinite dimensional, 6, 46
 real, 3
 real n-space ($\mathbb{R}^n$), <u>4</u>, *me*
 subspace, <u>28</u>, *me*
 the space of $m \times n$ matrices, 13
Veselić, K., *290*
virtual memory, *see* hierarchical memory
void matrix, <u>9</u>
 in pseudocode, 83
von Neumann, J., *143*, *245*

Watkins, D. S., *79*
weak stability, <u>133</u>, *144*, *245*, *324*, *325*
 corrected seminormal equations, 304

Wedderburn, J. H. M., 80
Wedin, P.-Å., *325*, *366*
Wei, M., 77
Weierstrass, K., 78
Weierstrass approximation theorem, 7
weighting method, *see* constrained least squares
Weyl's theorem, 67
Weyl, H., 72, *76*
whitening noise, 366
Wiener, N., 75
Wilkes, M., *120*
Wilkinson, J. H., *79*, *141*, *143*, *180*, *182*, *219*, *224*, *225*, *225*, *237*, *245*, *246*, *247*, *262*, *290*, *292*, *324*, *325*, *326*
Wilkinson diagram, <u>10</u>
Wilson, R., *354*
Woodbury's formula, 328, *354*
 numerical difficulties, 329–330, 351, *354*
Woodbury, M. A., 328, *354*
Wright, M. H., *325*

Yoo, K., *354*
Young, G., *76*

Zelkowitz, M. V., *86*
zero matrix, <u>9</u>
zero vector, <u>4</u>
Zielke, G., *354*